Evolutionary Computation for Modeling and Optimization

Daniel Ashlock

Evolutionary Computation for Modeling and Optimization

With 163 Figures

 Springer

Daniel Ashlock
Department of Mathematics and Statistics
University of Guelph
Guelph, Ontario NIG 2W1
CANADA
dashlock@uguelph.ca

Mathematics Subject Classification (2000): 6801, 68T20, 68T40

ISBN 978-1-4419-1969-4   e-ISBN 978-0-387-31909-4

Printed on acid-free paper.

Printed in the United States of America.

9 8 7 6 5 4 3 2 1

springeronline.com

Daniel Ashlock

Evolutionary Computation for Modeling and Optimization

October 21, 2005

Springer

Berlin Heidelberg New York
Hong Kong London
Milan Paris Tokyo

To my wife, Wendy

Preface

This book is an adaptation of notes that have been used to teach a class in evolutionary computation at Iowa State University for eight years. A number of people have used the notes over the years, and by publishing them in book form I hope to make the material available to a wider audience.

It is important to state clearly what this book is and what it is not. It is a text for an undergraduate or first-year graduate course in evolutionary computation for computer science, engineering, or other computational science students. The large number of homework problems, projects, and experiments stem from an effort to make the text accessible to undergraduates with some programming skill. This book is directed mainly toward application of evolutionary algorithms. This book is *not* a complete introduction to evolutionary computation, nor does it contain a history of the discipline. It is not a theoretical treatment of evolutionary computation, lacking chapters on the schema theorem and the no free lunch theorem.

The key to this text are the experiments. The experiments are small computational projects intended to illustrate single aspects of evolutionary computation or to compare different methods. Small changes in implementation create substantial changes in the behavior of an evolutionary algorithm. Because of this, the text does not tell students what will happen if a given method is used. Rather, it encourages them to experiment with the method. The experiments are intended to be used to drive student learning. The instructor should encourage students to experiment beyond the stated boundaries of the experiments. I have had excellent luck with students finding publishable new ideas by exceeding the bounds of the experiments suggested in the book.

Source code for experiments, errata for the book, and bonus chapters and sections extending material in the book are available via the Springer website www. Springeronline.com or at

www.eldar.http://eldar.mathstat.uoguelph.ca/dashlock/OMEC/

The book is too long for a one-semester course, and I have never managed to teach more than eight chapters in any one-semester offering of the course. The diagrams at the end of this preface give some possible paths through the text with different emphases. The chapter summaries following the diagrams may also be of some help in planning a course that uses this text.

Some Suggestions for Instructors Using This Text

- Make sure you run the code for an experiment before you hand it out to the class. Idiosyncratic details of your local system can cause serious problems. Lean on your most computationally competent students; they can be a treasure.
- Be very clear from the beginning about how you want your students to write up an experiment. Appendix A shows the way I ask students to write up labs for my version of the course.
- I sometimes run contests for Prisoner's Dilemma, Sunburn, the virtual politicians, or other competitive evolutionary tasks. Students evolve competitors and turn them in to compete with each other. Such competitions can be very motivational.
- Assign and grade lots of homework, including the essay questions. These questions are difficult to grade, but they give you, the instructor, excellent feedback about what your students have and have not absorbed. They also force the students that make an honest try to confront their own ignorance.

Possible Paths Through the Text

The following diagrams give six possible collections of paths through the text. Chapters listed in parentheses are prerequisite. Thus 13.3(6) means that Section 13.3 uses material from Chapter 6.

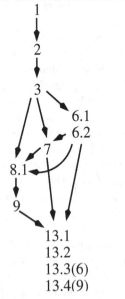

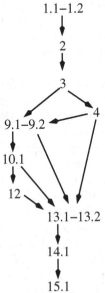

A course on using evolutionary algorithms for optimization.

A course on evolutionary algorithms using only simple string data structures.

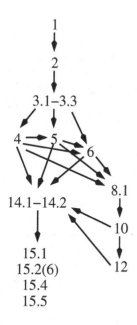

A course on using evolutionary algorithms for modeling.

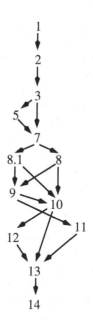

A broad survey of techniques in evolutionary computation.

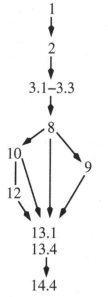

A course focused on genetic programming.

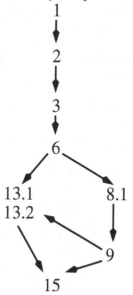

A course on evolutionary algorithms potentially useful in bioinformatics.

A Brief Summary of Chapters

Chapter 1 gives examples of evolutionary algorithms and a brief introduction to simple evolutionary algorithms and simple genetic programming. There is some background in biology in the chapter that may help a computational specialist understand the biological inspiration for evolutionary computation. There is also material included to help the instructor with students deeply skeptical of the scientific foundations of evolution. Chapter 1 can typically be skipped if there is a shortage of time. Most of the technically required background is repeated in greater detail in later chapters.

Chapter 2 introduces simple string data structure evolutionary algorithms. In this context, it surveys the "parts list" of most evolutionary algorithms including mutation and crossover operators, and selection and replacement mechanisms. The chapter also introduces two complex string problems: the Royal Road problem and the self-avoiding walk problem. These set the stage for the discussion of fitness landscapes in Chapter 3. The final section introduces a technical flourish and can easily be skipped.

Chapter 3 introduces real function optimization using a string of real numbers (array) as its representation. The notion of fitness landscape is introduced. The idea of niche specialization is introduced in Section 3.3 and may be included or skipped. Section 4 closely compares two fitness functions for the same problem. Section 5 introduces a simplified version of the circuit board layout problem. The fitness function is constant except where it is discontinuous and so makes a natural target for an evolutionary computation.

Chapter 4 introduces the idea of a model-based fitness function. Both the star fighter design problem (Sunburn) and the virtual politicians use a model of a situation to evaluate fitness. The model of selection and replacement is a novel one: gladiatorial tournament selection. This chapter is not deep, is intended to be fun, and is quite popular with students.

Chapter 5 introduces the programming of very simple artificial neural nets with an evolutionary algorithm in the context of virtual robots. These virtual robots, the symbots, are fairly good models of trophic life forms. This chapter introduces the problem of stochastic fitness evaluation in which there are a large number of fitness cases that need to be sampled. The true "fitness" of a given symbot is elusive and must be approximated. This chapter is a natural for visualization. If you have graphics-competent students, have them build a visualization tool for the symbots.

Chapter 6 introduces the finite state machine representation. The first section is a little dry, but contains important technical background. The second section uses finite state machines as game-playing agents for Iterated Prisoner's Dilemma. This section lays out the foundations used in a good deal of published research in a broad variety of fields. The third section continues on to other games. The first section of this chapter is needed for both GP automata in Chapter 10 and chaos automata in Chapter 15.

Chapter 7 introduces the permutation or ordered list representation. The first section introduces a pair of essentially trivial fitness functions for permutation genes. Section 2 covers the famous Traveling Salesman problem. Section 3 covers a bin-packing problem and also uses a hybrid evolutionary/greedy algorithm. The permutations being evolved control a greedy algorithm. Such hybrid representations enhance the power of evolutionary computation and are coming into broad use. Section 4 introduces an applied problem with some unsolved cases, the Costas array problem. Costas arrays are used as sonar masks, and there are some sizes for which no Costas array is known. This last section can easily be skipped.

Chapter 8 introduces genetic programming in its most minimal form. The variable (rather than fixed) sized data structure is the hallmark of genetic programming. The plus-one-recall-store problem, the focus of the chapter, is a type of maximum problem. This chapter tends to be unpopular with students, and some of the problems require quite advanced mathematics to solve. Mathematicians may find the chapter among the most interesting in the book. Only the first section is really needed to go on to the other chapters with genetic programming in them. The chapter introduces the practice of seeding populations.

Chapter 9 introduces regression in two distinct forms. The first section covers the classical notion of parameter-fitting regression, and the second uses evolutionary computation to perform such parameter fits. The third section introduces symbolic regression, i.e., the use of genetic programming both to find a model and fit its parameters. Section 4 introduces automatically defined functions, the "subroutine" of the genetic programming world. Section 5 looks at regression in more dimensions and can be included or not at the instructor's whim. Section 6 discusses a form of metaselection that occurs in genetic programming called bloat. Since the type of crossover used in genetic programming is very disruptive, the population evolves to resist this disruption by having individual population members get very large. This is an important topic. Controlling, preventing, and exploiting bloat are all current research topics.

Chapter 10 introduces a type of virtual robot, grid robots, with the Tartarus task. The robots are asked to rearrange boxes in a small room. This topic is also popular with students and has led to more student publications than any other chapter in the text. The first section introduces the problem. The second shows how to perform baseline studies with string-based representations. The third section attacks the problem with genetic programming. The fourth introduces a novel representation called the GP automaton. This is the first hybrid representation in the text, fusing finite state machines and genetic programming.

Chapter 11 covers a traditional topic: programming neural nets to simulate digital logic functions. While there are many papers published on this topic it is a little dry. The chapter introduces neural nets in a more complex form than Chapter 5. The chapter looks at direct representation of neural weights and at a way of permitting both the net's connectivity and weights to evolve, and finally attacks the logic function induction problem with genetic programming in its last section.

Chapter 12 introduces a novel linear representation for genetic programming called an ISAc list. ISAc lists are short pieces of simplified machine code. They use a form of goto and so can be used to evolve fully functioning programs with nontrivial flow of control. The chapter introduces ISAc lists in Section 1 and then uses them on the Tartarus problem in Section 2. Section 3 introduces a large number of new grid robot tasks. Section 4 uses ISAc lists as an inspiration to create a more powerful type of string representation for grid robot tasks. This latter section can be skipped.

Chapter 13 introduces a generic improvement to a broad variety of evolutionary algorithms. This improvement consists in storing the evolving population in a geography, represented as a combinatorial graph, that limits selection and crossover. The effect is to slow convergence of the algorithm and enhance exploration of the search space. The first section introduces combinatorial graphs as population structures. The second section uses the techniques on string-based representations. The third uses the graph-based population structure on more complex representations, such as finite state machines and ordered genes. The last section explores genetic programming on graphs. Other than the first section, the sections of this chapter are substantially independent.

Chapter 14 contains four extended examples of a generic technique: storing directions for building a structure rather than the structure itself. This type of representation is called a "cellular" representation for historical reasons. The first section uses a cellular representation to evolve two-dimensional shapes. The second introduces a cellular representation for finite state automata. The third introduces a novel editing representation that permits the evolution of a class of combinatorial graphs. The fourth section uses context free grammars to create a cellular encoding for genetic programming. This technique is quite powerful, since it permits transparent typing of genetic programming systems as well as the incorporation of domain-specific knowledge. The sections of this chapter may be used independently.

Chapter 15 gives examples of applications of evolutionary computation to bioinformatics. The first three sections are completely independent of one another. Section 1 gives an application of string-type genes to an applied (published) problem in bioinformatics. It both aligns and characterizes an insertion point of a type of genetic parasite in corn. Section 2 uses finite state machines to attempt to learn sets of PCR primers that work well and poorly. The finite state machines are intended as filters for subsequently designed primers. The third section introduces a hybrid evolutionary/greedy representation for a hard search problem, locating error-tolerant DNA tags used to mark genetic constructs. The last two sections give methods of visualizing DNA as a fractal.

Acknowledgments

I would like to thank my wife, Wendy, who has been a key player in preparing the manuscript and helping me get things done, and who has acted as a sounding board for many of the novel ideas contained in this text. I also owe a great deal to the students who supplied ideas in the book, such as John Walker, who thought up Sunburn and helped develop the symbots; Mark Joenks, the creator of ISAc lists and virtual politicians; Mark Smucker, whose ideas led to graph-based evolutionary algorithms; Warren Kurt vonRoeschlaub, who started the symbots and other projects; and Mike McRoberts, who coded up the first implementation of GP automata. I thank Jim Golden, who was a key participant in the research underlying the fractal visualization of DNA. I am also grateful to the numerous students who turned in edits to the manuscript over the years, including Pete Johnson, Steve Corns, Elizabeth Blankenship, and Jonathan Gandrud. The Bryden, Schnable, and Sheble labs at Iowa State have supplied me with many valuable students over the years who have asked many questions answered in this book. Mark Bryden, Pat Schnable, and Gerald Sheble all provided a valuable driving force toward the completion of this book.

Contents

1

An Overview of Evolutionary Computation

Evolutionary computation is an ambitious name for a simple idea: use the theory of evolution as an algorithm. Any program that uses the basic loop shown in Figure 1.1 could be termed evolutionary computation. In this text we will explore some of the many ways to fill in the details to the simple structure in Figure 1.1. Evolutionary algorithms operate on populations. We will choose data structures to represent the population, quality measures, and different ways to vary the data structures. We will need to decide how to tell when to stop. For any given problem there are many ways to implement an evolutionary computation system to attack the problem.

Generate a population of structures
Repeat
    Test the structures for quality
    Select structures to reproduce
    Produce new variations of selected structures
    Replace old structures with new ones
Until Satisfied

Fig. 1.1. The fundamental structure of an evolutionary computation.

The field of evolutionary computation has many founders and many names. A concise summary of the origins of evolutionary computation can be found in [8]. You may wonder how the notion of evolutionary computation could be discovered a large number of times without later discoverers noticing those before them. The reasons for this are complex and serve as a good starting point.

The simplest reason evolutionary computation was discovered multiple times is that techniques that cannot be applied yet are not remembered. During the Italian Renaissance, Leonardo da Vinci produced drawings for machines, such as the helicopter, that did not exist as working models for centuries. If he were not a genius and well remembered for other works, his version of the helicopter might well have faded into history. The idea of taking the techniques used by nature to produce diverse complex systems and use them as algorithms is a natural one. Fields like neural computation (computation with artificial neural nets) and fuzzy logic also draw inspiration from biology. The problem is that before the routine availability of large powerful computers these biologically derived ideas could not be implemented. Without big iron, even extremely simplified simulated biology is too slow for most applications.

Limited work with various levels of application and interest began in the 1950s. Sustained and widespread research in evolutionary computation began in the 1970s. By the late 1980s, computer power and human ingenuity combined to create an explosion of research. Vast numbers of articles can be found by searching the World Wide Web with any of the keys "Artificial Life," "Evolutionary Algorithms," "Genetic Algorithms"[29], "Evolutionary Programming"[23, 24], "Evolution Strategies"[13], or "Genetic Programming" [38, 39, 9]. To get a manageable-sized stack, you must limit these search keys to specific application or problem domains.

The second reason that evolutionary computation was discovered a large number of times is its interdisciplinary character. The field is an application of biological theory to computer science used to solve problems in dozens of fields. This means that different groups of people who *never* read one another's publications had the idea independently of using evolution as an algorithm. Early articles appear in journals as diverse as the *IBM Journal of Research and Development, the Journal of Theoretical Biology*, and *Physica D*. It is a very broad-minded scholar who reads journals that many floors apart in the typical university library. The advent of the World Wide Web has lowered, but not erased, the barriers that enabled the original multiple discoveries of evolutionary computation. Even now, the same problem is often attacked by different schools of evolutionary computation with years passing before the different groups notice one another.

The third source of the confused origins of evolutionary computation is the problem of naming. Most of the terminology used in evolutionary computation is borrowed from biology by computational scientists with essentially no formal training in biology. As a result, the names are pretty arbitrary and also annoying to biologists. People who understand one meaning of a term are resistant to alternative meanings. This leads to a situation in which a single word, e.g., "crossover," describes a biological process and a handful of different computational operations. These operations are quite different from one another and linked to the biology only by a thin thread of analogy: a perfect situation for confusion over who discovered what and when they did so. If you are interested in the history of evolutionary computation, you should read *Evolutionary Computation, the Fossil Record* [22]. In this book, David Fogel has compiled early papers in the area together with an introduction to evolutionary computation. The book supplies a good deal of historically useful context in addition to collecting the early papers.

As you work through this text, you will have ideas of your own about how to modify experiments, new directions to take, etc. Beware of being overenthusiastic: someone may have already had your clever idea; check around before trying to publish, patent, or market it. However, evolutionary computation is far from being a mature field, and relative newcomers can still make substantial contributions. Don't assume that your idea is obvious and must have already been tried. Being there first can be a pleasant experience.

1.1 Examples of Evolutionary Computation

Having warned you about the extended and chaotic beginnings of evolutionary computation, we will now look at some examples of applications of the discipline. These examples require no specific technical knowledge and are intended only to give the flavor of evolutionary computation to the novice.

1.1.1 Predators Running Backward

In Chapter 5, we will be looking at a variation of an experiment [12] reported in the first edition of the journal *Adaptive Behavior*. In this experiment, the authors use evolutionary computation to evolve a simple virtual robot to find a signal source. The robot's brain is an artificial neural net. The robot's sensors are fed into two of the neurons, and the robot's wheels are driven by two of the neurons. Evolution is used to select the exact weight values that specify the behavior of the neural net. Imagine the robot as a wheeled cart seeking a light source. Robots that find (stay near) the light source are granted reproductive rights. Reproduction is imperfect, enabling a search for robot brains that work better than the current best ones.

A student of mine, Warren Kurt vonRoeschlaub, attempted to replicate this experiment as a final project for an artificial intelligence class. Where the published experiment used a robot with six nonlinear artificial neurons, vonRoeschlaub created a robot with eight linear neurons. Without going into the technicalities, there is good reason to think that linear neurons are a good deal less powerful as a computational tool. In spite of this, the robots were able to find the simulated light source.

VonRoeschlaub, who had a fairly bad case of "hacker," modified the light source so that it could drift instead of holding still. The robots evolved to stay close to the light source. Not satisfied with this generalization, he then went on to give the light source its own neural net and the ability to move. At this point, the light source became "prey," and the simulation became a simulation of predator and prey. In order to generalize the simulation this way, he had to make a number of decisions.

He gave the predator and prey robots the same types of sensors and wheels. The sensors of prey robots could sense predators only in a 60-degree cone ahead of the robot; likewise, predator robots could sense prey only in a 60-degree cone ahead of them. Both species had two sensors pointing in parallel directions on one side of their body. Neither could sense its own type (and robots of the same type could pass through one another). If a predator and prey intersected, then the prey was presumed to have been eaten by the predator. At this point, several things happened simultaneously. The prey robot was deleted. The oldest surviving prey animal fissioned into two copies, one of which had its neural net modified. The predator also fissioned, and one of the two copies was also modified, and the oldest predator was deleted. This system exhibits many of the features of natural evolution. Reproduction requires that a creature "do a good job," either by eating or avoiding being eaten for the longest possible time. Children are similar but not always identical to their parents.

VonRoeschlaub ran this system a number of times, starting over each time with new randomly initialized neural nets for his predator and prey robots. In all the populations, the creatures wandered about almost at random initially. In most of the populations, a behavior arose in which a predator would leap forward when the signal strengths of its sensors became equal. Unless multiple prey are both in front of and near the predator, this amounts to the sensible strategy of leaping at prey in front of you.

Once this predator behavior had arisen, the prey in many cases evolved an interesting response. When their predator sensors returned similar strengths, they too leaped forward. This had the effect of generating a near miss in many cases. Given that the predators can see only in front of them and have effectively no memory, a near miss is the cleanest possible form of getaway. The response of some of the predators to this strategy was a little startling. The average velocity of predators in three of the populations became negative. At first, this was assumed to be a bug in the code. Subsequent examination of the movement tracks of the predator and prey robots showed that what was happening was in fact a clever adaptation. The predators, initially of the "leap at the prey" variety, evolved to first lower their leaping velocity and later to actually run backward away from prey.

Since the prey were leaping forward to generate a near miss, leaping more slowly or even running backward actually meant getting *more* prey. We named the backward predators the "reverse Ferrari lions," and their appearance illustrates the point that evolutionary computation can have surprising results. There are a number of other points worth considering about this experiment. Of about 40 initial populations, only three gave rise, during the time the experiment was run, to backward-running predators. Almost all generated the forward leaping predator, and many produced the near-miss-generating prey. It may be that all three of these behaviors would have been discovered in every simulation, if only it had been run long enough. It may also be that there were effective predator and prey behaviors that evolved that do not include these three detected behaviors. These alternative evolutionary paths for the robot ecology could easily have gone unnoticed by an experimenter with limited time and primitive analysis tools. It is important to remember that a run of simulated evolution is itself a sample from a space of possible evolutionary runs. Typically, random numbers are used to generate the starting population and also in the process of imperfect reproduction. Different outcomes of the process of evolution have different probabilities of appearing. The problem of how to tell that all possible outcomes have been generated is unsolved. This is a feature of the technique: sometimes a flaw (if all solutions must be listed) and sometimes a virtue (if alternative solutions have value).

The experiment that led to the backward-running predators is one offshoot of the original paper that evolved neural nets to control virtual robots. Chapter 5 of this text is another. The experiments that led to Chapter 5 were motivated by the fact that on the robotic light-seeking task, eight linear neurons outperformed six nonlinear neurons. The linear neurons were removed in pairs until minimum training time was found. Minimum training time occurred at *no* neurons. Readers interested in this subject will find a rich collection of possible experiments in [14].

This example, a predator–prey system, is absolutely classical biology. The advantage of adding evolutionary computation to the enterprise is twofold. First, it permits the researcher to sample the space of possible strategies for the predators and prey, rather than designing or enumerating them. Second, the simulation as structured incorporates the vagaries of individual predators and prey animals. This makes the simulation an *agent-based* one.

There are many different ways to derive or code biological models. Agent-based models follow individual animals (agents) through their interactions with the simulated environment. Another sort of model is a statistical model. These are usually descriptive models, allowing a researcher to understand what is typical or atypical behavior. Yet another sort of model is the equation-based model. Predator–prey models are usually of this type. They use differential equations to describe the impact of prey on the growth rate of the predator population as well as the impact of predators on the growth rate of the prey [44, 46]. Equation-based models permit the theoretical derivation of properties of the system modeled, e.g., that one should observe cyclic behavior of predator and prey population sizes. Each of these types of model is good for solving different types of problems.

1.1.2 Wood-Burning Stoves

Figure 1.2 shows the plan of a type of wood-burning stove designed in part using evolutionary computation. In parts of Nicaragua, stoves such as these or

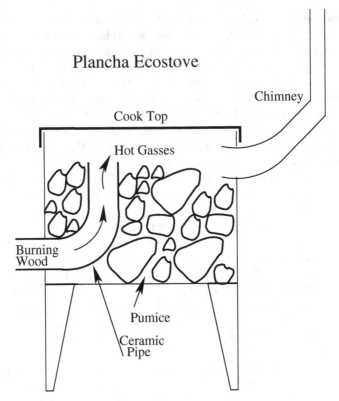

Fig. 1.2. Side cutaway view of a wood-burning Plancha EcoStove.

open hearths are found in most households. People spend as much as a quarter of their income on fuel wood for cooking. Indoor fires are often not vented, and thus they put smoke into the home causing respiratory problems and infant blindness. A not uncommon cause of injury is the "skirt fire." Improving the design of these stoves has the potential to dramatically improve the quality of people's lives. The stove design in Figure 1.2 has many advantages over an open hearth and other stove designs in current use. It uses a good deal less wood to cook the same amount of food, has a chimney to vent the fire, and can be made with inexpensive local materials by local craftsmen. The role of evolutionary computation in the project was to make a greater part of the cooktop useful for cooking by finding a way to spread the heat out more evenly.

The original design for the stove used a cooktop that was a square metal box, open on one side, that fit over the rest of the stove. Hot air would flow from the ceramic pipe where the wood was burned directly to the chimney. Over the ceramic pipe the cooktop was hot enough to boil water. Along the path from the hot spot to the chimney was a usefully hot cooking surface.

The rest of the cooktop was not really hot enough for cooking. A solution to the problem is to weld small, flat pieces of metal (baffles) to the underside of the cooktop to break up the flow of the hot gas. The question is, How many baffles and where do they go?

The stove's original designers found a twelve-baffle solution that yielded useful cooking heat over most of the cooktop. The field team in Nicaragua, however, was unhappy with the design. In order to weld baffles, they had to place hooks connected to cables over unshielded power lines running by the town dump where the stoves were manufactured. The line current was then used to charge a coil salvaged from an automobile. The coil was discharged though the baffle and cooktop, via alligator clips like those on automobile jumper cables, to create a spark-weld. This is not a rapid, or safe, process. Twelve baffles per cooktop was, in the opinion of the field team, too many. An example of a top view of a twelve-baffle design is shown in Figure 1.3.

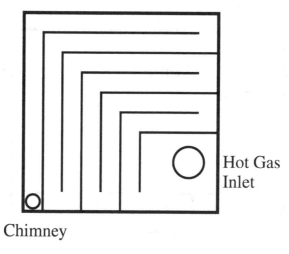

Hot Gas Inlet

Chimney

Fig. 1.3. A top view of a 12-baffle cooktop.

Figuring out where burning gases will go is itself a tricky problem. Computational fluid dynamics (CFD) is the field of predicting what a fluid will do next, given current conditions of momentum, pressure, etc. The equations that describe fluid dynamics are difficult to solve, and in some cases they are theoretically impossible to solve. The typical method of finding an approximate solution requires large computers. Another problem is that the solutions are often unintuitive. Turbulence, vortices, and the mechanical properties of the fluid interact to generate unpredictable behavior. Placing baffles with common sense does not lead to effective designs. The design in Figure 1.3 will force the hot gasses around and even the heating of the cooktop, but it may slow the air flow enough to lead to inefficient combustion and, as noted, requires too much welding.

The stove design team decided to use evolutionary computation to search for a good set of baffles. The evolving population in this case was a collection of baffle designs. Complete details of the experiment are available in [57]. The downtime of a large array of computers normally used to run a virtual reality system was salvaged to evaluate individual stove designs. A "stove design" in this case is a selection of a set of baffles. Some numerical experimentation showed that designs with 3 to 5 baffles achieved the same evenness of heating as designs with more baffles. The structure used by the evolutionary algorithm is a list of three baffles. Each baffle is given a starting position (x, y), a direction (up, down, left, right), a length, and a depth (how far the baffle projects below the cooktop). If two baffles intersect, they are counted as three baffles because one would have to be welded on in two pieces. The final baffle design found with the evolutionary computation system is shown in Figure 1.4. Notice that the design uses only three baffles, *nine* fewer than the design worked out by common sense.

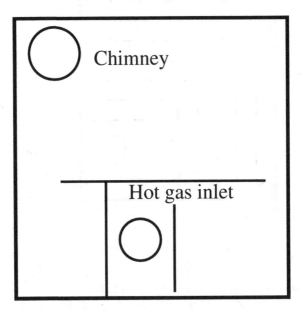

Fig. 1.4. The efficient three-baffle design located with evolutionary computation (the horizontal baffle is quite deep, reaching almost to the pumice, the vertical baffles are about half as deep).

At the beginning of the project, it took about a processor-hour to evaluate each stove design using a commercial CFD software package. The evolutionary computation system generated an input file and called the commercial package, examined the output file, and computed a figure-of-merit for evenness of heating of the cooktop. Generating new baffle designs from old ones used techniques like those in Chapter 3 and required an essentially trivial fraction of the total computation time.

Experimentally lowering the resolution of the CFD to where it was just accurate enough to compare designs lowered the evaluation time for a stove design to about five minutes. In a standard evolutionary computation run, the number of cases examined varies from hundreds of thousands to tens of millions. A five-minute evaluation for quality is exceedingly expensive. Even with an entire array of large processors available most evenings, six months of computation was required to produce the design shown in Figure 1.4. Techniques of the sort described in Chapter 13 were applied to make the process more efficient.

The final stove design was run through the commercial CFD software on very high resolution to numerically validate the design, and then Mark Bryden, the project leader, made a trip to Nicaragua to field-validate the design. At the time of this writing, several hundreds of these stoves are in operation in Nicaragua. Designing stoves in this fashion is an extreme case of evolutionary computation in which only a few tens of thousands of designs are examined with a very long, expensive fitness evaluation for each design.

1.1.3 Hyperspectral Data

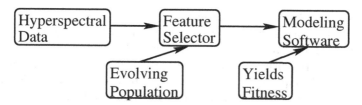

Fig. 1.5. Block diagram of an evolutionary feature-selection system.

A major application of evolutionary computation is modeling data. In Chapter 9, we will look at a number of methods for modeling data. If the data you are modeling are simple, then you won't need evolutionary computation. If the data you are modeling are complex and noisy, then evolutionary computation, properly applied, can be a big help.

In the early days of remote sensing, an aircraft or satellite might collect data on 3 to 7 frequencies. Depending on the application, these frequencies might be visible light or in the near infrared. Near infrared light contains a great deal of useful agricultural information. Corn that is about to shed pollen and diseased plants are both examples of interesting phenomena with detectable infrared signatures. Modern remote sensing platforms can collect data on thousands of frequencies. Such sensors are said to be *hyperspectral*. The data might be collected by a satellite, by aircraft overflight, or by sending light across a conveyor belt in a processing plant through the material on the belt. Even smokestacks are sometimes equipped with hyperspectral sensors that collect data on stack emissions by looking at what frequencies of light the smoke absorbs and transmits.

When you are collecting data on a small number of frequencies, you tend to spread them out across the spectrum. This means that each frequency contains fairly different information. If your sensors return data for thousands of frequencies, the frequencies will be closer together. Two similar frequencies tend to contain similar information. This means that a hyperspectral sensor array tends to return a lot of redundant data. Since sensors are always at least a little noisy, this redundancy (if you can find it) can be used to reduce the impact of noise or random measurement errors.

Evolutionary computation can reduce a huge block of noisy data to useful information while exploiting redundancy to compensate for noise. This is not an easy problem, especially given that some of the real information in the data may not be relevant to the phenomenon we are currently trying to detect. One possible approach is the pipeline shown in Figure 1.5. The useful items abstracted from the hyperspectral data are called *features*. The problem of picking good features is the *feature selection problem*. The strategy is to make a selection of features the data structure for the evolving population. The fitness of a selection of features is the quality of the data model derived from those features. An example of an evolutionary algorithm that performs this sort of data model selection appears in [53].

In this case, 321 frequencies of infrared data were available in the form of measured reflection off the canopy of a cornfield. Adjacent frequencies often take on quite similar values, and so this is exactly the sort of data set under discussion. The potential features were statistical abstractions of ranges of data. A block of frequencies had a statistical operator from the following list applied to them: average, median value, maximum value, minimum value, variance, and slope of a regression line through the data. A single member of the evolving population specified 47 such range abstractions. An individual range abstraction was specified by a starting frequency, a length (number of frequencies in the block), and a choice of one of the six operations.

The modeling software used to fit the model and check the fitness of the choice of range abstractions was a commercial *partial least squares* (PLS) package. This package normalizes all its input variables into the range [0, 1] before using them and so eliminates problems with, for example, the units of median and slope being different. To evaluate a given choice of range abstractions, the PLS software was used to generate a model from 80% of the data, and then the quality of the model was checked for error on the remaining 20% of the data. This procedure, intended to avoid overfitting, in essence memorizing the data instead of modeling it, is called *cross validation*. The dependent variable being fit was the degree to which the corn had tasseled, as assessed by human experts on the ground at the time the data was acquired.

This technique is an example of a *range operator enhanced* evolutionary algorithm. This particular example generated good models, better than using PLS on the raw data. This isn't the end of the story. Some natural questions that arise are, Do you need all six range operators? and Which parts of the spectra are used the most? By looking carefully at the sets of features that appeared in the evolved populations, the researchers could address these questions. Average and slope were the most useful operations, but all of them were used. There were definitely parts of the spectrum that were used more often, and a few regions appeared in almost every fit creature.

This example illustrates some valuable strategies that can be used in evolutionary computation. The first is using hybrid techniques. An evolutionary algorithm that tried to find the data model on its own, rather than handing off its feature selection to a modeling package, would not have functioned as well. The model-fitting package does not perform feature selection, and evolutionary algorithms are not good at fitting precise models quickly. This division of labor produced superior results.

A second notion is *data mining*. Data mining is the process of locating useful information buried in voluminous data. The feature selection that was the explicit purpose of the evolutionary algorithm in the example is one type of data mining. Looking at the evolved populations to see which parts of the spectrum and which range operators were useful is another. This latter data mining can be used to direct which spectral data are collected next time and also to simplify the evolutionary algorithm by reducing the set of range operators used.

Finally, this process of evolving abstractions and passing them off to modeling software, as shown in Figure 1.5, has the potential to be applied well beyond the example given here.

1.1.4 A Little Biology

A better understanding of biological evolution is the most fundamental application of evolutionary computation as well as its inspiration. The theory of evolution is central to both the discipline and to this text. Evolution itself is dead simple and widely misunderstood. The theory of evolution is subtle, complex, and widely misunderstood. Misunderstanding of evolution and the theory that describes evolution flows not from the topic's subtlety and complexity, though they help, but from active and malicious opposition to the theory. Because of this, we stop at this point for a review of the broad outline of the biology that inspires the techniques in the rest of the text.

The first thing we need is some definitions. If you don't know what DNA is or want a lot more detail on genes, look in any standard molecular biology text, e.g., [41]. A *gene* is a sequence of DNA bases that code for a trait, e.g., eye color or ability to metabolize alcohol. An *allele* is a value of a trait. The eye color gene could have a blue allele or a hazel allele in different people.

Definition 1.1 Evolution *is the variation of allele frequencies in populations over time.*

This definition is terse, but it is the definition accepted by most biologists. The term *frequency* means "fraction of the whole," in this case. Its precise meaning is the one used in statistics. Each time any creature is born or dies, the allele frequencies in its population change. When a blond baby is born, the fraction of blond alleles for some hair color gene goes up. When a man who had black hair in his youth dies, the frequency of black hair alleles drops. Clearly, evolution happens all the time.

Why, then, is there any controversy? The controversy exists partly because most people who oppose evolution have never even heard the definition given here. Try asking people who say they dislike evolution what the definition of evolution is. If you do this, try to figure out where (and from whom) the person to whom you are talking learned his definition of evolution.

The main reason for the controversy surrounding evolution is that people dislike the logical conclusions that follow from the above definition juxtaposed with a pile of geological, paleontological, molecular, and other evidence. It is not evolution, but the theory of evolution, that they dislike. The *theory of evolution* is the body of thought that examines evidence and uses it to deduce the consequences of the fact that evolution is going on all the time. In science, a theory means "explanation" not "tentative hypothesis." Scientific theories can be anywhere from entirely tentative to well supported and universally accepted. Within the scientific community, the theory of evolution is viewed as well supported and universally accepted. However, you do not need to accept the theory of evolution in biology to do evolutionary computation. Evolutionary computation uses the ideas in the theory of evolution, asserting nothing about their validity in biology. If you find some of the proceeding material distressing, for whatever reason, I offer the following thought. The concept of evolution exists entirely apart from the reality of evolution. Even if biological evolution is a complete fantasy, it is still the source from which the demonstrably useful techniques of evolutionary computation spring. We may set aside controversy, or at least wait and discuss it over a mug of coffee, later.

Why mention this in what is, essentially, an interdisciplinary computer science text? Because of the quite vigorous opposition to the teaching of evolution, most students come into the field of evolutionary computation in a state much worse than ignorance. Many have heard only myths, falsehoods, and wildly inaccurate claims about evolution. A wonderful essay on this problem is [17]. Since we will attempt to reforge evolution into an algorithm, fundamental misunderstandings about evolution are a handicap. So examine closely what you have been taught about evolution. There are a few key concepts important for evolutionary computation: reproduction, variation, and selection.

Evolution produces new forms over time, as can be seen from examining the fossil record and from looking at molecular evidence or "genetic fossils."

This ability to produce new forms, in essence to innovate without outside direction other than the imperative to have children that live long enough to have children themselves, is the key feature we wish to reproduce in software.

How does evolution produce new forms? There are two opposing forces that drive evolution: variation and selection. Variation is the process that produces new alleles and, more slowly, genes. Variation can also change which genes are or are not expressed in a given individual. The simplest method of doing this is sexual reproduction with its interplay of dominant and recessive genes. Selection is the process whereby some alleles survive and others do not. Variation builds up genetic diversity; selection reduces it.

In biology, the process of variation is quite complex and operates mostly at the molecular level. At the time of this writing, biologists are learning about whole new systems for generating variation at the molecular level. Biological selection is better understood than biological variation. Natural selection, the survival of forms better adapted to their current environment, has been the main type of biological selection. Selective breeding, such as that which produced our many breeds of dogs, is another example of biological selection.

Evolutionary computation operates on populations of data structures. It accomplishes variation by making random changes in these data structures and by blending parts of different structures. These two processes are called *mutation* and *crossover*, and together are referred to as *variation operators*. Selection is accomplished with any algorithm that favors data structures with a higher fitness score. There are many different possible selection methods.

Some ideas work very differently in biology and evolutionary computation. Consider the concept of fitness. The following is a common belief about evolution: "Evolution is the result of survival of the fittest." How do you tell who is fit? Clearly, the survivors are the most fit. Who survives? Clearly, the most fit are those that survive. This piece of circular logic both obscures the correct notion of fitness in biological evolution and makes it hard to understand the differences between biological evolution and the digital evolution we will work with in this text. In biology, the only reasonable notion of fitness is related to reproductive ability. If you have offspring that live long enough to have offspring of their own, then you are fit. A Nobel prize–winning Olympic triple medalist who never has children is completely unfit, by the simplest biological definition of fitness. Consider a male praying mantis. As part of his mating ritual, he gets eaten. He does not survive. The female that eats him goes on to lay hundreds of eggs. A male praying mantis is, thus, potentially a highly fit nonsurvivor. A better way to phrase the material in quotes might be "The results of evolution follow from differential survival and reproduction which, themselves, are the correct measure of fitness."

Oddly enough, "evolution is the result of survival of the fittest" is a pretty good description of many evolutionary computation systems. When we use evolutionary computation to solve a problem, we operate on a collection (population) of data structures (creatures). These creatures will have explicitly computed fitnesses used to decide which creatures will be partially or completely copied by the computer (have offspring). This fundamental difference in the notion of fitness is a key difference between biological evolution (or models of biological evolution) and most evolutionary computation. Some sorts of evolutionary computation do use computer models of the biological notion of fitness, but they are a minority.

Mutations of data structures can be "good" or "bad." A good mutation is one that increases the fitness of a data structure. A bad mutation is one that reduces the fitness of a data structure. Imagine, for the sake of discussion, that we view our data structures as living on a landscape made of a vast flat plane with a single hill rising from it. The structures move at random when mutated, and fitness is equivalent to height. For structures on the plane, any mutation that does not move them to the hill is neither good nor bad. Mutations that are neither good nor bad are called *neutral mutations*. Most of these mutations are neutral.

Let's focus on structures on or near the hill. For structures at the foot of the hill, slightly over half the mutations are neutral and the other half are good. The average effect of mutations at the foot of the hill is positive. Once we are well off the plane and onto the slope of the hill, mutations are roughly half good with a slightly higher fraction being bad. The net effect of mutation is slightly negative. Near or at the top of the hill, almost all movements result in lower fitness; almost all mutations are bad. Using this palette of possibilities, let's examine the net effect of mutation during evolution.

Inferior creatures, those not on the hill, cannot be harmed by mutation. Creatures on the hill but far from the top see little net effect from mutation. Good creatures are affected negatively by mutation. If mutation were operating in a vacuum, creatures would end up mostly on the hill with some bias toward the top. Mutation does not operate in a vacuum, however. Selection causes better structures to be saved. Near the top of the hill, those structures that leap downhill can be replaced and more tries can be made to move uphill from the better structures. The process of selection permits us to cherry-pick better mutations.

Biological mutations, random changes in an organism's DNA, are typically neutral. Much DNA does not encode useful information. The DNA that does encode useful information uses a robust encoding so that many single-base changes do not change what the gene does. The network of interaction among genes is itself robust with multiple copies of some genes and multiple different genes capable of performing a specific task. Biological organisms are often "near the top of the hill" in the sense of their local environment, but the hilltops are usually large and fairly flat. In addition, life has adapted over time to the process of evolution. Collections of genes that are adapted to other hilltops lie dormant or semifunctional in living organisms. Studying these adaptations, the process of "evolving to evolve" is fascinating but well beyond the scope of this text.

Since biological reality and evolutionary computation are not inextricably intertwined, we can harvest the blind alleys of biological science as valid avenues for computational research. Consider the idea of Lamarck that acquired characteristics can be inherited. In Lamarck's view, a muscular child can result from having parents work out and build muscle tissue prior to conceiving. Lamarck's version of evolution would have it that a giraffe's neck is long because of stretching up to reach the high branches. We are certain sure this is not how biology works. However, there is no reason that evolutionary computation cannot work this way, and in fact, some types of it do. The digital analogue to Lamarckian evolution is to run local optimization on a data structure and save the optimized version: acquired characteristics are inherited.

Issues to Consider

In the remainder of this text, you should keep the following notions in mind.

- The *representation* used in a given example of evolutionary computation is the data structure used together with the choice of variation operators. The data structure by itself is the *chromosome* (or *gene*) used in the evolutionary computation.
- The *fitness function* is the method of assigning a heuristic numerical estimate of quality to members of the evolving population. In some cases, it decides only which of two structures is better without assigning an actual numerical quality.

The choice of representation and fitness function can have a huge impact on the way an evolutionary computation system performs.

This text presumes no more familiarity with mathematics than a standard introduction to the differential and integral calculus. Various chapters use solid calculus, graph theory, Markov chains, and some statistics. The material used from these disciplines appears, in summary form, in the appendixes. Instructors who are not interested in presenting these materials can avoid them without much difficulty: they are in specific chapters and sections and not foundational to the material on evolutionary computation presented. The level of algorithmic competence needed varies substantially from chapter to chapter; the basic algorithms are nearly trivial qua algorithms. Genetic programming involves highly sophisticated use of pointers and dynamic allocation. Students whose programming skills are not up to this can be given software to use to perform experiments.

Problems

Problem 1. In the wood-burning stove example of evolutionary computation, the representation of the stove design was a list of baffles, each baffle being described by position, orientation, length, and depth. When a pair of baffles intersected, as shown below, one of the baffles was split into two baffles. If the representation specifies three baffles that must be vertical or horizontal (no diagonal baffles), then what is the largest number of baffles that can occur after splitting?

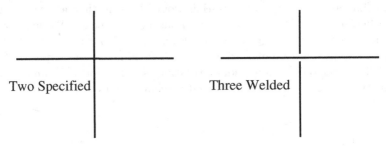

Two Specified

Three Welded

Problem 2. Read Problem 1. Suppose the representation specifies n baffles, then how many welded baffles can result? You answer will be a function of n.

Problem 3. Consider a system in which the chance of a good mutation is 10%, the chance of a bad mutation is 50%, and the chance of a neutral mutation is 40%. The population has two creatures. It is updated by copying the better creature over the worse and then mutating the copy. A good mutation adds one point of fitness and a bad mutation subtracts one point of fitness. If we start with two creatures that have fitness zero, compute the expected fitness of the best creature as a function of the number of population updatings.

Hill Function ――――

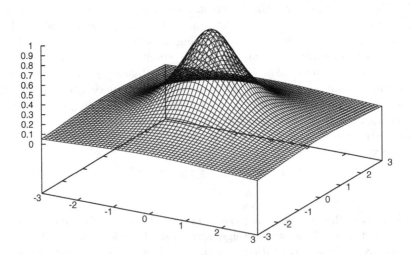

Problem 4. The function

$$f(x, y) = \frac{1}{x^2 + y^2 + 1}$$

is graphed above. It is a single hill with its peak at $(0,0)$. Suppose we have a data structure holding real values (x, y) with fitnesses $f(x, y)$. Mutation consists in moving a distance of exactly 1 in a direction selected uniformly at random.

(i) Give a minimal-length sequence of mutations that take the point $(2, 2)$ to the point $(0, 0)$ without ever lowering the fitness.
(ii) Prove that every point in the plane has a sequence of mutations that take it to the top of the hill.
(iii) Give a point (x, y) that cannot be taken by a sequence of mutations to $(0, 0)$ without lowering the fitness along the way.
(iv) Compute the minimal number of mutations needed to take (x, y) to $(0, 0)$ as a function of x and y.
(v) For which points (x, y) can the paths found in (iv) avoid a step in which fitness goes down?

Problem 5. Essay. Some genes generate traits fairly directly: if you block the gene, that trait goes away and the organism is otherwise unchanged. Other genes are more like control points. Knocking out a control gene can turn whole complexes of other genes off (or on). Which of these two sorts of genes are better targets for selective breeders? Imagine, for example, trying to breed high-yield corn or a dog with an entirely new appearance.

Problem 6. Essay. Consider the following animals: rabbit, box turtle, and deer. All three are herbivores living in North America. Do your best to assess, or at least discuss, the relative fitness of these creatures.

Problem 7. Essay. Compare and contrast North American deer, African antelopes, and Australian kangaroos. Do these animals live in similar environments? Do they do similar "jobs?" Is there a best way to be a large herbivore.

1.2 Evolutionary Computation in Detail

We already know that evolutionary computation uses algorithms that operate on populations of data structures by selection and variation. Figure 1.1 gave a very simple version of the basic loop for an evolutionary algorithm.

In an evolutionary algorithm, the first step is to create a population of data structures. These structures may be filled in at random, designed to some standard, or be the output of some other algorithm. A fitness function is used to decide which solutions deserve further attention. In the main loop of the algorithm, we pick solutions so that *on average* more fit solutions are chosen.

This is the process of selection. The selected solutions are copied over other solutions. The solutions slated to die may be selected at random or with a bias toward worse solutions. The copied solutions are then subjected to variation. This variation can be in the form of random tweaks to a single structure or exchange of material between structures. Changing a single structure is called *unary variation* or *mutation*. Exchanging material between structures is called *binary variation* or *crossover*.

The main loop iterates this process of population updating via selection and variation. In line with a broad outline of the theory of evolution, this should move the population toward more and more fit structures. This continues until you reach an optimum in the space of solutions defined by your fitness function. This optimum may be the best possible place in the entire fitness space, or it may merely be better than all structures "nearby" in the data structure space. Adopting the language of optimization, we call these two possibilities a *global optimum* and a *local optimum*. Unlike many other types of optimizer, an evolutionary algorithm can jump from one optimum to another. Even when the population has found an optimum of the fitness function, the population members scatter about the peak of that optimum. Some population members can leak into the area near another optimum. Figure 1.6 shows a fitness function with several optima. A well-scattered population on the left peak may still be able to discover the right peak. This breadth of search is a property of evolutionary algorithms that is both desirable for some search problems and a source of inefficiency for others.

Keep in mind that during the design process an evolutionary algorithm operates on a population of candidate solutions rather than on a single solution. Not all members of the population need to be fit for a population-based search algorithm to function well. For many types of problems, it is important that low-fitness members of the population exist: they can break through into new areas of the search space. The low-fitness individuals in generation 60 may be the ancestors of the highest-fitness members of generation 120.

It is important to keep in mind that not all problems have solutions. Evolutionary algorithms can also be used to solve problems that provably fail to have optimal solutions. Suppose that the task at hand is to play a game against another player. Some games, like tic-tac-toe, are futile, and you cannot learn to win them when playing against a competent player. Other games, like chess, may have exact solutions, but finding them in general lies beyond the computational ability of any machine envisioned within our current understanding of natural law. Finally, games like Iterated Prisoner's Dilemma, described in Robert Axelrod's book *The Evolution of Cooperation* [7], are intransitive: for every possible way of playing the game in a multiplayer tournament, there is another way of playing it that can tie or beat the first way. Oddly, this does not put Prisoner's Dilemma in a class with tic-tac-toe, but rather makes it especially interesting. A simpler game with this same intransitivity property is rock-paper-scissors. Every possible strategy has another that beats it.

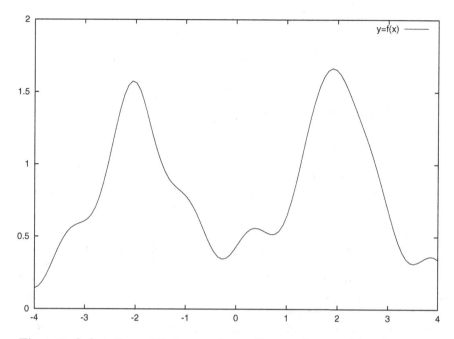

Fig. 1.6. A function with two major optima and several local optima.

Many real-world situations have strategies that work well in some contexts and badly in others. The "best strategy" for Prisoner's Dilemma varies depending on whom you are playing and also on who is playing in general. Political science and evolutionary biology both make use of Prisoner's Dilemma as a model of individual and group interaction. Designing a good fitness function to evolve solutions to these kinds of problems is less straightforward. We will treat Prisoner's Dilemma in greater depth in Chapter 6.

Genetic algorithms are, perhaps, the best-known type of evolutionary algorithm. Genetic algorithms are evolutionary algorithms that operate on a fixed-sized data structure and that use both mutation and crossover to accomplish variation. It is problem- and context-dependent whether crossover helps an evolutionary algorithm locate new structures efficiently. We will explore the utility of crossover in later chapters. There is a large variety of different types of crossover, even for fixed data structures. An example of crossover of two 6-member arrays of real numbers and of two 12-character strings is shown in Figure 1.7.

1.2.1 Representation

A central issue in evolutionary computation is the representation issue. Suppose, for example, that you are optimizing a real function with 20 variables. Would it be more sensible to evolve a gene that is an array of 20 real numbers

Parent 1	3.2 5.6 1.4 7.6 6.7 3.3
Parent 2	1.4 6.7 6.8 9.2 2.1 4.3
Child 1	3.2 5.6 6.8 9.2 2.1 4.3
Child 2	1.4 6.7 1.4 7.6 6.7 3.3

Parent 1	a a a b b b c c c d d d
Parent 2	A A A B B B C C C D D D
Child 1	a a a b B B B C C C d d d
Child 2	A A A B b b c c c D D D

Fig. 1.7. An example of crossover of data structures consisting of 6 real numbers and of 12 characters. (Crossover occurs after gene position 2 for the real-number structures and between positions 5 and 9 for the strings.)

or a gene that is a 960-bit string that codes for real numbers in some fashion? Should the crossover in the algorithm respect the boundaries of the real numbers or be allowed to split the structure in the middle of a real number? What about problems more complex than real function optimization? What data structure works best for them?

The representation question "What is the correct data structure and set of variation operators for the current problem?" is a thorny one. While there is theory in connection with some specific representations, there is not yet a good general theory of representation. Creating a new representation for a problem is an act of imagination. This is most easily seen when one is trying to design a software framework for evolutionary computation. The framework designers send a beta version of the framework to their test users. Within days, the test users will turn up a conference paper, journal article, or idea of their own that cannot be implemented smoothly in the framework because it involves a novel structure, population structure, or variation operator.

Consider the n-queens problem. The goal is to place n chess queens on an $n \times n$ board so that no two attack each other. Queens attack along rows, columns, and diagonals. If we use, as a representation, the coordinates of the queens on the board, then we must satisfy the obvious rule "one queen in each row and column" as part of our fitness evaluation. Most random placements of n queens will fail to satisfy this simple constraint. Vast amounts of time will be spent evaluating configurations that are obviously wrong. If instead we store an assignment of rows to columns (a bijection of the set of n rows with the n available columns), then the row–column mapping places the queens so that they do not attack along rows and columns. If, for example, we were working on the 3-queens problem, then the row–column assignment "row 1 is assigned to column 1, row 2 to column 3, and row 3 to column 2" produces the configuration in Figure 1.8. This configuration happens to solve the problem; in general, this is not true of row–column assignments. Now, is the row–column assignment a superior representation to listing queen positions (x, y)?

Probably: it satisfies many of the constraints of the problem intrinsically. Where is there room for doubt? The possible variation operators for row–column assignments encoded as *permutations* are more complex than those for simple lists. This issue is discussed at some length in Chapter 7.

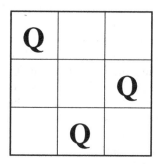

Fig. 1.8. A solution to the 3-queens problem.

Much of the recent progress in evolutionary computation has resulted from changing the representation used. Choosing a representation, as can be seen from the 3-queens example, is a way of placing problem-specific knowledge into the design of the evolutionary algorithm. The famous *no free lunch* theorem [60] demonstrates that there is no efficient general-purpose search algorithm. Evolutionary computation is quite general-purpose: this implies that it must be inefficient *unless* it is specialized to the problem at hand. Careful design of representation is a point at which such specialization can take place. We will examine these issues with experiments, problems, and examples in later chapters. This text introduces a broad variety of representations.

1.2.2 Evolution and Coevolution

Another important concept in evolutionary computation is coevolution. In his paper on evolving sorting networks [34], W. Daniel Hillis built an evolving system in which both the population of sorting networks and the collection of test cases being used to evaluate fitness were allowed to evolve. The solutions were judged to have fitness in proportion to the number of test cases they solved, while the test cases were judged to have fitness in proportion to the number of solutions they fooled. As Hillis's sorters got better, the problems they were tested on became harder (or at least focused on the weaknesses of the current crop of sorters). The biological idea that inspired Hillis was parasitism; a biologist might more properly term the Hillis technique coevolution of competing species. (The fact that a biologist might not like Hillis's analogy does not invalidate Hillis's technique: exactness of biological analogy is not only not required but may not really be possible.) The use of coevolving test problems did indeed enhance the performance of his search algorithm over that observed in earlier runs with a fixed set of test cases. By transforming a system that evolved to one that coevolved, Hillis enhanced performance.

There are two broad classes of evolutionary software that we will call *evolving* and *coevolving* in this text. An evolving population has members whose fitness is judged by some absolute and unchanging standard, e.g., smallness of the dependent variable when a function is minimized. The smaller the value of the evaluation function a given creature in an evolving system has found, the more fit it is. In a coevolving population, the fitness of a member of the evolving population is found by a context-dependent standard. A data structure may be quite fit at one time, unfit later in time, and then later return to being very fit. For example, when we evolve creatures to play Prisoner's Dilemma, the fitness of a creature will depend on the exact set of strategies in the current population. The intransitivity of Prisoner's Dilemma makes every strategy suboptimal in some situation.

Another example of transformation of an evolving system into a coevolving system appears in David Goldberg's classic *Genetic Algorithms in Search, Optimization, and Machine Learning* [29]. He suggests reducing the fitness of a member of a population in proportion to the number of other solutions that are essentially the same. In a real function optimizer, this might be the number of solutions that are close by in the domain space. The effect of this is to make solutions less good once they have been discovered by several members of the population. This reduces the accumulation of solutions onto a good, but suboptimal, solution found early on in the search. This technique is called *niche specialization* and is inspired by the notion of biological niches. The kangaroo in Australia, the deer in North America, and the gazelle in Africa are in the same biological niche. In theory, once a niche is filled, it becomes hard for new species to enter the niche. This is because the existing residents of the niche are already using the resources it contains.

Notice that niche specialization is a transformation from evolution to coevolution. The standard of fitness changes from an absolute one—the function being optimized—to one in which the current membership of the population is also relevant. This example, while coevolutionary, is in some sense closer to being evolutionary than the Prisoner's Dilemma example. There is not a strict dichotomy between evolution and coevolution. Rather, there is a spectrum of intermediate behaviors.

1.2.3 A Simple Type of Evolutionary Computation

Definition 1.2 *A* **string evolver** *is an evolutionary algorithm that tries to match a reference string starting from a population of random strings. The* underlying character set *of the string evolver is the alphabet from which the strings are drawn.*

String evolvers often serve as a baseline or source of reference behavior in evolutionary algorithm research. An evolutionary algorithm for a string evolver functions as follows. Start with a reference string and a population of random strings. The fitness of a string is the number of positions in which it has the same character as the reference string. To evolve the population, split it into small random groups called *tournaments*. Copy the most fit string (break ties by picking at random among the most fit strings) over the least fit string in each tournament. Then, change one randomly chosen character in each copy (mutation). Repeat until an exact match with the reference string is obtained. Typically, one records the number of tournaments, called *generations*, required to find a copy of the reference string.

A word of warning to student and instructor alike. The string evolver problem is a *trivial* problem. It is a place to cut your teeth on evolutionary algorithms, *not* an intrinsically interesting problem. It is an odd feature of the human mind that people immediately think of fifty or sixty potential improvements as soon as they hear a description of your current effort. If you have a cool idea about how to improve evolutionary algorithms, then you might try it out on a string evolver. However, bit-twiddling improvements that are strongly adapted to the string evolver problem are probably not of much value. An example of a string evolver's output is given in Figure 1.9.

Best String	Fitness	Appeared in Generation
HadDe Q'/--<jlm'	3	5
HadDe.em3m/<Ijm-	4	52
HadDe,em3m/<Ijm-	5	54
HadDm,ex3m/#Ijmj	6	73
HadDm,eI8m/#Ijmj	7	86
HadDm,eI8m[Ajjmt	8	118
HadDm,UI8m[Ajjm.	9	135
MadDm,zI8m4AJ1m.	10	154
Madam,zIXm4AJ1m.	11	163
Madam, InmqAJym.	12	256
Madam, I'mqArHm.	13	327
Madam, I'm AC~m.	14	473
Madam, I'm APam.	15	512
Madam, I'm Adam.	16	647

Fig. 1.9. The output of a string evolver operating over the printable ASCII characters with a population of 60 strings. (The reference string is "Madam, I'm Adam." Shown are each string that achieves a new best fitness together with its generation of appearance.)

Problems

Problem 8. Write and debug your own version of the string evolver described following Definition 1.2. Let your population contain 60 strings and set the size of the tournaments to $n = 2$. Run the algorithm 50 times and report the mean, standard deviation, and maximum and minimum number of generations needed to find solutions.

Problem 9. For the string evolver in Problem 8, what is the best value of n for the tournament size? In this case, "best" means "minimizes time-to-solution."

Problem 10. Following the discussion in Section 1.2.1, construct and defend a representation for the n-queens problem. Give the data structure and variation operators. State the advantages and disadvantages of your representation.

Problem 11. Assume that we are evolving strings of a fixed length l. Prove that the amount of time it takes the string evolver to converge is independent of the choice of characters in the reference string.

Problem 12. For tournament size 2, estimate mathematically and/or experimentally (consult your instructor) the dependence of the number of generations needed *on average* to find the reference string on the length l of the reference string. If you are taking the experimental route, give a careful description of your experiment. If you are estimating mathematically, the information on Markov chains in Appendix B may be helpful.

Problem 13. Modify the string evolver from Problem 8 to have crossover. Use tournament size $n = 4$. For each group of four strings, let the two with highest fitness cross over to produce two children that replace the two strings with lowest fitness. With probability m for each child, randomly change (mutate) one character of the child. The probability m is termed the *mutation rate*. To do crossover, as in the first part of Figure 1.7, select a random position in the parental strings and exchange the suffixes starting at that position to obtain the crossed-over strings of the children. Do 50 runs for $m = 0.4$ and $m = 0.8$ and compare the two mutation rates.

Problem 14. Implement an evolutionary algorithm that can find the maximum or minimum of a real function of n real variables. The real function can be hard-coded into your algorithm, and the number of variables should be something you can easily change. The data structures will be arrays of real numbers whose dimension is equal to the number of variables in the function you are optimizing. The fitness function is the function you are maximizing (minimizing) with the functional values interpreted appropriately. Crossover is done as in the string evolver, Problem 13, treating individual real numbers as if they were characters. Mutation consists in adding a uniformly distributed real number in the range $-0.2 \le x \le 0.2$ to some one random position in the creature's gene. Test your program on the following functions:

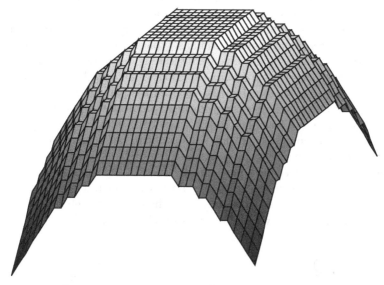

Fig. 1.10. $f(x, y) = 18 - \lfloor x^2 \rfloor - \lfloor y^2 \rfloor$, $-3 \le x, y \le 3$.

(i) Minimize $f(x, y, z) = x^2 + y^2 + z^2$, $-2 \le x, y, z \le 2$.

(ii) Maximize $f(x, y, z) = \frac{1}{x^2 + y^2 + z^2 + 1}$, $-2 \le x, y, z \le 2$.

(iii) Maximize $f(x, y) = \frac{\cos(\sqrt{x^2 + y^2})}{x^2 + y^2 + 1}$, $-2\pi \le x, y \le 2\pi$.

(iv) Maximize $f(x, y) = 18 - \lfloor x^2 \rfloor - \lfloor y^2 \rfloor$, $-3 \le x, y \le 3$.

The fourth function is shown in Figure 1.10 to aid your intuition. The symbol $\lfloor\ \rfloor$ means "floor" or round down to the nearest integer.

1.3 Genetic Programming

Genetic programming is an example of the use of a variable-sized data structure in evolutionary computation. We will explore genetic programming and compare the technique with other evolutionary algorithms in Chapters 8–15. In simple evolutionary algorithms, the data structure storing a member of the evolving population of solutions is of fixed size. This means that care must be taken to write a data structure that is general enough that it has the potential to contain a solution. For real function optimization this isn't a terribly difficult task: an array of real numbers sized to match the number of variables in the function suffices.

In solving more subtle problems, having a sufficiently general data structure can be a significant problem. An approach to this problem was invented by John Koza and David Rice and is called *genetic programming*

[38, 36, 39, 37, 9, 40]. Genetic programming (abbreviated GP) is, in spirit, the same as other evolutionary algorithms. The major difference is that the solutions are stored in variable-sized structures, most commonly in parse trees. These parse trees represent general formulas, typically with an upper bound on the size of the formulas. Operations are internal nodes of the trees; constants and variables are leaves (called *terminals*). Taken together, the operations and terminals of a parse tree are called the *nodes* of the parse tree.

Since almost any imaginable computational problem with a solution can be solved with one or more formulas, possibly involving iterative operations, this gives a general solution space. In fact, the problem becomes one of having a gigantic solution space, large enough to be quite difficult to search efficiently. Some example parse trees are given in Figure 1.11. To save space, we will usually give parse trees in a LISP-like notation in which a node and all its descendants are simply listed between parentheses, recursively. In LISP-like notation, we replace $f(x)$ with $(f\ x)$ and $(a + b)$ with $(+\ a\ b)$.

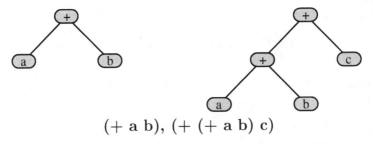

$(+\ a\ b),\ (+\ (+\ a\ b)\ c)$

Fig. 1.11. Some parse trees, together with their LISP-like form.

Genetic programming is a good technique that often works well, but nothing's perfect. It sometimes changes a problem from being like searching for a needle in a haystack to being like searching for a needle in Nebraska; the search space gets quite a lot larger. Several well known techniques exist for searching a haystack for a needle: sitting in the haystack and sifting through it, burning it and sorting through the ash with a magnet, etc. There are no known techniques for efficiently locating a needle randomly placed in the state of Nebraska. If you use genetic programming to evolve formulas that solve your problem, then you are in effect searching the space of formulas. This solution space is extra large, and the search will take forever unless it is done sensibly. You have to narrow your search to the right parts of Nebraska. This involves writing special-purpose languages for the formulas you evolve and using heuristics to bias the initial population of formulas.

Suppose, for example, that we are using genetic programming to find an efficient formula for encoding and interpolating a data set with two independent variables and one dependent variable. In addition, no pair of independent variables occurs more than once. (In other words, the data set describes a function.) One possible special-purpose language for this genetic programming task consists of real constants, the variables x and y, and the arithmetic operators $+$, $-$, $\times$, and $\div$. The fitness of a given parse tree is the sum over the data set of the square of the difference between the function the parse tree represents and the values given in the data set. In this case, we are minimizing the fitness function. This special-purpose language has no iterative operations. Every "program" (formula) returns a real-number result. It may be *just* complex enough to represent data sets that don't have too much wrong with them. An example of a function that could be encoded by this language is

$$f(x, y) = x^2 + y^2,$$

shown in parse tree form in Figure 1.12.

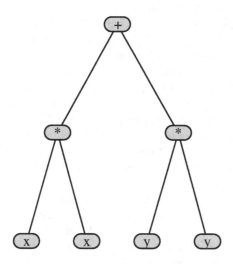

Fig. 1.12. $f(x, y) = x^2 + y^2$.

Notice that every possible formula and subformula in the proceeding example returns the same data type (real). This is done to simplify crossover and mutation operators and the creation of "random" parse trees for the initial population. The recursive calls to generate subtrees are easier to write if they need not do type-checking. Mutation in genetic programming consists in picking a random node in the parse tree, deleting the subtree of which it is the root, and then generating a new random subtree of about the same size to replace it. Crossover consists in locating a random node in each parent tree and then exchanging the subtrees of which they are the root. Examples of crossover and mutation are given in Figure 1.13. Paradoxically, it turns out that in genetic programming, crossover is computationally much easier than mutation; it is just a simple exchange of pointers.

Parent 1	(+(+ a b) c)
Parent 2	(* (* a (+ x y)) (+ a x))
Child 1	(+ (+ a (* a (+ x y))) c)
Child 2	(* b (+ a x))
	Crossover at b and (* a (+ x y))
Mutation	(+ (+ a (* (+ x y) (+ x y))) c)
(of child 1)	mutation takes a to (+ x y)

Fig. 1.13. Crossover and mutation.

One feature many GP implementations have is subroutines or *automatically defined functions* (ADFs). The number of ADFs may be fixed (often at one) or variable. When a creature has an ADF, its gene contains additional parse trees, one for each ADF. There is an operation for each ADF, available only outside the parse tree for which that ADF is defined, that is computed by executing the ADF parse tree. There are terminals in each ADF parse tree that give it access to the arguments from the calling parse tree. This chunking of the functionality of the program into subroutines (ADFs) is useful for many of the same reasons it is useful in standard programming. It also allows us to draw on a powerful biological paradigm: evolution by subsumption.

In the cells of your body, there are many organelles: ribosomes, mitochondria, etc. Some of these organelles have their own genetic code *different* from the genetic code used by the cell nucleus. It is thought that these subcellular organelles are descended from free living organisms that eons ago joined into colonial association with the single-celled ancestors of the type of cells that make up our body. This process, forming a colonial association that allows diverse types of organisms to share functionality that they evolved independently, is called *evolution by subsumption*. A variation operator that splices together the parse trees of ADFs and the main part of a GP creature allows this powerful sort of evolution to take place in a GP environment.

The preceding discussion suggests that GP has a more acute case of the representation problem than even other branches of evolutionary computation. In addition to selecting population size and structure, crossover and mutation type, mutation rate, and the plethora of other relevant parameters, a genetic programmer must select the parts of the special-purpose GP language.

Another issue that arises in genetic programming is that of *disruption*. In a string evolver, the positions in the data structure each have a simple mission: to match a character in the reference string. These missions are completely independent. When we modified the string evolver to be a real function optimizer, the mission specificity of each position in the array remained, but the independence was lost. Parse trees don't even have positions, except maybe "root node," and so they completely lack mission specificity of their entries. A node on the left side of an ancestor may end up on the right side of a descendant.

Crossover in the real function optimizer could break apart blocks of variables that had values that worked well together. This is what is meant by *disruption*. It is intuitively obvious and has been experimentally demonstrated that the crossover used in genetic programming is far more disruptive than the crossover used in algorithms with data structures organized as arrays of strings. Thus, the probability of crossover reducing fitness is higher in genetic programming. Oddly enough, this means that evolving to evolve is easier to observe in genetic programming. Contemplate the tree fragment **(* 0 T)**, where **T** is some subtree. This tree returns a zero no matter what. If **T** is large, then there are many locations for crossover that will not change the fitness of the tree. This sort of crossover resistance has been observed in many genetic programming experiments. It leads to a phenomenon called *bloat*, in which trees quickly grow as large as the software permits.

Problems

Problem 15. Suppose we are working in a genetic programming environment in which the language consists of the constant 1 and the operation +. How many nodes are in the smallest parse tree that can compute n? As an example, computing 3 can be done with the tree **(+ (+ 1 1) 1)**. This tree has 5 nodes.

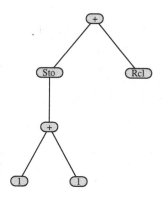

Fig. 1.14. Computing 4 for Problem 16.

Problem 16. Start with the same setup as Problem 15. Add to the language store (STO) and recall (RCL) instructions. The STO instruction takes a single argument and puts it into an external storage location and returns its value. The RCL instruction recalls the value in the storage location. Assume that the left argument of a + is executed before the right one. Find the smallest trees that compute the numbers $n = 3, 4, 5, 6, 7, 8, 9, 10, 11, 12$. The parse tree in Figure 1.14 computes 4, for example, using 6 nodes. (We would like to ask how many nodes are in the smallest parse tree that can compute n, but solving this problem in full generality is extremely hard.)

Problem 17. Using the language described in Problem 16, find the largest number that can be computed by a parse tree with k nodes for $k = 3, 4, 5, 6, 7, 8, 9, 10, 11, 12$. Advanced students with a great deal of time might want to solve this problem for general k.

Problem 18. Assume you are in a GP environment that has the operations $+, -, *, /$, a variable x, and integer constants. Give an example of a parse tree that can compute $x^5 + 3x^3 - 4x^2 - 4x + 1$ with as few nodes as possible. Advanced students should prove that they are using the smallest possible number of nodes.

Problem 19. The STO operation given in Problem 16 is mathematically the identity function; it simply returns its argument. It is valuable not for the computations it performs but rather because of its side effect of storing its argument into a memory location for later access with RCL. Using some sort of node with a side effect, define a set of operations and terminals that will let you compute the solutions to

$$ax^2 + bx + c = 0.$$

Give a single parse tree that returns the first solution the first time it is called and the second solution the second time it is called. The tree should also deal in a reasonable fashion with cases in which there are not two roots. You may use real or complex arithmetic as your base type.

Problem 20. One important difference between the string and tree crossover operators given in this chapter is that the string crossover operator changes nothing when it crosses over two identical structures, while the tree crossover operator can create children quite unlike a pair of identical parents. Give an example of two identical parents and crossover points that yield children unlike the parents.

Problem 21. Essay. Suppose we have a large number of data structures available and want to test a crossover operator for disruptiveness. Give and defend a scheme for such testing.

Problem 22. Essay. In computer science, one of the famous and foundational results concerns the *halting problem*. The main point of interest is that it is, in principle, impossible to separate the set of all computer programs into those that will eventually stop and those that will run forever. There are some programs that obviously fit into each category, but some interesting programs are utterly unclassifiable. Explain why the function approximation scheme given in this section produces parse trees that always halt, and then discuss methods for avoiding the halting problem even if the special purpose language used in a given instance of genetic programming is rich enough to allow general computer programs.

Problem 23. Essay. Discuss how a genetic program could be used to enhance the source code of other software. Be sure to discuss what notion of fitness would be used by the genetic programming environment. You may want to discuss trade-offs between accuracy and speed in the evolved code.

2

Designing Simple Evolutionary Algorithms

The purpose of this chapter is to show you how an evolutionary algorithm works and to teach you how to design your own simple ones. We start simply, by evolving binary character strings, and then try evolving more complex strings. We will examine available techniques for selecting which population members will breed and which will die. We will look at the available crossover and mutation operators for character strings; we will modify the string evolver to be a real function optimizer; and we will examine the issue of population size. We will then move on to more complex problems using string evolvers: the Royal Road problem and self-avoiding walks. The chapter concludes with a discussion of the applications of roulette selection beyond the basic algorithm, including a technique for performing a valuable but computationally difficult type of mutation (probabilistic mutation) efficiently. An example of a binary string evolver applied to a real world problem is given in Section 15.1. The experiments with various string evolvers continue in Chapter 13. Figure 2.1 lists the experiments in this chapter and shows how they depend on one another.

Evolutionary algorithms are a synthesis of several techniques: genetic algorithms, evolutionary programming, evolutionary strategies, and genetic programming. In this chapter, there is a bias toward genetic algorithms [29], because they were designed around the manipulation of binary character strings. The terminology used in this book comes from many sources; arbitrary choices were necessary when several terms exist for the same concept.

Figure 2.2 is an outline for a simple evolutionary algorithm. It is more complex than it seems at first glance. There are five important decisions that factor into the design of the algorithm:

What data structure will you use? In the string evolver and real function optimizer in Section 1.2, for example, the data structures were a string and an array of real numbers, respectively. This data structure is often termed the *gene* of the evolutionary algorithm. You must also decide how many genes will be in the evolving population.

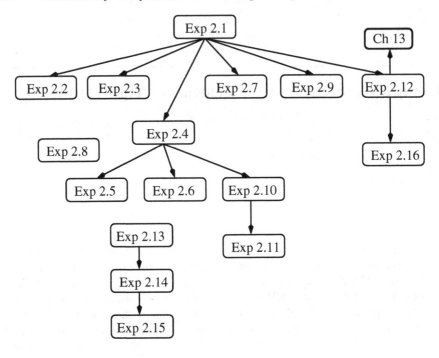

1   Basic string evolver.
2   Change replacement fraction.
3   Steady-state algorithm.
4   One- and two-point crossover.
5   Uniform crossover.
6   Adaptive crossover.
7   With and without mutation.
8   Basic real function optimizer.
9   Experimentation with population size.
10 Royal road function.
11 Royal Road with probabilistic mutation.
12 Introduce self avoiding walks.
13 The stochastic hill climber.
14 Stochastic hill climbing with more mutation.
15 Stochastic hill climbing with lateral movement.
16 Self-avoiding walks with helpful mutation derived from the stochastic hill climber.

Fig. 2.1. The topics and dependencies of the experiments in this chapter.

Create an initial population.
Evaluate the fitness of the population members.
Repeat
    Select pairs from the population to be parents, with a fitness bias.
    Copy the parents to make children.
    Perform crossover on the children (optional).
    Mutate the resulting children (probabilistic).
    Place the children in the population.
    Evaluate the fitness of the children.
Until Done.

Fig. 2.2. A simple evolutionary algorithm.

What fitness function will you use? A fitness function maps the genes
onto some ordered set, such as the integers or the real numbers. For the
string evolver, the fitness function has its range in the natural numbers; the
fitness of a given string is the number of positions at which it agrees with
a reference string. For the real function optimizer, the fitness function
is simply the function being optimized (when maximizing) or its negative
(when minimizing).

What crossover and mutation operators will you use? Crossover
operators map pairs of genes onto pairs of genes; they simulate sexual
reproduction. Mutation operators make small changes in a gene. Taken
together, these are called *variation operators*.

**How will you select parents from the population, and how will you
insert children into the population?** The only requirement is that the
selection method be biased toward "better" organisms. There are many
different ways to do this.

What termination condition will end your algorithm? This could be
after a fixed number of trials or when a solution is found.

Our prototype evolutionary algorithm will be the string evolver (as in
Definition 1.2 and Problem 13). Our data structure will be a string of characters, and our fitness function will be the number of agreements with a fixed
reference string. We will experiment with different variation operators and
different ways of picking parents and inserting children.

2.1 Models of Evolution

Definition 2.1 *The method of picking parents and the method of inserting
children back into the population, taken together, are called the* **model of
evolution** *used by an evolutionary algorithm.*

The model of evolution used in Problem 13 is called *single tournament
selection.* In single tournament selection, the population is shuffled randomly

and divided into small groups. The two most fit individuals in each small group are chosen to be parents. These parent strings are crossed over and the results possibly mutated to provide two children that replace the two least fit members of the small group.

Single tournament selection has two advantages. First, for small groups of size n, the best $n - 2$ creatures in the group are guaranteed to survive. This ensures that the maximum fitness of a group (with a deterministic fitness function) cannot decline as evolution proceeds. Second, no matter how fit a creature is compared to the rest of the population, it can have at most one child in each generation. This prevents the premature loss of diversity in a population that can occur when a single parent has a large number of children, a perennial problem in evolutionary algorithms of all sorts. When a creature that is relatively fit in the initial population dominates the early evolved population, it can prevent the discovery of better creatures by leading the population into a local optimum.

Definition 2.2 *A* **global optimum** *is a point in the fitness space whose value exceeds that of any other value (or is exceeded by every other value if we are minimizing). A* **local optimum** *is a point in the fitness space that has the property that no chain of mutations starting at that point can go up without first going down.*

Making an analogy to a mountain range, the global optimum can be thought of as the top of the highest mountain, while the local optima are the peaks of every mountain or foothill in the range. Even rocks will have associated local optima at their high points. Note that the global optimum is one of the local optima. Also, note that there may be more than one global optimum if two mountains tie for highest.

When the members of a population with the highest fitness are guaranteed to survive in an evolutionary algorithm, that algorithm is said to exhibit *elitism*. Those members of the population guaranteed to survive are called the *elite*. Elitism guarantees that a population with a fixed fitness function cannot slip back to a smaller maximum fitness in later generations, but it also causes the current elite to be more likely to have more children in the future causing their genes to dominate the population. Such domination can impair search of the space of genes, because the current elite may not contain the genes needed for the best possible creatures. A good compromise is to have a small elite. Single tournament selection has an elite of size 2. Half the population survives, but only two creatures, the two most fit, *must* survive. Other creatures survive only if they have the good luck to be put in a group with creatures less fit than they.

In single tournament selection, the selection of parents and the method for inserting children are wedded to one another by the picking of the small groups. This need not be the case; in fact, it is usually not the case. There are several other methods of selecting parents.

In *double tournament selection*, with tournament size n, you pick a group of n creatures and take the single most fit one as a parent, repeating the process with a new group of n creatures to get a second parent. Double tournament selection may also be done *with replacement* (the *same* parent can be picked twice) or *without replacement* (the same parent cannot be picked twice, i.e., the first parent is excluded during the selection of the second parent).

Roulette wheel selection, also called called roulette selection, chooses parents in direct proportion to their fitness. If creature i has fitness f_i, then the probability of being picked as a parent is f_i/F, where F is the sum of the fitness values of the entire population.

Rank selection works in a fashion similar to roulette wheel selection except that the creatures are ordered by fitness and then selected by their rank instead of their fitness. If creature i has rank f_i, then the probability of being picked as a parent is f_i/F, where F is the sum of the ranks of the entire population. Note: the *least* fit creature is given a rank of 1 so as to give it the smallest chance of being picked.

In Figure 2.3, we compare the probabilities for rank and roulette selection. If there is a strong fitness gradient, then roulette wheel selection gives a stronger fitness bias than rank selection and hence tends to take the population to a nearly uniform type faster. The utility of faster fixation depends on the problem under consideration.

Creature #	Fitness	Rank	P(chosen) Roulette	P(chosen) Rank
1	2.1	1	0.099	0.048
2	3.6	5	0.169	0.238
3	7.1	6	0.333	0.286
4	2.4	2	0.113	0.095
5	3.5	4	0.164	0.190
6	2.6	3	0.122	0.143

Fig. 2.3. Differing probabilities for roulette and rank selection.

A model of evolution also needs a child insertion method. If the population is to remain the same size, a creature must be removed to make a place for each child. There are several such methods. One is to place the children in the population at random, replacing anyone. This is called *random replacement*. If we select creatures to be replaced with a probability inversely proportional to their fitness, we are using *roulette wheel replacement* (also called roulette replacement). If we rank the creatures in the opposite order used in rank selection and then choose those to be replaced with probability proportional to their rank, we are using *rank replacement*. In another method, termed *absolute fitness replacement*, we replace the least fit members of the population with the children. Another possible method is to have children replace their parents only if they are more fit. In this method, called *locally elite replacement*, the two parents and their two children are examined, and the two most fit are put into the population in the slots occupied by the parents. In *random elite replacement*, each child is compared to a randomly selected member of the population and replaces it only if it is at least as good.

With all of the selection and replacement techniques described above you must decide how many pairs of parents to select in each generation of your evolutionary algorithm. At one extreme, you select enough pairs of parents to replace your entire population; this is called a *generational* evolutionary algorithm. At the other extreme, a *steady-state* evolutionary algorithm, you count each act of selecting parents and placing (or failing to place) the children in the population as a "generation." Such single-mating "generations" are usually called *mating events*.

Generational evolutionary algorithms were first to appear in the literature and were considered "standard." Steady-state evolutionary algorithms are described very well by Reynolds [49] and were discovered independently by Syswerda [54] and Whitley [59].

Experiment 2.1 *Write or obtain software for a string evolver (defined in Section 1.2). For each of the listed models of evolution, do 100 trials. Use 20-character strings of printable ASCII characters and a 60-member population. To stay consistent with single tournament selection in number of crossover events, implement all other models of evolution so that they replace exactly half the population. This updating of half the population will constitute a generation. For this experiment, use the type of crossover used in the first part of Figure 1.7 and Problem 13 in which the children are copies of the parents with their gene loci swapped after a randomly generated crossover point. For mutation, change a single character in each new creature at random.*

(i) Single tournament selection with small groups of size 4.
(ii) Roulette selection and locally elite replacement.
(iii) Roulette selection and random replacement.
(iv) Roulette selection and absolute fitness replacement.
(v) Rank selection and locally elite replacement.
(vi) Rank selection and random replacement.
(vii) Rank selection and absolute fitness replacement.

Write a few paragraphs explaining the results. Include the mean and standard deviation of the solution times (measured in generations) for each model of evolution. (A population is considered to have arrived at a "solution" when it contains one string that matches the reference string.) Compare your results with those of other students. Pay special attention to trials done by other students with identical models of evolution that give substantially different results.

Experiment 2.2 *Use the version of the code from Experiment 2.1 with roulette selection and random replacement. Compute the mean and standard deviation of time-to-solution of 100 trials in each of 5 identical populations in which you replace* $1/5, 1/3, 1/2, 2/3,$ *and* $4/5$ *of the population in each generation. Measure time in generations and in number of crossovers; discuss which measure of time is more nearly a fair comparison of the different models of evolution.*

Experiment 2.3 *Starting with the code from Experiment 2.1, build a steady-state evolutionary algorithm. For each of the following models of evolution, do 20 different runs. Give the mean and standard deviation of the number of mating events until a maximum fitness creature is located. Cut off the algorithm at 1,000,000 mating events if no maximum fitness creature is located. Assume that the double tournament selection is with replacement.*

(i) Single tournament selection with tournament size 4.
(ii) Single tournament selection with tournament size 6.
(iii) Double tournament selection with tournament size 2.
(iv) Double tournament selection with tournament size 3.

Problems

Problem 24. Assume that we are running an evolutionary algorithm on a population of 12 creatures, numbered 1 through 12, with fitness values of 1, 4, 7, 10, 13, 16, 19, 22, 25, 28, 31, and 34. Compute the expected number of children each of the 12 creatures will have for the following parent selection methods: (i) roulette selection, (ii) rank selection, and (iii) single tournament selection with tournament size 4. (The definition of *expected value* may be found in Appendix B.) Assume that both parents can be the same individual in the roulette and rank cases.

Problem 25. Repeat Problem 24 (i) and (ii), but assume that the parents must be distinct.

Problem 26. Compute the numbers that would appear in an additional column of Figure 2.3 for P(chosen) using single tournament selection with small groups of size 3.

Problem 27. Compute the numbers that would appear in an additional column of Figure 2.3 for P(chosen) using double tournament selection with small groups of size 4 and with replacement.

Problem 28. First, explain why the method of selecting parents, when separate from the method of placing children in the population, cannot have any effect on whether a model of evolution is elitist or not. Then, classify the following methods of placing children in the population as elitist or nonelitist. If it is possible for a method to be elitist or not depending on some other factor, e.g., fraction of population replaced, then say what that factor is and explain when the method in question is or is not elitist.

(i) random replacement.
(ii) absolute fitness replacement.
(iii) roulette wheel replacement.
(iv) rank replacement.
(v) locally elite replacement.
(vi) random elite replacement.

Problem 29. Essay. Aside from the fact that we already know the answer before we run the evolutionary algorithm, the problem being solved by a string evolver is very simple in the sense that all the positions in the creature's gene are independent. In other words, the degree to which a change at a particular location in the gene is helpful, unhelpful, or detrimental depends in no way on the value of the gene in other locations. Given that this is so, which of the possible models of evolution that you could build from the various parent selection and child placement methods, including single tournament selection, would you expect to work best and worst? Advanced students should support their conclusions with experimental data.

Problem 30. Give a sketch or outline of an evolutionary algorithm and a problem that together have the property that fitness in one genetic locus *can* be bought at the expense of fitness in another genetic locus.

Problem 31. Invent a model of evolution not described in this section that you think will be more efficient than any of those given for the string evolver problem. Advanced students should offer experimental evidence that their method beats both the models *single tournament selection* and *roulette selection with random replacement.*

Problem 32. Essay. Describe, as best you can, the model of evolution used by rabbits in their reproduction. One important difference between rabbits and a string evolver is that most evolutionary algorithms have a constant population whereas rabbit populations fluctuate somewhat. Ignore this difference by assuming a population of rabbits in which births and deaths are roughly equal per unit time.

Problem 33. Essay. Repeat Problem 32 for honeybees instead of rabbits. Warning: this is a hard problem.

Problem 34. Suppose that we modify the model of evolution "single tournament selection with group size 4" on a population of size $4n$ as follows. Instead of selecting the small groups at random, we select them in rotation as shown in the following table of population indices.

Generation	Group 1	Group 2 $\cdots$	Group n
1	0123	4567 $\cdots$	$(4n-4)(4n-3)(4n-2)(4n-1)$
2	$(4n-1)012$	3456 $\cdots$	$(4n-5)(4n-4)(4n-3)(4n-2)$
3	$(4n-2)(4n-1)01$	2345 $\cdots$	$(4n-6)(4n-5)(4n-4)(4n-3)$
4	$(4n-3)(4n-2)(4n-1)0$	1234 $\cdots$	$(4n-7)(4n-6)(4n-5)(4n-4)$
		etc.	

Call this modification *cyclic single tournament selection*. One of the qualities that makes single tournament selection desirable is that it can retard the rate at which the currently best gene spreads through the population. Would cyclic single tournament selection increase or decrease the rate of spread of a gene with relatively high fitness? Justify your answer.

Problem 35. Explain why double tournament selection of tournament size 2 without replacement and locally elite replacement is *not* the same as single tournament selection with tournament size 4. Give an example in which a set of 4 creatures is processed differently by these two models of evolution.

Problem 36. For double tournament selection with tournament size n with replacement and then without replacement, compute the expected number of mating events that the best gene participates in if we do one mating event for $n = 2, 3$, or 4 in a population of size 8.

2.2 Types of Crossover

Definition 2.3 *A* **crossover operator** *for a set of genes G is a map*

$$\text{Cross} : G \times G \to G \times G$$

or

$$\text{Cross} : G \times G \to G.$$

The points making up the pairs in the domain space of the crossover operator are termed **parents**, *while the points either in or making up the pairs in the range space are termed* **children**. *The children are expected to preserve some part of the parents' structure.*

In later chapters, we will study all sorts of exotic crossover operators. They will be needed because the data structures being operated on will be more complex than strings or arrays. Even for strings, there are a number of different types of crossover. The crossover used in Experiment 2.1 is called *single-point crossover*. To achieve a crossover with two parents, randomly generate a locus, called the *crossover point*, and then copy the loci in the genes from the parents to the child so that the information for each child comes from a different parent before and after the crossover point.

There is a problem with single-point crossover. Loci near one another in the representation used in the evolutionary algorithm are kept together with a much higher probability than those that are farther apart. If we are evolving strings of length 20 to match a string composed entirely of the character "A," then a creature with an "A" in positions 2 and 19 must almost be cloned during crossover in order to pass both good loci along. A simple way of reducing this problem is to have *multiple-point crossover*. In *two-point crossover*, as shown in Figure 2.4, two random loci are generated, and then the loci in the children are copied from one parent before and after the crossover points and from the other parent in between the crossover points. This idea generalizes in many ways. One could, for example, generate a random number of crossover points for each crossover or specify fixed fractions of usage for different sorts of crossover.

Parent 1	aaaaaaaaaaaaaaaaaaaa
Parent 2	bbbbbbbbbbbbbbbbbbbb
Child 1	aaaabbbbbbbbbaaaaaaa
Child 2	bbbbaaaaaaaaabbbbbbb

Fig. 2.4. Two-point crossover.

Experiment 2.4 *Modify the version of the code from Experiment 2.1 that does roulette selection with random elite replacement to work with different sorts of crossover. Run it as a steady-state algorithm for 100 trials. Use 20-character strings and a 60-member population. Measuring time in number of crossovers done, compare the mean and standard deviation of time-to-solution for the following crossover operators:*

(i) one-point,
(ii) two-point,
(iii) half-and-half one- and two-point.

When writing up your experiment, consult with others who have done the experiment and compare your trials to theirs.

Another kind of crossover, which is computationally expensive but eliminates the problem of representational bias, is *uniform crossover*. This crossover operator flips a coin for each locus in the gene to decide which parent contributes its genetic value to which child. It is computationally expensive because of the large number of random numbers needed, though clever programming can reduce the cost.

This raises an issue that is critical to research in the area of artificial life. It is easy to come up with new wrinkles for use in an evolutionary algorithm;

it is hard to assess their performance. If uniform crossover reduces the average number of generations, or even crossovers, to solution in an evolutionary algorithm, it may still be slower because of the additional time needed to generate the extra random numbers. Keeping this in mind, try the next experiment.

Experiment 2.5 *Repeat Experiment 2.4 with the following crossover operators:*

(i) one-point,
(ii) two-point,
(iii) uniform crossover.

In addition to measuring time in crossovers, also measure it in terms of random numbers generated and, if possible, true time by the clock. Discuss the degree to which the measures of time agree or fail to agree and frame and defend a hypothesis as to the worth of uniform crossover in this situation.

In some experiments, different crossover operators are better during different phases of the evolution. A technique to use in these situations is *adaptive crossover*. In adaptive crossover, each creature has its gene augmented by a *crossover template*, a string of 0's and 1's with one position for each item in the original data structure. When two parents are chosen, the crossover template from the first parent chosen is used to do the crossover. In positions where the template has a 0, items go from first parent to the first child and the second parent to the second child. In positions where the template has a 1, items go from the first parent to the second child and from the second parent to the first child. The parental crossover templates are themselves crossed over and mutated with their own distinct crossover and mutation operators to obtain the children's crossover templates. The templates thus coevolve with the creatures and seek out crossover operators that are currently useful. This can allow evolution to focus crossover activity in regions where it can help the most. The crossover templates that evolve during a successful run of an evolutionary algorithm may contain nontrivial useful information about the structure of the problem.

Example 1. Suppose we are designing an evolutionary algorithm whose gene consists of 6 real numbers. A crossover template would then be a string of six 0's and 1's, and crossover would work like this:

	Gene	Template
Parent 1	1.2 3.4 5.6 4.5 7.9 6.8	010101
Parent 2	4.7 2.3 1.6 3.2 6.4 7.7	011100
Child 1	1.2 2.3 5.6 3.2 7.9 7.7	010100
Child 2	4.7 3.4 1.6 4.5 6.4 6.8	011101

The crossover operator used on the crossover templates is single-point crossover (after position 3).

Adaptive crossover can suffer from a common problem called a *two-time-scale problem*. The amount of time needed to efficiently find those fit genes that are easy to locate with a given crossover template can be a great deal less than that needed to *find* the crossover template in the first place. For some problems this will not be the case, for some it will, and intuition backed by preliminary data is the best tool currently known for telling which problems might benefit from adaptive crossover. If a problem must be solved over and over for different parameters, then saving crossover templates between runs of the evolutionary algorithm may help. In this case, the crossover templates are being used to find good representations, relative to the crossover operator, for the problem in general while solving specific cases.

Experiment 2.6 *Repeat Experiment 2.4 with the following crossover operators:*

(i) one-point,
(ii) two-point,
(iii) adaptive crossover.

For the variation operators for the crossover templates, use one-point crossover together with a mutation operator that flips a single bit 50% of the time. When comparing solution times, attempt to compensate for the additional computational cost of adaptive crossover. Using real time-to-solution would be one good way to do this.

The last crossover operator we wish to mention is *null crossover*. In null crossover there is no crossover; the children are copies of the parents. Null crossover is often used as part of a mix of crossover operators or when debugging an algorithm. We conclude with a definition that will become important when we return to studying genetic programming.

Definition 2.4 *A crossover operator is called* **conservative** *if the crossover of identical parents produces children identical to those parents.*

Problems

Problem 37. Assume that we are working with a string evolver. If the reference string is

$$1111111111111111111111$$

then what is the expected fitness of the children of

$$11111111000000000000$$

and

$$00000000000011111111$$

under:

(i) one-point crossover,

(ii) two-point crossover,

(iii) uniform crossover.

Problem 38. Assume that we are maximizing the real function $f(x,y) = \frac{1}{x^2+y^2+1}$ with the technique described in Problem 14. Find a pair of parents (x_1, y_1), (x_2, y_2) such that neither parent has a fitness of more than 0.1 but one of their potential crossovers has fitness of at least 0.9. Crossover in this case would consist simply in taking the x coordinate from one parent and the y coordinate from the other. Fitness of a gene (a, b) is $f(a, b)$.

Problem 39. Usually we require that a crossover operator be conservative. Give a nonconservative crossover operator for use in the string evolver that you think will improve performance and show why the lack of conservation might help.

Problem 40. Essay. Taking the point of view that evolution finds pieces of a solution and then puts them together, explain why conservative crossover operators might be a good thing.

Problem 41. Suppose that we keep track of which pairs of parents have high- or low-fitness children by simply tracking the average fitness of all children produced by each pair of parents. We use these numbers to bias the selection of a second parent after the first is selected with a pure fitness bias. If this technique is used in a string evolver, will there be a two-time-scale problem? Explain what two separate process are going on in the course of justifying your answer. Hint: what is the average number of children a given member of the population has?

Problem 42. Prove that for the string evolver problem, all of the conservative crossover operators given in this section conserve fitness in the following sense: if we have a crossover operator take parents (p_1, p_2) to children (c_1, c_2), then the sum of the fitness of the children equals the sum of the fitness of the parents.

Problem 43. Read Problem 42. Find a problem that does not have the conservation property described. Prove that your answer is correct.

Problem 44. Essay. In the definition of the term "crossover operator" there were two possibilities, producing one or two children. If we transform a crossover operator that produces two children into an operator that produces one by throwing out the least fit child, then do we disrupt the conservation property described in Problem 42? Do you think this would improve the average performance of a string evolver or harm it?

Problem 45. Suppose we are running a string evolver with a 20-character reference string, a crossover operator producing two children, and no mutation operator. What condition must be true of the original population for there to be any hope of eventual solution? Does the condition that allows eventual solution ensure it? Prove your answers to both these questions. Estimate theoretically or experimentally the population size required to give a 95% chance of satisfying this condition.

2.3 Mutation

Definition 2.5 *A* **mutation operator** *on a population of genes G is a function*

$$\text{Mute} : G \to G$$

that takes a gene to another similar but different gene. Mutation operators are also called unary variation operators.

Crossover mixes and matches disparate creatures; it facilitates a broad search of the space of data structures accessible to a given evolutionary algorithm. Mutation, on the other hand, makes small changes in individual creatures. It facilitates a local search and also a gradual introduction of new ideas into the population. The string evolvers we have studied use a single type of mutation: changing the value of the string at a single position. Such a mutation is called a *point mutation*. More complex data structures might have a number of distinct types of minimal changes that could serve as point mutations. Once you have a point mutation, you can use it in a number of ways to build different mutation operators.

Definition 2.6 *A* **single-point mutation** *of a gene consists in generating a random position within the gene and applying a point mutation at that position.*

Definition 2.7 *A* **multiple-point mutation** *consists in generating some fixed number of positions in the gene and doing a point mutation at each of them.*

Definition 2.8 *A* **probabilistic mutation with rate** α *operates by going through the entire gene and performing a point mutation with probability α at each position. Probabilistic mutation is also called uniform mutation.*

Definition 2.9 *A* **Lamarckian mutation of depth** k *is performed by looking at all possible combinations of k or fewer point mutations and using the one that results in the best fitness value.*

Definition 2.10 *A* **null mutation** *is one that does not change anything.*

Any mutation operator can be made *helpful* by comparing the fitnesses of the gene before and after mutation and saving the better result. (Lamarckian mutation is already helpful, since "no mutations" is included in "k or fewer point mutations.")

Any mutation operator may be applied with some probability, as was done in several of the experiments in this chapter so far. The following experiment illustrates the use of mutations.

Experiment 2.7 *Modify the standard string evolver software used in Experiment 2.1 as follows. Use roulette wheel selection and random elite replacement. Use two-point crossover and put in an option to either use or fail to use single-point mutation in a given run of the evolutionary algorithm. When used, the single-point mutation should be applied to every new creature. Compute the average time-to-solution, cutting off the algorithm at generation 3000 if it has not found a solution yet. Report the number of runs that fail and the mean solution time of those that do find a solution. Explain the differences the mutation operator created.*

Definition 2.11 *A* **mode** *of a function is informally defined as a high point in the function's graph. Formally, a* **point mode** *is a point p in the domain of f such that there is a region R, also in the domain of f, about that point for which, for each $x \neq p \in R$, it is the case that $f(x) < f(p)$. Another type of mode is a contiguous region of points all at the same height in the graph of f, such that all points around the border of that region are lower than the points in the region. Figure 2.5 shows a function with two modes.*

The string evolver problem is what is called a *unimodal* problem; that is to say, there is one solution and an uphill path from any place in the gene space of the problem to the solution. For any given string other than the reference string, there are single character changes that improve the fitness.

Note: single-character changes (no matter whether they help, hurt, or fail to change fitness) induce a notion of distance between strings. Formally, the distance between any two strings is the smallest number of one-character changes needed to transform one into the other. This distance, called *Hamming distance* or *Hamming metric*, makes precise the notion of similarity in Definition 2.5. Mutation operators on any problem induce a notion of distance, but rarely one as nice as Hamming distance.

When designing an evolutionary algorithm, you need to select a set of mutation operators and then decide how often each one will be used. The probability that a given mutation operator will be used on a given creature is called the *rate* or *mutation rate* for that operator. The expected number of point mutations to be made in a new creature is called the *overall mutation rate* of the evolutionary algorithm. For helpful and Lamarckian mutation operators, computation of the overall mutation rate is usually infeasible; it depends on the composition of the population.

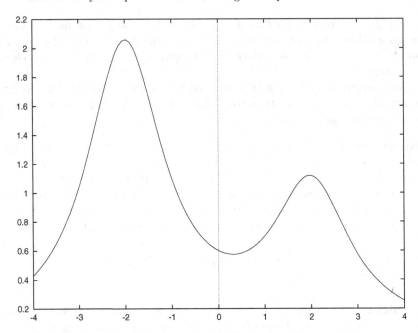

Fig. 2.5. A function with two modes.

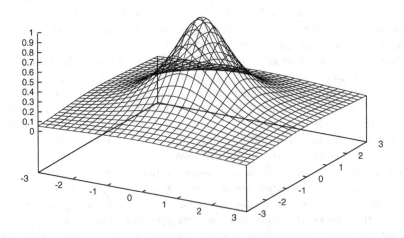

Fig. 2.6. Fake bell curve $f(x,y) = \frac{1}{1+x^2+y^2}$ in two dimensions.

To explore the effects of changing mutation rates, we will shift from string evolvers to function optimizers. It is folklore in the evolutionary algorithms community that an overall mutation rate equal to the reciprocal of the gene length of a creature works best for moving between nearby optima of nearly equal height when doing function optimization. Experiment 2.8 will test this notion.

The *fake bell curve in n dimensions* is given by the function

$$\mathcal{B}_n(x_1, x_2, \ldots, x_n) = \frac{1}{1 + \sum_{i=1}^{n} x_i^2}. \tag{2.1}$$

This function has a single mode at the origin of n-dimensional Euclidean space, as shown for $n = 2$ in Figure 2.6 (it is unimodal). By shifting and scaling this function we can create all sorts of test problems, placing optima where we wish, though some care is needed, as shown in Problem 48. Figure 2.5 was created in exactly this fashion.

Experiment 2.8 *Write or obtain software for a function optimizer with no crossover operator that uses probabilistic mutation with rate α. Use rank selection with random replacement. Use strings of real numbers of length n, where n is one of 4, 6, 8, and 10. Use overall mutation rates of r/n where r is one of 0.8, 0.9, 1, 1.1, 1.2. (Compute the α that yields the correct overall mutation rate: $r/n = n \cdot \alpha$, so $\alpha = r/n^2$.) Run the algorithm to optimize $f_n = \mathcal{B}_n(x_1, x_2, \ldots, x_n) + \mathcal{B}_n(x_1 - 2, x_2 - 2, \ldots, x_n - 2)$. Use a population of 200 creatures all initialized to $(2, 2, \ldots, 2)$. Do 100 runs. Compute the average time for a creature to appear that is less than 0.001 in absolute value in every locus.*

In Figures 2.6 and 2.5 we give examples of functions with one and two modes, respectively. The clarity of these examples relies on the smooth, continuous nature of the real numbers: both these examples are graphs of a continuous function from a real space to a real space. Our fundamental example, the string evolver, does not admit nice graphs. A string evolver operating on strings of length 20 would require a 20-dimensional graph to display the full detail of the fitness function. In spite of this, the string evolver fitness function is quite simple.

Problems

Problem 46. Suppose we modify a string evolver so that there are two reference strings, and a string's fitness is taken to be the number of positions in which it agrees with either of the reference strings. If the strings are of length l over an alphabet with k characters, then how many strings in the space exhibit maximum fitness? Hint: your answer will involve the number q of characters on which the two reference strings agree.

Problem 47. Is the fitness function given in Problem 46 unimodal? Prove your answer and describe any point or nonpoint modes.

Problem 48. Examine the fake bell curve, Equation 2.1, in 1 dimension, $f(x) = \frac{1}{1+x^2}$. If we want a function with two maxima, then we can take $f(x) + f(x - c)$ but *only if c is big enough*. Give the values of c for which $g(x) = f(x) + f(x - c)$ has one maximum, and those for which it has two.

Problem 49. Essay. Explain why it is difficult to compute the overall mutation rate for a Lamarckian or helpful mutation. Give examples.

Problem 50. Construct a continuous, differentiable (these terms are defined in calculus books) function $f(x, y)$ such that the function has three local maxima with the property that the line segment P (in x–y space) from the origin through the position of the highest maximum intersects the line segment Q joining the other two maxima, with the length of P at least twice the length of Q. Hint: multiply, don't add.

Problem 51. Suppose that we modify a string evolver to have two reference strings, but, in contrast to Problem 46, take the fitness function to be the maximum of the number of positions in which a given string matches one or the other of the reference strings. This fitness function can be unimodal, or it can have more than one mode. Explain under what conditions the function is uni- or multimodal.

Problem 52. Suppose that we are looking at a string evolver on strings of length 4 with underlying alphabet $\{0, 1\}$. What is the largest number of reference strings like those in Problem 51 that we could have and have as many modes as strings?

2.4 Population Size

Definition 2.12 *The* **population size** *of an evolutionary algorithm is the number of data structures in the evolving population.*

In biology it is known that small populations are likely to die out for lack of sufficient genetic diversity to meet environmental changes or because all members of the population share some defective gene. As we saw in Problem 45, analogous effects are possible even in simple evolutionary optimizers like the string evolver. On the other hand, a random initial population is usually jammed with average creatures. In the course of finding the reference string, we burn away a lot of randomness at some computational cost. There is thus a tension between the need for sufficient diversity to ensure solution and the need to avoid processing a population so large that it slows time-to-solution. Let's experiment with the string evolver to attempt to locate the sweet region and break-even point for increasing population size.

Experiment 2.9 *Modify the standard string evolver operating on 20-character strings as follows: Use roulette wheel selection, random elite replacement, and one-point mutation applied with probability one. Use a steady-state evolutionary algorithm and change the underlying alphabet to be $\{0, 1\}$. Put into the code the ability to change the population size. Measure the time-to-solution in crossover events, averaged over 100 runs, for populations of size* $20, 40, 60, 80, 100, 110,$ *and* $120.$ *Approximate the best size and do a couple of additional runs near where you suspect the best size is. Graph the results as part of your write-up.*

Problems

Problem 53. Essay. Larger populations, having higher initial diversity, should present less need to preserve diversity. Would you expect larger populations to be of more value in preserving diversity in a unimodal or polymodal problem as compared to diversity preservation techniques like single tournament selection?

Problem 54. Give a model of evolution that can process a large population more efficiently (for the string evolver problem) than any of the ones given in this chapter. Hint: concentrate on small subsets of the population without completely ignoring anyone.

Problem 55. Essay. There is no requirement in the theory of evolutionary algorithms that we have one population. In fact, when we do 100 experimental runs, we are using 100 different populations. Give a specification, like those in the text, for an experiment that will test into how many small populations 600 creatures should be divided for an arbitrary problem. It should explore reasonably between the extremes of running one population of 600 creatures and 600 populations of one creature each.

2.5 A Nontrivial String Evolver

An unfortunate feature of the string evolver is that it solves a trivial problem. It is possible to build very difficult string evolution problems by modifying the way in which fitness is computed. The standard example of this is the *Royal Road* function (defined by John Holland), which is defined over the alphabet $\{0, 1\}$. This function assumes a reference string of length 64, but blocks of 8 adjacent characters in positions 1–8, 9–16, ..., 57–64 are given special status. For each such block made entirely of 1's, the string's fitness is incremented by 8. Blocks with only some 1's give no fitness. This function is quite difficult to optimize and is a good test function for evolutionary optimization systems of difficult unimodal problems. The length of 64 and block size of 8 are traditional, but varying these numbers yields many possibly interesting test problems.

Definition 2.13 *Define the* **Royal Road function of length** l **and block size** b, *where* b *divides* l *evenly, to be a fitness function for strings where fitness is assessed by dividing the string into* l/b *pieces of length* b *and then giving a fitness of* b *for each piece on which a string in an evolving population exactly matches the reference string.*

Experiment 2.10 *Take the software you used for Experiment 2.4 and modify it to work on the Royal Road function with reference string "all ones" and alphabet* $\{0,1\}$ *with* $l = 16$ *and* $b = 1, 2, 4, 8$. *Report the mean and deviation time-to-solution over 100 runs for a population of 120 creatures, cutting off an unsuccessful run at 10,000 generations (do not include the cutoff runs in the mean and deviation computations). If you have a fast enough computer, obtain higher-quality data by increasing the cutoff limit. Use two-point crossover and single-point mutation (with probability one). In addition to reporting and explaining your results, explain why cutoff is probably needed and is a bad thing. What is the rough dependence of time-to-solution on* b?

Experiment 2.11 *Modify the software from Experiment 2.10 so that it uses probabilistic mutation with rate* α. *For* $l = 16$ *and* $b = 4$ *make a conjecture about the optimum value for* α *and test this conjecture by finding average time-to-solution over 100 runs for* $80\%, 90\%, 100\%, 110\%,$ *and* 120% *of your conjectured* α. *Feel free to revise your conjecture and rerun the experiment.*

Problems

Problem 56. Compute the probability of even one creature having nonzero fitness in the original population of n genes in a string evolver on the alphabet $\{0,1\}$ when the fitness function is the Royal Road function of length l and block size b for the following values:

(i) $n = 60$, $l = 36$, $b = 6$,
(ii) $n = 32$, $l = 49$, $b = 7$,
(iii) $n = 120$, $l = 64$, $b = 8$,
(iv) $n = 20$, $l = 120$, $b = 10$.

Problem 57. Essay. Suppose we are running a string evolver with the classical Royal Road fitness function ($l = 64$, $b = 8$). Which of the mutation operators in this section would you expect to be most helpful and why? Clearly, Lamarckian mutation with a depth of 8 would guarantee a solution, but it is computationally very expensive. Keeping this example in mind, factor computational cost into your discussion.

Problem 58. Essay. Single tournament selection does not perform well relative to roulette selection with random elite replacement on the basic string evolver. If possible, experimentally verify this. In any case, conjecture why this is so and tell whether you would expect this also to be so with the classical Royal Road fitness function ($l = 64$, $b = 8$). Support your argument with experimental data if it is available.

Problem 59. Read Problem 57. How many sets of point mutations must be checked in a single Lamarckian mutation of depth 8?

Problem 60. Consider a string evolver over the alphabet $\{0, 1\}$ using a Royal Road fitness function with $l = 4$ and a population of 2 creatures. The evolver proceeds by copying a single-point mutation of the best creature onto the worst creature in each generation. Estimate mathematically or experimentally the time-to-solution for $b = 1$, 2, 4 if the reference string is 1111 and the population is initialized to be all 0000. Appendix B, on probability theory, may be helpful.

Problem 61. Is the classical Royal Road fitness function unimodal?

2.6 A Polymodal String Evolver

In this chapter so far we have experimented with a number of evolutionary algorithms that work on unimodal fitness functions. In addition, we have worked, in Experiment 2.8, with a constructively bimodal fitness function. In this section, we will work with a highly polymodal fitness function. This polymodal fitness function is one used to locate *self-avoiding walks* that cover a finite grid.

Definition 2.14 *A* **grid** *is a collection of squares, called* **cells***, laid out in a rectangle (like graph paper).*

Definition 2.15 *A* **walk** *is a sequence of moves on a grid between cells that share a side. If no cell is visited twice, then the walk is* **self-avoiding***. If every cell is visited, then the walk is* **optimal***.*

From any cell in a grid, then, there are four possible moves for a walk: up, down, right, and left. We will thus code walks as strings over the alphabet $\{$**U, D, L, R**$\}$, which will be interpreted as the successive moves of a walk. Some examples of walks are given in Figure 2.7.

To evolve self-avoiding walks that cover a grid, we will permit the walks to fail to be self-avoiding, but we will write the fitness function so that the best score can be obtained only by a self-avoiding walk. Definition 2.16 gives such a function. If we think of self-avoiding walks as *admissible* configurations and walks that fail to avoid themselves as *inadmissible*, then we are permitting our evolutionary algorithm to search an entire space while looking for islands of admissibility. When a space is almost entirely inadmissible, attempting to search only the admissible parts of it is impractical. It is thus an interesting question, treated in the Problems, what fraction of the space is admissible.

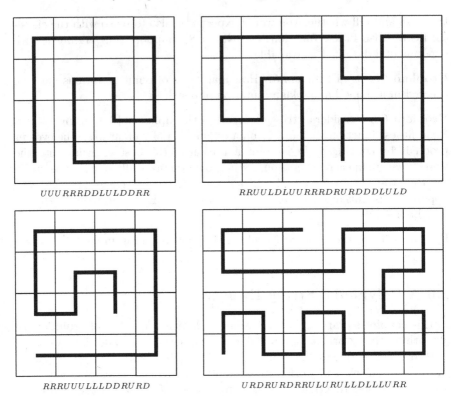

UUURRRDDLULDDRR

RRUULDLUURRRDRURDDDLULD

RRRUUULLLDDRURD

URDRURDRRULURULLDLLLURR

Fig. 2.7. Optimal self-avoiding walks on 4×4 and 4×6 grids that visit every cell. (The walks are traced as paths starting in the lower left cell and shown in string form beneath the grids with U=up, D=down, R=right, and L=left.)

Definition 2.16 *The* coverage fitness *of a random walk of length $NM - 1$ on an $N \times M$ grid is computed as follows: Begin in the lower left cell of the grid, marking it as visited. For each of the moves in the random walk, make the move (if it stays on the grid) or ignore the move (if it attempts to move off the grid). Mark each cell reached during the walk as visited. The fitness function returns the number of cells visited.*

Notice that this fitness function requires that the walk have exactly one fewer move than there are cells, so each move must hit a new cell. The examples given in Figure 2.7 have this property.

Experiment 2.12 *Modify the basic string evolver software to work on a population of n strings with two-point crossover and k-point mutation. Use size-4 tournament selection applied to the entire population. Make sure that changing n and k is easy. Run 400 populations each using the coverage fitness on 15-character strings over the alphabet {U, D, R, L} on a 4×4 grid for $n = 200, 400$ and $k = 1, 2, 3$. This is 2400 runs and will take a while on even* *a fast computer. Stop each individual run when a solution is found (this is a success) or when the run hits 1000 generations. Tabulate the number of successes and the fraction of successes. Discuss whether there is a clearly superior mutation operator and discuss the merits of the two population sizes (recalling that the larger one is twice as much work per generation).*

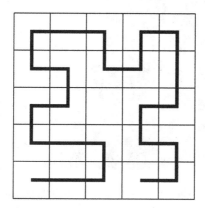

Fig. 2.8. A slightly suboptimal walk.

If the code used for Experiment 2.12 does a running trace of the best fitness, then it is easy to see that the search "gets stuck" sometimes. If you save time-to-solution for the runs that terminate in fewer than 1000 generations, you will also observe that solution is often rapid, much faster than 1000 generations. This suggests that not only are there many global optima (Figure 2.7 shows a pair of global optima for each of two different grid sizes), but there is probably a host of local optima. Look at the walk shown in Figure 2.8. It has a coverage fitness of 24; the optimal is 25. It is also several point mutations from any optimal gene. Thus, this walk forms an example of a local optimum.

As we will see in the Problems, each optimal self-avoiding walk has a unique encoding, but local optima have a number of distinct codings that in fact grows with their degree of suboptimality. As we approach an optimum, the fragility of our genetic representation of the walk grows. More and more of the loci are such that changing them materially decreases fitness. Let's take a look at how fitnesses are distributed in a random sample of strings coding for walks. Figure 2.9 shows how the fitnesses of 10,000 genes generated uniformly at random are distributed. Given that our evolutionary algorithms can find solutions to problems of this type, clearly the evolutionary algorithm is superior to mere random sampling. Our next experiment is intended to give us a tool for documenting the presence of a rich collection of local optima using the coverage fitness function.

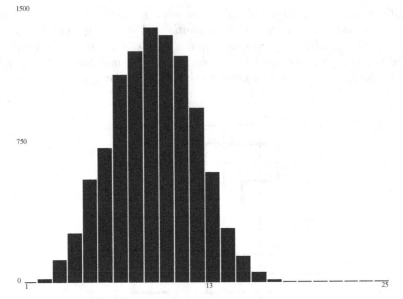

Fig. 2.9. A histogram of the covering fitness of 10,000 strings of 24 moves on a 5×5 grid. (The most common fitness was 10, attained by 1374 of the strings. The largest fitness obtained was 20.)

Definition 2.17 *A* **stochastic hill climber** *is an algorithm that repeatedly modifies an initial configuration, saving the new configuration only if it is better (or no worse).*

Experiment 2.13 *Write or obtain software for a stochastic hill climber that requires that new results be better for length-24 walks on a 5×5 grid starting in the lower left cell. Use single-point mutation to perform modifications. Run the hill climber for 1000 steps each time you run it, and run it until you get 5 walks of fitness 20 or more. Make pictures of the walks, pooling results with anyone else who has performed the experiment.*

Figure 2.10 shows four walks generated by a stochastic hill climber. The coverage fitnesses of these walks are 20, 16, 18, and 19, respectively. All four fail to self-avoid, and all four arose fairly early in the 1,000-step stochastic hill climb. If these qualities turn out to be typical of the walks arrived at in Experiment 2.13, then it seems that a stochastic hill climber is not the best tool for exploring this fitness landscape. In the interest of fairness, let us extend the reach of our exploration of stochastic hill climber behavior with an additional experiment.

Experiment 2.14 *Modify the stochastic hill climber from Experiment 2.13 to use two-point mutation. In addition to this change, perform 10,000 rather than 1000 mutations. (This is probably more than necessary, but it should*

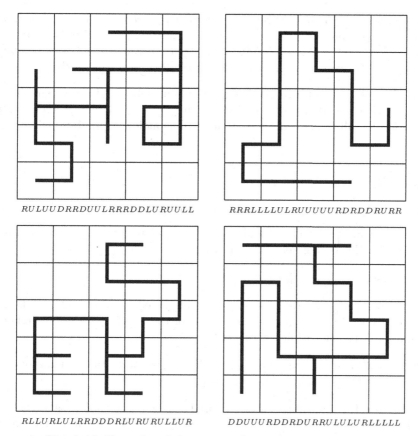

RULUUDRRDUULRRRDDLURUULL

RRRLLLLULRUUUUURDRDDRURR

RLLURLULRRDDDRLURURULLUR

DDUUURDDRDURRULULURLLLLL

Fig. 2.10. Examples of the output of a stochastic hill climber.

be computationally manageable.) Run both the old and new hill climbers 100 *times and compare histograms of the resulting fitnesses.*

The stochastic hill climbers in Experiments 2.13 and 2.14 require that new results be better, so they will make a move only if it leads uphill. Taking the mutated string only if it was *no worse* may tend to let the search move more, simply because "sideways" moves are permitted. Let's see what we can learn about the effect of these sideways moves.

Experiment 2.15 *Modify the stochastic hill climbers from Experiments 2.13 and 2.14 so that they accept mutated strings that are no worse. Repeat Experiment 2.14 with the modified hill climbers. Compare the results.*

A stochastic hill climber can be viewed as repeated application of a helpful mutation operator to a single-member population. After having done all this work on stochastic hill climbing, it might be interesting to see how it works within the evolutionary algorithm.

Experiment 2.16 *Modify the software from Experiment 2.12 to use helpful mutation operators part of the time. Rerun the experiment for $n = 200$ and $k = 1, 2$ with 50% and 100% helpful mutation. Compare with the corresponding runs from Experiment 2.12. Summarize and attempt to explain the effects.*

We conclude this phase of our exploration of polymodal fitness functions. We will revisit this fitness function in Chapter 13, where a technique for structurally enhancing evolutionary algorithms at low computational cost is explored.

Problems

Problem 62. For a 3×3 grid and walks of length 8 moves, give examples of:

(i) An optimal self-avoiding walk *other* than UURRDDLU (which is given later in this section as part of a problem).
(ii) A non-self-avoiding walk.
(iii) A self-avoiding nonoptimal walk.

Notice that you will have to waste moves at the edge of the grid (which are not moves at all) in order to achieve some of the answers. Be sure to reread Definition 2.16 before doing this problem.

Problem 63. Give an example of a self-avoiding walk that cannot be extended to an optimal self-avoiding walk. You may pick your grid size.

Problem 64. Make a diagram, structured as a tree, showing all self-avoiding walks on a 3×3 grid that start in the lower left cell, excluding those that waste moves off the edge of the grid. These walks will vary in length from 1 to 8. This is easy as a coding problem and a little time-consuming by hand. While there are only 8 optimal self-avoiding walks, there are quite a few self-avoiding walks.

Problem 65. Prove that the coverage fitness function given in Definition 2.16 awards the maximum possible fitness only to optimal self-avoiding walks.

Problem 66. Give an exact formula for the number of optimal self-avoiding walks on a $1 \times n$ and on a $2 \times n$ grid as a function of n. Assume that the walks start in the lower left cell.

Problem 67. Draw all possible optimal self-avoiding walks on a 3×3 grid and a 3×4 grid. Start in the lower left cell.

Problem 68. Give an exact formula for the number of optimal self-avoiding walks on a $3 \times n$ grid as a function of n. Assume that the walks start in the lower left cell. (This is a very difficult problem.)

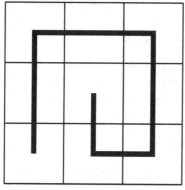

UURRDDLU

Problem 69. Review the discussion of admissible and inadmissible walks at the beginning of this section. For the length-8 walk given above, how many of the one-point mutants of the walk are admissible? Warning: there are 3^8 one-point mutants of this walk; you need either code or cleverness to do this problem.

Problem 70. Suppose that instead of wasting moves that move off the grid, we wrap the grid at the edges. Does this make the problem harder or easier to solver via evolutionary computation?

Problem 71. Prove that all single-point mutations of a string specifying an optimal self-avoiding walk are themselves nonoptimal.

Problem 72. Find a walk with a coverage fitness one less than the maximum on a 3×3 grid and then enumerate as many strings as you can that code for it (at least 2).

Problem 73. On a 5×5 grid, make an optimal self-avoiding walk and find a point mutation such that the fitness decrease caused by the point mutation is as large as possible.

Problem 74. Construct a string for the walk shown in Figure 2.8 that ends in a downward move off the grid (there is only one such string). Now find the smallest sequence of point mutations you can that makes the string code for an optimal self-avoiding walk.

Problem 75. Modify the software for Experiment 2.13 to record when the hill climber, in the course of performing the stochastic hill climb, found its best answer. Give the mean, standard deviation, and maximum and minimum times to get stuck for 1000 attempts.

Problem 76. Read the description of Experiments 2.13 and 2.14. Explain why a stochastic hill climber using two-point mutation might need more trials per hill climbing attempt than one using one-point mutation.

Problem 77. Given that we start in the lower left cell of a grid, prove that there are never more than three choices of a way for a walk to leave a given grid cell in a self-avoiding fashion.

Problem 78. Based on the results of Problem 77, give a scheme for coding walks starting in the lower left cell of a grid with a ternary alphabet. Find strings that result in the walks pictured in Figure 2.7.

Problem 79. Prove that the fraction of genes that encode optimal self-avoiding walks is less than $\left(\frac{3}{4}\right)^{NM-1}$ on an $N \times M$ grid.

Problem 80. Essay. Based on the sort of reencoding needed to answer Problem 78 (use your own if you did the problem), try to argue for or against the proposition that the reencoding will make the space easier or harder to search with an evolutionary algorithm. Be sure to address not only the size of the search space but also the ability of the algorithm to get caught. If you are feeling gung ho, support your argument with experimental evidence.

2.7 The Many Lives of Roulette Selection

In Section 2.1, we mentioned roulette selection as one of the selection techniques that can be used to build a model of evolution. It turns out that the basic roulette selection code, given in Figure 2.11, can be used for several tasks in evolutionary computation. The most basic is to perform roulette selection, but there are others. Let us trace through the roulette selection code and make sure we understand it first.

The routine takes, as arguments, an array of positive fitness values f and an integer argument n that specifies the number of entries in the fitness array. It returns an integer, the index of the fitness selected. The probability that a given index i will be selected is in proportion to the fraction of the total fitness in f at $f[i]$. Why? The routine first totals f, placing the resulting total in the variable ttl. It then multiplies this total by a random number uniformly distributed in the interval $[0, 1]$ to select a position in the range $[0, \text{Total Fitness}]$, which is placed in the variable $dart$. (The variable name is a metaphor for throwing a dart at a virtual dart board that is divided into areas that are proportional to the fitnesses in f.) We then use iterated subtraction, of successive fitness values from the dart, to find out where the dart landed. If the dart is in the range $[0, f[0])$, then subtracting $f[0]$ from the dart will drive the total negative. If the dart is in the range $[f[0], f[0] + f[1])$, then the iterated subtraction will go negative once we have subtracted both $f[0]$ and $f[1]$. This pattern continues with the effect that the probability that the iterated subtraction will drive the dart negative at index i is exactly $f[i]/ttl$. We thus return the value of i at which the iterated subtraction drives the dart negative.

```
//Returns an integer in the range 0 to (n-1) with probability of i
//proportional to f[i]/Sum(f[i]).

int RouletteSelect(double *f;int n){  //f holds positive fitness
values
                                      //n counts entries in f

int i; double ttl,dart;

   ttl=0;
   for(i=0;i<n;i++)ttl+=f[i];  //compute the total fitness
   dart=ttl*random01();        //generate randomly
                               //0<=dart<=(total fitness)
   i=-1;
   do {
      dart-=f[++i];  //subtract successive fitnesses;
   } while(dart>=0); //the one that takes you negative is
                     //where the dart landed

   return(i);  //tell the poor user what the decision is
}
```

Fig. 2.11. Roulette selection code.

Now that we have code for roulette selection, let's figure out what else we can do with it. It is often desirable to select in direct proportion to a function of the fitness. If, for instance, we have fitness values in the range $0 < x < 1$ but we want some minimal chance of every gene being selected, then we might use $x + 0.1$ as the "fitness" for selection. This could be coded by simply preprocessing the fitness array f before handing it off to the *RouletteSelect* routine. In general, if we want to select in proportion to $g(fitness)$, then we need only apply the function $g(x)$ to each entry of f before using it as the "fitness" array passed to *RouletteSelect*. It is, however, important for correct functioning of both evolution and the selection code that $g(x)$ be a monotone function, i.e., $a < b \rightarrow g(a) < g(b)$.

The other major selection method in Section 2.1 was *rank selection*. There we gave the most fit of n creatures rank n, the next most fit rank $n-1$, etc., and then selected creatures to be parents in proportion to their rank. Rank is thus nothing more than a monotone function of fitness. This means that the roulette selection code is also rank selection code as long as we pass an array of ranks. If we compute the ranks in reverse fashion, with the most fit creature's rank at 1, then the roulette selection code may be used to do the selection needed for rank replacement. Roulette replacement is also achieved by a simple modification of f. Let us now consider an application of roulette selection to the computational details of mutation.

The Poisson Distributions and Efficient Probabilistic Mutation

When we place a probability distribution on a finite set, we get a list of probabilities, each associated with one member of the finite set. Typically, a programming language comes equipped with a routine that generates random integers in the range $0, 1, \ldots, n - 1$ and with another routine that generates random floating point numbers in the range $(0, 1)$. As long as a uniform distribution on an interval is all that is required, an affine transformation $g(x) = ax + b$ can transform these basic random numbers into the integer or floating point distribution required. Computing nonuniform distributions can require a good deal of mathematical muscle. In Chapter 3 we will learn to transform uniform 0-1 random numbers into normal (also called *Gaussian*) random numbers. Here we will adapt roulette selection to nonuniform distributions on finite sets and then give an application for efficiently performing probabilistic mutation.

By now, the alert reader will have noticed that if we know a probability distribution on a finite set, then the roulette selection routine can generate probabilities according to that distribution if we simply hand it that list of probabilities in place of the fitness array. If, for example, we pass $f = \{0.5, 0.25, 0.25\}$ to the routine in Figure 2.11, then it will return 0 with probability 0.5, 1 with probability 0.25, and 2 with probability 0.25. In the course of designing simulations and search software in later chapters, it will be useful to be able to select random numbers according to any distribution we wish, but at present we want to concentrate on a particular distribution, the Poisson distribution.

In Appendix B, the *binomial distribution* is discussed at some length. When we are doing n experiments, each of which can either succeed or fail, the binomial distribution lets us compute the probability of k of the experiments succeeding. The canonical example of this kind of experiment is flipping a coin with "heads" being taken as a success. Now imagine we were to flip 3000 (very odd) coins, and that the chance of getting a head was only one in 1,500. Then, on average, we would expect to get 2 heads, *but* if we wanted to compute explicitly the chance of getting 0 heads, 1 head, etc., numbers like 3000! (three-thousand factorial) would come into the process, and our lives would become a trifle difficult. This sort of situation, a very large number of experiments with a small chance of success, comes up fairly often. A statistician examining data on how many Prussian cavalry officers were kicked to death by their horses (a situation with many experiments and few "successes") discovered a short cut.

As long as we have a very large number of experiments with a low probability of success, the *Poisson distribution*, Equation 2.2, gives the probability of k successes with great accuracy:

$$P(k \text{ successes}) = \frac{e^{-m} \cdot m^k}{k!} \tag{2.2}$$

The parameter m requires some explanation. It is the average number of successes. For n experiments with probability α of success we have $m = n\alpha$. In Figure 2.12 we give an example of the initial part of a Poisson distribution, both listed and plotted. How does this help us with probabilistic mutation?

When we perform a probabilistic mutation with rate α on a string with n characters, we generate a separate random number for each character in the string. If the length of the string is small, this is not too expensive. If the string has 100 characters, this can be a very substantial computational expense. Avoiding this expense is our object. Typically, we keep the expected number of mutations, $m = n\alpha$, quite small by keeping the string length times the rate of the probabilistic mutation operator small. This means that the Poisson distribution can be used to generate the number of mutations r, and then we can perform an r-point mutation.

There is one small wrinkle. As stated in Equation 2.2, the Poisson distribution gives a positive probability to each integer. This means that if we fill an n-element array with the Poisson probabilities of $0, 1, \ldots, n-1$, the array will not quite sum to 1 and will hence not quite be a probability distribution. Looking at Figure 2.12, we see that the value of the Poisson distribution drops off quite quickly. This means that if we ignore the missing terms after $n-1$ and send the not-quite-probability distribution to the routine $RouletteSelect(f, n)$, we will get something very close to the right numbers of mutations, so close, in fact, that it should not make any real difference in the behavior of the mutation operator. In the Problems, we will examine the question of when it is worth using a Poisson distribution to simulate probabilistic mutation.

Problems

Problem 81. The code given in Figure 2.11 is claimed to require that f be an array of *positive* fitness values. Explain why this is true and explain what will happen if (i) some zero fitness values are included and (ii) negative fitness values creep in.

Problem 82. The code given in Figure 2.11 returns an integer value without explicitly checking that it is in the range $0, 1, \ldots, n-1$. Prove that if all the fitness values in f are positive, it will return an integer in this range.

Problem 83. Modify the $RouletteSelect(f, n)$ routine to work with an array of integral fitness values. Other than changing the variable types, are any changes required? Why or why not?

Problem 84. If C is not your programming language of choice, translate the routine given in Figure 2.11 to your favored language.

Problem 85. Explicitly explain, including the code to modify the entries of f, how to use the $RouletteSelect(f, n)$ code in Figure 2.11 for roulette replacement. This, recall, selects creatures to be replaced by new creatures with probability inversely proportional to their fitness.

P(0)=0.135335
P(1)=0.270671
P(2)=0.270671
P(3)=0.180447
P(4)=0.0902235
P(5)=0.0360894
P(6)=0.0120298
P(7)=0.00343709
P(8)=0.000859272
P(9)=0.000190949
P(10)=3.81899e-05
P(11)=6.94361e-06

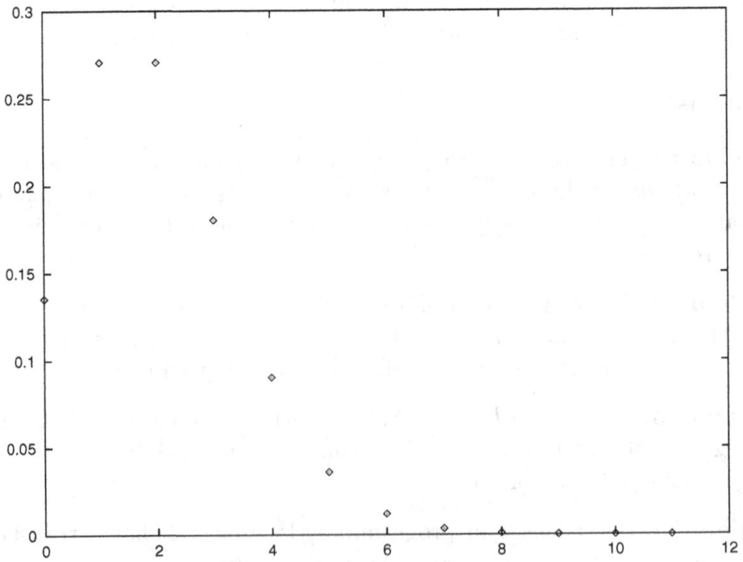

. . .

Fig. 2.12. A listing and plot of the Poisson distribution with a mean of $m = 2$.

Problem 86. Give the specialization of Equation 2.2 to a mean of $m = 1$ and compute for which k the probability of k successes drops to no more than 10^{-6}.

Problem 87. Suppose we wish to perform probabilistic mutation on a 100-character string with rate $\alpha = 0.03$. Give the Poisson distribution of the number of mutations and give the code to implement efficient probabilistic mutation as outlined in the text. Be sure to design the code to compute the partial Poisson distribution only once.

Problem 88. For an n-character string gene being modified by probabilistic mutation with rate α, compute the number of random numbers (other than those required to compute point mutations) needed to perform efficient probabilistic mutation. Compare this to the number needed to perform probabilistic mutation in the usual fashion. From these computations derive a criterion, in terms of n and α, for when to use the efficient version of probabilistic mutation instead of the standard one.

Problem 89. Suppose we are using an evolutionary algorithm to search for highly fit strings that fit a particular criterion. Suppose also that all good strings, according to this criterion, have roughly the same fraction of each character but have them arranged in different orders. If we know a few highly fit strings and want to locate more, give a way to apply $RouletteSelect(f, n)$ to generate initial populations that will have above average fitness. (Starting with these populations will let us sample the collection of highly fit strings more efficiently.)

Problem 90. Suppose we have an evolutionary algorithm that uses a collection of several different mutation operators. For each, we can keep track of the number of times it is used and the number of times it enhances fitness. From this we can get, by dividing these two numbers, an estimate of the probability each mutation operator has of improving a given gene. Clearly, using the most useful mutation operators more often would be good. Give a method for using $RouletteSelect(f, n)$ to probabilistically select mutation operators according to their estimated usefulness.

Problem 91. Essay. Read Problem 90. Suppose we have a system for estimating the usefulness of several mutation operators. In Problem 90, this estimate is the ratio of applications of a mutation operator that enhanced fitness to the total number of applications of that mutation operator. It is likely that the mutation operators that help the most with an initial, almost random, population will be different from those that help the most with a converged population. Suggest and justify a method for estimating the *recent* usefulness of each mutation operator, such as would enhance performance when used with the system described in Problem 90. Discuss the computational complexity of maintaining these moving estimates and try to keep the computational cost of your technique low.

3

Optimizing Real-Valued Functions

This chapter will expand the concept of string evolver, changing the alphabet from the character set to the set of real numbers. This will lead to our first problem of interest outside of the evolutionary computation community: maximization and minimization of real-valued functions of real variables. In this chapter we will explore different types of functions, from the continuous ones that can be optimized with classical methods like the calculus to more difficult functions that are constant except where they are discontinuous. This latter class of functions may sound artificial to someone whose mathematical education has been rich in the beautiful theory of real analysis, but such functions arise naturally in tasks like printed circuit board layout. The most difficult issue we will deal with in this chapter is that of mutation. Real variables take on continuous rather than discrete values, and so our mutation operators will become probability distributions. The set of "mutants" of one structure is no longer a finite set that can be enumerated with mere computer power, but rather a set which is theoretically infinite and in practice gigantic that must be handled with care.

In this chapter we will first create a basic evolutionary algorithm for optimizing real-valued functions. With this software in hand we will discuss one of the primary metaphors of evolutionary computation, the *fitness landscape*. We will then explore niche specialization. This is a modification inspired by the specialization of living organisms to different niches in nature. We will then explore two illustrative examples, finding a minimal-length path through the unit square and minimizing the crossing number of a combinatorial graph. Figure 3.1 lists the experiments in this chapter and shows how they depend on one another.

While optimizing real-valued functions we will call strings of real numbers by the more standard name, *arrays* of real numbers. A knowledge of calculus is helpful, and a few pertinent facts are included in Appendix C. We have already previewed this area in Problems 14 and 38 and in Experiment 2.8. Using real functions gives us some additional machinery. For example, we can use a much more efficient Lamarckian mutation when the function is differentiable. This

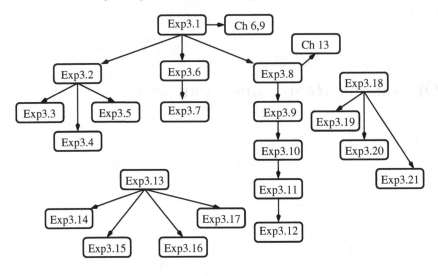

1 Basic real function optimizer.
2 Exploring types of mutation.
3 Real Lamarckian mutation.
4 Exploring types of crossover.
5 Comparing direct and binary representations.
6 Deceptive optima experiment.
7 The effects of space warp on deception.
8 Domain niche specialization, three optima.
9 Niche specialization, restricted initial range.
10 Niche specialization with nonpoint optima.
11 Testing range niche specialization.
12 Student's choice domain and range niche specialization.
13 Introducing the path length problem.
14 Path length with Gaussian mutation.
15 Path length with decreasing mutation.
16 Path length with alternative fitness.
17 Path length with alternating fitness functions.
18 Introducing the crossing number function.
19 Crossing number with Gaussian mutation.
20 Crossing number with niche specialization.
21 Crossing number on sparser graphs.

Fig. 3.1. The topics and dependencies of the experiments in this chapter.

mutation operator is explained in Appendix C. In this chapter, we will have our first nontrivial representation issues: comparing atomic and nonatomic representations of real numbers. (Recall that *atom* is from the Greek *a* (not) *tomos* (cut), meaning uncuttable.)

Some terminology concerning optima will be useful in this chapter. An *optimum* is a minimum or maximum of a function. An optimum is said to be *local* if it is larger (respectively smaller) than all nearby points. An optimum is *global* if it is larger (respectively smaller) than every other value the function takes on over its domain space. Since minimizing f is equivalent to maximizing $-f$, we will speak in the remainder of the chapter as if we were maximizing functions and as if our optima were maxima. Following terminology in statistics, optima are also sometimes called *modes*, and a function with only one optimum, e.g., $f(x) = 1 - x^2$, is said to be *unimodal*. Compare these with the definitions of mode and unimodal and local and global optimum from Chapter 2.

3.1 The Basic Real Function Optimizer

The crossover operators on strings, given in Section 2.2, carry over directly as crossover operators on arrays of real numbers. In one representation used for optimization of real-valued functions, real numbers are represented as strings of characters (representing bits), and the crossover operators are allowed to split real numbers "in the middle." This practice causes rather bizarre behavior in the split real numbers, and so we will avoid it for the most part by forcing our crossover operators to respect real boundaries. In the last chapter a point mutation consisted in replacing a character with another randomly generated character; a point mutation for a real number will consist in adding or subtracting small values to some locus of the gene.

Definition 3.1 *For a real number ϵ, we define a* **uniform real point mutation** *of a gene consisting of an array of real numbers to be addition of a uniformly distributed random number in the range $[-\epsilon, \epsilon]$ to a randomly chosen locus in the gene. The number ϵ is called the* maximum mutation size *of the mutation operator.*

A uniform real point mutation with maximum mutation size ϵ causes one of the loci in a gene to jump to a new value within ϵ of its current value. All numbers that can be reached by the mutation are equally likely. This definition of point mutation gives us an uncountable suite of mutation operators by varying the maximum mutation size continuously. Usually the problem under consideration suggests reasonable values for ϵ. Once we have a point mutation, we can build from it the one, two, and k-point mutations, probabilistic mutations, and the helpful mutations described in Section 2.3. The "all" clause in

the definition of Lamarckian mutation makes it impossible to use that definition with the definition of point mutation we have above, but in Appendix C
we define a derivative-based Lamarckian mutation for differentiable functions.

Definition 3.2 *For a real number σ, we define a* **Gaussian real point mu-**
tation *of a gene consisting of an array of real numbers to be addition of a
normally distributed random number with mean zero and standard deviation
σ to a randomly chosen locus in the gene.*

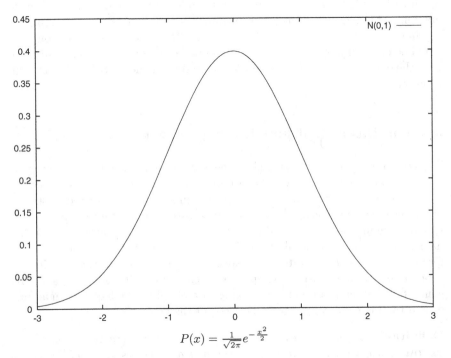

$$P(x) = \frac{1}{\sqrt{2\pi}}e^{-\frac{x^2}{2}}$$

Fig. 3.2. A graph of the probability density function of the Gaussian distribution
with mean zero and deviation 1. (Note that the tails of the distribution continue
indefinitely.)

A Gaussian real point mutation shares with the uniform point mutation
the property that its average is zero. Therefore, increasing the value of the
number hit with the mutation is symmetric to decreasing its value. The Gaussian mutation differs in being able to go farther. Where all possible numbers
are equally likely with a uniform mutation, the chance of reaching a number
drops off rapidly with a Gaussian mutation. The normal distribution has infinite tails: there is a positive probability of getting *any number at all* back
from a Gaussian mutation. Uniform mutation insists on local search; Gaussian mutation permits distant search, but usually performs local search. The

shape of the Gaussian distribution is shown in Figure 3.2. Generating Gaussian random variables with mean 0 and standard deviation 1 can be done with the formula given in Equation 3.1:

$$N(0,1) = \cos(2\pi \cdot R) \times \sqrt{-2 \cdot S},\qquad(3.1)$$

where R and S are independently generated uniform random variables in the range $[0, 1]$. In order to get Gaussian random numbers with standard deviation σ, simply multiply the results by σ.

In the rest of this chapter, except for Section 3.5, we will be testing real function optimizers on problems with known answers. We will say that we have "found" an answer when a member of our evolving population of candidate solutions is within a specified Euclidian distance, called a *tolerance*, of the true answer. Suppose we were optimizing, for testing purposes, the function $f(x) = x^2$. If we were working with a tolerance of 0.01, then any population member whose genes code for a number in the range $-0.01 \le x \le 0.01$ would serve as a witness that the evolving population had "solved" the problem of finding the minimum at $x = 0$.

Since models of evolution are independent of the problem being solved, we have available all of the models of evolution discussed so far. Selecting a basic real function optimizer requires that we go through the following steps. First, pick a fitness function. This function will be the function being optimized or its negative. Second, select a population size. The population will be made of n-dimensional arrays of reals, where n is the number of variables in the function being optimized. Third, select a suite of variation operators. You may want a mix of mutations that includes different maximum mutation sizes so that some mutations do a broad search, and others do a close search. The smallest maximum mutation size in any of your mutation operators should be small enough that you are unlikely to jump entirely past a nearby optimum with one mutation. (The fact that your mutations can only be as small as the real-number resolution of the computer you're working with keeps this from becoming an urgent concern.) Fourth, select a model of evolution. Fifth and last, come up with a stopping condition.

Step five is a killer. Any good stopping condition that can say, "Yes, this is it! We have found the true global optimum!" practically has to contain the answer to the optimization problem in order to work as advertised. When you are testing a real function optimizer on functions with known optima, optima you yourself have constructed within the function, an omniscient stopping condition is available. Normally, however, you have to make do with a second-rate, substitute stopping condition. A couple of demonstrably bad but widely used stopping conditions are (i) to stop if there has been no change in a long time and (ii) to run your real function optimizer as long as you can. Both of these stopping conditions can be marginally improved by rerunning the algorithm on different populations and seeing what sort of optima pop out. The values you get can be used to create new stopping conditions or restarting conditions. If, for example, you know of a much better optimum than the one you are in, and no progress has been made for several generations, it might well be profitable to restart evolution with a new random population. You might even share time among several evolving populations and delete the ones that appear to be going nowhere, replacing them with copies of the populations that are doing better. This is, in essence, an evolutionary algorithm whose individual creatures are evolutionary algorithm populations. This sort of thing is called a *metaselection algorithm* and is complex enough to be a technique of last resort.

In practical or applied problems, a stopping condition is often implied by the bounds implicit in reality. If your optimizer is stuck and has found a solution that brings you in under budget, you can just accept that solution and go home for the night. An applied problem usually has obvious bounds on optima: the speed of light, the size of the federal budget, the amount of zinc in North America, or whatever. Finally, in many optimization problems there is an extensive literature containing proven or estimated bounds that you can use as stopping conditions. If you are intending to use evolutionary algorithms for optimization, then you should become familiar with the problems you are solving to the greatest extent possible and use that knowledge to make the best stopping condition you can.

In his doctoral thesis, Kenneth DeJong proposed five standard test functions, f_1 through f_5, for evolutionary algorithms that optimize real functions. We do not use all of these functions in our experiments; the test suite is included for completeness and because of its historical importance in real function optimization with evolutionary algorithms. We generalize the test bed to the extent of making the number of variables arbitrary in some cases. Many of the functions in the test bed have optima that are very special points, e.g., the origin. It is considered good practice to shift the optima to points that do not have special status in the representation used. Here are DeJong's five functions:

$$f_1(x_1, \ldots, x_n) = \sum_{i=1}^{n} x_i^2 \tag{3.2}$$

with $-5.12 \le x_i \le 5.12$. This function is to be minimized.

$$f_2(x, y) = 100(x^2 - y^2)^2 + (1 - x)^2 \tag{3.3}$$

with $-2.048 \le x, y \le 2.048$. This function is to be minimized.

$$f_3(x_1, \ldots, x_n) = \sum_{i=1}^{n} [x_i] \tag{3.4}$$

with $-5.12 \le x_i \le 5.12$, where $[x]$ is the greatest integer in x. This function may be minimized or maximized.

$$f_4(x_1, \ldots, x_n) = \sum_{i=1}^{n} i \cdot x_i^4 + \text{Gauss}(0,1) \qquad (3.5)$$

with $-1.28 \leq x_i \leq 1.28$ and where Gauss(0,1) is a Gaussian random variable with a mean of zero and a standard deviation of 1. Recall that the formula for Gaussian random numbers is given in Equation 3.1. The Gaussian random variable is added each time the function f_4 is called. This function is to be minimized.

$$f_5 = 0.002 + \sum_{j=1}^{25} \frac{1}{j + \sum_{i=1}^{n}(x_i - a_{i,j})^6}, \qquad (3.6)$$

with $-65.536 \leq x_i \leq 65.536$ and where the $a_{i,j}$ are coordinates of a set of points within the function's domain generated at random and fixed before optimization takes place. This function is to be maximized.

Let us look qualitatively at these functions. The function f_1 has a single mode at the origin and is the simplest imaginable real function that has an optimum. The function f_2 is still not hard, but has two areas that look like optima locally with only one being a true optimum. The function f_3 is unimodal, but where it is not constant, it is discontinuous. It thus serves as a good example of a function for which the techniques of calculus are useless. The function f_4 is also unimodal with a mode at the origin, but it is flatter near its optimum than f_1 and has random noise added to it. The function f_5 simply has a large number of extremely tall, narrow optima with differing heights at the positions $(a_{i,j} : i = 1, \ldots, n$ and $j = 1, \ldots, 25)$. It could be used in a destruction test for mutation operators that made too large jumps.

DeJong's function test bed is a standard for testing evolutionary computation systems. Other functions can be used as building blocks to construct your own test functions. The fake bell curve, Equation 2.1, is one. Another is the *sombrero function*

$$\cos\left(\sqrt{x_1^2 + x_2^2 + \cdots + x_n^2}\right). \qquad (3.7)$$

To see where this function got its name, look at Figure 3.3, which shows the two-dimensional version on a square of side length 4π centered at the origin.

Experiment 3.1 *Write or obtain software for a real function optimizer for use on functions 1 through 5 given below. Use tournament selection with tournament size 4, two-point crossover, and uniform real single-point mutation with $\epsilon = 0.2$ on each new creature. Stop after 100 generations on functions 1 through 3; stop within a Euclidian distance of 0.001 of the true optimum on functions 4 and 5. Use a population size of 120 individuals. For functions 1 through 3, do 30 runs and give the solutions found. See whether they cluster in any way. For functions 4 and 5, find the mean and standard deviation of time-to-solution, using a tolerance of 0.001, averaged over 30 runs. Be sure your*

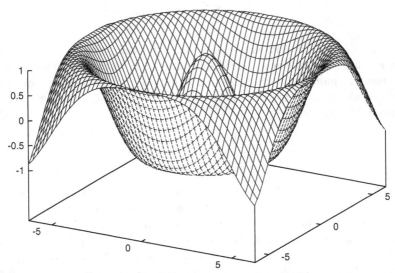

Fig. 3.3. The sombrero function in 2-D, $-2\pi \leq x, y \leq 2\pi$.

definition of two-point crossover does not choke in 2 dimensions; it should act like one-point crossover in this case.
(1) Minimize

$$x^4 + 2x^2y^2 + y^4 - 6x^3 - 10x^2y - 6xy^2 - 10y^3 + 37y^2 + 37x^2 - 12x - 20y + 70, \quad -10 \leq x, y \leq$$

(2) Maximize

$$\frac{\cos\left(5 \cdot \sqrt{x^2 + y^2}\right)}{x^2 + y^2 - 6x - 2y + 11}, \quad 0 \leq x, y \leq 5.$$

(3) Maximize and minimize

$$\frac{3xy - 2}{4x^2 - 4xy + y^2 + 4x - 2y + 2}, \quad -5 \leq x, y \leq 5.$$

(4) DeJong function f_1, Equation 3.2, in 10 dimensions.
(5) fake bell curve, Equation 2.1, in 4 dimensions, with $-2 \leq x_i \leq 2$.

In Experiment 3.1, functions 1 to 3 have optima at fairly weird points. The functions avoid "working out even," and the optima are hard to find even by staring at a graph of the functions, unless you play games with vertical scale. Functions 4 and 5 commit the sin of having optima at special points—the origin—and are the functions you should probably use when debugging your code.

Experiment 3.2 *Redo functions 4 and 5 of Experiment 3.1, replacing single-point mutation with (i) a probabilistic mutation operator that yields the same overall mutation rate as the one-point mutation used in Experiment 3.1; (ii) two-point mutation applied to each new creature; (iii) a probabilistic mutation that yields the same overall mutation rate as two-point mutation; and (iv) Gaussian mutation with standard deviation 0.1. In your write-up, compare the performance by drawing confidence intervals of one standard deviation about the mean solution times and checking for overlap.*

Experiment 3.3 *Redo Experiment 3.2, using single-point mutation and also replacing the single-point mutation with the Lamarckian mutation operator from Appendix C in 10% of the mutations. Compare the performance with the algorithm used in Experiment 3.2.*

Experiment 3.4 *Redo Experiment 3.2, using only the mutation operator you found to work best (this may be different for different functions). Rewrite the evolutionary algorithm to use one- or two-point crossover or uniform crossover. Compare the performance of the three different types of crossover, measuring time-to-solution in both crossover events and real numbers generated.*

In the next experiment, we will explore a representational issue: how does representing points in the domain space of a function with real numbers compare to a discrete representation of character strings coding for bits?

Experiment 3.5 *Redo Experiment 3.2 with two-point crossover, but this time radically modify your data structure be strings of 0's and 1's, twelve per real variable. Interpret these as real numbers by taking the characters in blocks of 12 as binary integers and then mapping those integers uniformly into the specified range of the function. These integers are in the range $0 \leq x \leq 4095$; so to map them onto a real y in the interval (r, s), set $y = (s - r) \cdot x / 4095 + r$. Let the mutation operator be single-point mutation. Compare the performance of this evolutionary algorithm with that of Experiment 3.2. Discuss the connection, if any, between this experiment and the Royal Road function with block size 12.*

We conclude by reminding you of a standard and possibly useful formula.

Definition 3.3 *The **Euclidian distance** between two points $p = (p_1, p_2, \ldots, p_n)$ and $q = (q_1, q_2, \ldots, q_n)$ in $\mathbb{R}^n$ is given by the formula*

$$d(p, q) = \sqrt{(p_1 - q_1)^2 + (p_2 - q_2)^2 + \cdots + (p_n - q_n)^2}.$$

Problems

Problem 92. Essay. Explain in detail why the unmodified Lamarckian mutation operator from Chapter 2 cannot be reasonably applied with the definition of point mutation given in this section.

Problem 93. Suppose we are optimizing some real-valued function in 4 dimensions with an evolutionary algorithm. If we are using single-point mutation with maximum mutation size ϵ, then what is the expected distance that 1, 2, 3, 4, or 5 single-point mutations will move us? Your answer should be in terms of ϵ.

Problem 94. Do Problem 93 experimentally for $\epsilon = 0.2$. In other words, generate $k = 1, 2, 3, 4$, or 5 single-point mutations of the gene $(0, 0, 0, 0)$ and check the average resulting difference for a large number of trials (e.g., thousands). Use the results to check your theoretical predictions.

Problem 95. A *neighborhood of radius* r of a point p in $\mathbb{R}^n$ is the set of all points within a distance r of p. Give an example of a continuous, nonconstant function from $\mathbb{R}^2$ to $\mathbb{R}$ and a point p such that for all choices of small r, neighborhoods of p contain a point q other than p such that $f(p) = f(q)$ and for no x in the neighborhood is $f(x) > f(p)$.

Problem 96. This problem requires you to compute functional gradients as described in Appendix C. For each of the following functions and points, compute a unit vector in the direction of maximum increase at the specified point:

(i) $f(x) = \cos(\frac{x}{x^2+1})$ at $x = 2$.

(ii) $f(x, y) = \frac{xy}{x^2+y^2+1}$ at $(1, 2)$.

(iii) $f(x, y, z) = \frac{\cos(x)\sin(y)}{z^2+1}$ at $(\pi/3, \pi/6, 1)$.

Problem 97. This problem requires you to compute functional gradients as described in Appendix C. Compute and display the gradient vectors within the square $-5 \leq x, y \leq 5$ of the function

$$f(x, y) = \cos\left(\sqrt{x^2 + y^2}\right).$$

You should compute the gradient at 25 points within the square. Choose each point from a different square in a regular 5×5 grid.

Problem 98. Give a function with an infinite number of local optima with distinct values on the interval $-1 \leq x \leq 1$. Can such a function be made to have a global optimum? Can such a function be made not to have a global optimum? For both questions give an example if the answer is yes and a proof if it is no.

Problem 99. The fake bell curve, Equation 2.1, produces values in the range $0 \leq f(x, y) \leq 1$. Assume we are optimizing this curve. Give the range of changes that are possible between the maximum fitness of a pair of parents and the maximum fitness of the children resulting from two-point crossover of those parents. In other words, what possible fitness difference can result from best parent to best child as a result of crossover? Assume that this is the type of crossover that treats creatures as strings of indivisible reals.

Problem 100. Do the results of Problem 99 change if we switch the data representation to that used in Experiment 3.5?

3.2 Fitness Landscapes

Evolutionary algorithms operate on populations of structures. The structures within a population are not all the same (if they were, there would be no point in having them). Over algorithmic time, the population changes as better structures are generated and selected. The space in which the population lives is called the *fitness landscape*. The fitness landscape is a metaphor that helps us to understand the behavior of evolutionary algorithms. Let us come up with a formal definition sufficiently general to permit its use in almost any evolutionary computation system.

Definition 3.4 *Let G be the space of all the data structures that could appear as members of the population in some evolutionary algorithm. Then, the* **fitness landscape** *of that evolutionary algorithm is the graph of the fitness function over the space given by G.*

As with many of the definitions in this text, Definition 3.4 cheats. It refers to the space of all data structures upon which the evolutionary algorithm in question might work. To generate an organized graph, there must be some sort of relationship among the data structures. In this chapter this isn't a problem, because all the data structures are points in $\mathbb{R}^n$. We are used to doing graphs over $\mathbb{R}^n$. When our evolutionary computation shifts to discrete structures, as in genetic programming, the structure of the domain space of the fitness landscape becomes far more complex.

For the type of evolutionary algorithms described in Section 3.1, the graph of the fitness function is the fitness landscape. In the next couple of sections of this chapter, we will attempt to enhance the performance of our evolutionary algorithms by modifying the fitness landscape or making the landscape dynamic. We will now examine the notion of fitness landscapes.

Examine Figure 3.4. The fake bell curve in one dimension is extremely simple to optimize, and as we see, the population rapidly moves from a disordered random state to a pileup beneath the optimum. Since the evolutionary algorithm running under Figure 3.4 had a relatively small mutation size and

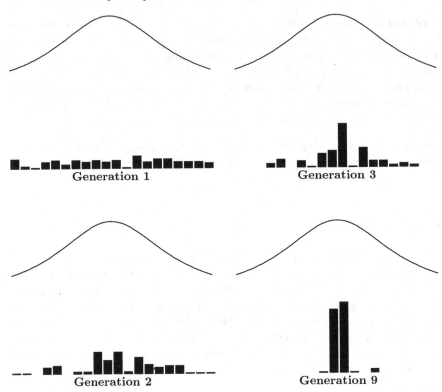

Fig. 3.4. Graph of the fake bell curve with one variable together with a histogram of the distribution of the population in the range $[-3, 3]$ at various times during evolution. (The evolutionary algorithm that generated these pictures used roulette selection on a population of 100 data structures with uniform single-point mutation of size 0.03 and no crossover. The algorithm is elitist, saving its best two genes.)

no crossover, this pileup is probably due mostly to reproduction and selection. Since the fake bell curve has a single mode, nothing can go wrong in optimizing it with an evolutionary algorithm. Selection and elitism ensure that good genes remain, a "fitness ratchet," and genes near the best genes are generated at random. This means that the population will rapidly pile up near the unique optimum. There is no way to trap the population away from the optimum: all roads are uphill toward the unique maximum. Let's look at a less friendly function.

In Figure 3.5, we see that when there is more than one optimum, the evolutionary algorithm can find different optima in different "runs." The difference between the left- and right-hand diagrams is caused by the choice of initial population. The function

$$f(x) = \frac{3.1}{1 + (40x - 4)^2} + \frac{3.0}{1 + (3x - 2.4)^4} \tag{3.8}$$

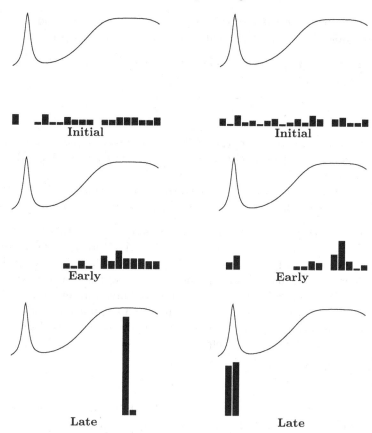

Fig. 3.5. Graph of $f(x) = \frac{3.1}{1+(40x-4)^2} + \frac{3.0}{1+(3x-2.4)^4}$ in the same style as Figure 3.4. (The results of running two distinct initial populations are shown.)

has a pair of maxima. The left-hand optimum is slightly higher, while the right-hand optimum is much wider and hence easier to find. An interesting question is: What is the probability of finding each of the optima?

Experiment 3.6 *Take the code from Experiment 3.1 and modify it to use roulette selection and random replacement with an elite of size 2, to run without crossover (on only one variable), and to use uniform mutation with size 0.01. Use this code with a population size of 50 to optimize Function 3.8 on the interval $[0, 1]$ and report how many times the algorithm "found" each of the two optima. In your write-up, be sure to give a clear definition of what it means for the population to have "found" an optimum.*

This experiment gives us a way of measuring the way the "converged" (final) population of an evolutionary algorithm samples between two optima. It is clear that if we have a large number of distinct optima floating around, the evolutionary algorithm will have some sort of probability distribution (see

Appendix B) on the optima of the function. This means that modifying the way we sample to produce the population may help. Let's try it.

Experiment 3.7 *Repeat Experiment 3.6, except that before evaluating the fitness of a point x, replace x with $g(x)$, where*

$$g(x) = \frac{x}{6 - 5x}.$$

Again: do everything as before, except that the fitness of a population member x is $f(g(x))$ instead of $f(x)$. For each run, report x, $g(x)$, and $f(g(x))$ for the best population member. The $g(x)$ values are the ones used to judge which of the two maxima the population has converged to. In your write-up, state the effect of the modification on the chance of locating the left-hand optimum.

We will return to the issue of fitness landscapes many times in the subsequent text. The idea of fitness landscapes is helpful for speculating on the behavior of evolutionary algorithms, but should never be taken as conclusive unless experiments or mathematical proof back the intuition generated. As the dimension of a problem increases, the quality of human intuition degrades. If we cease to consider functions of real variables, as we will in a majority of subsequent chapters, the intuition generated by the nice smooth graphs presented in this section will be even less helpful.

Problems

Problem 101. Prove that the function $g(x)$ in Experiment 3.7 leaves population members that start in the range $[0, 1]$ in that range.

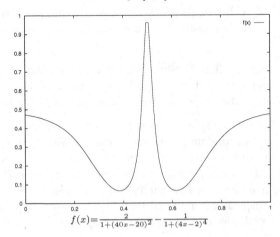

$$f(x) = \frac{2}{1+(40x-20)^2} - \frac{1}{1+(4x-2)^4}$$

Problem 102. Examine the function above. This function is *deceptive* in the sense that, unlike the fake bell curve, there are many points from which uphill is away from the global maximum. Compute the probability that an algorithm that picks a point uniformly at random in the interval $[0, 1]$ and then heads uphill will find the true optimum of this function. Hint: this involves computing the position of the two minima, which you may do numerically if you wish.

Problem 103. Find an explicit formula for some $f(x, y)$ that generalizes the function given in Problem 102 to two dimensions. Having done this, compute the probability of successful hill climbing for that function for points chosen uniformly at random in the unit square. Will the problem get harder as the dimension increases?

Problem 104. Explain, analytically, what the function $g(x)$ in Experiment 3.7 does. Hint: start by graphing it.

Problem 105. Assuming that everything worked as intended, Experiment 3.7 modified the evolutionary algorithm to find the narrower but taller optimum more often. The function $g(x)$ in that experiment is, in some sense, "tailored" to Function 3.8. Examine, for positive α, the functions

$$g_\alpha(x) = \frac{x}{(\alpha + 1) - \alpha x} \qquad (3.9)$$

and

$$h_\alpha(x) = \frac{(\alpha + 1)x}{1 + \alpha x}. \qquad (3.10)$$

Suppose we are going to run an evolutionary algorithm, like the one in Experiment 3.7, a large number of times. Using the functions $g_\alpha(x)$ and $h_\alpha(x)$, come up with a scheme for running a different sampling of the space each time. Explain what roles the functions play. Assume that we have no idea where the optima are in the functions we are trying to optimize. You may want to give your answer as pseudocode.

Problem 106. Prove that the functions $g_\alpha(x)$ and $h_\alpha(x)$ from Problem 105 have the following properties on the interval $[0, 1]$:

(i) $g_\alpha(h_\alpha(x)) = h_\alpha(g_\alpha(x)) = x$.
(ii) The functions are monotone for all choices of α. (A function $q(x)$ is *monotone* if $a < b$ implies that $q(a) < q(b)$.)
(iii) Both functions take on every value in the range $[0, 1]$ exactly once each.

Problem 107. Generalize the scheme you worked out in Problem 105 to multivariate functions. Suggestion: Think of $g_\alpha(x)$ and $h_\alpha(x)$ as *distortions*; it may be good to have different distortions for each dimension.

Problem 108. Essay. Speculate as to the effect of changing the population size in Experiment 3.6. Advanced students should actually redo the experiment with different population sizes to support their speculation.

Problem 109. Essay. Speculate as to the effect of changing the mutation size in Experiment 3.6. Advanced students should actually redo the experiment with different mutation sizes to support their speculation.

Problem 110. Essay. In Section 2.5, the Royal Road function is described. What is the fitness landscape for the evolutionary algorithm described in Experiment 2.10?

3.3 Niche Specialization

In this section we will tinker with the fitness function of our real function optimizer to try to enhance performance. The idea of *niche specialization* for use in evolutionary algorithms was proposed by David Goldberg [29]. It is inspired by a biological notion with the same name, discussed in Section 1.2. The basic idea is simple: reduce or divide the fitness of a member of an evolving population by a function of the number of other essentially similar members of the population. In order to do niche specialization one needs to create a similarity measure. In real function optimization at least two obvious similarity measures are available. The population we operate on in our evolutionary algorithm for optimizing real functions is just a collection of points in $\mathbb{R}^n$, and so proximity in $\mathbb{R}^n$ is one possible similarity measure. Call this *domain niche specialization*. Let r be the *similarity radius* and define a new fitness measure as follows. Where before we used the value of the function f being optimized as a fitness function, we will now take the value of the function divided by a penalty based on the number of members of the population nearby. Let m be the number of members of the population at distance r or less from a particular population member v. Let $q(m)$ be the *penalty function* and set

$$\text{Fitness}(v) := f(v)/q(m). \qquad (3.11)$$

The effect of this will be to make an optimum less attractive as more creatures find it. Since we are going to use it quite a lot, the function $q(m)$ needs to be inexpensive to compute but also needs to avoid throwing creatures out of optima before they can explore them and find their true depth. One very easy to compute function is $q(m) = m$, but this function may well be far too harsh a penalty to the fitness. One might try a simple modification of the identity function like

$$q(m) = \begin{cases} 1, & m \leq 4, \\ m/4, & m > 4. \end{cases} \qquad (3.12)$$

The next experiment tests domain niche specialization on a function with three modes.

Experiment 3.8 *Recall that $\mathcal{B}_n$ is the fake bell curve, Equation 2.1. Use the following function with three modes,*

$$f(x, y, z) = \mathcal{B}_3(x, y, z) + 2 \cdot \mathcal{B}_3(x - 2, y - 2, z - 2) + 3 \cdot \mathcal{B}_3(x - 4, y - 4, z - 4),$$

and then modify the software from Experiment 3.1 to test niche specialization on $f(x, y, z)$ as follows. Use an initial population of 60 creatures all of the form $(0, 0, 0)$. Test the time for a creature to get within distance 0.05 of $(4, 4, 4)$ in real time and in generations. Compute the mean and standard deviation of these times over 50 runs in each of 3 sets of trials, as follows. The first set of trials should be the basic real function optimizer. The second should be one with domain niche specialization with similarity radius 0.05 and $q(m) = m$. The third should be like the second, except that $q(m) = \sqrt{m}$. Recall that $m \geq 1$ because the creature is close to itself.

Since the computation of m and $q(m)$ are potentially quite expensive, the measurement of real time taken is very important, a more objective measure of efficiency than the generations or crossovers to solution. Notice that in Experiment 3.8 we place the initial population in a very bad place, inside a local optimum and with another local optimum between it and the unique global optimum. This is very different from having an initial population uniformly placed throughout the domain space.

Experiment 3.9 *Redo Experiment 3.8 with a random initial population placed uniformly on the domain $-1 \leq x, y, z \leq 5$. Emphasize the differences in performance resulting from initial population placement in your write-up.*

The similarity measure used in domain niche specialization seems to assume that the optima are zero-dimensional, that is, that all directions away from a local optimum result in points with lower fitness, as measured by the function being optimized. Functions that attain their optimum on some non-point region might prove a challenge for domain niche specialization. Instead of forcing the creatures to move away from an optimum and find new optima, domain niche specialization might well force them into different parts of the "same" local optimum. In Problem 95, we asked you to construct a function that had this property, and in fact Equation 3.7, the sombrero function, is just such a function. All but one of its local optima are $(n - 1)$-dimensional spheres centered at the origin. A problem with the sombrero function is that all the local optima are global optima, but we can get around this by combining the sombrero function with the fake bell curve. Examine the following pair of equations:

$$f(x_1, x_2, \ldots, x_n) = \frac{\cos\left(\sqrt{x_1^2 + x_2^2 + \cdots + x_n^2}\right)}{x_1^2 + x_2^2 + \cdots x_n^2 + 1}; \qquad (3.13)$$

$$f(x_1, x_2, \ldots, x_n) = \frac{\cos\left(\sqrt{x_1^2 + x_2^2 + \cdots + x_n^2}\right)}{\left(4 - \sqrt{x_1^2 + x_2^2 + \cdots + x_n^2}\right)^2 + 1}. \qquad (3.14)$$

The first, Equation 3.13, is a sombrero function multiplied by a fake bell curve, so that there is a single global optimum at the origin and sphere-shaped local optima at each radius $2\pi k$ from the origin. The second, Equation 3.14, is a sombrero function multiplied by a fake bell curve that has been modified to have a single spherical global optimum at a radius of 4 from the origin, placing the global optimum of the function in a spherical locus at a distance of nearly 4 from the origin. Parts of the graphs of these functions are shown in Figure 3.6.

Experiment 3.10 *Redo Experiment 3.9 using Equations 3.13 and 3.14. If you can, for the experiment in two dimensions dynamically plot the location of each population member. Do the creatures spread out along the spherical local optimum? Given that we have crossover available, is that a bad thing? In any case, check the mean and standard deviation of time-to-solution to some part of the global optimum in $2, 3,$ and 4 dimensions. Use a tolerance of 0.05 for stopping. Since testing tolerance is harder with nonpoint optima, carefully document how you test the stopping condition.*

The second sort of similarity measure we want to examine in this section is based on the position a creature codes for in the function's range space. We retain the similarity radius r from the preceding discussion as well as the penalty function $q(m)$ but change the method of computing m, the number of essentially similar creatures. If f is the function we are optimizing, then let m for a creature y be the number of creatures x such that $|f(x) - f(y)| \leq r$. In other words, we are just computing the number of creatures that have found roughly the same functional value. We call niche specialization on this new similarity measure *range niche specialization*. Range niche specialization avoids some of the potential pitfalls of domain niche specialization. It can compensate for the funny spherical local optima that come from the sombrero function. Imagine a function that jumps around quite a lot in a very small region. Range niche specialization wouldn't tend to drive creatures out of such a region before it was explored. On the other hand, if there was a local optimum with a steep slope leading into it, then range niche specialization might create a population living up and down the sides of the optimum.

Experiment 3.11 *Redo Experiment 3.10 with range niche specialization. Does it help, hurt, or make very little difference?*

Since the two types of niche specialization we have discussed have different strengths and weaknesses and both also differ from the basic algorithm, it might be worthwhile to use them intermittently. Given that they have non-trivial computational costs, using them intermittently is a computationally attractive option.

Experiment 3.12 *Pick a function that has been difficult for both sorts of niche specialization and modify your optimizing software as follows. Run the basic real function optimizer for 20 generations and then one or the other sort*

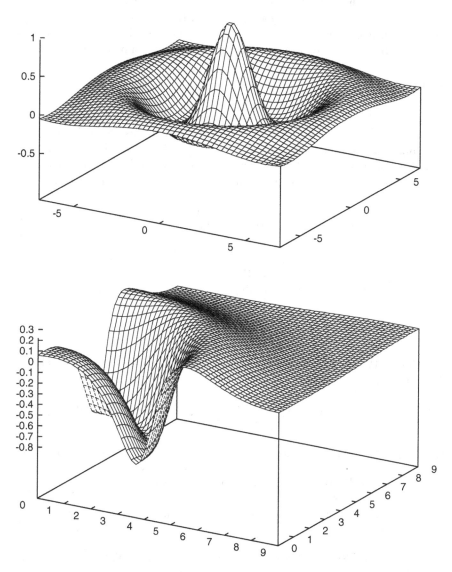

Fig. 3.6. Graphs of Equations 3.13 and 3.14.

of niche specialization for 10 generations, and repeat. Also try alternating the types of niche specialization. Do you observe any improvement? Remember to measure time-to-solution with both a "stopwatch" and in generations. Since the directions given for this experiment are fuzzy and imprecise, you should give a detailed write-up of exactly what you did and why.

Range niche specialization may be of more use in situations in which the functions being optimized are not continuous. In the next section we will see some examples of functions that are constant where they are not discontinuous. These functions may be a more fruitful area for use of niche specialization.

Problems

Problem 111. Suppose we are optimizing the fake bell curve in two dimensions with an 8 member population. If the creatures' genes are as shown in the following table, compute the modified fitness of each creature for domain niche specialization with similarity radius $r = 0.1$ and penalty function $q(m) = \sqrt{m}$. Do you think the use of niche specialization will help in this instance?

x	y
0.042	0.043
0.047	0.048
0.051	0.066
0.121	0.136
0.077	0.081
0.166	0.135
0.042	0.055
0.211	0.056

Problem 112. Do Problem 111 over but for range niche specialization with similarity radius $r = 0.01$.

Problem 113. In Experiment 3.8 it was asserted that the function being optimized,

$$f(x, y, z) = \mathcal{B}_3(x, y, z) + 2 \cdot \mathcal{B}_3(x - 2, y - 2, z - 2) + 3 \cdot \mathcal{B}_3(x - 4, y - 4, z - 4),$$

has 3 modes. In Problem 48 we saw that if fake bell curves are put too close together, the modes blend. Derive from $f(x, y, z)$ a single-variable function that will have the same number of modes as f and graph it to verify that f has 3 modes.

Problem 114. Essay. Construct a function that you think will be more easily optimized by a basic real function optimizer than by that same optimizer modified by domain niche specialization. Discuss how you construct the function and give your reasons for thinking that domain niche specialization might run afoul of your function. Consult your instructor as to the degree of precision he demands in specifying a function. Advanced students should support their conclusions experimentally.

Problem 115. Essay. Do Problem 114 for range niche specialization.

Problem 116. Carefully graph

$$f(x, y) = \frac{C}{\left(d - \sqrt{x^2 + y^2}\right)^2 + 1}$$

for $C = 1$ and $d = 3$ and discuss its relationship to Equation 3.14 and the fake bell curve, Equation 2.1. This function is called the *crater* function.

Problem 117. Essay. Assume that we are using the stopping condition that says "stop if there has been no change in the gene of the most fit creature for 20 generations." The crater function, introduced in Problem 116, creates an annoying optimum that takes on its maximum value at an infinite number of points. Since all these points are true global optima, this creates a problem for our stopping criterion. Discuss what would happen to the optimization process if we added a fake bell curve or other bounded, easily computed curve to the crater function. Assume that the curve added has a very small maximum compared to the maximum of the curve actually being optimized. In particular, would adding this other "bias" function ease the problem afflicting the stopping criterion? What if the bias function was instead composed of a fixed set of random noise of small amplitude? Remember to discuss not only the effects on the population structure but the computational costs.

Problem 118. Essay. Suppose we are optimizing real functions with either domain or range niche specialization. Try to characterize the set of continuous functions that you think would be more likely to be optimized more efficiently under each of these two specialization techniques.

Problem 119. Explain why the following are desirable properties of a penalty function $q(m)$, $m \varepsilon \{1, 2, 3, \ldots\}$:
   (i) $q(1) = 1$,
   (ii) $q(m) \geq 1$ for all m, and
   (iii) $q(m + 1) \geq q(m)$ for all m.

Problem 120. Essay. One potential problem with niche specialization is that it may drive parts of a population out of a niche before it is explored. Discuss how the choice of the penalty function can be used to combat this problem without removing all the benefits of niche specialization, e.g., by making $q(m) = 1$. Give at least three penalty functions and discuss their advantages and disadvantages both in terms of the population structures they encourage and their computational cost. The *unpenalized local population size* is the size of a population that can exist within a single optimum smaller than the similarity radius without any fitness decrease. For each penalty function you discuss, compute the unpenalized local population size. How could this size be zero and why would that be bad?

3.4 Path Length: An Extended Example

In this section we will work with minimizing the length or cost of a path (like a garden tour in which you are given a list of addresses that you are to visit in order). This is a rich class of minimization problems with some interesting properties. An interesting thing about this class of problems is that there are two different "natural" fitness functions that produce different fitness landscapes, but which both have a global optimum that solves the problem. We note that an algorithm called *dynamic programming*, which appears in the robotics literature, can be used to solve the type of problems treated in this section far more efficiently than an evolutionary algorithm: we are using path length optimization to illuminate some of the features of evolutionary computation.

A *path* with n points is a sequence $\mathcal{P} = \{(x_0, x_1), (x_1, y_1), \ldots, (x_{n-1}, y_{n-1})\}$ of points to be traversed in order.

The *length* of a path is the sum

$$Len(\mathcal{P}) = \sum_{k=0}^{n-2} \sqrt{(x_k - x_{k+1})^2 + (y_k - y_{k+1})^2}. \qquad (3.15)$$

Our paths will live in the unit square:

$$\mathcal{U} = \{(x, y) : 0 \leq x, y \leq 1\}.$$

For the nonce we will anchor our paths at $(0, 0)$ and $(1, 1)$ by insisting that $(x_0, y_0) = (0, 0)$ and $(x_{n-1}, y_{n-1}) = (1, 1)$.

An optimal solution for minimizing the length of a path is to place all of the points in the path in order on the diagonal. There are a large number of optimal solutions, since moving a point anywhere between its adjacent points on the diagonal does not change the length of the path. This means that if we were to use Equation 3.15 as a fitness function on a population of paths, there would be a nontrivial flat space at the bottom (remember, we are minimizing) of the fitness landscape. When there is only one unanchored point, the path length is a function of its position (illustrated in Figure 3.7). Note that the flat space is a line corresponding to placing the unanchored point on the line segment joining $(0, 0)$ with $(1, 1)$.

For those familiar with the notion of dimension, the flat space, while larger than a point optimum, is a lower-dimensional subset of the fitness landscape. In this case, the fitness landscape is a two-dimensional surface, while the optimal solutions all lie on a one-dimensional line segment.

Experiment 3.13 *Write or obtain software for a real-function optimizer evolutionary algorithm that can optimize the length of a path of the sort described above. The function to be optimized is Equation 3.15 with the independent variables being the x and y positions of the points other than the anchor*

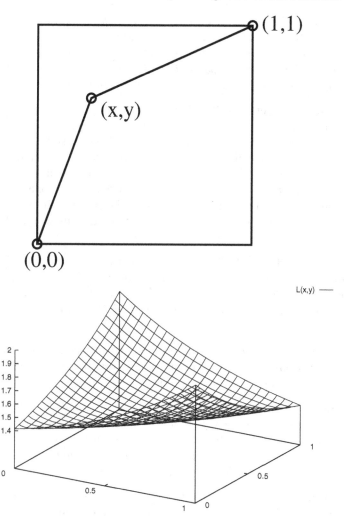

Fig. 3.7. A path with one nonanchored point and the graph of the length $L(x, y)$ of that path as a function of the unanchored point. (Note the trough-shaped set of minima along the diagonal from $(0,0)$ to $(1,1)$.)

points. Make the number of points $n \geq 3$ a parameter that you can change easily.

Your algorithm should operate on a population of 100 paths. Use roulette selection and random replacement with an elite of 2 paths. Use two-point crossover and uniform two-point mutation with a size of 0.1. Run the algorithm 50 times for $n = 5$ points (the 5 includes the 2 anchors). Save the best (shortest) path from each run and plot the best, worst, and 3 others of these "best of run" paths. Run until the path is within 5% of the true minimum length or for 1000 generations, whichever is shortest.

Now repeat the experiment, but instead use a mutation operator that mutates both coordinates of one point. Plot the same collection of paths and, in your write-up, comment on the effect of the new mutation operator.

The point of Experiment 3.13 is to compare two mutation operators, one of which "knows something" about the problem. The gene used in Experiment 3.13 has more obvious structure than the ones we've used so far. It is organized into pairs of points. We said nothing in the experiment about how to organize the variables. There are 720 possible orders for them, but in some sense, only a few natural orders. We could group the x coordinates into one block and the y coordinates into another, or we could group the variables in pairs giving the position of each point. The next experiment compares uniform and Gaussian mutations.

Experiment 3.14 *Modify the software from Experiment 3.13 to use Gaussian mutations in place of uniform ones. Compute a standard deviation for the Gaussian mutation that is the same as that of the uniform mutation used before. In your write-up, compare uniform and Gaussian mutations for this problem.*

So far, we have tinkered with the action of mutation on points and with the shape of mutations. The next experiment suggests another potential avenue for improvement, but you'll have to hunt around for a way to make it work.

Experiment 3.15 *Modify the software from Experiment 3.13 to let the size of the mutations decrease as a function of the number of generations. Repeat the experiment with at least two different schemes for decreasing the mutation size as time passes. In your write-up, document whether this mutation-size shrinking helped and speculate as to why or why not.*

We are now ready to look at the second fitness function mentioned at the beginning of the section. In Equation 3.15 we compute the length of the line segments making up the path $\mathcal{P}$. The square root function, which appears numerous times in this computation, is a monotone function, and so removing it will still permit us to minimize path length in some fashion. (Think about why this is desirable.)

$$\text{Len } 2(\mathcal{P}) = \sum_{k=0}^{n-2} (x_k - x_{k+1})^2 + (y_k - y_{k+1})^2 \tag{3.16}$$

Experiment 3.16 *Modify the software from Experiment 3.13 to use Equation 3.16 as a fitness function and perform the experiment again using the mutation operator acting on both coordinates of a point. Compare both the time required to get within 5% of optimum and the character of the "best of run" paths.*

Having two different fitness functions for the same problem helps us avoid traps in the fitness landscape. As we will see in the problems, the fitness landscape for each function does contain traps, in the form of tortuous (mutational) paths rather than direct paths to the optima. If we view the population as moving across the fitness landscape, then one function may have a trap where the other does not. Thus, alternating between the fitness functions may permit more rapid convergence than the use of either fitness function alone.

Experiment 3.17 *Modify the software from Experiment 3.13 to use either Equation 3.15 or Equation 3.16 as a fitness function and perform the experiment again using the mutation operator acting on both coordinates of a point. Try alternating the use of the two fitness functions both every other generation and every fifth generation. In your write-up, compare the time required for each version of the algorithm to get within 5% of optimum.*

Problems

Problem 121. In Figure 3.7, there is a graph of Equation 3.15 for the case of a path with one mobile point. Copy this graph and make a graph for both Equation 3.16 and for the difference of Equation 3.15 and Equation 3.16. Comment on the difference between their minima.

Problem 122. Prove that all optima of Equation 3.16 are also optima of Equation 3.15, and give an example that shows that the reverse is not true.

Problem 123. Construct an example of a path with 5 total points (including the 2 anchored points) and a mutation of both coordinates of a point such that (i) the mutation makes the fitness of the path worse, and (ii) the mutation moves the point to a place where it would be in an optimal solution.

Problem 124. Construct an example of a nonoptimal path and a mutation of both coordinates of a point that does not change the fitness of the path.

Problem 125. Essay. Reread Problem 106. Does the use of the functions $g_\alpha(x)$ and $h_\alpha(x)$ remove traps in the fitness landscape, as we hope the alternation of fitness functions will do in Experiment 3.17, or does it do something else? Explain.

Problem 126. Essay. Suppose that instead of alternating fitness functions, as in Experiment 3.17, we instead added a small random number to the fitness evaluation of a path. First of all, would this help us out of traps, and second, what other benefits or problems might this cause? Give the design of an experiment to test this hypothesis.

3.5 Optimizing a Discrete-Valued Function: Crossing Numbers

The third function in DeJong's test suite, Equation 3.4, has the property that it is discontinuous everywhere it is not constant. When looking at the definition of this function, one may be inspired to ask, "do functions like that ever come up in practice?" The answer is a resounding yes! and in this section we will define and optimize a class of such functions.

Our motivation for this class of functions will be circuit layout. Circuit layout is the configuration of the components of an electric circuit on a printed circuit board. Efficiency will be measured by the number of jumpers needed. A printed circuit board is a nonconducting board with the wiring diagram of a circuit printed on the board in copper. The parts are soldered into holes punched inside the copper-coated regions. In laying out the circuit board, it may be that two connections must cross one another on the two-dimensional surface of the board without any electrical connection. When this happens we need a jumper. Where the crossing happens, one of the two connections in the copper to be printed on the board is broken, and an insulated wire is soldered in to bridge the gap. It turns out that some circuit layouts require far fewer bridges than others, so locating such good layouts is desirable.

Since representing actual circuit diagrams in the computer would be a little tricky, we will formalize circuit layouts as drawings of *combinatorial graphs*. The elementary theory of such graphs is discussed in Appendix D. Briefly, a graph is a collection of points (called *vertices*) and lines (called *edges*) joining some pairs of points. A *drawing* of a graph is simply a placement of the vertices of the graph into the plane together with a representation of the edges as simple curves joining their endpoints. A drawing is said to be *rectilinear* if all the edges are line segments. The *crossing number* of a drawing of a graph is the number of pairs of edges that intersect (other than at their endpoints). The *crossing number* of a graph is the minimum crossing number attained by any drawing. The *rectilinear crossing number* of a graph is the minimum crossing number of any rectilinear drawing of that graph. It is known that for some graphs, the rectilinear crossing number is strictly larger than the crossing number. A theorem of graph theory says that any graph with a crossing number of zero also has a rectilinear crossing number of zero. Graphs with a crossing number of zero are said to be *planar*.

The *complete graph on n vertices*, denoted by K_n, is the graph with all possible edges. A small amount of work will convince the reader that K_n is planar only when $n \leq 4$. In Figure 3.8, we show a standard presentation of K_5 and a rectilinear drawing of K_5 that has crossing number one.

At the time of this writing, the crossing number and rectilinear crossing number of K_n are known only for $n \leq 9$. The known values for rectilinear crossing numbers of K_n are given in Figure 3.9.

The first step in designing software to approximate the rectilinear crossing number is to figure out when two lines cross. The specific problem is, given the

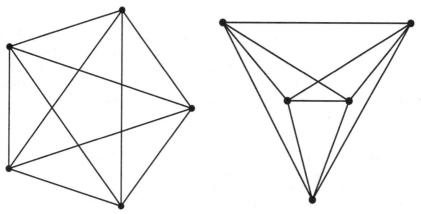

Fig. 3.8. The complete graph K_5 drawn with five crossings and with one crossing.

n	2	3	4	5	6	7	8	9
Crossing Number	0	0	0	1	3	9	19	36

Fig. 3.9. Known rectilinear crossing numbers for K_n.

endpoints of two line segments, how do you decide whether those line segments cross other than at their endpoints? First, if two lines emanate from the same point, then they do not contribute to the crossing number and need not be compared. Such checks should be done by your algorithm before any question of lines intersecting is treated. Suppose that we have two line segments with endpoints $(x_1, y_1), (x_2, y_2)$ and $(r_1, s_1), (r_2, s_2)$ respectively, with $x_1 \leq x_2$ and $r_1 \leq r_2$. Then, using simple algebra to compute the slopes m_1 and m_2 of the lines respectively, we find that the x-coordinate of the intersection of the two lines is

$$x_i = \frac{m_1 x_1 - y_1 - m_2 r_1 + s_1}{m_1 - m_2}, \tag{3.17}$$

and so we can check if $x_1 \leq x_i \leq x_2$ and $r_1 \leq x_i \leq r_2$.

There are some problems with this test: a Boolean predicate that computes the intersection of lines would need to check for $m_1 = m_2$ or $m_i = \infty$, $i = 1, 2$, and deal with such situations appropriately. It is also possible for things to get very confusing if 3 vertices are co-linear. In the event that 3 vertices are co-linear, treat the structure as a single intersection. The task of avoiding all these pitfalls is left for you in Problem 134. In addition, numerical precision issues bedevil anyone who wishes to compute whether two line segments intersect in their interior. These issues are beyond the scope of this text and are the business of the field of *computational geometry*.

Assuming that we can compute when two line segments intersect in their interior, we are ready to modify any of the real-function optimizers that we've developed so far to estimate crossing numbers. The function we are optimizing will have $2n$ variables for an n-vertex graph, where these variables are the x- and y-coordinates of the vertices for the drawing of the graph. In the case of the complete graph, K_n, we will be looking at all possible pairs of line segments. In general, we will need an *adjacency matrix* for the graph as defined in Appendix D. The fitness function to minimize is the following: for each pair of edges in the graph that intersect but do not share a vertex, add 1 to the fitness function.

Experiment 3.18 *Write or obtain software for a real function optimizer that estimates the rectilinear crossing numbers of complete graphs. Test this optimizer on K_4, K_5, and K_6. Use tournament selection with tournament size 4, single-point mutation with mutation size $\epsilon = 0.1$ applied to each new creature, and two-point crossover. Force your starting drawing to lie in the unit square but allow mutation to take the vertices outside the unit square if it wishes. Compute the mean and standard deviation of time-to-solution over 30 runs for each graph. Turn in the visually most pleasing drawings you obtain for $K_4, K_5,$ and K_6.*

Experiment 3.19 *Modify Experiment 3.18 to use Gaussian point mutation with the same variance as in Experiment 3.18 and compare with the results from uniform mutation.*

Experiment 3.20 *Take the software from Experiment 3.18 and modify it to use domain and range niche specialization. Taking the known answer for K_6, compare the raw algorithm with both domain and range niche specialization. The comparison of time-to-solution should be made by comparing mean times with a one standard deviation error bar. Use a similarity radius of 0.05 for the domain niche specialization. Range niche specialization will consider creatures the same if they have the same fitness. For both sorts of niche specialization use the penalty function $q(m)$ given in Equation 3.12.*

Notice that in Experiment 3.20, the range niche specialization uses a similarity radius of zero. This is sensible because we are optimizing an integer-valued function. Zero isn't the only possible sensible value in this case, but for real-valued functions it is never sensible, since it might never happen in practice that two creatures had the same fitness. One corollary of the fitness function being integer-valued with real domain is that a great deal of information must be lost. In other words, there are many, many genes that map to the same fitness value.

Experiment 3.21 *Modify the basic program from Experiment 3.18 so that the algorithm incorporates a (possibly hard-coded) adjacency matrix for a graph G and modify the algorithm so that the algorithm estimates the crossing number for G. Try your optimizer on 12 vertices connected in a cycle and on the graph with the following adjacency matrix:*

$$\begin{bmatrix} 0 & 1 & 0 & 0 & 1 & 1 & 0 & 0 & 0 & 0 \\ 1 & 0 & 1 & 0 & 0 & 0 & 1 & 0 & 0 & 0 \\ 0 & 1 & 0 & 1 & 0 & 0 & 0 & 1 & 0 & 0 \\ 0 & 0 & 1 & 0 & 1 & 0 & 0 & 0 & 1 & 0 \\ 1 & 0 & 0 & 1 & 0 & 0 & 0 & 0 & 0 & 1 \\ 1 & 0 & 0 & 0 & 0 & 0 & 0 & 1 & 1 & 0 \\ 0 & 1 & 0 & 0 & 0 & 0 & 0 & 0 & 1 & 1 \\ 0 & 0 & 1 & 0 & 0 & 1 & 0 & 0 & 0 & 1 \\ 0 & 0 & 0 & 1 & 0 & 1 & 1 & 0 & 0 & 0 \\ 0 & 0 & 0 & 0 & 1 & 0 & 1 & 1 & 0 & 0 \end{bmatrix}$$

Problems

Problem 127. Verify the formula given in Equation 3.17 showing your work clearly. Obtain a similar formula for the y-coordinate of the intersection.

Problem 128. Do the evolutionary algorithms discussed in this section place upper or lower bounds on crossing numbers? Justify your answer in a sentence or two.

Problem 129. What is the maximum crossing number possible for a rectilinear drawing of K_n? Advanced students should offer a mathematical proof of their answer.

Problem 130. Essay. Would you expect domain or range niche specialization to be more effective at improving the performance of an evolutionary algorithm that estimated rectilinear crossing number? Justify your answer with logic and with experimental evidence if it is available. You may wish to examine Problem 131 before commencing.

Problem 131. Shown below are two drawings of K_4. Suppose we add a vertex somewhere and connect it to all the other vertices so as to get a rectilinear drawing of K_5. First, give a proof that the crossing number of the new drawing depends only on the region of the drawing in which the added vertex is placed. Second, redraw these pictures and place the resulting crossing number in each of the regions.

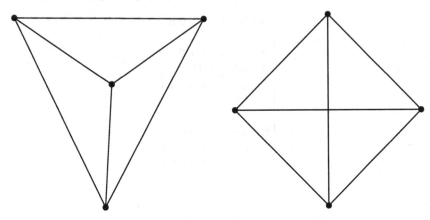

Problem 132. Problem 131 suggests a novel mutation operator for use in computing crossing numbers. Discuss it briefly and outline an algorithm that would implement such a mutation operator.

Problem 133. There are an infinite number of possible drawings of a graph in the plane. (Given a drawing, just move a vertex of the drawing any distance in any direction to get another drawing.) A little thought shows that this isn't a useful definition of the word "different" when one is attempting to compute crossing numbers. Following Problem 131, come up with an alternative definition of when two drawings are the same or different and explain why your definition is useful. You may want to glance at a book on topology to get some of your terms. Consider the role of vertices, edges, and intersections of edges in your definition of "different." Is it easy to compute when two graphs are different under your definition?

Problem 134. Give the code for a function that accepts two line segments and returns true if they intersect in their interior. Intersection at the endpoints or no intersection is returned as false. Make sure your code deals well with parallel lines and lines with infinite slope.

Problem 135. Write a function such that given the number n of vertices in a graph, two real arrays x and y that specify the vertex coordinates, and a matrix A with $a_{i,j} = 1$ if vertices i and j are joined by an edge and $a_{i,j} = 0$ otherwise, the function returns the rectilinear crossing number of the graph. It should use the code from Problem 134.

Problem 136. Essay. In a practical implementation of an evolutionary algorithm to estimate rectilinear crossing number, there will be problems with almost parallel and almost vertical lines. Discuss hacks to avoid having numerical precision problems. You may wish to consider modifying the fitness function to penalize certain slopes or pairs of slopes, moving vertices periodically to avoid these sorts of problems, or moving vertices exactly when a vertex is apparently causing such problems. Be sure to discuss the computational cost of such changes as well as their efficiency in preventing numerical precision problems. Would putting the vertices of the graph onto a discrete grid help?

Problem 137. In this section, we developed evolutionary algorithms to estimate rectilinear crossing numbers of graphs. Working with the true crossing number would require that we somehow manipulate fully general curves in the plane, which is very hard. Suppose that instead of representing edges as line segments, we represent them as multiple line segments. This would bring the rectilinear crossing number closer to the true crossing number. Find a graphical construction that transforms a graph G into a graph G' by drawing the edges of G using k consecutive line segments so that the rectilinear crossing number of G' is an estimate of the crossing number of G.

Problem 138. Suppose we have a drawing of a graph in which one vertex lies *on* a line segment joining two others. In the formal graph theory that computes crossing numbers, this is viewed as a pathological case, repaired by making a tiny change in the position of the vertex causing the problem. In this chapter, this is counted as a crossing in order to get on with things. Our motivating example was a circuit board. Argue from physical reality that a vertex in the middle of an edge should not be counted as a crossing at all. Give an example.

Problem 139. Using software that computes the crossing numbers of drawings experimentally, estimate and graph the crossing number of randomly drawn k-gons for $k = 4, 5, \ldots, 12$. Also estimate the probability of getting a correct solution in the initial population for the 12-cycle half of Experiment 3.21.

4

Sunburn: Coevolving Strings

Sunburn is a model for designing a simplified video-game spacecraft. The model was proposed by John Walker. The model has a fitness function that is predicated on winning battles with other spacecraft. The fitness function explores the space of designs for spaceships without using actual numbers; it just uses win, lose, or tie results. The genes for spaceships are character strings augmented by an integer.

There are two important new ideas in this chapter. First, the model of evolution used for Sunburn involves a strange variant on tournament selection, termed *gladiatorial tournament selection*. Second, the strings used as genes in Sunburn are tested against other strings. This means that Sunburn is a *coevolving* evolutionary algorithm.

Somewhat surprisingly, for many choices of parameters, the Sunburn model behaves like a unimodal optimizer rather than a coevolving system; evolution does not appear to be contingent on the initial conditions. This suggests that, for those choices of parameters, the system is quite likely to be very close to a unimodal optimization problem.

Coevolving systems depend more strongly on the choices made in the design of the system than the fixed-fitness evolving systems we have studied thus far. In this chapter we will explore variations of the coevolving Sunburn system and so gain some knowledge and experience in designing and controlling the behavior of such systems. Variations of the Sunburn system are explored via experimentation. The final section of the chapter introduces a computationally similar evolutionary algorithm called VIP for VI(rtual) P(olitician), which is entirely nonviolent. This system was invented by Mark Joenks. Figure 4.1 gives the dependencies of the experiments in this chapter.

4.1 Definition of the Sunburn Model

The gene for a spacecraft in the Sunburn model has a fixed number of slots for ships' systems (the character string) together with a desired distance the

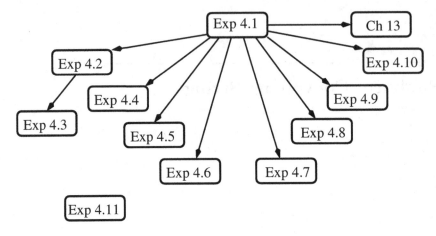

1 Basic Sunburn evolver.
2 Exploring weapons' effectiveness curves.
3 Varying initial range.
4 Student exploration of weapons' effectiveness curves.
5 Finding the true measure of weapons' effectiveness.
6 Generational gladiatorial tournament selection.
7 Exploring the shield multiplier.
8 Weapons that don't always fire.
9 Randomized hit location.
10 Exploring multiple shield types.
11 VIrtual Politicians.

Fig. 4.1. The topics and dependencies of the experiments in this chapter.

spacecraft wishes to be from an opponent during a battle (the integer). There are five types of systems: guns, lasers, missiles, drives, and shields. Figure 4.2 shows a typical gene. We will use a value of 20 system slots and choose ranges between 1 and 20.

Systems	Preferred Range
GLMDSSMGLSDSMLGMLSDS	16

Fig. 4.2. A Sunburn gene.

In biology, the map from genes to creature can be quite complex; in most artificial life systems, it is quite direct. The process of transforming a genome into a creature is called the *developmental biology* of a creature. In Sunburn we have our first nontrivial developmental biology. The sense behind this developmental biology will become apparent as we describe the rules for combat, and it is discussed explicitly in Section 4.3.

To derive a spaceship from a Sunburn gene, you count the number of loci in which an S appears and then put 3 times that many shields on the front of the ship. The remaining systems, G(uns), L(asers), M(issles), and D(rives), are placed behind the shields in the order in which they appear in the gene. Subscript the ship with its preferred range. The ship described by the gene in Figure 4.2 looks like

$$SSSSSSSSSSSSSSSSSSGLMDMGLDMLGMLD_{16},$$

with the front of the ship being to the left. The *shield multiplication factor* is the number of shields a ship receives for each genetic locus in which it has an S. In this example and in most of this chapter, the shield multiplication factor is 3.

Combat between two ships is conducted as follows: Combat is initiated at a starting range. (We will explore different methods of generating starting ranges.) Once combat has started, the ships iteratively go through turns consisting of shooting and then moving. This loop continues until a winner and loser are found or until a draw occurs. The two ships shoot simultaneously using a gun, missile, or laser system once each turn. Each of the three types of weapons has an *effectiveness curve* that specifies its probability of scoring a hit at each possible range. These effectiveness curves are a very important feature of the Sunburn model. Examples of gun, laser, and missile effectiveness curves are shown in Figure 4.3.

The design of the effectiveness curves was inspired by imagining how space weapons might work. A gun emits a physical missile fired at very high velocity but with no guidance beyond initial efforts at aiming. The chance of hitting a target with a gun thus drops off with distance, and guns are most effective at short range. A laser moves at an incredible velocity (the speed of light) in a very straight line and so is equally effective at all ranges where a target can be detected. It does not have a high energy content compared to the kinetic energy locked in a physical projectile like a missile or shell and so has a relatively low kill rate. A missile, like the shells fired by a gun, is a physical projectile. Unlike a shell, it picks up speed throughout its attack run and has target-seeking qualities. At short range (low velocity), missiles are relatively easy to intercept. Their effectiveness thus climbs as they approach top velocity. This conception is helpful in framing the initial analysis of the behavior of the system. Other conceptions and sets of weapons' effectiveness curves that match them are certainly possible.

Each hit scored removes one system from the front of a ship. After shooting, the ships move. Each drive enables a ship to move a distance of one. The ships take turns moving a distance of one toward their preferred ranges, dropping out when they run out of drives.

In order to win a combat, a ship must destroy all the systems on the other ship and have at least one drive left itself. In the event that neither ship has

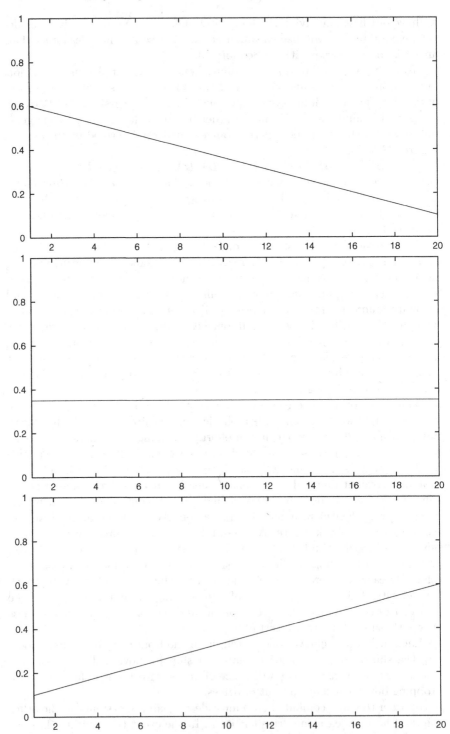

Fig. 4.3. From the top, gun, laser, and missile weapons' effectiveness curves.

remaining drives or neither ship has remaining weapons systems, the fight is a draw. If two ships have not fought to a draw or victory in 100 combat turns, their combat is a draw by fiat. (Sometimes ships appear that take forever to resolve a fight, slowing down evolution without substantial benefit. This happens, for example, when the ship's preferred range is badly mismatched with its weapons mix.)

One point that may need emphasis is that the genes of the creatures are separate from the ships built from them. When a ship loses systems, its gene remains intact. When a ship wins, its *gene*, not itself, is used in reproduction.

Now that we have the gene, developmental biology, and combat described, we can define *gladiatorial tournament selection*. In this model of evolution, instead of generations there are mating events that asynchronously update the population. This is a form of steady-state genetic algorithm. A mating event is performed by conducting combat on randomly selected pairs of distinct ships, returning pairs that draw to the population untouched, until two winners and two losers are obtained. The losers are deleted from the population, and the winners reproduce as in the algorithms in previous chapters. The two children they produce replace the losers.

The fact that there is a group of 4, in which the two best mate and produce children that replace the two worst, makes this sort of selection reminiscent of tournament selection, but there are some important differences. The measure of fitness is extremely local: all you really know is that a ship was good enough to beat one other randomly selected ship in one particular combat. The measure of fitness is not a heuristic estimation of quality, but entirely objective within the bounds of luck. All that Sunburn ships do is fight other Sunburn ships, and we are testing them on exactly this task. Since ships are tested against other ships, Sunburn is clearly a coevolutionary genetic algorithm.

The gladiatorial tournament selection model of evolution can be used whenever two creatures are playing some sort of game against one another. Since it grants fitness for beating one other creature, it is a model of evolution that rewards skill in pairwise interactions, unaffected by other pairwise actions in the population. One might expect that it would be very bad for the emergence of cooperation in a group; we shall test this expectation in Chapter 6 when we study Iterated Prisoner's Dilemma.

Problems

Problem 140. With gladiatorial tournament selection, a creature may survive indefinitely by simply not being chosen for combat. Compute the probability that a creature remains untested after k mating events in a population of $n = 2m$ creatures. Compute also the expected waiting time until a test.

Problem 141. Reread Problem 140. Suppose that we want to know the time until a creature undergoing evolution with gladiatorial tournament selection breeds or dies. Compute the expected time to death or reproduction in terms of the given probability p that any given encounter is a draw.

Problem 142. The gladiatorial tournament selection described in Section 4.1 uses groups of size 4. Give a reasonable description for how to do gladiatorial tournament selection with larger groups. Avoid situations in which some of the creatures are not at risk.

Problem 143. Gladiatorial tournament selection, as portrayed in this section, is a steady-state evolutionary algorithm. Give a selection algorithm for a generational version in which every ship engages in combat each generation. Relative to your algorithm, identify at least two details that could have been done in a different way and justify your choice.

Problem 144. Essay. One of the advantages of tournament selection in general is its ability to preserve population diversity. If we define a generation to be m mating events, where there are $n = 2m$ creatures in the population, is gladiatorial tournament selection better or worse at preserving diversity than single tournament selection? Justify your answer logically or experimentally. Be sure to fix a problem domain in which to make your judgment as to what is better. The general answer is likely to be "it depends," which is both correct and uninformative. Possible problem domains are the various example targets of evolutionary computation in the earlier chapters of the book.

Problem 145. Short Essay. Is gladiatorial tournament selection elitist when it is used in Sunburn? If the answer were yes or no, this would not be an essay question.

Problem 146. Come up with one or more techniques for converting an optimization that uses tournament selection to one that uses gladiatorial tournament selection. For each such scheme, answer the following questions: Is the scheme elitist? If there are $n = 2m$ creatures in the population, what is the expected number of children a creature will have in a mating event as a function of its rank in the population?

Problem 147. Essay. In gladiatorial tournament selection one strategy that avoids death is to fight to a draw every time. While this prevents you from having children, it would seem to be a survival strategy. Under what sorts of circumstances is it an effective survival strategy?

Problem 148. Suppose that we wish to redesign the Sunburn genetic algorithm to evolve ships that beat a fixed, existing design. Give an outline of such a genetic algorithm. Explain the degree to which the algorithm is evolutionary or coevolutionary.

4.2 Implementing Sunburn

Now that Sunburn is defined, we can play with it. What is the effect of changing the weapons' effectiveness curves? How does the value of the initial range affect the outcome? Which kinds of weapons are most effective? Which model of evolution works best with Sunburn?

Experiment 4.1 *Write or obtain software for a Sunburn evolutionary algorithm using gladiatorial tournament selection on a population of* 200 *randomly generated ship designs with* 20 *systems (G, L, M, S, or D) and a preferred range ranging from 1 to 20. Use the weapons' effectiveness curves from Figure 4.3, which are given by the following formulas. At a range r the probabilities of hitting for guns, lasers, and missiles are*

$$P_G(r) = 0.6 - (r-1)/38,$$
$$P_L(r) = 0.35,$$
$$P_M(r) = 0.1 + (r-1)/38.$$

Use two-point crossover with the preferred range treated as the last character in the string making up the ship design. Use single-point mutation which replaces a ship's system locus with a new random system, or increments or decrements the preferred range by 1. Always use a starting range of 20.

In place of generations, do 100 mating events. After each 100 mating events, report the fraction of the population's genes devoted to each type of system, the fraction of ships that have a drive in the last position (something the rules favor), and the ratio of shields (genetic loci that are "S") to weapons. In addition, report the mean and standard deviation of the preferred ranges and the fraction of combats that are draws. Run each population for 100 "generations," i.e., 10,000 mating events, and run 20 populations with distinct initial populations for comparison.

If you have implemented Experiment 4.1 correctly, it should often converge to a design we call the "starbase" fairly quickly. A "starbase" has lots of missiles, few other weapons, a respectable number of shields, and a single drive on the end of the ship. Preferred ranges will tend to be large, but are essentially irrelevant because of the speed with which combat is resolved. Let us now make a foray into exploring other weapons' effectiveness curves.

Experiment 4.2 *Modify the software from Experiment 4.1 to use the following weapons' effectiveness curves:*

$$P_G(r) = \frac{0.7}{(r-3)^2 + 1},$$

$$P_L(r) = \frac{0.6}{(r-6)^2 + 1},$$

$$P_M(r) = \frac{0.5}{(r - 9)^2 + 1}.$$

Gather the same data as before. These weapons' effectiveness curves should provide some small incentive for ships to have drive systems other than because of the technicality that a drive system is required to win. Do they? Why or why not? In addition to answering those questions, report the various statistics mentioned in Experiment 4.1. A graph of these weapons' effectiveness curves is shown in Figure 4.4.

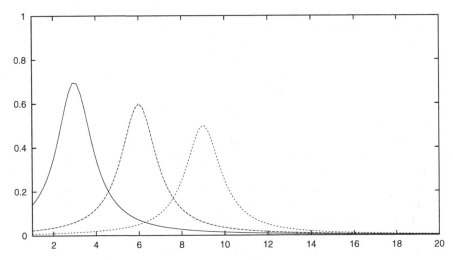

Fig. 4.4. Modified weapons' effectiveness curves.

Next, we will experimentally explore the effect of varying the initial range. Intuitively, starting at long range favors starbases that can launch barrages of highly effective missles. A very short starting range should create a sort of opposite of a starbase, bristling with guns, but otherwise similar to a starbase. We will test the effects of intermediate starting ranges where it is not clear which weapons are best.

Experiment 4.3 *Use the software from Experiment 4.1. Take either (consult your instructor) of the two sets of weapons' effectiveness curves we have examined so far and change the starting range to be normally distributed with a mean of 10 and a standard deviation of 2. You will need to round to the nearest integer and throw out values larger than 20 or smaller than 1. Report the same statistics as in Experiment 4.1. Does changing the starting range have an effect on evolution? (If you have forgotten the formula to generate normally distributed random numbers, refer to Equation 3.1.)*

The dynamics of selection in Sunburn are strongly controlled by the weapons' effectiveness curves. In the next experiment, we have a more than usually open-ended setup in which you are sent off into the wilderness to find interesting weapons' effectiveness curves.

Experiment 4.4 *If evolution is* contingent, *the fractions of various types of systems at the end of the evolution should vary nontrivially from run to run. Using the software from Experiment 4.1 with appropriate modifications, try to find a set of weapons' effectiveness curves for which evolution is more strongly contingent. State in your report how you will document contingency and state the degree to which you found contingency. Did initial findings cause you to revise your method of documenting contingency?*

One issue in the Sunburn system is the question of what makes evolution favor a given type of weapon. The *initial effectiveness* of a weapon is its expected probability of hitting at the initial range. The *total effectiveness* of a weapon is the area under its weapons' effectiveness curve. The *maximal effectiveness coefficient* of a weapon is the fraction of the possible ranges where the weapon is more likely to hit than any other. The *range-weighted effectiveness* is the sum over possible ranges r of the weapons' effectiveness curve multiplied by a range weighting function $\omega(r)$. All of these measures are related to effectiveness, but it isn't very clear which one is closest to being the thing optimized for by evolution. When reason does not speak to a question, experimentation may help.

Experiment 4.5 *Let* $\omega(r) = \frac{1}{(r-s)^2+1}$, *where s is the mean starting range for ships. Give and logically justify a hypothesis about which of initial effectiveness, total effectiveness, maximal effectiveness coefficient, and range-weighted effectiveness the Sunburn genetic algorithm is in fact optimizing. Then design and perform an experiment to test that hypothesis for a variety of weapons' effectiveness curves and starting ranges. Advanced students may also wish to test other possible choices of* $\omega(r)$.

One of the themes that appears over and over in this book is that the choice of model of evolution can have a substantial effect on the results. We will now experimentally test the effect of changing the model of evolution on Sunburn.

Experiment 4.6 *If you have not done Problem 143 yet, do so. Now modify the code from Experiment 4.1 to use a generational version of gladiatorial tournament selection. Perform the experiment again, keeping the number of mating events roughly constant (within two times). In your write-up, discuss the changes in the dynamics of evolution and the difference in final designs, if any.*

There are a *lot* of other experiments we could perform. Sunburn has been a rich source of interesting student projects over the years (see Section 4.4). We will cut off the parade of experiments with the basic Sunburn model here and look at less-than-basic models in the next section.

Problems

Problem 149. There are several ways to detect the outcome of a combat without fighting to the bitter end. Give at least one set of circumstances where a combat must end in a draw and one where it is clear which ship will win even though both ships still have working systems.

Problem 150. Suppose we have a simplification of the Sunburn model in which there are only missiles and shields. Ships don't move, and the winner is the last ship with any systems. If the probability that a missile hit will destroy a system is p, then what is the optimal number of shields in a gene with 20 systems slots and a shield multiplication factor of 3? Attack this problem by thinking of the ships as pairs of numbers that add to 20 and saying that one pair dominates another if there is a probability of more than 0.5 that the corresponding ship will win a fight. This gives you a domination relationship on these pairs of numbers that you can compute by hand or machine. If there are ships not dominated by any other, they are the optimal designs; otherwise, there is no optimum. If the problem seems too hard, try working with a smaller number of system slots until you get a feel for the problem.

Problem 151. In a given turn, a ship in the Sunburn genetic algorithm may fire several weapons that all have the same probability p of hitting. Each weapon needs its own random number. Each such event is a *Bernoulli trial* as defined in Appendix B. If you have several, say more than 3, independent Bernoulli trials, then they are very close to being normally distributed with mean np and standard deviation $\sqrt{np(1-p)}$. This means that we could substitute a single normal random variable for a large number of uniform random variables. In order to get a normal random variable with mean μ and standard deviation σ, denoted by $G(\mu, \sigma)$, you set

$$G(\mu, \sigma) = \sigma \cdot G(0, 1) + \mu,$$

where the formula for $G(0, 1)$ is given after Equation 3.5.

Assuming that we wish to replace any set of 4 or more weapon resolutions of weapons of the same type with a normal random variable, give pseudocode for doing so for missiles, lasers, and guns. Truncate the normal random variable when it produces unreasonable values. Advanced students may wish to experimentally measure the amount of time saved by this technique.

Problem 152. Please compute the total effectiveness, maximal effectiveness coefficient, and range-weighted effectiveness of missiles, lasers, and guns for the weapons' effectiveness curves in Experiments 4.1 and 4.2. Assume that $\omega(r) = \frac{1}{(r-20)^2+1}$.

Problem 153. Come up with a better algorithm than the one given in the text for figuring out the change in position of Sunburn ships during the movement part of a turn. You should produce pseudocode that accepts current positions, desired ranges, and number of drives for two ships and returns their new positions.

4.3 Discussion and Generalizations

Why does the developmental biology of Sunburn have a shield multiplication factor? Well, shields do nothing except soak up damage. If one shield locus in the gene corresponded to one shield in the finished ship, then evolution would quickly replace the shields with a system that did something. This system, whatever it was, could still absorb one hit and the ship would get some additional use out of it. If a shield gene can soak up more than one hit, it has some nontrivial usefulness compared to other systems, and there is reason for a gene to contain some nontrivial number of shield loci.

Experiment 4.7 *Using the software from Experiment 4.1 with appropriate modifications, repeat the experiment with shield multiplication factors of 2 and 4. Compare the results to the original experiment with a shield multiplication factor of 3.*

Drives, which are far less useful in the basic Sunburn model than the designer thought they would be, are used to move ships to where their weapons are more effective. A large part of the reason drives, other than the one required for victory, are not useful is that ships quickly evolve to be maximally effective at the mean initial range. Experiment 4.2 tries to prevent this by making all weapons woefully ineffective at the starting range. If you are interested, you could try a more extreme tactic by making weapons completely useless at the initial range.

The three sorts of weapons in the initial Sunburn work done by John Walker were thought of as effective at short range (guns), effective at long range (missiles), and somewhat effective at all ranges (lasers), which is reflected in the weapons' effectiveness curves shown in Figure 4.3. Changing the weapons' effectiveness curves can radically affect the outcome of evolution. Also somewhat surprising is the apparent unimodality of the coevolving space of designs for Sunburn for the weapons' effectiveness curves tried thus far. For these curves the evolution of the resulting populations seems to close in on a single optimal design with a few ineffective forays into dead ends. A few choices of weapons' effectiveness curves do yield contingent evolution, but these sets of curves seem rare.

It is obvious that one could increase or decrease the number of types of weapons in Sunburn to get different systems. A model with a single type of weapon is easier to analyze theoretically, as we saw in Problem 150. In the remainder of this section we will suggest other possible modifications of the Sunburn model that make good term projects.

Limited Shots

One completely unrealistic feature of the Sunburn model is the unlimited use of the ships' weapons. The ships fire as many shots as they like, one per turn per weapons system. Removing this implausible feature leads to several variations. First one might try simply imposing a maximum on the total uses of each weapon type, perhaps with less-effective weapon types having more shots. Choose these limits carefully. If they are too high, they do not change the behavior of the model at all. If they are too low, they cause a high number of stalemates.

It is plausible that limiting weapon use would encourage the creatures to learn to move to effective ranges as fast as possible before those weapons were used up. The need to use weapons effectively so as not to waste shots also suggests another possible addition to the Sunburn model: a way of deciding when to fire. The technology for such a decision maker could be quite simple, e.g., a real number for each weapon type that serves as a probability threshold. The creature fires a given type of weapons system only when the range is such that the probability of hitting equals or exceeds the threshold.

Experiment 4.8 *Suppose we are doing a Sunburn model with a single type of weapon and that we have augmented the genetics with the probability threshold described above. Running several populations with starting range 20 and weapons' effectiveness curve*

$$f_w(r) = \frac{0.7}{(r - 10)^2 + 1},$$

ascertain whether there is a single optimum weapons' threshold or whether the weapons' threshold is contingent on the initial conditions. Describe carefully the design of trials you use to settle this question. Does it matter how you incorporate the threshold into the data structure used to hold the spaceship design?

Another natural choice for a decision-to-fire device is a small neural net. The net might take as inputs range, remaining charges, distance from the weapons system to the front of the ship, hit probability at current range, or number of shields remaining. Its output would be a fire/don't fire decision. Even a very small neural network (see Section 11.1) with one to three neurons could make fairly complex decisions. The ship should have three neural nets, one for each weapon type, rather than one net per weapons system. Ships without a given type of weapon could have vestigial neural control systems for that weapon type. The connection weights for these neural nets would need to be incorporated into the data structure holding a spaceship design.

Another control technology that would be interesting but much more difficult to implement is genetic programs (see Section 1.3). As with neural control structures, there should be one GP parse tree for each weapon type that takes integer parameters and outputs a fire/no fire decision. Since the constant function "fire" is a tolerable program for some parts of the parameter space, there is room for evolution to take an extremely simple control strategy and improve it. This is typically an evolution-friendly situation in which an evolutionary algorithm can shine.

For each of the three possible control technologies we mention above and any other that you think up on your own, it may be best to start with a single type of weapon until the control-strategy-evolving software is working properly. Only after this software is working in a predictable fashion should you diversify the weapons mix. Keep in mind that the "always fire" weapons' control of the basic Sunburn model, when using one weapon type, is an invitation for evolution to solve the optimization problem given in Problem 150. With the control systems in place, however, nontrivial evolutionary behavior is possible even with a single weapons' type.

Another variation that could make the Sunburn model more plausible is to add a type of ship's system called a magazine, which contains some fixed number of missiles, laser charges, and shells for guns. This way, the evolutionary algorithm would have to figure out how many magazines there are and where they should be placed. Taking this variation further, there could be three types of magazine: one for missiles, one for laser charges, and one for shells. Other variations might include being able to use magazines only if they are adjacent to appropriate weapons' types or destroying systems adjacent to magazines when the magazine is hit.

Sensible Movement

The movement in the original Sunburn model is as simple as possible and could easily be made more realistic and difficult. One could change the drives from devices that churn out distance to devices that churn out acceleration. This would probably require placing the ships on a real number line. Acceleration requires three variables: position (s), velocity (v), and acceleration (a). A simple algorithm for acceleration is as follows:

```
For each drive available do
Begin
    If we are too close a := a-1 else a := a+1
end;
v := v+a;
s := s+v;
```

This model of acceleration is very primitive. When farther away than its preferred range, the ship simply accelerates toward the other ship; otherwise, it accelerates away. This will result in very little time spent at optimum range and, interestingly, with the ships having their highest velocity when they are near their desired range. This doesn't have to be a bad thing. Our semantic interpretation of the number we call the ship's "preferred range" is the range the ship wants to be at during combat. With the more complex movement system in place, evolution may well find another use for that number, in essence changing the "meaning" of the number. Instead of using this number as a range to maintain, it will become the useful mark for starting to decelerate. It is not unintuitive that this range be chosen so that the ship is slowing down and turning around at a point where its weapons are most effective.

From our knowledge of basic physics, we can guess that a much larger space will be needed for ships with drives that generate acceleration than for ones that generate just position changes. Once you have chosen your larger board size, you may wish to make moving beyond maximum range a condition that leads to stalemate.

As with the weapons systems, it may also be desirable to have a neural net or genetic program that decides whether it wants to use the ship's drives in a given turn. The inputs to this net or GP could include current range, number of ship's systems left, number of drives left, number of weapons left, and current velocity. This opens up the possibility that a ship will try to run away and achieve stalemate when victory has become impossible, an action that will require some revision to the methods of early detection of stalemates. The more sophisticated drive controls could be used in either the position- or acceleration-generating environments. Given that a severely damaged ship would want to flee, one also might simply add the following rule to the basic Sunburn model: when a ship is out of weapons systems, it will use all remaining drives to increase range. The thought here is that the ship is attempting to achieve stalemate.

Finally, if you implement a more complex model of movement, it might be sensible to factor the ship's relative velocity into the weapons' effectiveness curves. A faster-moving ship might be harder to hit (recall that velocity is a relative *not* an absolute quantity). Some thought should be given to determining how speed affects weapons' effectiveness. It may be that it affects different weapons to different degrees, and it is almost certainly not linear.

Variations in Taking Damage

It is implausible that a ship's systems would be destroyed *in order*. The position of systems would provide a strong bias for the order of destruction but not utterly control it. One simple method of dealing with this is to place a distribution on the systems that favors the front. Pick some probability p and then, each time a ship is hit, do the following: with probability p, the first system remaining is destroyed; with probability $1-p$, you skip the first system and go on to the rest of the ship. If $p = \frac{1}{2}$, then the probability of systems being destroyed is, starting from the front, $\frac{1}{2}, \frac{1}{4}, \frac{1}{8}, \ldots$.

Experiment 4.9 *Modify the software from Experiment 4.1 to use the probabilistic hit evaluation described above. For $p = 0.75$ and $p = 0.5$, rerun the* *experiment and report on the differences that this type of hit location assessment causes. In particular, does the number of drives go up and does the ratio of shields to weapons change?*

Another implausible feature of damage assessment in Sunburn is that the same generic shield can stop three very different sorts of weapons. This implausibility can be remedied in a fashion that will also make the evolution more complex and perhaps more interesting. Instead of a single type of shield, there will be one type of shield per type of weapons system. In the original Sunburn model this would mean a missile shield, a laser shield, and a gun shield. These three types of shield would be represented in the genome by distinct letters. The developmental biology is modified as follows: take the pattern of shield genes in the order they appear in the gene; place this pattern of shields on the front of the ship repeated a number of times equal to the shield multiplication factor.

Example 2. If we had a Sunburn simulation with multiple shield types, S_M, S_L, and S_G, and the following gene,

$$GLMDS_M S_M MGLS_L DS_G MLGMLS_M DS_L \ 16,$$

then the resulting ship would look like

$$S_M S_M S_L S_G S_M S_L S_M S_M S_L S_G S_M S_L S_M S_M S_L S_G S_M S_L GLMDMGLDMLGMLD_{16}.$$

Here is one possible way to modify the combat rules in order to accommodate multiple shield types. In combat, a shield is transparent to the type of weapons it is not intended for and is still destroyed by them. Suppose we had a ship with 3 missile shields and then a laser shield as its front 4 systems. If a laser hit the ship, then those 3 missile shields would be destroyed without doing anything, and the laser shield would be destroyed stopping the laser. Notice that having different types of shields makes the order in which various attacks hit important. Assume that the enemy's hits are generated from the front to the back of his ship. If his forward missile launcher and aft laser turret are hit, the missile hit is processed first. This gives the ordering of weapons systems a new importance.

Experiment 4.10 *Modify the software from Experiment 4.1 to have multiple shield types as described above. What differences do the multiple shield types make?*

There are many, many other modifications one could make to the Sunburn model, but we leave these to your inventiveness.

Problems

Problem 154. There is at least one obvious and stupid method of choosing weapons' effectiveness curves that will force the fraction of the gene devoted to each type of weapons system to be a contingent feature of evolution. Find an example of such a method.

Problem 155. Give a choice of weapons' effectiveness curves and a fixed starting range for a Sunburn model that will coerce the use of drives. Make your example substantially different from those discussed in the text.

Problem 156. For p equal to each of 0.9, 0.75, 0.5, and 0.3, graph on the same axes the probability of systems being destroyed starting from the front of the ship using the probabilistic hit location system described in Section 4.3 (Variations in Taking Damage).

Problem 157. In Example 2, we repeat the pattern of shields in the gene 3 times. Why is this better or worse than taking the pattern once and triplicating each shield as we go? Give an example.

Problem 158. Essay. Describe a Sunburn variation in which there are two or more weapon types and for which weapons, shields, and drives all draw off a common reserve of energy. Make predictions about the behavior of your system. Advanced students should test these predictions experimentally.

4.4 Other Ways of Getting Burned

One objection that has been raised to the Sunburn model as an example of coevolving strings is that it is violent. The author, being a fairly typical recovering video and war game addict, simply did not imagine that abstract warfare between character strings was offensive. Once made aware that Sunburn was sufficiently violent to offend some portion of the public, the author put it as a challenge to his students to take the basic notion of Sunburn and find a nonviolent fitness function that could be used to experiment with a similar set of issues. This section is inspired by one of the better attempts to meet the challenge invented by Mark Joenks as a term project for the class.

The essence of Sunburn is to have strings that compete in some fashion and whose genes spell out how that competition is approached by the entities derived from those genes. In this section, we will develop such a model for individuals participating in political campaigns. It is for you to decide whether this is more or less offensive than simulated warfare.

In Sunburn, the character strings had five letters: D, G, L, M, and S. These character strings were developed into abstract models of fighter craft. For our model of political campaigns we need a new alphabet as follows: **A**dopting popular programs, **B**ribing, **D**oing what your opponent did in the last move of the campaign, **F**und-raising, **L**aying low, **N**egative campaigning, **P**andering, and having a **S**candal. The artificial agents in this simulation of politics will be called **virtual** politicians, or *VIPs*.

Of prime importance in evaluating a character string representing a VIP is our model of the behavior of the electorate. Our fitness evaluation, still used for gladiatorial tournament selection, will run through the campaign season. At regular intervals, the genes of each of our virtual politicians will be expressed, one location at a time, in order. A series of variables will change value during the campaign, depending on the actions of the competitors. At the end, the electorate will vote, probabilistically, deciding the contest. The state variables stored for each candidate are given in Table 4.1.

Credentials	C	the candidate's standing with his single-issue voters
Credibility	R	the candidate's perceived ability to serve competently
Name Recognition	N	related to being recognized by a voter
Scandal Factor	S	the degree to which the candidate is tainted by scandal
Finances	B	powers everything else

Table 4.1. State variables for VIPs.

The way we initialize the candidates' state variables describes how the candidates start the campaign season. We update the variables according to the following rules, as we scan down the candidate's gene strings.

Rule 1. Credentials, credibility, name recognition, and scandal factor all undergo exponential decay. At the beginning of each period of the campaign, they are multiplied by r_C, r_R, r_N, and r_S, all smaller than 1. Since the voters have short memories for anything complicated, we insist that

$$r_N > r_R > r_C > r_S.$$

Rule 2. Finances grow exponentially. At the beginning of each period of the campaign, a candidate's money is multiplied by r_F, bigger than 1. This represents fundraising by the candidate's campaign organization.

Rule 3. Adopting a popular program adds 2 to a candidate's credibility and subtracts 1 from his credentials. If he has at least half as much money as his opponent, it adds 2 to his credibility and subtracts 1 from his credentials. Otherwise, it adds 2 to his credibility and subtracts 1 from his credentials (he swiped the idea).

Rule 4. Bribing either subtracts 5 from a candidate's finances or cuts them in half if his total finances are less than 5. Bribing adds 5 to his credentials, 2 to his scandal factor, and 1 to his name recognition.

Rule 5. Doing what a candidate's opponent did last time is just what it sounds like. On the first action, this action counts as laying low.

Rule 6. Fundraising adds 3 to a candidate's finances and 1 to his name recognition. It represents a special, personal effort at fundraising by the candidate.

Rule 7. Laying low has no effect on the state variables.

Rule 8. Negative campaigning subtracts 1 from a candidate's credibility and credentials and adds 1 to the other candidate's credentials. If he has at least half as much money as his opponent, then this goes his way. Otherwise, it goes the other candidate's way.

Rule 9. Pandering adds 5 to a candidate's credentials, 1 to his name recognition, and subtracts 1 from his credibility.

Rule 10. Scandal adds 4 to a candidate's name recognition and subtracts 1 from his credentials and credibility.

Once we have used the rules to translate the VIP's genes into the final version of the state variables, we have an election. In the election, we have 25 special-interest voters aligned with each candidate and 50 unaligned voters. Each voter may choose to vote for a candidate or refuse to vote at all. The special-interest voters will vote for their candidate or not vote. For each voter, check the following probabilities to tally the vote.

A special-interest voter will vote for his candidate with probability

$$P_{\text{special}} = \frac{e^{C-S}}{2 + e^{C-S}}. \tag{4.1}$$

An unaligned voter will choose a candidate first in proportion to name recognition. He will vote for the first candidate with probability

$$P_{\text{unaligned}} = \frac{e^{R-S}}{3 + e^{R-S}}. \tag{4.2}$$

If not voting for the first candidate, the voter will consider the second candidate using the same distribution. If he still has not voted, then he will repeat this procedure two more times. If, at the end of 3 cycles of consideration, he has still not picked a candidate, the voter will decline to vote. The election (and the gladiatorial tournament selection) are decided by the majority of voters picking a candidate. If no one votes, then the election is a draw.

Experiment 4.11 *Using the procedure outlined in this section, create an evolutionary algorithm for VIPs using gladiatorial tournament selection on a population of 200 VIPs. Use two-point crossover on a string of 20 actions with two-point mutation. Set the constants as follows: $r_N = 0.95$, $r_R = 0.9$, $r_C = 0.8$, $r_S = 0.6$, and $r_F = 1.2$. Use uniform initial conditions for the VIPs with the state variables all set to 0, except finances, which is set to 4. Perform 100 runs lasting for 20,000 mating events each. Document the strategies that arise. Track average voter turnout and total finances for each run.*

There is an enormous number of variations possible on the VIP evolutionary algorithm. If you find one that works especially well, please send the author a note.

Problems

Problem 159. Essay. Compare and contrast the Sunburn and VIP simulators as evolving systems.

Problem 160. The choices of constants in this section were pretty arbitrary. Explain the thinking that you imagine would lead to the choices for the four decay constants in Experiment 4.11.

Problem 161. Explain and critique the rules for the VIP simulator.

Problem 162. In terms of the model, and referring to the experiment if you have performed it, explain how scandals might help a candidate. At what point during the campaign might they be advantageous?

Problem 163. Essay. The VIPs described in this section have a preprogrammed set of actions. Would we obtain more interesting results if they could make decisions based on the state variables? Outline how to create a data structure that could map the state variables onto actions.

Problem 164. Cast your mind back to the most recent election in your home state or country. Write out and justify a VIP gene for the two leading candidates.

Problem 165. The VIP simulator described in this section is clearly designed for a two-sided contest. Outline how to modify the simulator to run a simulation of a primary election.

Problem 166. We have the electorate divided 25/50/25 in Experiment 4.11. Outline the changes required to simulate a 10/50/40 population in which one special-interest group outnumbers another, but both are politically active. Refer to real-world political situations to justify your design.

Problem 167. Analyze Equations 4.1 and 4.2. What are the advantages and disadvantages of those functions? Are they reasonable choices given their place in the overall simulation? Hint: graph $f(x) = \frac{e^x}{c + e^x}$ for $c = 1, 2, 3$.

Problem 168. Should the outcome of some actions depend on what the other candidate did during the same campaign period? Which ones, why, and how would you implement the dependence?

5

Small Neural Nets : Symbots

In this chapter, you will learn to program a very simple type of neural net with evolutionary algorithms. These neural nets will be control systems for virtual autonomous robots called *symbots*, an artificial life system developed by Kurt vonRoeschlaub, John Walker, and Dan Ashlock. These neural nets will have no internal neurons at first, just inputs and outputs. The symbots are a radical simplification of a type of virtual robot investigated by Randall Beer and John Gallagher [12]. Beer and Gallagher's neural robot has six neurons. It performs well on the task of finding a single signal-emitting source. In an attempt to discover the "correct" number of neurons, a series of experiments was performed in which neurons were removed. Training time declined with each neuron removal, leading to the zero-neuron symbot model. Chapter 11 explores the evolution of more complex neural nets.

Symbots have two wheels and two sensors, as shown in Figure 5.2. The sensors report the strength of a signal field at their position. The sensors can be thought of as eyes, nostrils, Geiger counters, etc.; the field could be light intensity, chemical concentration, the smell of prey, whatever you wish to model. The symbot's neural net takes the sensor output and transforms it into wheel drive strengths. The wheels then cause the symbot to advance (based on the sum of drive strengths) and turn (based on the difference of drive strengths). Computing a symbot's motion requires simple numerical integration. A model, called a kinematic model, of the forces acting on the symbot is what the numerical integration operates on to produce the symbot's motion.

In the course of the chapter, we will introduce a new theoretical concept, the *lexical product* of fitness functions, which is used to combine two fitness functions in a fashion that allows one to act as a helper for the other. The lexical product is of particular utility when the fitness function being maximized gives an initial value of zero for almost all creatures. Figure 5.1 gives the dependencies of the experiments in this chapter.

We will start with the problem of finding a single signal source as in Beer and Gallagher's initial work. We will then train symbots to find and eat

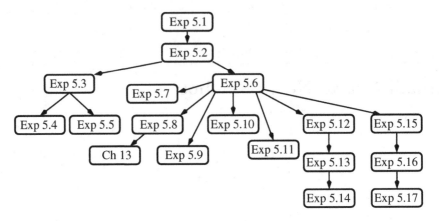

1 Basic symbot motion loop.
2 Evolving symbots.
3 Alternative fitness function.
4 Exploring symmetry.
5 Introduction of lexical product fitness.
6 Multiple signal sources.
7 Using roulette selection.
8 Exploring different types of walls.
9 Evolving sensor locations.
10 Evolving body size.
11 Symbots with three sensors.
12 Moving sensors with a control neuron.
13 Two neuron symbots.
14 Alternative neural transition function.
15 Clear-the-board fitness.
16 Polysymbots: multiple symbots per gene.
17 Coevolving populations of symbots.

Fig. 5.1. The topics and dependencies of the experiments in this chapter.

multiple sources. It is in this context that we explore lexical fitness. We then examine allowing the symbots to evolve their bodies and their control nets. We conclude by attempting to involve groups of symbots that work together (or at least avoid obstructing one another).

In the more complex virtual worlds explored here, like the worlds with multiple signal sources, symbots display what appears to be complex behavior. Since this behavior is the result of simply numerically integrating a linear control model, it follows that the complexity is being drawn from the symbot's environment. The extreme simplicity of the symbots permits them to evolve quickly to have a desired behavior. Likewise, symbot movement is computationally trivial to manage, and so swarms of symbots can be used in a given application.

5.1 Basic Symbot Description

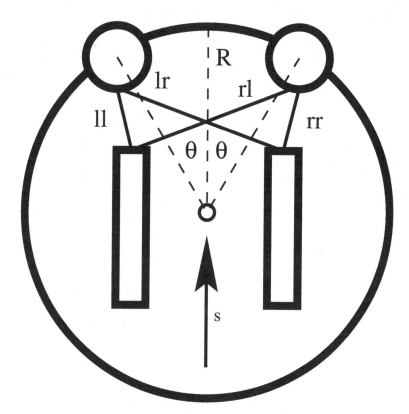

Fig. 5.2. Basic symbot layout.

Symbots live on the unit square: the square with corners $(0,0)$, $(0,1)$, $(1,0)$, and $(1,1)$. A basic symbot, shown in Figure 5.2, is defined by a radius R, an angular displacement of sensors from the symbot's centerline θ, and 5 real parameters that form the symbot's control net. The symbot is controlled by a neural net. A *neural net* is a structure which is programmed by changing the connection weights between its neurons. In this case, the neurons are the the left and right sensors and the left and right wheels. The input of each neuron is the sum of the output of each neuron connected to it times its connection weight to that neuron. The symbots parameters are the connection weights from the right and left sensors to the right and left wheels, and the *idle speed*. The idle speed is the symbot's speed when it is receiving no stimulation from its sensors. We will denote these five real parameters by ll, lr, rl, rr, and s. (The two-letter names have as their first character the sensor ((l)eft or (r)ight) with which they are associated and as the second the wheel with which they are associated.)

The symbot's neural net uses the sensors as the input neurons and the wheels as the output neurons. Each sensor reports the strength of some field; these strengths are multiplied by the appropriate connection weights and then summed to find out how hard and in which direction the wheels push the symbot. The symbot's motion is simulated by iterating the algorithm given in Figure 5.3, called the *basic symbot motion loop*. The code given in this loop is an Euler's method integration of a kinematic motion model. The step size of the integration is controlled by the constant C_s.

```
Begin
    x₁ := x + R · cos(τ + θ);              //left sensor position
    y₁ := y + R · sin(τ + θ);
    x₂ := x + R · cos(τ − θ);              //right sensor position
    y₂ := y + R · sin(τ − θ);
    dₗ := f(x₁, y₁) · ll + f(x₂, y₂) · rl;   //find wheel
    dᵣ := f(x₂, y₂) · rr + f(x₁, y₁) · lr;   //drive strengths
    ds := Cₛ · (dₗ + dᵣ + s · R/2);          //change in position
    If |ds| > R/2 then ds := sgn(ds) · R/2;  //truncate at R/2
    dτ := 1.6 · Cₛ/R · (dᵣ − dₗ);            //change in heading
    If |dτ| > π/3 then dτ := sgn(dτ) · π/3;  //truncate at π/3
    x := x + ds · cos(τ);                    //update position
    y := y + ds · sin(τ);
    τ := τ + dτ;                             //update heading
end;
```

Fig. 5.3. Basic symbot motion loop for symbot of radius R and idle speed s at position (x, y) with heading τ. (The function $f(x, y)$ reports the field strength; $\mathrm{sgn}(x)$ is $-1, 0$, or 1 for x negative, zero, or positive; C_s controls the step size of integration.)

The basic symbot motion loop computes, for one time slice, the symbot's response to the inputs felt at its left and right sensors. This response consists in updating the symbot's position (x, y) and heading τ. The loop in Figure 5.3 is completely inertialess and contains a constant C_s that is used to scale the symbot's turning and motion.

The basic intuition is as follows: Compute the current position of both sensors and get the strength of the field at their position with the function $f(x, y)$. Multiply by the connection weights of sensors to wheels, obtaining the drive of each wheel. The forward motion of the symbot is the sum of the wheel drives plus the idle speed. The change in heading is the difference of the drives of the left and right wheels. Both the motion and change in heading are truncated if they are too large. (Notice that the symbot's position (x, y) is the *center* of its circular body.)

The function $f(x, y)$ reports the field strength at position (x, y). This function is the primary part of the description of the symbot's world and the definition of its task. If, for example, the symbot is a scavenger, the function f would be a diffusion process, possibly modeled with a cellular automaton, that spreads the smell of randomly placed bits of food. If the symbot is attracted to light sources, the field strength would be computed from the standard inverse square law. If the symbot's task is to follow a line on the floor, the field strength would simply be a binary function returning the color of the floor: 0 for off the line and 1 for on it.

We said that the symbot lives on the unit square. What does this mean? What if it tries to wander off? There are many ways to deal with this problem; here are four suggestions: In a *wall-less world*, the symbot lives in the Cartesian plane, and we simply restrict interesting objects to the unit square, e.g., lines to be followed, inverse square law sources, and food. In a *lethal-wall world*, we simply end the symbot's fitness evaluation when it wanders off the unit square. (In a lethal-wall world, the fitness function should be a nondecreasing function of the time spent in the world. If this is not the case, evolution may select for hitting the wall.) In a *reflective world*, the walls are perfect reflectors, and we simply modify the symbot's heading as appropriate to make the angle of incidence equal to the angle of reflection. In a *stopping world*, we set the symbot's forward motion ds to 0 for any move that would take it beyond the boundaries of the unit square. This does not permanently stop the symbot, since it still updates its heading and can turn away from the wall.

Because the symbot's world is highly idealized, we have to be careful. Suppose we are generating the field strength from inverse square law sources. For a single source at (a, b), the pure inverse square law says that its field is

$$f(x, y) = \frac{C_f}{(x - a)^2 + (y - b)^2},$$ (5.1)

where C_f is a constant that gives the intensity of the source. The problem with this is that a symbot that has a sensor near (a, b) experiences an awesome signal and as a result may suddenly shoot off at a great speed or spin through an angle so large, relative to the numerical precision of the computer you are using, that it is essentially a random angle. To avoid this, we assume that the inverse square law source is not a point source, but rather has a radius r_c with a constant field strength inside the source equal to the value the inverse square law would give at the boundary of the source. Call such inverse square law sources *truncated inverse square law sources*. The equation for a truncated inverse square law source with radius r_c at position (a, b) is given by

$$f(x,y) = \begin{cases} \frac{C_f}{(x-a)^2+(y-b)^2} & \text{if } (x-a)^2 + (y-b)^2 \geq r_c^2 \\ \frac{C_f}{r_c^2} & \text{if } (x-a)^2 + (y-b)^2 < r_c^2. \end{cases} \qquad (5.2)$$

The following experiment implements basic symbot code without imposing the additional complexity of evolution. Later, it will serve as an analysis tool for examining the behavior of evolved symbots. Keeping this in mind, you may wish to pay above-average attention to the user interface, since you will use this code with symbots that you evolve later in the chapter.

Experiment 5.1 *Write or obtain software to implement the basic symbot motion loop. You should have a data structure for symbots that allows the specification of radius R, angular displacement θ of sensors from the axis of symmetry, the four connection weights ll, lr, rl, rr, and the idle speed s. Use the basic symbot motion loop to study the behavior of a single symbot placed in several locations and orientations on the unit square. Define field strength using a truncated inverse square law source with radius $r_c^2 = 0.001$ at position (0.5, 0.5) with $C_f = 0.1$ and $C_s = 0.001$. Test each of the following symbot parameters and report on their behavior:*

Symbot	R	θ	ll	lr	rl	rr	idle
1	0.05	π/4	−0.5	0.7	0.7	−0.5	0.3
2	0.05	π/4	−0.2	1	1	−0.2	0.6
3	0.05	π/4	−0.5	0.5	0.7	−0.7	0.4
4	0.1	π/4	1	0	0	1	0.3
5	0.05	π/2	−0.5	0.7	0.7	−0.5	0.3

Characterize how each of these symbots behaves for at least four initial position/orientation pairs. Use a wall-less world. It is a good idea to write your software so that you can read and write symbot descriptions from files, since you will need this capability later.

With Experiment 5.1 in hand, we can go ahead and evolve symbots. For the rest of this section, we will set the symbots the task of finding truncated inverse square law sources. We say that a symbot has *found* a source if the distance from the source to the symbot's center is less than the symbot's radius. There is the usual laundry list of issues (model of evolution, variation operators, etc.), but the most vexing problem for symbots is the fitness function. We need the fitness function to drive the symbot toward the desired behavior. Also, it is desirable for the fitness function to be a nondecreasing function of time, since this leaves open the possibility of using lethal walls (see Problem 169). In the next few experiments, we will use evolution to train symbots to find a single truncated inverse square law source at (0.5, 0.5). (If you have limited computational capacity, you can reduce population size or number of runs in the following experiments.)

Experiment 5.2 *Use the same world as in Experiment 5.1. Fix symbot radius at R = 0.05 and sensor angular displacement at θ = π/4. Build an* *evolutionary algorithm with the gene of the symbot being the five numbers ll, lr, rl, rr, and s, treated as indivisible reals. Use tournament selection with tournament size 4, one-point crossover, and single-point real mutation with mutation size 0.1. Evaluate fitness as follows. Generate three random initial positions and headings that will be used for all the symbots. For each starting position and heading, run the symbots forward for 1000 iterations of the basic symbot motion loop. The fitness is the sum of $f(x, y)$ across all iterations, where (x, y) is the symbot's position. Evolve 30 populations of 60 symbots for 30 generations.*

Report the average and maximum fitness and the standard deviation of the average fitness. Save the best design for a symbot from the final generation for each of the 30 runs. Characterize the behavior of the most fit symbot in the last generation of each run. (This is not as hard as it sounds, because the behaviors will fall into groups.)

Define finding the source to be the condition that exists when the distance from the symbot's nominal position (x, y) to the source is at most the symbot's radius R. Did the symbots do a good job of finding the source? Did more than one technique of finding the source arise? Do some of the evolved behaviors get a high fitness without finding the source?

Are some of the behaviors physically implausible, e.g., extremely high speed spin? Explain why the best and average fitnesses go up and down over generations in spite of our using an elitist model of evolution.

Some of the behaviors that can arise in Experiment 5.2 do not actually find the source. In Figures 5.4 and 5.5 you can see the motion traces of symbots from our version of Experiment 5.2. If we wish to find the source, as opposed to spending lots of time fairly near it, it might be good to tweak the fitness function by giving a bonus fitness for finding the source. There are a number of ways to do this.

Experiment 5.3 *Write a short subroutine that computes when the symbot has found the source at $(0.5, 0.5)$, and then modify Experiment 5.2 by replacing the fitness function with a function that counts the number of iterations it took the symbot to find the target for the first time. Minimize this fitness function. Report the same results as in Experiment 5.2.*

The results may be a bit surprising. Run as many populations as you can and examine the symbot behaviors that appear.

The fitness function in Experiment 5.3 is the one we really want if the symbot's mission is to find the source. However, if this function acts in your experiments as it did in ours, there is a serious problem. The mode fitness of a random creature is zero, and unless the population size is extremely large, it is easy to have *all* the fitnesses in the initial population equal to zero for most test cases. How can we fix this? There are a couple of things we can try.

Experiment 5.4 *Redo Experiment 5.3, but in your initial population generate 3 rather than 5 random numbers per symbot, taking ll = rr and lr = rl.*

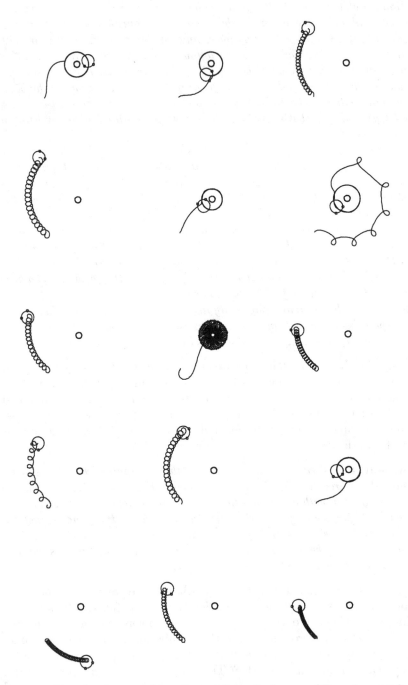

Fig. 5.4. Plots for the most fit symbot at the end of a run of Experiment 5.2, runs 1–15.

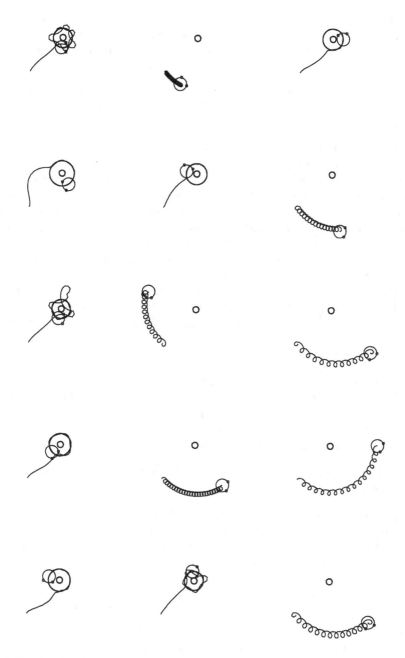

Fig. 5.5. Plots for the most fit symbot at the end of a run of Experiment 5.2, runs 16–30.

The effect of this is to make the initial symbots symmetric. Do two sets of runs:

(i) *Runs where the condition $ll = rr$ and $lr = rl$ is maintained under mutation: if one connection weight changes, change the other.*

(ii) *Runs in which the evolutionary operations are allowed to change all five parameters independently.*

Do 100 runs. Does evolution tend to preserve symmetry? Does imposed symmetry help? How often do we actually get a symbot that reliably finds the source?

The key to Experiment 5.4 is restriction of the space the evolutionary algorithm must search. From other work with symbots, it is known that there are very good solutions to the current symbot task that have symmetric connection weights. More importantly, the probability of a symmetric symbot being a good solution is higher than that probability for an asymmetric symbot. The symmetry restriction makes the problem easier to solve. Keep in mind that Experiment 5.4 doesn't just demonstrate the value of symmetry but also checks the difference between a 3-parameter model (i) and a 5-parameter model with a few restrictions on the initial conditions (ii).

The question remains: can we solve the original 5-parameter problem more efficiently without cooking the initial values? One technique for doing so requires that we introduce a new type of fitness function. The fitness functions we have used until now have been maps from the set of genes to an ordered set like the real numbers.

Definition 5.1 *The **lexical product** of fitness functions f and g, denoted by f lex g, is a fitness function that calls a gene x more fit than a gene y if $f(x) > f(y)$ or $f(x) = f(y)$ and $g(x) > g(y)$. In essence, g is used only to break ties in f. We say that f is the dominant function. (This terminology helps us remember which function in a lexical product is the tie-breaker.)*

With the notion of lexical product in hand, we can do Experiment 5.2 a different way.

Experiment 5.5 *Modifying the fitness evaluation techniques used in Experiments 5.2 and 5.3, evolve symbots with a fitness function that is the lexical product of (i) the number of iterations in which a symbot has found the source with (ii) the sum of the field strength at (x, y) in all iterations. Let the number of iterations in which the symbot has found the source be the dominant function. Do 30 runs on a population of size 60 for 30 generations and compare to see whether using the lexical product gives an improvement on the problem of minimizing the number of iterations it took the symbot to find the target.*

The motivation for the lexical product of fitness functions is as follows: Imagine a case in which the the fitness function you want to satisfy has a fitness landscape for which almost all random creatures have the same rotten fitness (so much so that random initial populations tend to be uniformly unfit). When this happens, evolution needs a secondary heuristic or fitness function to be used when the first gives no information.

Maximizing function (ii) from Experiment 5.5, the sum of field strengths over iterations, biases the symbot toward approaching the source. Once the symbots tend to approach the source, the probability that some will actually run over it is much higher, and evolution can proceed to optimize the ability to find the source (function (i)). Notice that the sum-of-field-strength function almost always distinguishes between two symbots. With similar symbots, it may do so capriciously, depending on the initial positions and directions selected in a given generation. The quality of being virtually unable to declare two symbots equal makes it an excellent tie breaker. Its capriciousness makes it bad as a sole fitness function, as we saw in Experiment 5.1.

Next, we will change the symbot world. Instead of a single source at a fixed location, we will have multiple, randomly placed sources. An example of a symbot trial in such a world is shown in Figure 5.6.

Experiment 5.6 *Write or obtain software for an evolutionary algorithm with a model of evolution and variation operators as in Experiment 5.2. Use a world without walls. Implement routines and data structures so that there are k randomly placed sources in the symbot world. When a symbot finds a source, the source should disappear and a new one be placed. In addition, the same random locations for new sources should be used for all the symbots in a given generation to minimize the impact of luck. This will require some nontrivial information-management technology. In this experiment let $k = 5$ and test two fitness functions, to be maximized:*

(i) Number of sources found.
(ii) Lexical product of the number of sources found with $\frac{1}{d+1}$, where d is the closest approach the symbot made to a source it did not find.

Use populations of 32 symbots for 60 generations, but only do one set of 1500 iterations of the basic symbot motion loop to evaluate fitness (the multiple-source environment is less susceptible to effects of capricious initial placement). Run 30 populations with each fitness function. Plot the average and maximum score of each population and the average of these quantities over all the populations for both fitness functions. Did the secondary fitness function help? If you have lots of time, rerun this experiment for other values of k, especially 1. Be sure to write the software so that it can save the final population of symbots from each run to a file for later use or examination.

If possible, it is worth doing graphical displays of the "best" symbots in Experiment 5.6. There is a wide variety of possible behaviors, many of which are amusing and visually appealing: symbots that move forward, symbots that move backward, whirling dervishes, turn-and-advance, random-looking motion, a menagerie of behaviors, etc.

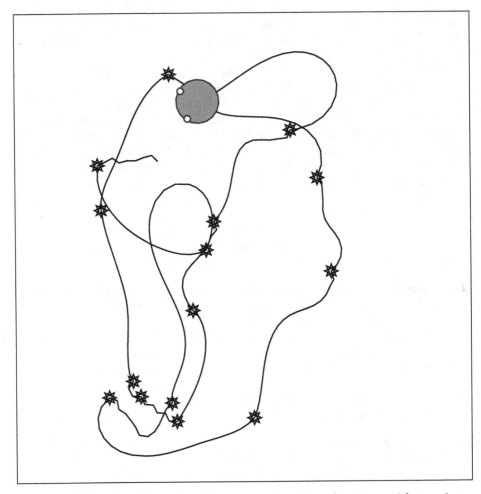

Fig. 5.6. A symbot, its path, and sources captured in a $k = 5$ run with stopping walls. Note that the symbot has evolved to avoid the walls, saving the time required to turn.

Experiment 5.7 *Redo Experiment 5.6 with whichever fitness function exhibited superior performance, but replace tournament selection with roulette selection. What effect does this have? Be sure to compare graphs of average and best fitness in each generation.*

Problems

Problem 169. When we defined lethal walls, the statement was made that "in a lethal wall world the fitness function should be a nondecreasing function of the time spent in the world." Explain why in a few sentences. Give an example of a fitness function that decreases with time and has evil side effects.

Problem 170. Essay. Explain in colloquial English what is going on in the basic symbot motion loop depicted in Figure 5.3. Be sure to say what each of the local variables does and explain the role of C_s. What does the constant 1.6 say about the placement of the wheels?

Problem 171. In a world with reflecting walls, a symbot is supposed to bounce off of the walls so that the angle of incidence equals the angle of reflection. Give the formula for updating a symbot's heading θ when it hits a wall parallel to the x-axis and when it hits a wall parallel to the y-axis. Hint: this is really easy. You may want to give your answer as a modification of the basic symbot motion loop, given in Figure 5.3.

Problem 172. Carefully graph the field that results from

(i) an inverse square law source at $(0.5, 0.5)$,
(ii) a truncated inverse square law source at $(0.5, 0.5)$ with radius 0.1,
(iii) two truncated inverse square law sources with radius 0.1 at $(0.25, 0.25)$ and $(0.75, 0.75)$.

Problem 173. Essay. Think about a light bulb. Why is there no singularity in the field strength of the light emitted by the bulb? The inverse square law is a good description of the bulb's field at distances much greater than the size of the bulb. Is it a good description close up?

Problem 174. Essay. Suppose we are running an evolutionary algorithm with a lexical fitness function f lex g (f dominant). If f is a real-valued, continuous, nonconstant function, how often will we use g? Why is it good, from the perspective of g being useful, if f is a discretely valued function? An example of a discretely valued function is the graph-crossing-number function explored in Section 3.5.

Problem 175. Essay. In Chapter 3, we used niche specialization to keep a population from piling up at any one optimum. Could we use a lexical product of fitness functions to do the same thing? Why or why not? More to the point, for which sorts of optimization problems might the technique help and for which would it have little or no effect.

Problem 176. Think about what you know about motion from studying physics. Rewrite the basic symbot motion loop so that the symbots have mass, inertia, and rotational inertia. Advanced students should give the symbots' wheels rotational inertia as well.

Problem 177. If you are familiar with differential equations, write out explicitly the differential equations that are numerically solved by the basic symbot motion loop. Discuss them qualitatively.

Problem 178. Essay. In which of the experiments in Section 5.1 are we using a fixed fitness function and in which are we using one that changes? Can the varying fitness functions be viewed as samples from some very complex fixed fitness function? Why or why not?

Problem 179. Short Essay. Are the fitness functions used to evolve symbots polymodal or unimodal? Justify your answer with examples and logic.

Problem 180. Suppose we have two symbots with sensors $\pi/4$ off their symmetry axes, radius of 0.05, and connection strengths as follows:

LL	1	0
LR	0	1
RL	0	1
RR	1	0
Idle	0.5	0.5

There is a single inverse truncated square law source at $(0.5, 0.5)$. For each symbot, compute its direction of motion (forward/backward) and current turn direction (left/right) at $(0.25, 0.25)$, $(0.25, 0.75)$, and $(0.75, 0.75)$ assuming that it is facing in the positive y direction and then, again, assuming that it is facing in the positive x direction. Do we need C_s and C_f to do this problem?

5.2 Symbot Bodies and Worlds

In this section, we will explore various symbot worlds, free up the parameters that define the symbot's body, and allow evolution to attempt to optimize details of the symbot body plan. At its most extreme, this will involve modifying the basic symbot body plan to allow asymmetry and additional sensors.

In Section 5.1, we defined reflecting, lethal, and stopping walls but did not use them. Our first experiment in this section explores these other possible symbot worlds. The experiment asks you to report on what differences result from changing the symbot world. Before doing this experiment, you should discuss in class what you *expect* to happen when you change the world. Write down your predictions both before and after the discussion and compare them with the actual outcome of the experiment.

Experiment 5.8 *Modify the software from Experiment 5.6 so that the walls may be optionally lethal, reflecting, stopping, or nonexistent. Using whichever fitness function from Experiment 5.6 gave the best performance, run 20 ecologies in each sort of world. Do different behaviors arise in the different worlds? How do average scores differ? Do the symbots merely deal with or do they actually exploit the reflective and stopping walls?*

In experiments with lethal walls, it is interesting to note that the symbots can actually learn where the walls are, even though they have no sensors that directly detect them. If you have the time and inclination, it is instructive to recode Experiment 5.2 to work with lethal walls. In Experiment 5.2, the placement of the source gives reliable information about the location of the walls, and hence the symbot can learn more easily where the walls are.

In Section 5.1, we had symbots with sensors that were at an angular displacement of $\pi/4$ from the symbot's axis of symmetry. This choice was an aesthetic one, it makes the symbots look nice. We also know that the symbots were able to show a good level of performance with these fixed sensor locations. There is, however, no reason to think that fixing the sensors at $\pi/4$ off the axis of symmetry is an optimal choice, and we will now do an experiment to see whether, in fact, there are better choices.

Experiment 5.9 *Modify the evolutionary algorithm used in Experiment 5.6 so that it operates on a gene that contains two additional loci, the displacements off the axis of symmetry of the left and right sensors in radians. Run 30 populations of size 60 for 75 generations with the displacements*

(i) equal but with opposite sign, and
(ii) independent.

That is to say, the sensors should be coerced to be symmetric in one set of runs and allowed to float independently in the other. What values for sensor displacements occur? How does the performance of evolution in this experiment compare with that in Experiment 5.6? When the two sensor locations float independently, you will need to make a small, obvious modification to the basic symbot motion loop. Include a discussion of this modification in your write-up.

In our version of Experiment 5.9, two common designs were Chameleon (sensors at $\pi/2$ off the axis of symmetry) and Cyclops (both sensors on the axis of symmetry, one in front and one in back). Note that Cyclops can occur only in the second set of runs. When writing up Experiment 5.9, be sure to note any designs that are substantially different from Cyclops and Chameleon.

So far, each symbot in our experiments has had a body size of 0.05, 1/20 the width of the unit square. Making a symbot larger would clearly benefit the symbot; even blundering movement would cover a greater area. A symbot with a radius of 1, for example, would cover all or most of the unit square and hence would "find" things quite efficiently. In addition, if we assume fixed sensor locations, then symbots that are larger have more resolution on their sensors. It is not clear whether this is good or bad. The farther apart their two sensors are, the more difference in the field strength they feel. If a symbot is big enough, it is often the case that one sensor is near one source and the other is near another. Such a symbot may have different design imperatives than a symbot that is small.

In the following experiment, we will explore the radius parameter R for symbots. We will use a new technique, called *population seeding*. In population seeding, an evolved population generated in the past is used as the starting population for a new evolution. Sometimes this is done just to continue the evolution, possibly multiple different times, to test for contingency or look for added progress toward some goal. However, it also gives you the opportunity to change the fitness function so as to approach some goal stepwise. If we start with a population of symbots that can already find sources efficiently, then evolution can concentrate on optimizing some other quality, in this case the symbot's radius.

A bit of thought is required to design an experiment to explore the utility of radius to a symbot. The area of a symbot is πR^2, while the cross section it presents in the direction of motion is $2R$. The symbot's area is the fraction of the unit square it covers, but since it moves, its leading surface might well be the "useful" or "active" part of the symbot. There is also the role of sensor separation in maneuvering to consider. Symbots that are too small feel almost no difference in their sensor strengths, while symbots that are too large can have inputs from distinct sources dominating each of their sensors. This means that symbot fitness might vary linearly as size, quadratically as size, or vary according to the average distance between sources. The truth is probably some sort of subtle combination of these and other factors. The following experiment places some bounds and can serve as the starting point for designing additional experiments.

Experiment 5.10 *Modify the software from Experiment 5.6 with fitness function (i) setting $k = 5$ to provide a source-rich environment. Modify the symbot gene so that radius, set initially to 0.05, is part of the evolving gene. Allow radii in the range $0.01 \leq R \leq 0.25$ only. Run 3 sets of 30 populations with population size 32 for 60 generations with the fitness function*

(i) unmodified,
(ii) divided by the symbot's diameter $2R$, and
(iii) divided by the symbot's area πR^2.

Instead of generating random initial creatures, use a population of evolved symbots from Experiment 5.6. Doing this will allow the use of the simpler fitness function: an already evolved population should not need the lexical fitness function boost to its early evolution.

For your write-up, plot the distribution of radii in the final population of each run. Write a few paragraphs that explain what this experiment has to say about the effect of radius on fitness. Did some sets of runs move immediately to the upper or lower boundary of permitted radius?

So far, the symbots we have examined have had two sensors and, with the exception of Experiment 5.9, bilateral symmetry. This is because they are modeled on biological creatures. The sensors are thought of as two eyes. Maybe three sensors would work better. Let's try it and see.

Experiment 5.11 *Rewrite the code from Experiment 5.6 so that the symbots have 3 genetic loci that give the angular position of 3 sensors, where 0 is the direction the symbot moves as the result of idle speed alone (the forward direction along its axis of symmetry). You will need to rewrite the basic symbot motion loop to involve 6 sensor/wheel connections, as in Problem 184. Run*

20 populations of 60 symbots for 75 generations saving average and maximum fitness and the sensor positions of the best symbot in the final generation of each population. What arrangements of sensors occur in your best symbots? How does fitness compare with the fitnesses in Experiments 5.6 and 5.9?

There are several hundred possible experiments to be done with symbots, just by using the elements of the experiments presented so far in this section. A modest application of imagination can easily drive the total into the thousands. The author urges anyone who thinks up *and performs* such experiments to contact him. Some additional suggestions: a symbot with a 2-segment body, segments joined by a spring; moving the wheels of the symbot around; adding noise to the symbot's sensors; implementing more realistic underlying physics for the symbots. In this book, our next step will be to give the symbots some modest additional control mechanisms.

Problems

Problem 181. Write out the new version of the basic symbot motion loop, given in Figure 5.3, needed by Experiment 5.9.

Problem 182. Often, a lexical product fitness function is used in evolving symbots. Explain why if we seed a population with evolved symbots and then continue evolution, such a lexical product is not needed.

Problem 183. Essay. Suppose we are running an evolutionary algorithm in which we found a lexical product of two fitness functions f and g with f dominant to be helpful. Discuss the pros and cons of using f lex g for only the first few generations and then shifting to f alone as the fitness function. Give examples.

Problem 184. Give pseudocode, as in Figure 5.3, for the basic symbot motion loop of a symbot with 3 sensors at angular positions τ_1, τ_2, and τ_3 counterclockwise from the direction of forward motion.

Problem 185. True or False? A symbot with a single sensor could find sources and evolve to higher fitness levels using the setup of Experiment 5.9.

5.3 Symbots with Neurons

The symbots we have studied so far have a feed-forward neural net with 2 or 3 input neurons (the sensors), 2 output neurons (the wheels), and no hidden layers or interneurons. The complexity of the symbot's behavior has been the result of environmental interactions: with the field, with the walls, and with the sources. In this section, we will add some neurons to the symbot's control structures.

A neuron has inputs that are multiplied by weights, summed, and then run through a transfer function. The name of a type of neuron is usually the name of its transfer function (hyperbolic tangent, arctangent, or Heaviside, for example). The underlying function for the neuron may be modified by vertical and horizontal shifting and stretching. These are represented by 4 parameters, so that with $f(x)$ being our transfer function, in

$$a \cdot f(b \cdot (x - c)) + d \qquad (5.3)$$

the parameter a controls the degree of vertical stretching; the parameter b controls the degree of horizontal stretching; the parameter c controls the horizontal shift; and the parameter d controls the vertical shift. To see examples of these sorts of shifts look at Figure 5.7.

In Experiment 5.9, we allowed evolution to explore various fixed locations for a pair of sensors. What if the symbot could change the spacing of its sensors in response to environmental stimuli? Let's try the experiment. We should design it so that it is *possible* for evolution to leave the sensors roughly fixed, in case that solution is superior to moving the sensors. In order to do this, we will take the basic symbot and make the symmetric sensor spacing parameter θ dynamic, controlled by a single neuron. Since $-\pi/2 \le \theta \le \pi/2$ is a natural set of possible sensor positions, we will choose an arctangent neuron. The neuron should use the sensors as inputs, requiring 2 connection weights, and will have 2 parameters that are allowed to vary, b and c from Equation 5.3 (a and d are set to 1).

Experiment 5.12 *Modify the software from Experiment 5.6 with fitness function (ii) and the basic symbot motion loop to allow the symmetric spacing of the sensors to be dynamically controlled by an arctangent neuron of the form*

$$\arctan(b \cdot (x - c)).$$

The parameters b and c as well as the connection strengths ln and rn of the left and right sensors to the neuron must be added as new loci in the symbot gene. Before iterating the basic symbot motion loop during fitness evaluation, initialize θ to $\pi/4$. Here is the modification of the basic symbot motion loop.

```
Begin
    x₁ := x + R · cos(τ + θ);                              //left sensor position
    y₁ := y + R · sin(τ + θ);
    x₂ := x + R · cos(τ − θ);                              //right sensor position
    y₂ := y + R · sin(τ − θ);
    dₗ := f(x₁, y₁) · ll + f(x₂, y₂) · rl;                 //find wheel
    dᵣ := f(x₂, y₂) · rr + f(x₁, y₁) · lr;                 //drive strengths
    θ = arctan(b · (ln · f(x₁, y₁) + rn · f(x₂, y₂) − c))  //new sensor spacing
    ds := Cₛ · (dₗ + dᵣ + s · R/2);                        //change in position
    If |ds| > R/2 then                                     //truncate at R/2
        If ds > 0 then ds := R/2 else ds := −R/2;
```

$d\tau := 1.6 \cdot C_s/R \cdot (d_r - d_l);$ //change in heading
If $d\tau > \pi/3$ then $d\tau := \pi/3;$ //truncate at $\pi/3$
$x := x + ds \cdot \cos(\tau);$ //update position
$y := y + ds \cdot \sin(\tau);$
$\tau := \tau + d\tau;$ //update heading
end;

The parameters b and c should be initialized to 1 when you are generating the initial population; the connection strengths ln and rn should start in the range $-1 \le x \le 1$. Do two sets of 20 runs on populations of 40 symbots for 75 generations. In the first set of runs, generate all the symbot genetic loci randomly. In the second set of runs get the parameters rr, rl, lr, ll, and s from an evolved population generated by Experiment 5.6. In addition to the usual fitness data, save the mean and standard deviation of the 4 neuron parameters and devise a test to see whether the symbots are using their neurons. (A neuron is said to be used if θ varies a bit during the course of a fitness evaluation.)

Recall that

$$\tanh(x) = \frac{e^x - e^{-x}}{e^x + e^{-x}}. \tag{5.4}$$

Now we will move to a 2-neuron net, one per wheel, in which we just put the neurons between the sensors and the wheels. We will use hyperbolic tangent neurons. Recall the reasons for truncating the inverse square law sources (Section 5.1): we did not want absurdly large signal inputs when the symbots had a sensor too near an inverse square law source. These neurons represent another solution to this problem. A neuron is *saturated* when no increase in its input will produce a significant change in its output. High signal strengths will tend to saturate the neurons in the modified symbots in the following experiment.

Experiment 5.13 *Take either Experiment 5.6 or Experiment 5.12 and modify the algorithm so that instead of*

$d_l := f(x_1, y_1) \cdot ll + f(x_2, y_2) \cdot rl;$
$d_r := f(x_2, y_2) \cdot rr + f(x_1, y_1) \cdot lr;$

we have

$d_l := R/2 \cdot \tanh(bl \cdot (f(x_1, y_1) \cdot ll + f(x_2, y_2) \cdot rl) + cl);$
$d_r := R/2 \cdot \tanh(br \cdot (f(x_2, y_2) \cdot rr + f(x_1, y_1) \cdot lr) + cr);$

where bl, cl, br, cr are new real parameters added to the symbot's gene. Initialize bl and br to 1 and cl and cr to 0. This will have the effect of having the

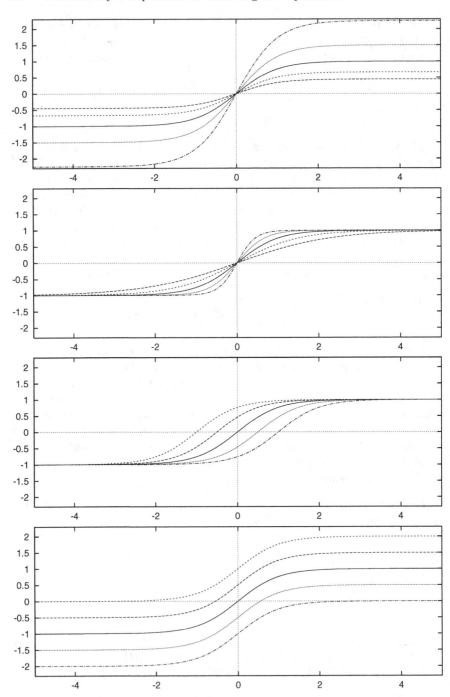

Fig. 5.7. Variations of vertical stretching, horizontal stretching, horizontal shift and vertical shift for the hyperbolic tangent.

neurons fairly closely mimic the behavior of the original network for small signal strengths. Seed the values of $ll, lr, rl, rr,$ *and* s *with those of an evolved population from Experiment 5.6.*

Run at least 10 *populations of* 60 *symbots for* 75 *generations. Document changes in the efficiency of evolution and comment on any new behaviors (things that did not happen in the other evolutions).*

The hyperbolic tangent neuron is computationally expensive, so we should see whether a cheaper neuron could work as well. The transfer function

$$f(x) = \begin{cases} -1 & \text{if } x \leq -1, \\ x & \text{if } -1 < x < 1, \\ 1 & \text{if } 1 \leq x, \end{cases} \tag{5.5}$$

is much cheaper to compute. Let us do an experiment to test its performance.

Experiment 5.14 *Repeat Experiment 5.13 replacing the* $\tanh(x)$ *function with Equation 5.5. Compare the performance of the final symbots and the speed with which the populations converge to their final form.*

Problems

Problem 186. Can the symbots in Experiment 5.12 set the parameters of their sensor positioning neuron so as to mimic symbots with fixed sensor positions? Give neuron parameters that yield fixed sensors or show why this cannot be done. In the event that only *some* fixed positions are possible, show which these are.

Problem 187. Assume that you are working in a computer language that does not have hyperbolic tangents as primitive functions. Unless you are using quite advanced hardware, computing exponentials is more expensive than multiplication and division, which are in turn more expensive than addition and subtraction. Assume that you have the function e^x available (it is often called $\exp(x)$). Find a way to compute $\tanh(x)$ (Equation 5.4) using only one evaluation of e^x and two divisions. You may use as many additions and subtractions as you wish.

Problem 188. Describe a way to *efficiently* substitute a lookup table with 20 entries for the function $\tanh(x)$ in Experiment 5.13. Give pseudocode. A lookup table is an array of real numbers together with a procedure for deciding which one to use for a given x. In order to be efficient, it must not use too many multiplications or divisions and only a moderate amount of addition and subtraction. Graph $\tanh(x)$ and the function your lookup table procedure produces on the same set of axes. Advanced students should also augment the lookup table with linear interpolation.

Problem 189. Essay. Examine the graph of $\tanh(x^3)$ as compared to $\tanh(x)$. Discuss the qualitative advantages and disadvantages of the two functions as neuron transfer functions. What about the shape of the first function is different from the second, and when might that difference be significant?

Problem 190. Essay. If we use hyperbolic tangent neurons as in Experiment 5.13, then large signal strengths are ignored by saturated neurons. Using experimental data, compare the minimal symbols that rely on truncating (Experiment 5.6) with the ones that have saturation available. Are the neuron-using symbols superior in terms of performance, "realism," or stability?

Problem 191. Essay. Explain the choices of a and d made in Experiment 5.12. Why might vertical shift and stretch be bad? How would you expect the symbots to behave if these parameters were allowed to vary?

5.4 Pack Symbots

In this section, we will examine the potential for coevolving symbots to work together. We will also try to pose somewhat more realistic tasks for the symbots.

To this end, we define the *Clear-the-Board* fitness function. Start with a large number of sources and place no new sources during the course of the fitness evaluation. Fitness is the lexical product of the number of sources found with $\frac{1}{d+1}$, where d is the closest approach the symbot made to a source it did not find (compare with Experiment 5.6).

We will distribute the large number of sources using one of three algorithms: uniform, bivariate normal, and univariate normal off of a line running through the fitness space. Think of the sources as spilled objects. The uniform distribution simulates a small segment of a wide-area spill. The bivariate normal distribution is the scatter of particles from a single accident at a particular point. The univariate normal off of a line represents something like peanuts spilling off of a moving truck.

Experiment 5.15 *Modify the software in Experiment 5.6, fitness function (ii), to work with a Clear-the-Board fitness function. If two symbots both clear the board, then the amount of time taken is used to break the tie (less is better). Change the symbots' radius to 0.01 and have $k = 30$ sources. Run 20 populations of 60 symbots for 50 generations on each of the 3 possible distributions:*

(i) Uniform,
(ii) Bivariate normal with mean (0.5, 0.5) and variance 0.2, and
(iii) Univariate normal with variance 0.1 off of a line.

See Problem 193 for details of distribution (iii). Seed the populations with evolved symbots from Experiment 5.6. When the normal distribution produces points not inside the unit square, simply ignore those points and generate new ones until you get enough points. Report mean and best fitness and say which distributions allowed the symbots to learn to clear the board most often. If it appears that the symbots could clear the board given a little more time, you might try increasing the number of iterations of the symbot motion loop. You should certainly terminate fitness evaluation early if the board is cleared.

Experiment 5.15 is intended to give you practice with the new fitness function and the new patterns of source distribution. With these in hand, we will move on to *pack symbots*. Pack symbots are symbots that learn to "work together" in groups. There are two ways to approach pack symbots: specify a set of symbots with a single gene, or evolve several populations of symbots whose fitness is evaluated in concert. For both approaches, there are many symbots present in the unit square simultaneously. It may be that the various symbots will learn to coevolve to do different tasks. One would hope, for example, that in the experiments with a bivariate normal source distribution, several symbots would intensively scour the center of the region while others swept the outer fringes.

A new problem that appears in multiple-symbot environments is that of symbot collision. Symbots realized in hardware might well not care too much if they bumped into one another occasionally, but it is not desirable that we evolve control strategies in which symbots pass through one another. On the other hand, realistic collisions are difficult to simulate. Aside from mentioning it, we will, for the present, ignore this problem of symbot collisions.

Experiment 5.16 *Modify the software from Experiment 5.15 so that a gene contains the description of m symbots. The resulting object is called a polysymbot. All m symbots are run at the same time with independent positions and headings. The fitness of a polysymbot gene is the sum of the individual fitnesses of the symbots specified by the gene. Run 20 populations of 60 polysymbots for 100 generations on one of the 3 possible distributions of sources for $m = 2$ and $m = 5$. Use $k = 30$ sources on the board.*

In addition to documenting the degree to which the symbots clean up the sources and avoid colliding with each other, try to document, by observing the motion tracks of the best cohort in the final generation of each run, the degree to which the symbots have specialized. Do a few members of the group carry the load, or do all members contribute?

In the next experiment, we will try to coevolve distinct populations instead of gene fragments.

Experiment 5.17 *Modify the software from Experiment 5.16, with $m = 5$ symbots per pack, so that instead of a gene containing 5 symbots, the algorithm contains 5 populations of genes that describe a single symbot. For each fitness* *evaluation the populations should be shuffled and cohorts of five symbols, one from each population, tested together. Each symbol is assigned to a new group of five in each generation. The fitness of a symbot is the fitness that its cohort, as a whole, gets. Do the same data-acquisition runs as in Experiment 5.16 and compare the two techniques. Which was better at producing coevolved symbots that specialize their tasks?*

Problems

Problem 192. Is the fitness function specified in Experiment 5.15 a lexical product? Check the definition of lexical products very carefully and justify your answer.

Problem 193. In the experiments in this section, we use a new fitness function in which the symbots attempt to clear the board of sources. To generate uniformly distributed sources, you generate the x and y coordinates as uniform random numbers in the range $0 \leq x, y \leq 1$. The bivariate normal distribution requires that you generate two Gaussian coordinates from the random numbers (the transformation from uniform to Gaussian variables is given in Equation 3.1). In this problem, you will work out the details of the Gaussian distribution of sources about a line.

(i) Give a method for generating a line uniformly selected from those that have at least a segment of length 1 inside the unit square.

(ii) Given a line of the type generated in (i), give a method for distributing sources uniformly along its length but with a Gaussian scatter about the line (with the line as the mean). Hint: use a vector orthogonal to the line.

Problem 194. Imagine an accident that would scatter toxic particles so that the particles would have a density distribution that was a Gaussian scatter away from a circle. Give a method for generating a field of sources with this sort of density distribution.

Problem 195. Give a method for automatically detecting specialization of symbots for different tasks, as one would hope would happen in Experiments 5.16 and 5.17. Logically justify your method. Advanced students should experimentally test the method by incorporating it into software.

Problem 196. Essay. Describe a baseline experiment that could be used to tell whether a polysymbot from either Experiment 5.16 or 5.17 was more effective at finding sources than a group of symbots snagged from Experiment 5.6.

6

Evolving Finite State Automata

In this chapter, we will evolve finite state automata. (For the benefit of those trained in computer science, we note that the finite state automata used here are, strictly speaking, finite state transducers: they produce an output for each input.) The practice of evolving finite state automata is a very old one, having started in the early 1960s with the foundational work of Larry Fogel [23, 25]. Finite state automata (or FSAs) are a staple of computer science. They are used to encode computations, recognize events, or as a data structure for holding strategies for playing games. The dependencies of the experiments in this chapter are given in Figure 6.1. Notice that there are two unconnected sets of experiments.

The first section of the chapter examines different methods of representing finite state automata in an array and introduces very simple fitness functions similar to the one-max problem for strings. In finite state automata, this simple function consists in predicting the next bit in a periodic input stream. This is a simple version of more complex tasks such as modeling the stock or commodity markets. There is also an excellent lexical fitness function that substantially improves performance.

The latter two sections explore the use of finite state automata as game-playing agents. One section deals entirely with Iterated Prisoner's Dilemma, on which numerous papers have been published. Some of these are referenced at the beginning of Section 6.2. The third section expands the exploration into other games, including the graduate student game and Divide the Dollar. This is just a hint of the rich generalizations of these techniques that are possible.

In Section 6.1, we start off with a very simple task: learning to predict a periodic stream of zeros and ones. In Section 6.2, we apply the techniques of artificial life to perform some experiments on Iterated Prisoner's Dilemma. In Section 6.3, we use the same technology to explore other games. We need a bit of notation from computer science.

Definition 6.1 *If A is an alphabet, e.g., $A = \{0,1\}$ or $A = \{L, R, F\}$, then we denote by A^n the set of strings of length n over the alphabet A.*

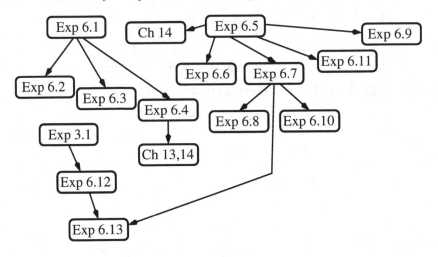

1 First string-prediction experiment.
2 Changing the crossover representation.
3 Self-driving length fitness function.
4 Using self-driving length as a lexical partner.
5 Evolving FSAs to play Iterated Prisoner's Dilemma.
6 Analysis of evolved strategies for IPD.
7 Experimenting with tournament size for selection.
8 Computing strategies' mean evolutionary failure time.
9 Looking for error-correcting strategies.
10 Variation: the Graduate School game.
11 Optional Prisoner's Dilemma.
12 Divide the Dollar.
13 Tournament size variation and comparison of DD with IPD.

Fig. 6.1. The topics and dependencies of the experiments in this chapter.

Definition 6.2 *A* **sequence** *over an alphabet A is an infinite string of characters from A.*

Definition 6.3 *By A^* we mean the set of all finite-length strings over A.*

Definition 6.4 *The symbol λ denotes the empty string, a string with no characters in it.*

Example 3.

$$\{0,1\}^3 = \{000, 001, 010, 011, 100, 101, 110, 111\},$$
$$\{a,b\}^* = \{\lambda, a, b, aa, ab, ba, bb, aaa, aab, aba, abb, \ldots\}.$$

Definition 6.5 *For a string s we denote by $|s|$ the length of s (i.e., the number of characters in s).*

Example 4.

$$|\lambda| = 0,$$
$$|heyHeyHEY| = 9.$$

6.1 Finite State Predictors

A finite state automaton consists of an input alphabet, an response alphabet, a collection of states, a transition function, a response function, an initial state, and an initial response. The states are internal markers used as memory, like the tumblers of a combination lock that "remember" whether the user is currently dialing in the second or third number in the combination. The transition function encodes how the automaton moves from one state to another. The response function encodes the responses produced by the automaton, depending on the current state and input.

An example may help make some of this clear. Consider a thermostat. The thermostat makes a decision every little while and must not change abruptly from running the furnace to running the air-conditioner and vice versa. The input alphabet for the thermostat is {hot, okay, cold}. The response alphabet of a thermostat is {air-conditioner, do-nothing, furnace}. The states are {ready, heating, cooling, just-heated, just-cooled}. The initial response is do-nothing. The initial state, transition function and response function are shown in Figure 6.2.

The thermostat uses the "just-cooled" and "just-heated" states to avoid going from running the air-conditioner to the furnace (or the reverse) abruptly. As an added benefit, the furnace and air-conditioner don't pop on and off; the "just" states slow the electronics down to where they don't hurt the poor machinery. If this delay were *not* needed, we wouldn't need to distinguish between states and responses. We could let the states be the set of responses and "do" whatever state we were in. A finite state automaton that does this is called a *Moore* machine. The more usual type of finite state automaton, with an explicitly separate response function, is termed a *Mealey* machine. In general, we will use the Mealey architecture.

Notice that the transition function t (shown in the second column of Figure 6.2) is a function from the set of ordered pairs of states and inputs to the set of states; i.e., t(state,input) is a member of the set of states. The response function r (in the third column) is a function from the set of ordered pairs of states and inputs to the set of responses, i.e., r(state,input) is a member of the response alphabet.

Colloquially speaking, the automaton sits in a state until an input comes. When an input comes, the automaton generates a response (with its response

Initial State: ready		
When current state and input are	make a transition to state	and respond with
(hot, ready)	cooling	air-conditioner
(hot, heating)	just-heated	do-nothing
(hot, cooling)	cooling	air-conditioner
(hot, just-heated)	ready	do-nothing
(hot, just-cooled)	ready	do-nothing
(okay, ready)	ready	do-nothing
(okay, heating)	just-heated	do-nothing
(okay, cooling)	just-cooled	do-nothing
(okay, just-heated)	ready	do-nothing
(okay, just-cooled)	ready	do-nothing
(cold, ready)	heating	furnace
(cold, heating)	heating	furnace
(cold, cooling)	just-cooled	do-nothing
(cold, just-heated)	ready	do-nothing
(cold, just-cooled)	ready	do-nothing

Fig. 6.2. A thermostat as a finite state automaton.

function) and moves to a new state (which is found by consulting the transition function). The initial state and initial response specify where to start.

For the remainder of this section, the input and response alphabets will both be $\{0, 1\}$, and the task will be to learn to predict the next bit of an input stream of bits.

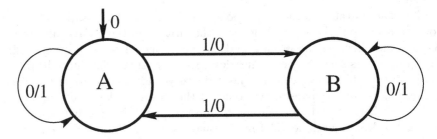

Fig. 6.3. A finite state automaton diagram.

A finite state automaton of this type is shown in Figure 6.3. It has two states: state A and state B. The transition function is specified by the arrows in the diagram, and the arrow labels are of the form input/response. The initial response is on an arrow that does not start at a state and that points to the initial state. This sort of diagram is handy for representing automata on paper. Formally: the finite state automaton's response function is $r(A, 0) = 1$, $r(A, 1) = 0$, $r(B, 0) = 1$, $r(B, 1) = 0$, and the initial response is 0. Its transition function is $t(A, 0) = A$, $t(A, 1) = B$, $t(B, 0) = B$, $t(B, 1) = A$. The initial state is A.

Initial response:0	
Initial state:A	

State	If 0	If 1
A	1→A	0→B
B	1→B	0→A

Fig. 6.4. A finite state automaton table.

Figure 6.4 specifies the finite state automaton shown in Figure 6.3 in tabular format. This is not identical to the tabular format used in Figure 6.2. It is less explicit about the identity of the functions it is specifying and much easier to read. The table starts by giving the initial response and initial state of the finite state automaton. The rest of the table is a matrix with rows indexed by states and columns indexed by inputs. The entries of this matrix are of the form response → state. This means that when the automaton is in the state indexing the row and sees the response indexing the column, it will make the response given at the tail of the arrow, and then make a transition to the state at the arrow's head.

You may want to develop a computer data structure for representing finite state automata (FSA). You should definitely build a routine that can print an FSA in roughly the tabular form given in Figure 6.4; it will be an invaluable aid in debugging experiments.

So that we can perform crossover with finite state automata, we will describe them as arrays of integers and then use the usual crossover operators for arrays. We can either group the integers describing the transition and response functions together, termed *functional* grouping, or we can group the integers describing individual states together, termed *structural* grouping. In Example 5, both these techniques are shown. Functional grouping places the integers describing the transition function and those describing the response function in contiguous blocks, making it easy for crossover to preserve large parts of their individual structure. Structural groupings place descriptions of individual states of an FSA into contiguous blocks, making their preservation easy. Which sort of grouping is better depends entirely on the problem being studied.

Example 5. We will change the finite state automaton from Figure 6.4 into an array of integers in the structural and functional manners. First we strip the finite state automaton down to the integers that describe it (setting $A = 0$ and $B = 1$) as follows:

Initial response:0		
Initial state:A		
State	If 0	If 1
A	1→A	0→B
B	1→B	0→A

0	
0	
1 0	0 1
1 1	0 0

To get the structural grouping gene we simply read the stripped table from left to right, assembling the integers into the array

$$0010011100. \tag{6.1}$$

To get the functional gene we note that the pairs of integers in the stripped version of the table above are of the form

$$\boxed{\text{response} \quad \text{state}}.$$

We thus take the first integer (the response) in each pair from left to right, and then the second integer (the transition state) in each pair from left to right to obtain the gene

$$0010100110. \tag{6.2}$$

Note that in both the functional and structural genes the initial response and initial state are the first two integers in the gene.

We also want a definition of point mutation for a finite state automaton. Pick at random any one of the initial response, the initial state, any transition state, and any response; replace it with a randomly chosen valid value.

Now that we know how to do crossover and mutation, we can run an evolutionary algorithm on a population of finite state automata. For our first such evolutionary algorithm, we will use a task inspired by a Computer Recreations column in *Scientific American*. Somewhat reminiscent of the string evolver, this task starts with a reference string. We will evolve a population of finite state automata that can predict the next bit of the reference string as that string is fed to them one bit at a time.

We need to define the alphabets for this task and the fitness function. The reference string is over the alphabet $\{0, 1\}$, which is also the input alphabet and the response alphabet of the automaton. The fitness function is called the *String Prediction fitness function*, computed as follows. Pick a reference string in $\{0, 1\}^*$ and a number of bits to feed the automaton. Bits beyond the length of the string are obtained by cycling back though the string again. Initialize fitness to zero. If the first bit of the string matches the initial response of the FSA, fitness is +1. After this, we use the input bits as inputs to the FSA, checking the response of the FSA against the next bit from the string; each time they match, fitness is +1. The finite state automaton is being scored on its ability to correctly guess the next bit of the input.

Example 6. Compute the String Prediction fitness of the finite state automaton in Figure 6.3 on the string 011 with 6 bits.

Step	FSM guess	String bit	State after guess	Fitness
0	0	0	A	+1
1	1	1	A	+1
2	0	1	B	-
3	0	0	A	+1
4	1	1	A	+1
5	0	1	B	-
			Total fitness:	4

The String Prediction fitness function gives us the last piece needed to run our first evolutionary algorithm on finite state automata.

Experiment 6.1 *Write or obtain software for randomly generating, printing, and handling file input/output of finite state automata as well as the variation operators described above. Create an evolutionary algorithm using size-4 tournament selection, two-point crossover, single-point mutation, and String Prediction fitness. Use the structural grouping for your crossover. Run 30 populations for up to 1000 generations, recording time-to-solution (or the fact of failure) for populations of 100 finite state automata with*

(i) *Reference string 001, 6 bits, 4-state FSA,*
(ii) *Reference string 001111, 12 bits, 4-state FSA,*
(iii) *Reference string 001111, 12 bits, 8-state FSA.*

Define "solution" to consist in having at least one creature whose fitness equals the number of bits used. Graph the fraction of populations that have succeeded as a function of the number of generations for all 3 sets of runs on the same set of axes.

Experiment 6.2 *Redo Experiment 6.1 with functional grouping used to represent the automaton for crossover. Does this make a difference?*

Let's try another fitness function. The *Self-Driving Length function* is computed as follows. Start with the finite state automaton in its initial state with its initial response. Thereafter, use the last response as the current input. Eventually, the automaton must simultaneously repeat both a state and response. The number of steps it takes to do this is its Self-Driving Length fitness.

Example 7. For the following FSAs with input and response alphabet $\{0, 1\}$, find the Self-Driving Length fitness.

Initial response:1	
Initial state:D	

State	If 0	If 1
A	1→B	0→B
B	1→A	0→B
C	1→C	0→D
D	0→A	0→C

Time-step by time-step:

Step	Response	State
1	1	D
2	0	C
3	1	C
4	0	D
5	0	A
6	1	B
7	0	B
8	1	A
9	0	B

So, in time step 9, the automaton repeats the response/state pair 0, B. We therefore put its Self-Driving Length fitness at 8.

Notice that in our example we have *all* possible pairs of states and responses. We can do no better. This implies that success in the Self-Driving Length fitness function is a score of twice the number of states (at least over the alphabet $\{0, 1\}$).

Experiment 6.3 *Rewrite the software for Experiment 6.1 to use the Self-Driving Length fitness function. Run 30 populations of 100 finite state automata, recording time to success (achieving the maximum possible fitness value) and cutting the automata off after 2000 generations. Graph the fraction of populations that succeeded after k generations, showing the fraction of failures on the left side of the graph as the distance below one. Do this experiment for automata with 4, 6, and 8 states. Also report the successful strings.*

It is easy to describe a finite state automaton that does not use some of its states. The Self-Driving Length fitness function encourages the finite state automaton to use as many states as possible. In Experiment 6.1 the string 001111, while possible for a 4-state automaton to predict, was difficult. The string 111110 would prove entirely impossible for a 4-state automaton (why?) and very difficult for a 6-state automaton.

There is a very large local optimum in Experiment 6.1 for an automaton that predicts the string 111110; automata that just churn out 1's get relatively high fitness. Of the automata that churn out only 1's, many use few states. The more states involved, the easier to have one that is associated with a response of 0, either initially or by a mutation. A moment's thought shows, in fact, that 1-making automata that use a large number of states are more likely to have children that don't, and so there is substantial evolutionary pressure to stay in the local optimum associated with a population of FSAs generating 1's, and using a small number of states to do so. This leaves only extremely low probability evolutionary paths to an automaton that predicts 111110.

Where possible, when handed lemons, make lemonade. In Chapter 5, we introduced the lexical product of fitness functions. When attempting to optimize for the String Prediction fitness function in difficult cases like 111110, the Self-Driving Length fitness function is a natural candidate for a lexical product; it lends much greater weight to the paths out of the local optimum described above. Let us test this intuition experimentally.

Experiment 6.4 *Modify the software from Experiment 6.1 to optionally use either the String Prediction fitness function or the lexical product of String Prediction with Self-Driving Length, with String Prediction dominant. Report the same data as in Experiment 6.1, but running 6- and 8-state automata with both the plain and lexical fitness functions on the reference string 111110 using 12 bits. In your write-up, document the differences in performance and give all reasons you can imagine for the differences, not just the one suggested in the text.*

Experiment 6.4 is an example of an evolutionary algorithm in which lexical products yield a substantial gain in performance. Would having more states cause more of a gain? To work out the exact interaction between additional states and the solutions present in a randomly generated population, you would need a couple of stiff courses in finite state automata or combinatorics. In the next section, we will leave aside optimization of finite state automata and proceed with coevolving finite state automata.

Problems

Problem 197. Suppose that A is an alphabet of size n. Compute the size of the set $\{s \in A^* : |s| \leq k\}$ for any nonnegative integer k.

Problem 198. How many strings are there in $\{0, 1\}^{2m}$ with exactly m ones?

Problem 199. Notice that in Experiment 6.1 the number of bits used is twice the string length. What difference would it make if the number of bits were equal to the string length?

Problem 200. If we adopt the definition of success given in Experiment 6.1 for a finite state automaton predicting a string, is there any limit to the length of a string on which a finite state automaton with n states can succeed?

Problem 201. Give the structural and functional grouping genes for the following FSAs with input and response alphabet $\{0, 1\}$.

(i)

Initial response: 1		
Initial state: B		
State	If 0	If 1
A	1→A	1→C
B	1→B	0→A
C	0→C	0→A

(ii)

Initial response: 1		
Initial state: A		
State	If 0	If 1
A	1→A	0→B
B	1→C	1→A
C	1→B	0→C

(iii)

Initial response: 0		
Initial state: D		
State	If 0	If 1
A	1→B	1→D
B	1→C	0→A
C	0→D	1→B
D	0→A	0→C

(iv)

Initial response: 0		
Initial state: D		
State	If 0	If 1
A	0→B	0→D
B	0→C	1→A
C	1→D	0→B
D	1→A	1→C

Problem 202. For each of the finite state automata in Problem 201, give the set of all strings the automaton in question would count as a success if the string were used in Experiment 6.1 with a number of bits equaling twice its length.

Problem 203. Prove that the maximum possible value for the Self-Driving Length fitness function of an FSA with input and response alphabet $\{0, 1\}$ is twice the number of states in the automaton.

Problem 204. Give an example that shows that Problem 203 does not imply that the longest string a finite state automaton can succeed on in the String Prediction fitness function is of length $2n$ for an n-state finite state automaton.

Problem 205. In the text it was stated that a 4-state automaton *cannot* succeed, in the sense of Experiment 6.1, on the string 111110. Explain irrefutably why this is so.

Problem 206. Problems 203, 204, and 205 all dance around an issue. How do you tell whether a string is too "complex" for an n-state finite state automaton to completely predict? Do your level best to answer this question, over the input and response alphabet $\{0, 1\}$.

Problem 207. Work Problem 203 over, assuming that the finite state automaton uses the input and response alphabets $\{0, 1, \ldots, n - 1\}$. You will have to conjecture what to prove and then prove it.

6.2 Prisoner's Dilemma I

The work in this section is based on a famous experiment of Robert Axelrod's concerning Prisoner's Dilemma [7]. The are many published works that use evolution to explore Iterated Prisoner's Dilemma. Much of the work on evolutionary computation and Prisoner's Dilemma was done by David Fogel. An early paper on using evolutionary programming to train Prisoner's Dilemma playing agents is [19]. An exploration of the impact of the parameters of the Prisoner's Dilemma game (changing the scoring system) on the course of evolution appears in [20]. A critical feature of Iterated Prisoner's Dilemma is the number of rounds in the game. In [21] the duration of an encounter is allowed to evolve.

Other representations besides finite state automata are used for evolutionary training of Iterated Prisoner's Dilemma agents. [42] uses a lookup table driven by past moves. One of the mutation operators is able to change the time horizon (number of past moves referenced by the lookup table). This representation is unique, and the variation of the time horizon is a clever idea, borrowed in Chapter 10 to create more effective string controllers for virtual robots. This line of research continues in [43], where the Prisoner's Dilemma players are placed on a spatial grid.

The representations used to code Prisoner's Dilemma playing agents include artificial neural nets (discussed in Chapter 11), genetic programming (discussed in Chapters 8, 9, and 10), ISAc lists (examined in Chapter 12), and also with Markov chains and other even more exotic structures. The exploration of Prisoner's Dilemma in this text is an entry, no more, to a rich field of research.

6.2.1 Prisoner's Dilemma Modeling the Real World

The original Prisoner's Dilemma was a dilemma experienced by two accomplices, accused of a burglary. The local minions of the law are sure of the guilt of the two suspects they have in custody, but have sufficient evidence to convict them only of criminal trespass, a much less serious crime than burglary. In an attempt to get better evidence, the minions of the law separate the accomplices and make the same offer to both. The state will drop the criminal trespass charge and give immunity from any self-incriminating statements made, if the suspect will implicate his accomplice. There are four possible outcomes to this situation:

(1) Both suspects remain mum, serve their short sentence for criminal trespass, and divide the loot.
(2) One suspect testifies against the other, going off scot-free and keeping all the loot for himself. The other serves a long sentence as an unrepentant burglar.
(3) Same as 2 except with the suspects reversed.

(4) Both suspects offer to testify against the other and receive moderate sentences because they are repentant and cooperative burglars. Each also keeps some chance at getting the loot.

In order to analyze Prisoner's Dilemma, it is convenient to arithmetize these outcomes as numerical payoffs. We characterize the response of maintaining silence as *cooperation* and the response of testifying against one's accomplice as *defection*. Abbreviating these responses as C and D, we obtain the payoff matrix for Prisoner's Dilemma shown in Figure 6.5. Mutual cooperation yields a payoff of 3, mutual defection a payoff of 1, and unilateral stabbing the other player in the back yields a payoff of 5 for the stabber and 0 for the stabbee. These represent only one possible set of values in a payoff matrix for Prisoner's Dilemma. Discussion of this and other related issues are saved for Section 6.3.

Player 2

C   D

Player 1
C (3,3) (0,5)
D (5,0) (1,1)

Fig. 6.5. Payoff matrix for Prisoner's Dilemma.

Prisoner's Dilemma is an example of a *game* of the sort treated by the field of *game theory*. Game theory was invented by John von Neumann and Oskar Morgenstern. Their foundational text, *The Theory of Games and Economic Behavior*, appeared in 1953. Game theory has been widely applied to economics, politics, and even evolutionary biology. One of the earliest conclusions drawn from the paradigm of Prisoner's Dilemma was somewhat shocking. To appreciate the conclusion von Neumann drew from Prisoner's Dilemma, we must first perform the standard analysis of the game.

Imagine you are a suspect in the story we used to introduce Prisoner's Dilemma. Sitting in the small, hot interrogation room you reflect on your options. If the other suspect has already stabbed you in the back, you get the lightest sentence for stabbing him in the back as well. If, on the other hand, he is maintaining honor among thieves and refusing to testify against you, then you get the lightest sentence (and all the loot) by stabbing him in the back. It seems that your highest payoff comes, in all cases, from stabbing your accomplice in the back. Unless you are altruistic, that is what you'll do.

At the time he and Morgenstern were developing game theory, von Neumann was advising the U.S. government on national security issues. A central European refugee from the Second World War, von Neumann was a bit hawkish and concluded that the game-theoretic analysis of Prisoner's Dilemma indicated that a nuclear first strike against the Soviet Union was the only rational course of action. It is, perhaps, a good thing that politicians are not especially respectful of reason. In any case, there is a flaw in von Neumann's reasoning. This flaw comes from viewing the "game" the U.S. and U.S.S.R. were playing as being exactly like the one the two convicts were playing. Consider a similar situation, again presented as a story, with an important difference. It was inspired by observing a parking lot across from the apartment the author lived in during graduate school.

Once upon a time in California, the police could not search a suspected drug dealer standing in a parking lot where drugs were frequently sold. The law required that they see the suspected drug dealer exchange something, presumably money and drugs, with a suspected customer. The drug dealers and their customers found a way to prevent the police from interfering in their business. The dealer would drop a plastic bag of white powder in the ornamental ivy beside the parking lot in a usual spot. The customer would, at the same time, hide an envelope full of money in a drain pipe on the other side of the lot. These actions were performed when the police were not looking. Both then walked with their best "I'm not up to anything" stride, exchanged positions, and picked up their respective goods. This is quite a clever system as long as the drug dealer and the customer are both able to trust each other.

In order to cast this system into a Prisoner's Dilemma format, we must decide what constitutes a defection and a cooperation by each player. For the drug dealer, cooperation consists in dropping a bag containing drugs into the ivy, while defection consists in dropping a bag of cornstarch or baking soda. The customer cooperates by leaving an envelope of federal reserve notes in the drain pipe and defects by supplying phony money or, perhaps, insufficiently many real bills. The arithmetization of the payoffs given in Figure 6.5 is still sensible for this situation. In spite of that, this is a new and different situation from the one faced by the two suspects accused of burglary.

Suppose the dealer and customer both think through the situation. Will they conclude that ripping off the other party is the only rational choice? No, in all probability, they will not. The reason for this is obvious. The dealer wants the customer to come back and buy again, tomorrow, and the customer would likewise like to have a dealer willing to supply him with drugs. The two players play the game many times. A situation in which two players play a game over and over is said to be *iterated*. One-shot Prisoner's Dilemma is entirely unlike Iterated Prisoner's Dilemma, as we will see in the experiments done in this section.

Iterated Prisoner's Dilemma is the core of the excellent book *The Evolution of Cooperation* by Robert Axelrod. The book goes through many real-life examples that are explained by the iterated game and gives an accessible mathematical treatment.

Before we dive into coding and experimentation, a word about altruism is in order. The game theory of Prisoner's Dilemma, iterated or not, assumes that the players are not altruistic: that they are acting for their own self-interest.

This is done for a number of reasons, foremost of which is the mathematical intractability of altruism. One of the major results of research on Iterated Prisoner's Dilemma is that cooperation can arise in the absence of altruism. None of this is meant to denigrate altruism or imply that it is irrelevant to the social or biological sciences. It is simply beyond the scope of this text.

In the following experiment we will explore the effect of iteration on play. A population of finite state automata will play Prisoner's Dilemma once, a small number of times, and a large number of times. A *round robin tournament* is a tournament in which each possible pair of contestants meet.

Experiment 6.5 *This experiment is similar to one done by John Miller. Write or obtain software for an evolutionary algorithm that operates on 4-state finite state automata. Use $\{C, D\}$ for the input and response alphabets. The algorithm should use the same variation operators as in Experiment 6.1. Generate your initial populations by filling the tables of the finite state automata with uniformly distributed valid values.*

Fitness will be computed by playing a Prisoner's Dilemma round robin tournament. To play, a finite state automaton uses its current response as the current play, and the last response of the opposing automaton as its input. Its first play is thus its initial response. Each pair of distinct automata should play n rounds of Prisoner's Dilemma. The fitness of an automaton is its total score in the tournament. Start each automata over in its initial state with its initial response with each new partner. Do not save state information between generations.

On a population of 36 automata, use roulette selection and absolute fitness replacement, replacing 12 automata in each generation for 100 generations. This is a strongly elitist algorithm with $\frac{2}{3}$ of the automata surviving in each generation. Save the average fitness of each population divided by 35n (the number of games played) in each generation of each of 30 runs.

Plot the average of the averages in each generation versus the generations. Optionally, plot the individual population averages. Do this for $n = 1$, $n = 20$, and $n = 150$. For which of the runs does the average plot most closely approach cooperativeness (a score of 3)? Also, save the finite state automata in the final generations of the runs with $n = 1$ and $n = 150$ for later use.

There are a number of strategies for playing Prisoner's Dilemma that are important in analyzing the game and aid in discussion. Figure 6.6 lists several such strategies, and Figure 6.7 describes five of these as finite state automata.

The strategies, Random, Always Cooperate, and Always Defect represent extreme behaviors, useful in analysis. Pavlov is special for reasons we will see later.

The strategy Tit-for-Tat has a special place in the folklore of Prisoner's Dilemma. In two computer tournaments, Robert Axelrod solicited computer strategies for playing Prisoner's Dilemma from game theorists in a number of academic disciplines. In both tournaments, Tit-for-Tat, submitted by Professor Anatole Rapoport, won the tournament. The details of this tournament are reported in the second chapter of Axelrod's book, *The Evolution of Co-operation*.

The success of Tit-for-Tat is, in Axelrod's view, the result of four qualities. Tit-for-Tat is *nice*; it never defects first. Tit-for-Tat is *vengeful*; it responds to defection with defection. Tit-for-Tat is *forgiving*; given an attempt at cooperation by the other player, it reciprocates. Finally, Tit-for-Tat is *simple*; its behavior is predicated only on the last move its opponent made, and hence other strategies can adapt to it easily. Note that not all these qualities are advantageous in and of themselves, but rather they form a good group. Always Cooperate has three of these four qualities, and yet it is a miserable strategy. Tit-for-Two-Tats is like Tit-for-Tat, but nicer.

Random The Random strategy simply flips a coin to decide how to play.

Always Cooperate The Always Cooperate strategy always cooperates.

Always Defect The Always Defect strategy always defects.

Tit-for-Tat The strategy Tit-for-Tat cooperates as its initial response and then repeats its opponent's last response.

Tit-for-Two-Tats The strategy Tit-for-Two-Tats cooperates for its initial response and then cooperates whenever its opponent's last two responses have not been cooperation.

Pavlov The strategy Pavlov cooperates as its initial response and then cooperates if its response and its opponent's response matched last time.

Fig. 6.6. Some common strategies for Prisoner's Dilemma.

Always Cooperate		
Initial response:C		
Initial state:1		
State	If D	If C
1	C→1	C→1

Always Defect		
Initial response:D		
Initial state:1		
State	If D	If C
1	D→1	D→1

Tit-for-Tat		
Initial response:C		
Initial state:1		
State	If D	If C
1	D→1	C→1

Tit-for-Two-Tats		
Initial response:C		
Initial state:1		
State	If D	If C
1	C→2	C→1
2	D→2	C→1

Pavlov		
Initial response:C		
Initial state:1		
State	If D	If C
1	D→2	C→1
2	C→1	D→2

Fig. 6.7. Finite state automaton tables for common Prisoner's Dilemma strategies.

Before we do the next experiment, we need a definition that will help cut down the work involved. The *self-play string* of a finite state automaton is the string of responses the automaton makes playing against itself. This string is very much like the string of responses used for computing the Self-Driving Length fitness, but the string is not cut off at the first repetition of a state and input. The self-play string is infinite.

Thinking about how finite state automata work, we see that a automaton might never repeat its first few responses and states, but that it will eventually loop. For any finite state automaton, the self-play string will be a (possibly empty) string of responses associated with state/input pairs that never happen again followed by a string of responses associated with a repeating sequence of state/input pairs. For notational simplicity, we write the self-play string in the form string1:string2 where string1 contains the responses associated with unrepeated state/input pairs and string2 contains the responses associated with repeated state/input pairs. Examine Example 8 to increase your understanding.

Example 8. Examine the following automaton:

Initial response:C		
Initial state:4		
State	If D	If C
1	D→2	C→2
2	C→1	D→2
3	D→3	D→4
4	C→1	C→3

The sequence of plays of this automaton against itself is as follows:

Step	Response	State
1	C	4
2	C	3
3	D	4
4	C	1
5	C	2
6	D	2
7	C	1
...	...	...

The self-play string of this finite state automaton is

CCD:CCD.

Notice that the state/response pairs (4,C), (3,C), and (4,D) happen exactly once while the state/response pairs (1,C), (2,C), and (2,D) repeat over and over as we drive the automaton's input with its output. It is possible for two automata with different self-play strings to produce the same output stream when self-driven.

In Experiment 6.6, the self-play string can be used as a way to distinguish strategies. Before doing Experiment 6.6, do Problems 215 and 216.

Experiment 6.6 *Take the final populations you saved in Experiment 6.5 and look through them for strategies like those described in Figures 6.6 and 6.7. Keep in mind that states that are not used or that cannot be used are unimportant in this experiment. Do the following:*

(i) For each of the strategies in Figure 6.6, classify the strategy (or one very like it) as occurring often, occasionally, or never.

(ii) Call a self-play string dominant if at least $\frac{2}{3}$ of the population in a single run has that self-play string. Find which fraction of the populations have a dominant strategy.

(iii) Plot the histogram giving the number of self-play strings of each length, across all 30 populations evolved with $n = 150$.

(iv) Plot the histogram as in part (iii) for 1080 randomly generated automata.

In your write-up, explain what happened. Document exactly which software tools you used to do the analyses above (don't, for goodness' sake, do them by hand).

Experiment 6.6 is very different from the other experiments so far in Chapter 6. Instead of creating or modifying an evolutionary algorithm, we are sorting through the debris left after an evolutionary algorithm has been run. It is usually much harder to analyze an evolutionary algorithm's output than it is to write the thing in the first place. You should carefully document and save any tools you write for sorting through the output of an evolutionary algorithm so you can use them again.

We now want to look at the effect of models of evolution on the emergence of cooperation in Iterated Prisoner's Dilemma.

Experiment 6.7 *Modify the software from Experiment 6.5 so that the model of evolution is tournament selection with tournament size 4. Rerun the experiment for $n = 150$ and give the average of averages plot. Now do this all over again for tournament size 6. Explain any differences and also compare the two data sets with the data set from Experiment 6.5. Which of the two tournament selection runs is most like the run from Experiment 6.5?*

A strategy for playing a game is said to be *evolutionarily stable* if a large population playing that strategy cannot be invaded by a single new strategy mixed into the population. The notion of invasion is relative to the exact mechanics of play. If the population is playing round robin, for example, the new strategy would invade by getting a higher score in the round robin tournament.

The notion of evolutionarily stable strategies is very important in game theory research. The location of such strategies for various games is a topic of many research papers. The intuition is that the stable strategies represent attracting states of the evolutionary process. This means that you would expect an evolving system to become evolutionarily stable with high probability once it had been going for a sufficient amount of time. In the next experiment, we will investigate this notion.

Both Tit-for-Tat and Always Defect are evolutionarily stable strategies for Iterated Prisoner's Dilemma in many different situations. Certainly, it is intuitive that a group playing one or the other of these strategies would be very difficult for a single invader to beat. It turns out that neither of these strategies is in fact stable under the type of evolution that takes place in an evolutionary algorithm.

Define the *mean failure time* of a strategy to be the average amount of time (in generations) it takes a population composed entirely of that strategy, undergoing evolution by an evolutionary algorithm, to be invaded. This number exists relative to the type of evolution taking place and is not ordinarily something you can compute. In the next experiment, we will instead approximate it.

Experiment 6.8 *Modify the software from Experiment 6.7, for size-4 tournaments as follows. Have the evolutionary algorithm initialize the entire population with copies of a single automaton. Compute the average score per play that automaton gets when playing itself, calling the result the* baseline score. *Run the evolutionary algorithm until the average score in a generation differs from the baseline by 0.3 or more (our test for successful invasion) or until 500 generations have passed. Report the time-to-invasion and fraction of populations that resisted invasion for at least 500 generations for 30 runs for each of the following strategies:*

(i) Tit-for-Two-Tats,
(ii) Tit-for-Tat,
(iii) Always Defect.

Are any of these strategies stable under evolution? Keeping in mind that Tit-for-Two-Tats is not evolutionarily stable in the formal sense, also comment on the comparative decay rates of those strategies that are not stable.

One quite implausible feature of Prisoner's Dilemma as presented in this chapter so far is the perfect understanding that the finite state automata have of one another. In international relations or a drug deal there is plenty of room to mistake cooperation for defection or the reverse. We will conclude this section with an experiment that explores the effect of error on Iterated Prisoner's Dilemma. We will also finally discover why Pavlov, not a classic strategy, is included in our list of interesting strategies. Pavlov is an example of an *error-correcting* strategy. We say that a strategy is error-correcting if it avoids taking too much revenge for defections caused by error. Do Problem 211 by way of preparation.

Experiment 6.9 *Modify the software for Experiment 6.5, with* $n = 150$, *so that responses are transformed into their opposites with probability* α. *Run 30 populations for* $\alpha = 0.05$ *and* $\alpha = 0.01$. *Compare the cooperation in these populations with the* $n = 150$ *populations from Experiment 6.5. Save the finite state automata from the final generations of the evolutionary algorithm and answer the following questions. Are there error-correcting strategies in any of the populations? Did Pavlov arise in any of the populations? Did Tit-for-Tat? Detail carefully the method you used to identify these strategies.*

We have barely scratched the surface of the ways we could explore Iterated Prisoner's Dilemma with artificial life. You are encouraged to think up your own experiments. As we learn more techniques in later chapters, we will revisit Prisoner's Dilemma and do more experiments.

Problems

Problem 208. Explain why the average score over some set of pairs of automata that play Iterated Prisoner's Dilemma with one another is in the range $1 \leq \mu \leq 3$.

Problem 209. Essay. Examine the following finite state automaton. We have named the strategy encoded by this finite state automaton *Ripoff*. It is functionally equivalent to an automaton that appeared in a population containing immortal Tit-for-Two-Tats automata. Describe its behavior colloquially and explain how it interacts with Tit-for-Two-Tats. Does this strategy say anything about Tit-for-Two-Tats as an evolutionarily stable strategy?

Initial response:D		
Initial state:1		
State	If D	If C
1	C→3	C→2
2	C→3	D→1
3	D→3	C→3

Problem 210. Give the expected (when the random player is involved) or exact score for 1000 rounds of play for each pair of players drawn from the following set:

{Always Cooperate, Always Defect, Tit-for-Tat, Tit-for-Two-Tats, Random, Ripoff}.

Ripoff is described in Problem 209. Include the pair of a player with itself.

Problem 211. Assume that we have a population of strategies for playing Prisoner's Dilemma consisting of Tit-for-Tats and Pavlovs. For all possible pairs of strategies in the population, give the sequence of the first 10 plays, assuming that the first player's response on round 3 is accidentally reversed. This requires investigating 4 pairs, since it matters which type of player is first.

Problem 212. Find an error-correcting strategy other than Pavlov.

Problem 213. Assume that there is a 0.01 chance of a response being the opposite of what was intended. Give the expected score for 1000 rounds of play for each pair of players drawn from the set {Always Cooperate, Always Defect, Tit-for-Tat, Tit-for-Two-Tats, Pavlov, Ripoff}. Ripoff is described in Problem 209. Include the pair of a player with itself.

Problem 214. Give a finite state automaton with each of the following self-play strings:

(i) :C,
(ii) D:C,
(iii) C:C,
(iv) CDC:DDCCDC.

Problem 215. Show that if two finite state automata have the same self-play string, then the self-play string contains the moves they will use when playing one another.

Problem 216. Give an example of 3 automata such that the first 2 automata have the same self-play string, but the sequences of play of each of the first 2 automata against the third differ.

Problem 217. In Problem 209, we describe a strategy called Ripoff. Suppose we have a group of 6 players playing round robin with 100 plays per pair. Players do not play themselves. Compute the scores of the players for each possible mix of Ripoff, Tit-for-Tat, and Tit-for-Two-Tats containing at least one of all 3 player types. There are 10 such groupings.

Problem 218. Essay. Outline an evolutionary algorithm that evolves Prisoner's Dilemma strategies that does not involve finite state automata. You may wish to use a string-based gene, a neural net, or some exotic structure.

Problem 219. For each of the finite state automata given in Figure 6.7 together with the automaton *Ripoff* given in Problem 209, state which of the following properties the strategy encoded by the automaton has: niceness, vengefulness, forgiveness, simplicity. These are the properties to which Axelrod attributes the success of the strategy Tit-for-Tat (see page 157).

6.3 Other Games

In this section, we will touch briefly on several other games that are easily programmable as artificial life systems. Two are standard modifications of Prisoner's Dilemma; the third is a very different game, called *Divide the Dollar*.

The payoff matrix we used in Section 6.2 is the classic matrix appearing on page 8 of *The Evolution of Cooperation*. It is not the only one that game theorists allow. Any payoff matrix of the form given in Figure 6.8 for which $S < Y < X < R$ and $S + R < 2X$ is said to be a payoff matrix for Prisoner's Dilemma. The ordering of the 4 payoffs is intuitive. The second condition is required to make alternation of cooperation and defection worth less than sustained cooperation. We will begin this section by exploring the violation of that second constraint.

The *Graduate School* game is like Prisoner's Dilemma, save that alternating cooperation and defection scores higher, on average, than sustained cooperation. The name is intended to refer to a situation in which both members of a married couple wish to go to graduate school. The payoff for going to school is higher than the payoff for not going, but attending at the same time causes hardship. For the iterated version of this game, think of two preschoolers with a tricycle. It is more fun to take turns than it is to share the tricycle, and both those options are better than fighting over who gets to ride. We will use the payoff matrix given in Figure 6.9.

For the Graduate School game, we must redefine out terms. Complete cooperation consists in two players alternating cooperation and defection. Par-

$$
\begin{array}{cc}
 & \text{Player 2} \\
 & \text{C} \qquad \text{D} \\
\text{Player 1} \quad \begin{array}{c} \text{C} \\ \text{D} \end{array} & \begin{array}{cc} \text{(X,X)} & \text{(S,R)} \\ \text{(R,S)} & \text{(Y,Y)} \end{array}
\end{array}
$$

Fig. 6.8. General payoff matrix for Prisoner's Dilemma. (Prisoner's Dilemma requires that $S < Y < X < R$ and $S + R < 2X$.)

$$
\begin{array}{cc}
 & \text{Player 2} \\
 & \text{C} \qquad \text{D} \\
\text{Player 1} \quad \begin{array}{c} \text{C} \\ \text{D} \end{array} & \begin{array}{cc} \text{(3,3)} & \text{(0,7)} \\ \text{(7,0)} & \text{(1,1)} \end{array}
\end{array}
$$

Fig. 6.9. Payoff matrix for the Graduate School game.

tial cooperation is exhibited when players both make the cooperative play together. Defection describes two players defecting.

Experiment 6.10 *Use the software from Experiment 6.7 with the payoff matrix modified to play the Graduate School game. As in Experiment 6.5, save the final ecologies. Also, count the number of generations in which an ecology has a score above 3; these are generations in which it is clear there is complete cooperation taking place. Answer the following questions.*

(i) Is complete cooperation rare, occasional, or common?
(ii) Is the self-play string histogram materially different from that in Experiment 6.6?
(iii) What is the fraction of the populations that have a dominant strategy?

A game is said to be *optional* if the players may decide whether they will or will not play. Let us construct an optional game built upon Iterated Prisoner's Dilemma by adding a third move called "Pass." If either player makes the play "Pass," both score 0, and we count that round of the game as not played. Call this game Optional Prisoner's Dilemma. The option of refusing to play has a profound effect on Prisoner's Dilemma, as we will see in the next experiment.

Experiment 6.11 *Modify the software from Experiment 6.5 with $n = 150$ to work on finite state automata with input and response alphabets $\{C, D, P\}$. Scoring is as in Prisoner's Dilemma, save that if either player makes the P move, then both score zero. In addition to a player's score, save the number of times he actually played instead of passing or being passed by the other player. First, run the evolutionary algorithm as before, with fitness equal to total score. Next, change the fitness function to be score divided by number of plays. Comment on the total level of cooperation as compared to the nonoptional game and also comment on the differences between the two types of runs in this experiment.*

At this point, we will depart radically from Iterated Prisoner's Dilemma to a game with a continuous set of moves. The game *Divide the Dollar* is played as follows. An infinitely wealthy referee asks two players to write down what fraction of a dollar they would like to have for their very own. Each player writes a bid down on a piece of paper and hands the paper to the referee. If the bids total at most one dollar, the referee pays both players the amount they bid. If the bids total more than a dollar, both players receive nothing.

For now, we will keep the data structure for playing Divide the Dollar simple. A player will have a gene containing 6 real numbers (yes, we will allow fractional cents). The first is the initial bid. The next 5 are the amount to bid if the last payout p (in cents) from the referee was 0, $0 < p \leq 25$, $25 < p \leq 50$, $50 < p \leq 75$, or $p > 75$, respectively.

Experiment 6.12 *Modify the software from Experiment 3.1 to work on the 6-number genome for Divide the Dollar given above. Set the maximum mutation size to be 3.0. Use a population size of 36. Replace the fitness function* *with the total cash a player gets in a round robin tournament with each pair playing 50 times. Run 50 populations, saving the average fitness and the low and high bids accepted in each generation of each population, for 60 generations. Graph the average, over the populations, of the per generation fitness and the high and low bids.*

Divide the Dollar is similar to Prisoner's Dilemma in that it involves cooperation and defection: high bids in Divide the Dollar are a form of defection; bids of 50 (or not far below) are a form of cooperation. Low bids, however, are a form of capitulation, a possibility not available in Prisoner's Dilemma. Also, in Divide the Dollar the result of one player cooperating (say bidding 48) and one defecting (say bidding 87) is zero payoff for both. From this discussion, it seems that single moves of Divide the Dollar do not map well onto single moves of Prisoner's Dilemma. If we define cooperation to be making bids that result in a referee payout, we can draw one parallel, however.

Experiment 6.13 *Following Experiment 6.7, modify the software from Experiment 6.12 so that it also saves the fraction of bids with payouts in each generation. Run 30 populations as before and graph the average fraction of acceptance of bids per generation over all the populations. Modify the software to use tournament selection with tournament size 6 and do the experiment again. What were the effects of changing the tournament size? Did they parallel Experiment 6.7?*

There is an infinite number of games we could explore, but we have done enough for now. We will return to game theory in future chapters once we have developed more artificial life machinery. If you have already studied game theory, you will notice that the treatment of the subject in this chapter differs radically from the presentation in a traditional game theory course. The approach is experimental (an avenue only recently opened to students by large, cheap digital computers) and avoids lengthy and difficult mathematical analyses. If you found this chapter interesting or entertaining, you should consider taking a mathematical course in game theory. Such a course is sometimes found in a math department, occasionally in a biology department, but most often in an economics department.

Problems

Problem 220. In the Graduate School game, is it possible for a finite state automaton to completely cooperate with a copy of itself? Prove your answer. Write a paragraph about the effect this might have on population diversity as compared to Prisoner's Dilemma.

Problem 221. Suppose we have a pair of finite state automata of the sort we used to play Prisoner's Dilemma or the Graduate School game. If the automata have n states, what is the longest they can continue to play before they repeat a set of states and responses they were both in before. If we were to view the pair of automata as a single finite state device engaged in self play, how many states would it have and what would be its input and response alphabets?

Problem 222. Take all of the one-state finite state automata with input and response alphabets $\{C, D\}$, and discuss their quality as strategies for playing the Graduate School game. Which pairs work well together? Hint: there are 8 such automata.

Problem 223. Essay. Explain why it is meaningless to speak of a single finite state automaton as coding a good strategy for the Graduate School game.

Problem 224. Find an error-correcting strategy for the Graduate School game.

Problem 225. Essay. Find a real-life situation to which Optional Prisoner's Dilemma would apply and write-up the situation in the fashion of the story of the drug dealer and his customer in Section 6.2.

Problem 226. Are the data structures used in Experiments 6.12 and 6.13 finite state automata? If so, how many states do they have and what are their input and response alphabets.

Problem 227. Is a pair of the data structures used in Experiments 6.12 and 6.13 a finite state automaton? Justify your answer carefully.

Problem 228. Essay. Describe a method of using finite state automata to play Divide the Dollar. Do *not* change the set of moves in the game to a discrete set, e.g., the integers 1–100, and then use that as the automaton's input and response alphabet. Such a finite state automaton would be quite cumbersome, and more elegant methods are available. It is just fine to have the real numbers in the range 0–100 as your response alphabet, you just cannot use them directly as input.

Problem 229. To do this problem you must first do Problem 228. Assume that misunderstanding a bid in Divide the Dollar consists in replacing the bid b with $(100 - b)$. Using the finite state system you developed in Problem 228, explain what an error-correcting strategy is and give an example of one.

7

Ordered Structures

The representations we have used thus far have all been built around arrays or vectors of similar elements, be they characters, real numbers, the ships' systems from Sunburn, or states of a finite state automaton. The value at one state in a gene has no effect on what values may be present at another location, except for nonexplicit constraints implied by the fitness function.

In this chapter, we will work with ordered lists of items called *permutations*, in which the list contains a specified collection of items once each. Just as we used the simple string evolver in Chapter 2 to learn how evolutionary algorithms worked, we will start with easy problems to learn how systems for evolving ordered genes work. The first section of this chapter is devoted to implementing two different representations for permutations: a direct representation storing permutations as lists of integers $0, 1, \ldots, n$ varying only the order in which the integers appear, and the *random key* representation, which stores a permutation as an array of real numbers. To test these representations, we will use them to minimize the number of reversals in a permutation, in effect to sort it, and to maximize the order of a permutation under composition.

The second section of the chapter will introduce the *Traveling Salesman* problem. This problem involves finding a minimum-length cyclic tour of a set of cities. The third section will combine permutations with a greedy algorithm to permit us to evolve packings of various sizes of objects into containers with fixed capacity; this is an interesting problem with a number of applications.

The last section will introduce a highly technical mathematical problem, that of locating Costas arrays. Used in the processing and interpretation of sonar data, some orders of Costas arrays are not known to exist. The author would be overjoyed if anyone finding one of these unknown arrays would inform him. The dependencies of the experiments in this chapter are given in Figure 7.1. Notice that there are several sets of experiments that do not share code.

The basic definition of a permutation is simple: an order in which to list a collection of items, no two of which are the same. To work with structures of this type, we will need a bit of algebra and a cloud of definitions.

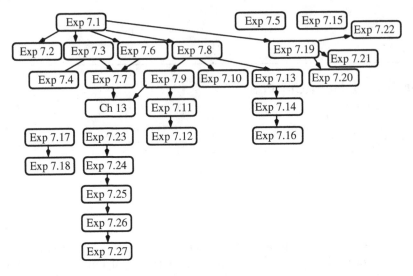

1   Evolving permutations to maximize reversals.
2   Explore the impact of permutation and population size.
3   Evolve permutations to maximize the permutation's order.
4   Change permutation lengths when maximizing order.
5   Stochastic hill-climber baseline for maximizing order.
6   Introducing random key encodings for reversals.
7   Maximizing order with random key encodings.
8   Introducing the Traveling Salesman problem.
9   Random-key-encoded TSP.
10 Adding more cities to the TSP.
11 Exploring the difficulty of different city arrangements.
12 Using random city coordinates.
13 Population seeding with the closest-city heuristic.
14 Population seeding with the random key encoding.
15 Closest-city and city-insertion heuristics.
16 Population seeding with the city-insertion heuristic.
17 Testing greedy packing fitness on random permutations.
18 Stochastic hill-climber for greedy packing fitness.
19 Evolving solutions to the Packing problem.
20 Exploring different cases of the Packing problem.
21 Problem case generator for the Packing problem.
22 Population seeding for the Packing problem.
23 Evolving Costas arrays.
24 Varying the mutation size.
25 Varying the crossover type.
26 Finding the largest size array you can.
27 Population seeding to evolve Costas arrays.

Fig. 7.1. The topics and dependencies of the experiments in this chapter.

Definition 7.1 *A **permutation** of the set $N = \{0, 1, \ldots, n-1\}$ is a bijection of N with itself.*

Theorem 1. *There are $n! := n \cdot (n-1) \cdot \cdots \cdot 2 \cdot 1$ different permutations of n items.*

Proof:

Order the n items. There are n choices of items onto which the first item may be mapped. Since a permutation is a bijection, there are $n - 1$ items onto which the second item may be mapped. Continuing in like fashion, we see the number of choices of destination for the ith item is $n - i + 1$. Since these choices are made independently of one another with past choice not influencing present choice among the available items, the choices multiply, yielding the stated number of permutations. □

Example 9. There are several ways to represent a permutation. Suppose the permutation f is: $f(0) = 0, f(1) = 2, f(2) = 4, f(3) = 1$, and $f(4) = 3$. It can be represented thus in **two-line** notation:

$$\begin{pmatrix} 0\ 1\ 2\ 3\ 4 \\ 0\ 2\ 4\ 1\ 3 \end{pmatrix}.$$

Two line notation lists the set in "standard" order in its first line and in the permuted order in the second line. **One line** notation,

$$(0\ 2\ 4\ 1\ 3),$$

is two-line notation with the first line gone. Another notation commonly used is called **cycle** notation. Cycle notation gives permutations as a list of disjoint cycles, ordered by their leading items, with each cycle tracing how a group of points are taken to one another. The cycle notation for our example is

$$(0)(1\ 2\ 4\ 3),$$

because 0 goes to 0, 1 goes to 2 goes to 4 goes to 3 returns to 1.

Be careful! If the items in a permutation make a single cycle, then it is easy to confuse one-line and cycle notation.

Example 10. Here is a permutation of the set $\{0, 1, 2, 3, 4, 5, 6, 7\}$ shown in one-line, two-line, and cycle notation.

Two line:

$$\begin{pmatrix} 0\ 1\ 2\ 3\ 4\ 5\ 6\ 7 \\ 2\ 3\ 4\ 7\ 5\ 6\ 0\ 1 \end{pmatrix}$$

One line:

$$(2\ 3\ 4\ 7\ 5\ 6\ 0\ 1)$$

Cycle:

$$(0\ 2\ 4\ 5\ 6)(1\ 3\ 7)$$

A permutation uses each item in the set once. The only real content of the permutation is the order of the list of items. Since permutations are functions, they can be composed.

Definition 7.2 Multiplication *of permutations is done by composing them.*

$$(f * g)(x) := f(g(x)). \tag{7.1}$$

Definition 7.3 *The permutation that takes every point to itself is the* **identity** *permutation. We give it the name e.*

Since permutations are bijections, it is possible to undo them, and so permutations have inverses.

Definition 7.4 *The* **inverse** *of a permutation $f(x)$ is the permutation $f^{-1}(x)$ such that*

$$f(f^{-1}(x)) = f^{-1}(f(x)) = x.$$

In terms of the multiplication operation, the above would be written

$$f * f^{-1} = f^{-1} * f = e.$$

Example 11. Suppose we have the permutations in cycle notation $f = (02413)$ and $g = (012)(34)$. Then

$$\begin{aligned}
f * g &= f(g(x)) = (0314)(2), \\
g * f &= g(f(x)) = (0)(1423), \\
f * f &= f(f(x)) = (04321), \\
g * g &= g(g(x)) = (021)(3)(4), \\
f^{-1} &= f^{-1}(x) = (03142), \\
g^{-1} &= g^{-1}(x) = (021)(34).
\end{aligned}$$

Cycle notation may seem sort of weird at first, but it is quite useful. The following definition and theorem will help you to see why.

Definition 7.5 *The* **order** *of a permutation is the smallest number k such that if the permutation is composed with itself k times, the result is the identity permutation e. The order of the identity is 1, and all permutations of a finite set have finite order.*

Theorem 2. *The order of a permutation is the least common multiple of the lengths of its cycles in cycle notation.*

Proof:

Consider a single cycle. If we repeat the action of the cycle a number of times less than its length, then its first item is taken to some other member of the cycle. If the number of repetitions is a multiple of the length of the cycle, then each item returns to its original position. It follows that, for a permutation, the order of the entire permutation is a common multiple of its cycle lengths. Since the action of the cycles on their constituent points is independent, it follows that the order is the least common multiple. $\square$

Definition 7.6 *The* **cycle type** *of a permutation is a list of the lengths of the permutation's cycles in cycle notation. The cycle type is an unordered partition of n into positive pieces.*

Example 12. If $n = 5$, then the cycle type of e (the identity) is 1 1 1 1 1. The cycle type of (01234) is 5. The cycle type of (012)(34) is 3 2.

n	Max Order	n	Max Order	n	Max Order
1	1	13	60	25	1260
2	2	14	84	26	1260
3	3	15	105	27	1540
4	4	16	140	28	2310
5	6	17	210	29	2520
6	6	18	210	30	4620
7	12	19	420	31	4620
8	15	20	420	32	5460
9	20	21	420	33	5460
10	30	22	420	34	9240
11	30	23	840	35	9240
12	60	24	840	36	13,860

Table 7.1. The maximum order of a permutation of n items.

Table 7.1 gives the maximum order possible for a permutation of n items ($n \leq 36$). The behavior of this number is sort of weird, growing abruptly with n sometimes and staying level other times. More values for the maximum order of a permutation of n items may be found in the *On Line Encyclopedia of Integer Sequences* [51].

Example 13. Here are some permutations in cycle notation and their orders.

Permutation	Order
(012)(34)	6
(01)(2345)(67)	4
(01)(234)(56789)	30
(0123456)	7

Definition 7.7 *A **reversal** in a permutation is any pair of items such that in one-line notation, the larger item comes before the smaller item.*

Example 14. Here are some permutations in one-line notation and their numbers of reversals.

Permutation	Reversals
(012345678)	0
(210346587)	5
(160584372)	17
(865427310)	31
(876543210)	36

Theorem 3. *The maximum number of reversals of a permutation of n items is*

$$\frac{n(n-1)}{2}.$$

The minimum number is zero.

Proof:

It is impossible to have more reversals than the number obtained when larger numbers strictly precede smaller ones. In that case, the number of reversals is the number of pairs (a, b) of numbers with a larger than b. There are $\binom{n}{2}$ such pairs, yielding the formula desired. The identity permutation has no reversals, yielding the desired lower bound. $\square$

Reversals and orders of permutations are easy to understand and simple to compute. Maximizing the number of reversals and maximizing the order of a permutation are simple problems that we will use to dip our toes into the process of evolving permutations.

Definition 7.8 *A **transposition** is a permutation that exchanges two items and leaves all others fixed, e.g., (12)(3)(4) (cycle notation).*

Theorem 4. *Any permutation on n items can be transformed into any other permutation on n items by applying at most $n - 1$ transpositions.*

Proof:

Examine the first $n - 1$ places of the target permutation. Transpose these items into place one at a time. By elimination, the remaining item is also correct. $\square$

7.1 Evolving Permutations

In order to evolve permutations, we will have to have some way to store them in a computer. We will experiment with more than one way to represent them. Our first will be very close to one-line notation.

Definition 7.9 *An array containing a permutation in one-line notation is called the* **standard representation**.

While the standard representation for permutations might seem quite natural, it has a clear flaw. If we fill in the data structure with random numbers, even ones in the correct range, it is easy to get a nonpermutation. This means that we must be careful how we fill in the array and how we implement the variation operators.

A typical evolutionary algorithm generates a random initial population. Generating a random permutation is a bit trickier than filling in random characters in a string, because we have to worry about having each list item appear exactly once. The code given in Figure 7.2 can be used to generate random permutations.

```
CreateRandomPerm(perm p,int n){//assume p is an array of n integers

int i,rp,sw;  //loop index, random position, swap

  for(i=0;i<n;i++)perm[i]=i;     //fill in the identity permutation

  for(i=0;i<n-1;i++){//for all entries
    rp=random(n-i)+i;   //get a random number in the range i..(n-1)

    sw=p[i];          //swap the
    p[i]=p[rp];       //current
    p[rp]=sw;         //and random entry

  }

}
```

Fig. 7.2. Code for creating a random permutation of the list $0, 1, \ldots, (n-1)$ of integers.

Now that we have a way of generating random permutations, we need to choose the variation operators. The choice of variation operators has a substantial impact on the performance of an evolutionary algorithm. Mutation is easy; we will simply use transpositions.

Definition 7.10 *A* **transposition mutation** *is the application of a transposition to a permutation. An n-transposition mutation is the application of n transpositions to a permutation.*

Crossover is a more challenging operation than mutation. If we try to apply our standard crossover operator to permutations, there is a high probability of destroying the property that each item appears only once in the list. The following is a standard crossover operation for permutations.

Definition 7.11 *A* **one-point partial preservation** *crossover of two permutations in the standard representation is performed as follows. A single crossover point is chosen. In each permutation, those items present at or before the crossover point are left untouched. Those items after the crossover point also appear after the crossover point, but in the order that they appear in the other permutation. This crossover operator produces two permutations from two permutations but is fairly destructive.*

Example 15. Let's look at some examples of one-point partial preservation crossover. All the permutations are given in one-line notation.

Parents	Crossover Point	Children
(01234)	3	(01243)
(14320)		(14302)
(03547612)	4	(03547126)
(73412065)		(73410562)
(012345)	1	(012345)
(123450)		(102345)

We now have sufficient machinery to perform a simple experiment with evolving permutations. We will be looking for permutations with a large number of reversals.

Experiment 7.1 *Either write or obtain code that can randomly generate permutations and can apply transposition mutations and one-point partial preservation crossover. Use the standard representation for permutations. Also obtain or write code for computing the number of reversals in a permutation. Using a population of 120 permutations write a steady-state evolutionary algorithm using single tournament selection with tournament size 7 to evolve permutations of 16 items to have a maximal number of reversals. The best possible fitness in this case is 120.*

Perform 100 runs each for probabilities of 0%, 50%, and 100% of using the crossover operator and for using 0 or 1, 0 to 2, or 0 to 3 transposition mutations on each new permutation, selecting the number of mutations uniformly at random. Cut a given run off after 10,000 mating events. Report the mean and standard deviation of time-to-solution, measured in mating events. Report the number of runs that timed out (did not find an optimal solution in 10,000 mating events). State any differences in the behavior of time-to-solution for the 9 different ways of using variation operators.

Maximizing reversals is a unimodal problem (see Problems 232 and 231). In fact, it is a very nice sort of unimodal problem in which not only is there only one hill, but there is a nonzero slope at every point. More precisely, any transposition must change the fitness of a permutation. The next experiment explores how much harder the problem of maximizing the number of reversals gets as n grows.

Experiment 7.2 *Run the software from Experiment 7.1 for all even permutation lengths from $n = 8$ to $n = 20$ using the best of the 9 settings for variation operators. Report the mean time-to-solution, the deviation of time-to-solution, and the number of failures. Perform the experiment again for population size 14 instead of 120. Discuss your results.*

Experiment 7.2 shows that the difficulty of maximizing reversals grows in a convex fashion, but not an unmanageable one. This is related to the very nice shape of the fitness landscape. Maximizing the order of a permutation has a less elegant fitness landscape. Let's repeat Experiment 7.1 for the problem of maximizing the order of a permutation.

Experiment 7.3 *Take the software from Experiment 7.1 and modify it to evolve permutations of maximal order. Evolve permutations of 20 items (which have a maximal order of 420 according to Table 7.1). Leave the other parameters of the algorithm the same. Report the mean time-to-solution and standard deviation of the time-to-solution for each of the 9 ways of using variation operators. Compare with the results of Experiment 7.1. Also, check the permutations that achieve the maximum and report the number of distinct solutions.*

Experiment 7.3 demonstrates that the problem of maximizing the order of a permutation interacts with the variation operators in a very different fashion from the problem of maximizing the number of reversals. Let's check and see whether the difficulty of the problem increases smoothly with the number of items permuted.

Experiment 7.4 *Run the software from Experiment 7.3 for all permutation lengths from $n = 20$ to $n = 30$, using the best of the 9 settings for variation operators. Report the mean time-to-solution, the standard deviation of time-to-solution, and the number of failures. Get the correct maxima from Table 7.1. Does the problem difficulty grow directly with the number of items permuted?*

The preceding experiment should convince you that the fitness landscape for the problem of maximizing a permutation's order changes in an irregular manner as the number of items permuted grows. In the next experiment, we will try to learn more about the structure of the fitness landscape.

176   Evolutionary Computation for Modeling and Optimization

Experiment 7.5 *Review Definition 2.17 of stochastic hill climbers. Build a stochastic hill climber for locating permutations of high order with single transposition mutations that accepts new configurations that are no worse. Run the algorithm 100 times for 100,000 mutations per trial on permutations of length 22, 24, and 27. Report the fraction of the time that the final configuration was of maximal order and make a histogram of the frequency with which each order located was found. Explain the results.*

Experiment 7.5 uses software that can climb only up (or around) a hill. It has no ability to go downhill or to jump to another nearby hill. This means that it can be used to locate local optima of the search space and give you some notion of how hard it is to fall into them. Unlike maximizing the number of reversals, the order maximization has places in which you can get stuck. In spite of this, Experiment 7.4 should have convinced you that it is not a very hard problem. We will explore this juxtaposition of qualities in the Problems.

Random Key Encoding

It has already been noted that performing crossover on two permutations as if they were strings often yields nonpermutations. One-point partial preservation crossover is a way of crossing over permutations that works but seems, at least intuitively, to be a fairly disruptive operator. The issue of how to code a permutation in order to evolve it is another example of the representation issue. James Bean [11] came up with a pretty clever way of coding permutations as an array of real numbers like those used in Chapter 3. His technique is called a *random key* encoding.

In this case the word *key* is used to mean sorting key. A sorting key is a field in a database used to sort the records of the database. You might sort a customer database by customer's last name for billing and later by total sales to figure out whom to have your salesmen visit. In the first instance, the name is the sorting key; in the second, total sales is the sorting key.

If we generate a list of n random real numbers, then there is some permutation of those n numbers that will sort the numbers into ascending order. The random list of numbers, treated as a list of sorting keys, specifies a permutation. So: why bother? If we specify our permutations as the sorting order of lists of real numbers, then we can use the standard variation operators for arrays of real numbers that we defined in Chapter 3. This, at least, yields a new sort of crossover operator for permutations, and in some cases may yield superior performance. Let's do some examples to explain this new encoding for permutations.

Example 16. For a selection of numbers placed somewhat randomly in an array, let's compute the permutation that sorts them into ascending order:

Random key:	0.7	1.2	0.6	4.5	−0.6	2.3	3.2	1.1
Permutation:	2	4	1	7	0	5	6	3

Notice that this permutation simply labels each item with its rank.

Definition 7.12 *The* **random key encoding** *for permutations operates by storing a permutation implicitly as the sorting order of an array of numbers. The array of numbers is the structure on which crossover and mutation operate. The random key encoding is another representation for permutations.*

This encoding for permutations has some subtleties. First of all, notice that if two numbers in the random key are equal, then we have an ambiguous result. We will break ties by letting the first of the two equal numbers have the smaller rank. If we use a random number generator that generates random numbers in the range $(0,1)$, then the probability of two numbers being equal is very close to zero, and so this isn't much of a problem.

A more important subtle feature of this encoding is its ability to have multiple inequivalent encodings for the same permutation. Notice that we can change the value of a key by a small amount and not change the permutation. We simply avoid changing the relative order of the numbers. Such a small change could change a number enough to change where it showed up in a new permutation after crossover. In other words, when we use random key encoding of permutations, there are mutations that do not change the permutation, but which do change the results when the permutation undergoes crossover. This is not good or bad, just different. For a given problem, random key encoding may be better or worse than the standard representation of permutations defined earlier in the chapter. Let's perform some comparisons.

Experiment 7.6 *Either write or obtain code that implements random key encoding for permutations. Use it instead of the standard representation with the software of Experiment 7.1. Initialize the random keys from the range $(0,1)$. Use two-point crossover of the random keys and uniform single-point mutation of size 0.1. Perform the experiment with the random key modifications and compare with the results obtained using the standard representation. Discuss the pattern of differences in your write-up.*

Since changing the representation to random key encoding should impact different problems to different degrees, we should test it on multiple problems.

Experiment 7.7 *Modify the software from Experiment 7.6 to evolve permutations of maximal order. Evolve permutations of 20 items (which have a maximal order of 420 according to Table 7.1), but leave the other parameters of the algorithm the same. Report the mean time-to-solution and the standard deviation of the time-to-solution for each of the 9 settings for variation operators. Compare with the results of Experiment 7.3. Also, check the permutations that achieve the maximum and report the number of distinct solutions. Did the shift in the encoding make a noticeable change?*

We will look again at random key encodings in the later sections of this chapter and compare their performance on applied problems. There is a large number of possible encodings for permutations, and we will examine another in the Problems.

Problems

Problem 230. What is the order of a permutation of n items that has a maximal number of reversals? Prove that the permutation in question, the one maximizing the number of reversals, is unique among permutations of $0, 1, \ldots, n-1$.

Problem 231. Perform the following study to help you understand the fitness landscape for the problem of maximizing the number of reversals in a permutation. Generate 10,000 random permutations of 12 items and perform the following steps for each permutation: compute the number of reversals, perform a single transposition mutation, and compute the change in the number of reversals. Make a histogram of the number of reversals times the number of changes.

Problem 232. Prove the following statements:

(i) A transposition must change the number of reversals by an odd number.
(ii) The maximum change in the number of reversals caused by a transposition applied to a permutation of n items is $2n - 3$.
(iii) The fitness landscape for Experiment 7.1 is unimodal. *Hint:* find an uphill path of transpositions from any permutation to the unique answer located in Problem 230.

Problem 233. Given the cycle type as an unordered partition $k_1, k_2, \ldots, k_m$ of n into positive pieces, find a formula for the number of permutations of that type. (This is a very hard problem.)

Problem 234. List all possible cycle types of permutations of $0, 1, \ldots, 5$ together with the number of permutations with each cycle type. Recall that the total number of permutations is $5!=120$.

Problem 235. Compute the number of cycle types for permutations of 10 items that yield permutations of order 6.

Problem 236. For $n=3$, 5, 7, 9, 11, 13, and 15, find a cycle type for a permutation that hits the maximum order given in Table 7.1.

Problem 237. Find the smallest n for which the cycle type of a permutation that hits the maximum order given in Table 7.1 is not unique, i.e., the smallest n for which there are multiple cycle types that attain the maximum order.

Problem 238. Is there a permutation that has the following properties?

(i) The permutation does not have the largest possible order.

(ii) Every application of a transposition to the permutation yields a permutation with a smaller order.

Discuss such a permutation in the context of Experiment 7.5 whether or not it exists.

Problem 239. Essay. Experiment 7.5 can be used to document the existence of local optima in the search space, when we are trying to find permutations of maximum order. In spite of the existence of these local optima, the evolutionary algorithm typically succeeds in finding a member of the global optima. Describe, to the best of your ability, the global optima of the search space. Hint: figure out why the maximum order is 420 for permutations of 19, 20, 21, and 22 items, keeping firmly in mind that there is not a unique permutation with maximum order.

Problem 240. Give an example of three random key encodings of permutations of 5 items such that the first two encode the same permutation, but when the first one is crossed over with the third , it yields a different permutation than when the second one is crossed over with the third (using the same two-point crossover).

Problem 241. Is it possible to cross over two random key encodings of the same permutation and get new random key encodings of a different permutation? If your answer is yes, give an example; if your answer is no, offer a mathematical proof.

Problem 242. Consider the type of mutation used on random key encodings in Experiment 7.6. It is possible for such a mutation to fail to change the permutation that the random key encodes. True or false: if it does change the permutation, that change will be a transposition. Offer a proof of your answer.

Problem 243. Suppose that we have two mutation operators for random key encodings that behave as follows. The first picks a location at random and then finds the next-smallest number to the one in that location, if it exists. If the number picked is the smallest, the mutation does nothing; otherwise, it decreases the number in the chosen location by half the difference between it and the next-smallest number. The second mutation is very like the first, save that it finds the next-largest number and increases the number in the chosen location by half the difference between it and the next-largest number. Notice that these mutations are neutral in the sense that they do not change the permutation encoded. What effect do they have?

Problem 244. Essay. As given in Experiment 7.6, the mutation operator permits the values of random keys to grow or shrink outside their initialized ranges. What effect does having a very large (or large negative) value have on the permutation the random key encodes? Discuss this in the context of the fitness functions of Experiments 7.6 and 7.7.

Problem 245. Essay. Suppose we do some large number of evolutionary algorithm runs for a problem using a random key encoding of permutations. Suppose that some positions within the permutation are uniformly large (or have large negative values) across all populations, while others have small mean values over all populations. Discuss what, if any, useful information can be gleaned from these observations about solutions to the problem in general.

Problem 246. Prove that the initialization for permutations used in Experiment 7.6 is fair in the sense that all permutations have an equal chance of being selected.

Problem 247. Prove that any permutation of the set $\{0, 1, \ldots, n-1\}$ is the product of transpositions of the form $(0\ i)$ for $1 \le i \le n-1$. Let those transpositions be represented by the symbols $a_1, a_2, \ldots, a_{n-1}$. If we use strings over this collection of symbols, can we then evolve permutations using a string evolver?

Problem 248. For the representation of permutations given in Problem 247, compute the length of string needed to permit any permutation to be represented.

Problem 249. Is the encoding defined in Problem 247 fair, in the sense the word is used in Problem 246? Prove your answer.

Problem 250. The set of transpositions used to create an $(n-1)$-symbol representation for permutations in Problem 247 is an instance of a more general method. True or false: there is a binary string representation for permutations that uses only two permutations. This would require that all permutations be generated as a product of those two permutations repeated in some order. Prove your answer.

7.2 The Traveling Salesman Problem

The Traveling Salesman problem is a classic optimization problem. It is typically phrased as follows: "Given a collection of cities for which the cost of travel between each pair of cities is known, find an ordering of the cities that lets a salesman visit the cities cyclically with minimum cost." If the cost of travel is dynamic, as with airline tickets, this problem can become difficult to even think about. The Traveling Salesman problem is usually abstracted into the following simpler form.

Definition 7.13 The Traveling Salesman Problem. *Given a set of points in the plane, find an order that minimizes the total distance involved in a cyclic tour of the points.*

Two examples of such tours, with integer-valued coordinates for the cities, are shown in Figure 7.3. Notice in the second example that some very close links between pairs of cities are not used. The Traveling Salesman problem is an NP-hard problem and one of substantial economic importance. Such problems are typical applications of evolutionary computation. In this section, we will work with small examples with 16 or fewer cities in which the evolutionary algorithm is only slightly better than a brute-force exhaustive search. In real-world applications, the number of cities can range up into the thousands.

City	Position	City	Position
A	(1, 93)	G	(12, 39)
B	(45, 75)	H	(47, 38)
C	(29, 18)	I	(8, 27)
D	(87, 18)	J	(88, 73)
E	(50, 5)	K	(50, 75)
F	(23, 98)	L	(98, 75)

City	Position	City	Position
A	(10, 90)	H	(32, 33)
B	(48, 35)	I	(28, 60)
C	(76, 50)	J	(98, 85)
D	(56, 35)	K	(10, 10)
E	(34, 20)	L	(34, 52)
F	(68, 52)	M	(1, 92)
G	(42, 28)		

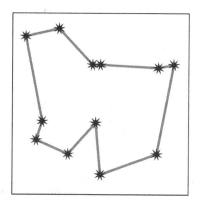 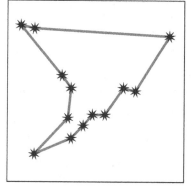

Fig. 7.3. Coordinates for two Traveling Salesman tours and pictures of their corresponding optimal tours. (The first has 12 cities that have an optimal Traveling Salesman tour with length 355.557; the second comprises 13 cities with optimal tour length 325.091.)

Different instances of the Traveling Salesman problem on the same number of cities can have varying difficulty. Compare the 13-city example in Figure 7.3 with one having the 13 cities arranged in a circle. The spike-like feature in the lower left of the picture of the tour contains cities that are close together and yet are not neighbors in the optimal tour. This means that a simple algorithm that starts at a random city and then jumps to the closest city not already part of the tour could get stuck on this example, while it would solve 13 cities in a circle correctly no matter where it started. Let's check and see whether these two cases look different to an evolutionary algorithm.

Before we build the evolutionary algorithm, there is one important point to make. For evaluating permutations as Traveling Salesman tours, we will use cycle notation, always with only one cycle. This looks very much like one-line notation, but it isn't. Using one-line notation would not be a good idea, because it would be so easy to accidently mutate a tour into one with multiple independent cycles. Keeping this in mind, let's do out first Traveling Salesman experiment.

Experiment 7.8 *Take the software from Experiment 7.1 and modify the fitness function to be the length of the Traveling Salesman tour a permutation produces on the 13-city example given in Figure 7.3. For each of the 9 ways of using variation operators, report a 95% confidence interval on the number of mutating events to solution. Selecting the best of these 9, rerun the software on a 13-city tour with the 13 cities placed equally around a circle of radius 45 centered at* (50, 50). *You must compute the length of this tour yourself. Using 95% confidence intervals, compare the time-to-solution for the two problem cases.*

When possible, it is good to make immediate checks to see whether representation has an impact on a given problem. Random key encoding is the "competing" encoding so far in this chapter. Let's check it on the current problem.

Experiment 7.9 *Modify the software from Experiment 7.8 to optionally use random key encoding. Using the combination of crossover probability and mutation type you found to work best, redo the experiment. Compare the 95% confidence intervals for the original and random key encodings. Is there a significant performance difference?*

In Experiments 7.2 and 7.4 we tried to estimate the degree to which a problem gets more difficult when made longer. For the Traveling Salesman problem, this is quite tricky, because the individual problems vary considerably in difficulty. Let's examine the degree to which the problem gets harder as it gets longer for a specific set of relatively easy instances: cities in a circle.

Experiment 7.10 *Use the software from Experiment 7.8 at its most efficient settings. For* $n = 6, 7, 8, 9, 10, 11, 12, 13, 14, 15,$ *and* 16, *arrange* n *cities so that they are equally spaced around a circle of radius 45 with center* (50, 50). *Note that computing the length of the tour is part of the work of setting up the experiment. Perform at least 100 runs per instance, more if you can. Compute the mean time-to-solution with a 95% confidence interval. Plot the confidence intervals on the same set of axes. Check your results against Figure 7.4.*

The difficulty of solving a given instance of the Traveling Salesman problem varies. Examine the solved examples shown in Table 7.2, which are visualized in Figure 7.5. Discuss in class, if possible, what sorts of traps might be lurking in the examples that would make them especially difficult.

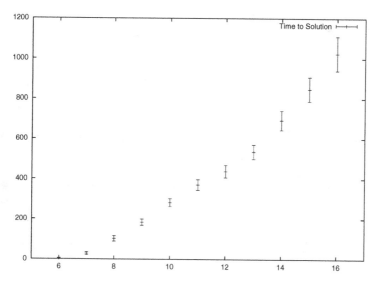

Fig. 7.4. Confidence intervals for time-to-solution for simple circular tours used in Experiment 7.10.

Experiment 7.11 *Before performing the computational portion of this experiment, predict which of the examples given in Table 7.2 are more or less difficult and your reason for thinking so. Using the software from Experiment 7.9 on its most efficient settings, perform 400 runs for each of the examples on 16 cities given in Table 7.2. Plot 95% confidence intervals for the time-to-solution. Discuss the result in light of your predictions.*

The issue of the level of difficulty of particular instances of the Traveling Salesman problem is a tricky one. What makes an instance difficult? The next experiment will generate some clues for answering this question.

Experiment 7.12 *In this experiment we will examine instance complexity (the difficulty of particular instances) of the Traveling Salesman problem. Ideally we would like to generate random instances and check their complexity; this requires we know the actual minimal tour length. We don't know it, so we will have to approximate it. Generate 10 random TSP problems with 12 cities. Let the coordinates of the cities be random in the range [5, 95]. Use the software from Experiment 7.11 to perform 100 runs each on these cities with the software set to run a full 20,000 mating events. Take the best final result as your estimate of the "true" minimum tour length. If a whole class is running the experiment, compile and exchange results to find best tour lengths. Now, run the software on the tours again, but this time use your estimates of the best lengths so that the software can stop when it finds the "correct" solution. Run the software 400 times in this second set of runs. Plot 95% confidence*

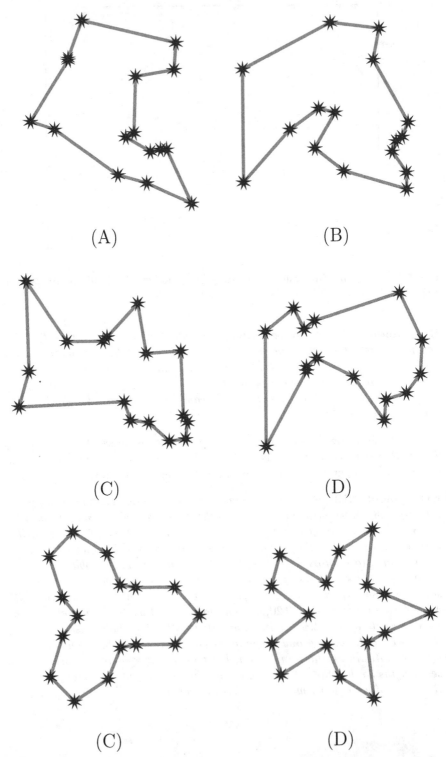

(A)   (B)

(C)   (D)

(C)   (D)

Fig. 7.5. Visualizations of tour solutions for the test problems given in Table 7.2.

Ex	Length	City coordinates			
A)	< 307	(57, 20)	(93, 6)	(33, 76)	(86, 83)
		(85, 70)	(73, 31)	(40, 94)	(71, 16)
		(66, 67)	(14, 46)	(65, 40)	(78, 32)
		(26, 42)	(81, 32)	(33, 75)	(61, 38)
B)	< 343	(85, 20)	(54, 21)	(5, 70)	(40, 32)
		(69, 74)	(72, 89)	(85, 12)	(83, 37)
		(86, 44)	(78, 30)	(42, 51)	(80, 35)
		(50, 49)	(48, 92)	(5, 16)	(28, 41)
C)	< 315	(66, 79)	(89, 14)	(88, 25)	(51, 63)
		(70, 55)	(87, 56)	(62, 23)	(90, 22)
		(31, 61)	(49, 61)	(71, 22)	(11, 90)
		(59, 32)	(81, 13)	(12, 47)	(7, 30)
D)	< 304	(91, 44)	(91, 44)	(15, 10)	(58, 43)
		(39, 70)	(81, 83)	(34, 66)	(29, 76)
		(84, 35)	(40, 52)	(73, 22)	(92, 60)
		(74, 32)	(35, 48)	(15, 65)	(35, 46)
E)	< 261	(95, 50)	(83, 64)	(64, 64)	(56, 65)
		(50, 80)	(33, 91)	(21, 79)	(28, 59)
		(35, 50)	(28, 41)	(21, 21)	(33, 9)
		(50, 20)	(56, 35)	(64, 36)	(83, 36)
F)	< 325	(95, 50)	(72, 59)	(64, 64)	(67, 91)
		(50, 80)	(44, 65)	(21, 79)	(17, 64)
		(35, 50)	(17, 36)	(21, 21)	(44, 35)
		(50, 20)	(67, 9)	(64, 36)	(72, 41)

Table 7.2. A collection of 16-city examples of the Traveling Salesman problem. (Solutions are visualized in Figure 7.5.)

intervals for the time to solution. Are there disjoint intervals? How did the random problem compare to the circle of 12 cities from Experiment 7.11?

Population seeding is a standard practice in evolutionary computation to generate smart initialization strategies. Population seeding consists in finding superior genes and incorporating them into the initial population. The superior genes are supposed to supply good pieces that can be mixed and matched by the algorithm. There is a real danger that population seeding will start the algorithm at a local optimum and get it stuck there. The superior genes may come from any source, and in this case, we will use greedy algorithms to generate them.

Definition 7.14 *A greedy algorithm is one that makes an immediate choice that is the best possible without looking at the overall situation.*

Algorithm 7.1 Closest-city heuristic

Input: A list of cities given as points in the plane and a starting city
Output: A Traveling Salesman tour on the cities

Details:
    Let the current end of the tour be the starting city.
    While cities remain:
        Find the city outside the tour closest to its end.
        Add that city to the tour as its new end.
    End(While)
    Complete the tour by connecting the first and last cities.

Experiment 7.13 *Modify the software from Experiment 7.8 to use population seeding. Generate the closest-city heuristic tour (Algorithm 7.1) for each of the 13 possible starting points and add those 13 tours into the starting population. Perform all 9 sets of runs for the different settings of variation operators. Did the population seeding help? Did a different combination of settings for variation operators work best?*

Random key encoding translates an array of random real numbers into a permutation. If we are going to use population seeding with random key encoding, we must be able to generate random key encodings that yield specific permutations. It is important that while doing this, we do not bias the random key. Recall that changes to the random key that do not change the permutation encoded can still have an effect (see Problems 240, 241). Before doing the next experiment, do Problem 258.

Experiment 7.14 *Modify the software from Experiment 7.13 to use random key encoding. Perform all 9 sets of runs for the different settings of variation operators. Compare your results with those from Experiments 7.8 and 7.13.*

A technique that is part of the standard toolkit for general problem-solving is to break a problem into parts and solve it one piece at a time. Algorithm 7.1 is an example of this technique: you make a sequence of simple decisions. With the Traveling Salesman problem, the attempt to break up the problem into pieces makes some globally bad decisions, a topic we will explore in the Problems. Let's look at another method of breaking up the problem.

Algorithm 7.2 City-insertion heuristic

Input: A list of cities given as points in the plane
Output: A Traveling Salesman tour on the cities
Details:
    Select 3 cities at random and make a triangular tour.
    While cities remain outside the tour:
        Pick a city at random.
        For each adjacent pair in the tour,
            compute the sum of distances from the pair to the selected city.
        Insert the city between the adjacent pair with least joint distance.
    End(While)

Before we do an evolutionary experiment with the second heuristic, let's compare the performance of the two heuristics.

Experiment 7.15 *For the 6 tours on 16 cities given in Table 7.2, compare the performance of Algorithms 7.1 and 7.2. Run Algorithm 7.1 once for each possible starting city, and run Algorithm 7.2 1000 times on each example. Report the mean and best results for each algorithm.*

Now, with a sense of their relative merit as stand-alone heuristics, let us compare the two heuristics as population seeders.

Experiment 7.16 *Modify the software from Experiments 7.13 and 7.14 to do their population seeding using Algorithm 7.2. Use the crossover and mutation settings that worked best (which may be different for the two experiments), and perform the experiments again. Seed 13, 30, and 60 tours produced with Algorithm 7.2 into the initial population. The 13-seed runs permit direct comparison. Does use of the new seeding heuristic have an impact? Does the number of seeded population members have an impact? Explain your results.*

The Traveling Salesman problem is an applied problem and a hard problem. As Experiment 7.12 shows, the Traveling Salesman problem has a fitness landscape that changes from problem instance to problem instance. There is an enormous body of research on the Traveling Salesman problem, and many publications on evolutionary algorithms have the Traveling Salesman problem as their main focus. Many other techniques besides evolutionary algorithms are used to attack the problem. If the brief introduction in this section interests you, there is endless reading ahead. A good resource is the TSPBIB Home Page assembled by Pablo Moscato (www.ing.unlp.edu.ar/cetad/mos/TSPBIB_home.html). Making a real contribution to the Traveling Salesman literature is difficult (because so many clever people are there ahead of you), but worthwhile, because improvement in the algorithm has practical applications that save people money.

Problems

Problem 251. Demonstrate that each optimal Traveling Salesman tour on n cities has $2n$ different versions in one-line notation. Give a normalization (a transformation of any of these tours into a unique representative) that picks out a unique way of writing the tour.

Problem 252. Essay. As we say in the experiments in this section, the case of cities arranged in a circle is a very easy one. Real cities often are located in river valleys. Suppose we have a branching system of rivers with cities only along the rivers. Would you expect this sort of problem to be about as easy as the circle example, or much harder? Would you expect this sort of problem to be easier than, on a par with, or much harder than random problems generated in a manner similar to that used in Experiment 7.12.

Problem 253. Suppose that we start with 4 cities in the corners of a square with side length 1 and coordinates $(0,0), (0,1), (1,0), (1,1)$. Consider a fifth city anywhere in the Cartesian plane with coordinates (x, y). Let $f(x, y)$ be the length of an optimal Traveling Salesman tour on the 5 cities. Answer the following questions:

- Is $f(x, y)$ continuous?
- Describe for which points $f(x, y)$ is differentiable and for which it is not.
- Sketch the set of points where $f(x, y)$ is not differentiable.

Problem 254. Build an exhaustive searcher that can locate the true minimal tour length for a given collection of cities. Hint: use the results of Problem 251. Generate 1000 random examples with 8 cities that have their coordinates chosen uniformly at random from the interval $[5, 95]$, and compute their correct minimal tour length by exhaustion. Plot the results as a histogram. What type of distribution is it?

Problem 255. Suppose we were to place cities uniformly at random within a circle of radius R. Give and defend a bound on the maximum length of the shortest tour involving both the radius and the number of cities. This is a funny problem in that we want to know the maximum, over all possible examples, of the minimum tour length. You may want to try some numerical examples, but your final answer must be defended by logic in a mathematical fashion. (This is a hard problem.)

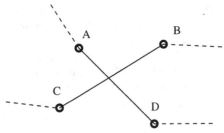

Problem 256. Suppose that 4 cities are configured as shown above as part of a Traveling Salesman tour. Prove that the other two ways to connect the 4 cities (A-B C-D and A-C B-D) yield shorter total distance for the two trip legs involving these 4 cities. Is it always possible to change to one of the other two connections of A, B, C, and D *without* breaking the tour into two pieces?

Problem 257. Suppose you have a predicate that can detect the crossing of two line segments. Describe a mutation operator based on Problem 256, and perform a cost-benefit analysis in a few sentences.

Problem 258. Verify that the following procedure for converting a permutation into a random key is correct and unbiased. Assume that we are creating a random key to encode a permutation $f(i)$ on n items. Generate an array of n uniformly distributed random numbers and sort it. The random key will use the values in this array. Going through the permutation, place the ith item of the sorted array into the $f(i)$th position in the random key. You must show two things: that this procedure generates a random key that encodes $f(i)$ (is correct), and that the key is chosen uniformly at random from among those that encode $f(i)$ (is unbiased). If either of these properties does not hold, document the flaw and repair the procedure.

Problem 259. Consider the following collection of points in the plane:

$$(0,0), (2,2), (4,0), (6,2), (8,0), (10,2), (12,0), (14,2).$$

Compute a tour resulting from applying Algorithm 7.1, and, also compute the optimal tour.

Problem 260. Implement Algorithm 7.2 and run it several times on the collection of points given in Problem 259. Compare with the optimal tour and the result of using Algorithm 7.1.

Problem 261. What is the largest number of distinct answers that Algorithm 7.1 produces for a given n-city tour selected at random?

Problem 262. Essay. Contrast and compare Algorithms 7.1 and 7.2. Address the following issues. Are there obvious mistakes that each of these algorithms makes? Which of these algorithms are deterministic? Which of these algorithms exhibits superior performance *on its own*? Does superior stand-alone performance lead directly to improved performance in a seeded evolutionary algorithm?

Problem 263. Essay. There is a saying that "you can't make money without spending money," and in fact, most methods of making a lot of money require investment capital of some sort. A counterexample to this principle is the fortune of J. K. Rowling, author of the fantastically popular *Harry Potter* series. At the time of this writing, the author's fortune exceeds that of the Queen of England, and yet the total investment in producing the first book was quite modest. Keeping these two different notions of wealth creation in mind, comment on the worth and impact of improving performance on the Traveling Salesman problem. There is economic impact from such improvements: what sort of capital investment do such improvements require? Is the process of improving algorithmic performance on the Traveling Salesman problem more like that of a venture capitalist or a successful author? Explain.

Problem 264. Create and test your own Traveling Salesman heuristic as a stand-alone tool using the solved examples in this section. Compare it with the two heuristics given.

7.3 Packing Things

This section will treat the problem of packing generic objects in generic containers. The packing will be performed by a greedy algorithm that is controlled by an evolved permutation. This problem is easier, on average, than the Traveling Salesman problem because an optimal solution often has many forms.

Let's describe the basic greedy algorithm that we will subsequently modify into a fitness function.

Algorithm 7.3 Greedy Packing Algorithm

Input: A set of n indivisible objects G of sizes $g_1, g_2, \ldots, g_n$ and a set C of k containers of sizes $c_1, c_2, \ldots, c_k$
Output: An assignment of objects to containers and a list of objects that fit in no container
Details:
Taking the objects in the order presented, assign an item to the first container with sufficient space remaining for it. When an object is assigned to a container, subtract the object's size from the container's capacity. Objects that cannot fit in a container, when their turn comes up to be assigned, are put in the group that fit in no container.

Definition 7.15 *The* **excess** *is the sum of the sizes of the objects that fit in no container.*

In order to make the algorithm clear, we will trace it in an example.

Example 17. Suppose we have 6 objects of sizes 20, 40, 10, 50, 40, and 40, and 2 containers of sizes 100 and 100. Then, as we assign the objects to the containers, the remaining capacity in the containers will behave in the following fashion:

Good number	Capacity 1	Capacity 2	Excess	Assigned to
-start-	100	100	0	
1	80	100	0	1
2	40	100	0	1
3	30	100	0	1
4	30	50	0	2
5	30	10	0	2
6	30	10	40	excess

Notice that the greedy algorithm did not do a good job. Since $20+40+40 = 100$ and $10 + 40 + 50 = 100$, all the objects could have been packed (and would have if they had been presented in a different order).

With this example in hand, it is possible to describe a fitness function for permutations that permits us to search for good assignments of objects to containers. In the example, we could have achieved a better packing by presenting the objects in a different order. Since permutations specify orders, we can improve the greedy algorithm by having a permutation reorder the objects before they are presented to it.

Definition 7.16 *The* **greedy packing fitness** *of a permutation σ for a given set of objects $g_1, g_2, \ldots, g_n$ and a given set of containers of sizes $c_1, c_2, \ldots, c_n$ is the excess that results when the algorithm is applied with the objects presented in the order $g_{\sigma(1)}, g_{\sigma(2)}, \ldots, g_{\sigma(n)}$. This fitness function is to be minimized for efficient packing.*

At this point, we need some example problems. A problem consists of a number of containers, the sizes of those containers, a number of objects, and the sizes of those objects. Table 7.3 gives 9 examples chosen for varying size and difficulty. In general, size and difficulty increase together, but not always. While there is no requirement that the objects available exactly fill the containers, all the examples given in Table 7.3 have this property. Let's start by getting a handle on the relative difficulty of these experimental cases.

Experiment 7.17 *For each of the 9 experimental cases given in Table 7.3, build or obtain software to randomly generate permutations until one is found with a greedy packing fitness of 0. One run of this software is called a trial. Write the software to stop if it must examine more than 20,000 permutations. If no permutation is found that has an excess of 0 after 20,000 permutations are examined, record the trial as a failure; otherwise, record it as a success. The number of permutations examined before a trial succeeds is its length. The length of failures is undefined. For each of the 9 cases given, perform 400 trials. Report the number of failures and the mean and standard deviation of the lengths of successful trials.*

Experiment 7.18 *Modify the code from Experiment 7.17 to perform trials in a different manner: as a stochastic hill climber (see Definition 2.17). Starting with a random permutation, the software should apply a mutation and save the result if it is no worse (has no larger excess) than the current permutation. For each of the cases given in Table 7.3, run 4 different stochastic hill climbers, 400 times each. The 4 stochastic hill climbers should use (i) the standard representation with single transposition mutation, (ii) the standard representation with double transposition mutation, (iii) the random key encoding with single-point mutation as in Experiment 7.6, and (iv) the random key encoding with two-point mutation. Record the number of successes and failures for each case and type of stochastic hill climber.*

As we saw in Section 2.6, the stochastic hill climber can be used to estimate the roughness of the "terrain" in a fitness landscape. Since the current

Case	Containers	Sizes	Objects	Sizes
1	3	100(3)	12	39, 37, 34, 33, 28, 25, 23, 22, 19, 17, 15, 8
2	4	100(4)	14	60, 40, 39, 37, 34, 33, 28, 25, 23, 22, 19, 17, 15, 8
3	5	100(5)	23	60, 55, 50, 45, 40, 30, 25(2), 20, 13(2), 12(2), 11(5), 9(5)
4	6	100(6)	20	56, 55, 54, 53, 52, 51, 49, 24(3), 23(3), 22(2), 9(5)
5	7	100(7)	22	60, 59, 58, 57, 56, 49, 48, 46, 36, 27, 26, 25, 24, 22, 21, 20, 19, 15, 14, 7, 6, 5
6	7	100(7)	22	60, 59, 58, 57, 56, 49, 48, 46, 36, 27, 26, 25, 24, 22, 21, 20, 19, 15, 14, 7, 6, 5
7	7	100(7)	22	65, 56(2), 55, 54, 53, 52, 47, 38, 25, 24(2), 23(3), 22, 13, 11, 9(4)
8	6	100(5), 200	22	65, 56(2), 55, 54, 53, 52, 47, 38, 25, 24(2), 23(3), 22, 13, 11, 9(4)
9	5	100(3), 200(2)	22	65, 56(2), 55, 54, 53, 52, 47, 38, 25, 24(2), 23(3), 22, 13, 11, 9(4)

Table 7.3. Examples of containers and objects for the Packing problem. (Numbers in parentheses indicate the number of containers or objects of a given size.)

permutation is never allowed to increase its excess, the hill climber tends to find the top of whatever hill it started on. The key observation is that the hills *do not* exist in the space of permutations alone. Rather, the hills are created by the connectivity induced by the mutation operators. This means that Experiment 7.18 may well have predictive value for the best representation and mutation operator to use in our evolutionary algorithms.

Experiment 7.19 *Modify the code from earlier experiments that evolve permutations using either the standard representation of permutations with one-point partial preservation crossover and one-point transposition mutation, or using the random key encoding with two-point crossover and single-point mutation as in Experiment 7.6. Using all 4 types of evolutionary algorithms on Cases 1 through 5 of the Packing problem given in Table 7.3, compute the mean time-to-solution, standard deviation of time-to-solution, and the number of times the algorithms failed to find a solution. Use initial populations of 200 permutations and evolve for at most 100,000 mating events using a steady-state algorithm. Use single tournament selection with tournament size 7. Compare the two methods. If you performed Experiment 7.18, check and see whether the stochastic hill climber was predictive of the behavior of the evolutionary algorithms.*

Cases 1 through 5 in Table 7.3 can all be solved in a reasonable amount of time by the stochastic hill climber (if it hits the right hill), and so are reasonable targets for any of the evolutionary algorithms in Experiment 7.19. Cases 6 through 9 are all variations of one basic problem that explore variation of the distribution of objects and sizes of containers.

Experiment 7.20 *Pick the evolutionary algorithm from Experiment 7.19 that turned in the best performance. Use that code on cases 6 through 9 of the Packing problem given in Table 7.3. Record the same data as in Experiment 7.19. Discuss the issue, "Is there variable hardness in a random problem case?"*

Experiment 7.20 demonstrates that there is a difference in the difficulty of different cases of the Packing problem. It would be interesting to see whether we can locate difficult and simple instances of the problem automatically. To do this, we need the ability to rapidly generate instances of the Packing problem. This is related to a well-known combinatorial problem of placing balls in bins. Suppose that we have n adjacent bins and want to place k indistinguishable balls in those bins. The problem is to enumerate the space of possible configurations. In a combinatorics course, we would simply count the configurations; here we want a representation that permits us to search (or even evolve) Packing problems.

Theorem 5. *There are $\binom{n+k-1}{k}$ ways of placing k indistinguishable balls into n adjacent bins. Each configuration corresponds to a binary word with $n-1$ ones and k zeros.*

Proof: The zeros represent balls. Zeros before the first one represent balls in the first bin. Each one represents the division between two bins. Zeros between two ones are all in the bin "between" those ones. The zeros after the last one in a given binary string are the balls placed in the last bin. The number of such strings is equal to the sum of the number of ones and zeros choose (in the sense of binomial coefficients) the number of zeros. □

Example 18. Suppose we have 9 balls and 3 bins. Then, the correspondence given in Theorem 5 maps the string **00100000100** to the following ball-bin configuration:

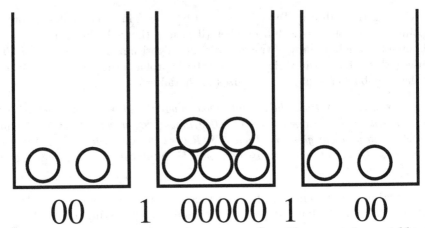

$$00 \qquad 1 \ \ 00000 \ \ 1 \qquad 00$$

The 1's divide the containers from one another. Three containers yield two boundaries between containers.

What does filling bins with indistinguishable balls have to do with creating examples of the Packing problem? We will use the ball-bin problem to help us generate a list of sizes for objects, given a list of sizes of containers. We can limit ourselves to the problem of finding a list of object sizes that exactly fill the containers. If we want examples that overfill the containers, we can add a few objects; if we want excess capacity, we can remove a few objects. Cases with exact solutions, thus, are the starting points from which we can generate any possible case.

Assuming that we have a list $c_1, c_2, \ldots, c_i, \ldots, c_k$ of container sizes, we treat finding the objects to go in each container as a single balls-in-bins problem. The number of balls is equal to the capacity of the container c_i, while the number of objects n is equal to the number of bins. Generating a random binary string with c_i zeros (representing balls) and $n - 1$ ones (separating the bins) thus specifies n objects to fill container i. Repeating this procedure for each container gives a set of object sizes that fills the container exactly. Generating a random binary string with a specified number of ones and zeros is similar to generating the standard representation of a permutation. Put the correct number of ones and zeros in an array, and then repeatedly swap randomly chosen pairs of items in the array.

Experiment 7.21 *Create or obtain code that implements the problem case generator described above. The generator should save not only the sizes of the objects, but the binary strings used in creating the case. Use 5 containers of size 100 with 4 objects per container.*

For a given case, count the number of permutations out of 100,000 selected at random that correctly solve it. This is its hardness *(low numbers are hard). Generate 2000 problem cases and give the distribution of hardnesses as a histogram with 20 bars, each with a width of 5000 hardness units. If you per-*

formed Experiment 7.19 or 7.20, run that software on the hardest and easiest problem case you generated in this experiment.

For the Traveling Salesman problem, we examined the effect of seeding the population with superior solutions generated by heuristics. It would be interesting to see whether that procedure helps solve harder cases of the Packing problem. We will use a very simple heuristic.

Experiment 7.22 *Pick the evolutionary algorithm from Experiment 7.19 that turned in the best performance. Modify the part of the code that creates the initial population, so that, rather than using 200 permutations generated randomly, you use the best 200 out of 400, 800, and 1200. Record the mean time-to-solution, the standard deviation of time-to-solution, and the number of times the algorithms failed to find a solution for cases 6 through 9 of the Packing problem given in Table 7.3, as well as the hardest case you located in Experiment 7.21 (if you performed that experiment). Does the population seeding help? If so, does it help enough to justify its cost? Is the effect uniform over the problem cases? Do the experiments suggest that the filtration of even larger initial groups would be a good idea for any of the problem cases?*

A class activity that is interesting is to compete to construct difficult cases of the Packing problem. Each student should turn in a case together with a solution (certificate of solvability). These cases should then be compiled by the instructor and given to all students (with the names removed). Each student then attempts to solve each of the problem cases with evolution code (any version). Students are scored on the number of other students that fail to solve their case.

Problems

Problem 265. How many different ways can you completely fill a container of size 100 with n objects? Clearly, there is only one way to do it with 1 object (the size of the object must be 100). If we assume that the objects have integer sizes, then there is only one way to fill the container with 100 objects: use objects of size 1. How many ways are there, if we do not care about the order in which the objects are put into the container, to fill it with $n = 2, 3$, or 4 integer-sized objects? Hint: don't do this by hand. (This is a type of mathematical problem called *partitioning*).

Problem 266. Enumerate all the orderings of the objects in Example 17 that permit all the objects to be packed by the greedy algorithm. An explicit listing may be rather long; be clever in how you enumerate.

Problem 267. In many of the examples in this section we have only one size of container. For examples that have multiple container sizes, would it help to permute the order in which the containers are filled? Prove that your answer is correct.

Problem 268. A case of the Packing problem is *tight* if the sum of the sizes of the objects exactly equals the sum of the sizes of the containers. Give an example of a tight problem, with 3 containers of size 100, that is impossible: there is no way to fit all the objects in the bins.

Problem 269. Reread Problem 268. Construct a tight problem with 3 containers of size 100 that has an excess of 0 for *every* permutation of the presentation order of its objects.

Problem 270. Assume that objects and containers have integer sizes. Prove that every Packing problem in which the sizes of the containers and objects have a common divisor larger than one is equivalent to a problem in which they do not.

Problem 271. Draw the balls-in-bins diagrams for the following binary strings in the same style as Example 18:

(i) **010**
(ii) **0010001000010**
(iii) **10001000100**
(iv) **001100**
(v) **0110100100110**

Problem 272. On page 194, there is an informal description, but not pseudocode, for an algorithm for generating random problem cases for the Packing problem. Carefully write out this pseudocode.

Problem 273. Generalize the random case generator to also generate, using a trinary string, the container capacities. Just as 1's were boundaries between bins, 2's become boundaries between containers. Be sure to give an interpretation for all possibilities, such as adjacent 1's or 2's.

Problem 274. It would be interesting to know, for a given case of the Packing problem, how many different solutions there are. A problem with zero solutions is, in some sense, maximally hard. A solvable problem is probably easier if there are many solutions. Give and defend a definition of *different solutions*. Answer the following questions as part of your defense. Is it enough for the permutations controlling the greedy packing algorithm to be different? Is it enough for a given object to be in a different container? Does being different require that the numbers added to fill a container not appear in the solution being judged to be different?

Problem 275. Essay. Suppose that we use a string evolver to attack the Packing problem. The alphabet will contain one character for each container and an additional character for the placement of an object in no container. A given string will have a length equal to the number of objects. The ith character gives the container assignment of the ith object. Both the excess (total size of objects assigned to no container) and the number of containers overfilled (containers in which no more objects will fit) are computed. When strings are compared, having fewer containers overfilled is more important than having less excess. These two numbers are used in a lexical product fitness function. Your task: write an essay that estimates the relative performance of this evolutionary algorithm and one of the algorithms used in this section. (Implementing the algorithm would provide the best form of evidence for your essay, but is not required.)

Problem 276. The containers used in this section have a pretty boring sort of geometry. They simply have a one-dimensional capacity that is used up as objects are placed in them. Suppose that we have containers that are $N \times M$ rectangles and objects that are $H \times K$ rectangles. Rewrite the greedy packing fitness for placing rectangular objects into rectangular containers so that they fill the containers efficiently. Modify the permutation data structure if you deem it necessary. Since the objects have two orientations in which they may be placed, you may need to augment your permutations with some sort of orientation switch. Write pseudocode for the new fitness function.

Problem 277. The hardness criterion given in Experiment 7.21 is rather slow to use as a fitness function for evolving hard (or easy) problem cases. Give and defend a better fitness function for evolving hard problem cases.

7.4 Costas Arrays

This section develops the theory needed to attack an unsolved mathematical problem, the existence of a Costas array of order 32.

Definition 7.17 *A* **Costas array** *is an $n \times n$ array with one dot in each row and column and the property that vectors connecting any two dots in the array do not occur more than once (i.e., all such vectors have different lengths or different slopes). The number n is the* **order** *of the array.*

Costas arrays were originally devised as signal processing masks by John P. Costas. Given that these arrays are useful for signal processing, the next mathematical question is "do they exist?" For infinitely many n, the answer is known to be yes, but for some n, the answer is not yet known. The $n < 100$ for which no Costas array is known to exist at the time of this writing are **32, 33, 43, 48, 49, 54, 63, 73, 74, 83, 84, 85, 89, 90, 91, 92, 93,** and **97**. The total number of Costas arrays of order n is a very odd function of n. Examine Table 7.4 to gain some sense of this oddity. Examples of Costas arrays of order 10 and 12 are given in Figure 7.6. Before continuing on with Costas arrays, we need to review some useful linear algebra.

If we have an $n \times n$ matrix M, then the process of multiplying a vector $\mathbf{v}$ by M creates a map from $\mathbb{R}^n$ to itself. If M is a matrix with a single one in

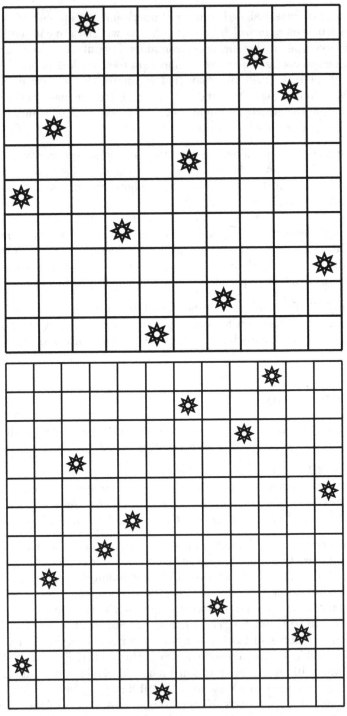

Fig. 7.6. Matrix presentations of Costas arrays of orders 10 and 12.

Dimension	Number	Dimension	Number
1	1	2	2
3	4	4	12
5	40	6	116
7	200	8	444
9	760	10	2160
11	4368	12	7852
13	12828	14	17258
15	19612	16	21104
17	18276	18	15096
19	10240	20	6464
21	3536	22	2052
23	872	24	? (> 1)

Table 7.4. Costas arrays of those sizes for which the total number of Costas arrays is known.

each row and column and all other entries zero, it acts on $\mathbb{R}^n$ by permuting the dimensions.

Definition 7.18 *A* **permutation matrix** *is a matrix with a single 1 in each row and column and all other entries 0.*

Assume that $\mathbf{v}$ is a row vector and M is a permutation matrix. To multiply M by $\mathbf{v}$, we could compute either $M \cdot \mathbf{v}^t$ or $\mathbf{v} \cdot M$. The resulting permutations are inverses of each other. Let's do an example.

Example 19. Suppose that

$$M = \begin{bmatrix} 0 & 1 & 0 \\ 0 & 0 & 1 \\ 1 & 0 & 0 \end{bmatrix}$$

and that $\mathbf{v} = (a, b, c)$. Then

$$M \cdot \mathbf{v}^t = (b, c, a),$$

while

$$\mathbf{v} \cdot M = (c, a, b).$$

The permutations (bca) and (cab), in one-line notation, are inverses of one another.

The reason for introducing permutation matrices at this point is that the "one-per-row-and-column" condition is common between Costas arrays and permutation matrices. To evolve Costas arrays, we will take the machinery already developed for evolving permutations, add a map from permutations to permutation matrices, and then search the space of permutation matrices for those that obey the additional condition needed to be Costas arrays. (To get Costas arrays, we then just substitute dots for 1's and blanks for 0's.)

Definition 7.19 *For a permutation* σ *of* $\{1, 2, \ldots, n\}$ *let the* **permutation matrix associated with** σ, M_σ, *be the permutation matrix with ones in position* $(i, \sigma(i))$, $i = 1, 2, \ldots, n$, *and zeros elsewhere.*

Example 20. If $\sigma = (24153)$ in one-line notation, then

$$M_\sigma = \begin{bmatrix} 0 & 1 & 0 & 0 & 0 \\ 0 & 0 & 0 & 1 & 0 \\ 1 & 0 & 0 & 0 & 0 \\ 0 & 0 & 0 & 0 & 1 \\ 0 & 0 & 1 & 0 & 0 \end{bmatrix}.$$

The map from permutations to potential Costas arrays is useful for displaying the final array in the form of a Costas array. The fundamental structure undergoing evolution, however, is the permutation, and the fitness function we will use is based on the permutation in one-line notation. Let's examine two permutations, one of which is a Costas array, in order to understand the additional property of vectors being unique.

Example 21. Shown below is a pair of 5×5 arrays with one dot in each row and column. The first is a Costas array: the 10 vectors joining the pairs of dots AB, AC, AD, AE, BC, BD, BE, CD, CE, and DE are all different in their x or y length or both. In the second array, the vectors joining BC and DE are identical. Notice that we do not distinguish between **BC** and **CB**.

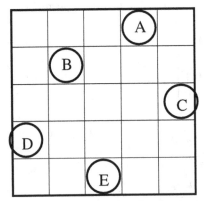 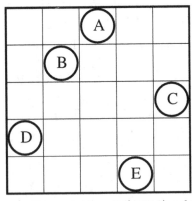

The permutation for the first array, in one-line notation, is (42513), while the second array is associated with the permutation (32514).

We need a fitness function that will favor Costas arrays over mere permutation matrices. Given that Costas arrays require vectors connecting dots in the matrix to be unique, there is a natural choice: count the number of reused vectors (and minimize).

Definition 7.20 *The* **number of violations of the Costas condition** *in a permutation matrix is the number of times that a vector is reused. A* *vector used 3 times is reused twice and so contributes 2 to the total. The* **VCC fitness function** *of a permutation σ is the number of violations of the Costas condition that appear in M_σ, the corresponding permutation matrix.*

The definition of the VCC fitness function does not immediately suggest a simple algorithm for computing it. If we list the permutations in one-line notation, then dots in adjacent columns of M_σ are adjacent in σ, dots two apart horizontally in M_σ are two apart in σ, and so on. So we can count violations of the Costas condition by looking for repetitions in the distance-i differences of the permutation.

Example 22. Compute the number of violations of the Costas condition in the permutation $\sigma = (\mathbf{456123})$.

$\sigma(i)$	4	5	6	1	2	3	Repeats
$\sigma(i+1) - \sigma(i)$	1	1	-5	1	1		3
$\sigma(i+2) - \sigma(i)$	2	-4	-4	2			2
$\sigma(i+3) - \sigma(i)$	-3	-3	-3				2
$\sigma(i+4) - \sigma(i)$	-2	-2					1
$\sigma(i+5) - \sigma(i)$	-1						0

Totaling the repetitions for each horizontal component length, we have $3 + 2 + 2 + 1 + 0 = 8$ violations. Notice that since the permutation does not wrap around, there are fewer vectors with long horizontal components than short ones. In a permutation of 6 numbers, there is only one vector with horizontal component 5 (in this case, its vertical component is -1), and so there is no chance of repetition there.

Experiment 7.23 *Using a random key encoding for permutations in one-line notation, build or obtain software for an evolutionary algorithm for locating Costas arrays. Initialize the random key to have its numbers in the range $[0, n]$, where n is the order of the Costas arrays being searched. Let your point mutation be uniform real mutation with size 0.5. Test the software with one-, two-, and three-point mutation and two-point crossover for Costas arrays of order 12. Use a population of 1000 permutations. Compute the mean time until an array is located and the standard deviation of this time over 400 trials. Have the algorithm stop after 200,000 fitness evaluations and declare an algorithm that times out a failure. Record the number of failures. Compare the 3 different mutation operators. Do the data indicate that trying four-point mutation might be advisable?*

The number of mutations is one possible parameter that governs the mutation-based part of the evolutionary search. Another is the size of the mutations.

Experiment 7.24 *Perform Experiment 7.23 again, only this time fix the number of mutations at whatever worked best. Rather than varying the number of mutations, vary their size. Use mutation sizes 0.25, 0.5, 1.0, and 2.0, pulling in the data for 0.5 from the previous experiment.*

The crossover operator also has some impact on the behavior of a system that evolves permutations, and so may be worth examining. Since Costas arrays are connected with unsolved problems, we will cast a fairly wide net, going beyond the obvious possibility of playing with the number of crossover points to the extremity of introducing a new type of crossover operator.

Definition 7.21 Nonaligned crossover *for a pair of genes organized as a string or an array is performed in the following manner. A length smaller than or equal to the length of the shorter of the two participating strings or arrays is chosen. Starting positions in both strings are chosen. Going down the strings, we then exchange characters in substrings or subarrays of the selected length, wrapping if we go off the end of the string.*

The standard crossover operator contains a tacit agreement that each position within the objects being crossed over has a particular meaning. When these meanings are distinct, using nonaligned crossover would mean comparing apples and oranges. With random key encoding, the relative rank of the numbers in the array is what matters, not their positions. This means that all meaning is between numbers, and so nonaligned crossover is, perhaps, more meaningful.

Experiment 7.25 *Perform Experiment 7.24 again, only this time fix the number and size of mutations at whatever worked best (two-point and 1.0, for example). Instead of testing different mutation operators, test one-point crossover and two-point crossover. Use both the standard and nonaligned forms of the crossover with the probability of nonaligned crossover being 0%, 10%, and 25% in different runs. This requires 6 different sets of runs. Does nonaligned crossover help? Might upping the percentage be a good thing? Warning: be sure your random key encoding software deals consistently with ties.*

Experiments 7.23–7.25 all tuned up our search system on order-12 Costas arrays. It is now time to see how far the system can be pushed.

Experiment 7.26 *Perform Experiment 7.25 again, using the collection of settings that you found to work best. Allocating runtime on available machines so as to avoid actually getting into trouble or breaking rules, find the largest order Costas array you can with your software. Run experiments on several* *orders to obtain a baseline behavior before doing the big run. Keep in mind that the absolute number of Costas arrays declines at order 17, even though the search space (or number of permutations) is growing at a faster than exponential rate (in fact, factorially). This means that time estimates for orders below 17 will have no meaning for those over 17. If you find a Costas array of one of the unknown orders, publish the result and send the author of this book a copy of the manuscript, please.*

On page 197, the claim was made that an infinite number of n are known for which there are Costas arrays. Such arrays may be useful in seeding initial populations (or they may not).

Definition 7.22 *For a prime number p, a number k (mod p) is said to be* **primitive** *(mod p), if every nonzero value (mod p) is a power of k.*

Example 23. Let's look at all powers of all nonzero numbers (mod 7). Recall that the powers cycle after the sixth power (and sooner in elements that are not primitive).

	Power					
k	1	2	3	4	5	6
1	1	1	1	1	1	1
2	2	4	1	2	4	1
3	3	2	6	4	5	1
4	4	2	1	4	2	1
5	5	4	6	2	3	1
6	6	1	6	1	6	1

As the table shows, the only two primitive numbers (mod 7) are 3 and 5. There are many, many patterns in the above table, and most books on group theory or number theory both generalize and explain these patterns.

It turns out that primitive numbers can be used to generate Costas arrays.

Theorem 6. *Let k be a primitive number (mod p) for some prime p. Then the permutation σ on $\{1, 2, \ldots, (p-1)\}$ for which $\sigma(i) = k^i$ (mod p) has the property that M_σ is a Costas array.*

Proof: left as an exercise, Problem 283.

Example 24. Notice that the construction given in Theorem 6 yields permutations of $p - 1$ items that don't include zero. Let's look at the Costas array generated by 3 (mod 7). Using Example 23, we see that σ makes the following assignments: $1 \to 3$, $2 \to 2$, $3 \to 6$, $4 \to 4$, $5 \to 5$, and $6 \to 1$. This corresponds to the following permutation matrix:

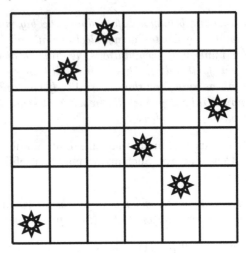

Let's conclude with an experiment in population seeding.

Experiment 7.27 *Perform Experiment 7.26 again at the maximum order for which you were able to find a Costas array or the next-smallest order, whichever is not one less than a prime. The initial population should be generated in the normal fashion and also by taking permutations of the sort described in Theorem 6 and randomly extending them to the desired order. Invent and describe the means you use to extend these permutations. Be very sure your extended items are still permutations. Does seeding the population in this fashion help?*

Problems

Problem 278. Prove that if we apply any of the 8 symmetries of the square to a Costas array, it remains a Costas array.

Problem 279. Can you devise a better fitness function for searching for Costas arrays? Find a function that is different from the one used in the chapter and document its strengths and weaknesses.

Problem 280. A tie in a random key encoding happens when two numbers in the array are equal. Does the use of nonaligned crossover increase the probability of ties within members of a population of random key encodings? If the answer is yes, give a technique for eliminating such ties.

Problem 281. For each of the following permutations in one-line notation, compute the number of violations of the Costas condition. Hint: using a computer may be faster and is almost surely more accurate.

(i) (5 1 7 4 3 9 6 8 2 0)
(ii) (0 3 6 2 1 9 7 8 4 5)
(iii) (8 7 4 2 9 1 3 5 0 6)
(iv) (2 7 1 8 6 5 9 0 3 4)
(v) (9 4 7 8 1 2 6 3 5 0)
(vi) (7 9 3 5 0 4 8 6 1 2)
(vii) (5 0 7 1 3 2 8 6 4 9)
(viii) (8 5 0 4 7 6 1 3 9 2)
(ix) (3 7 2 1 8 9 6 4 0 5)
(x) (6 8 4 3 1 2 0 7 5 9)
(xi) (6 7 0 5 1 4 2 8 9 3)
(xii) (2 7 4 3 0 6 5 1 9 8)

Problem 282. For each of the following 4 arrays, compute the number of violations of the Costas condition and, on a copy of the matrix presentation, show the vectors that repeat more than once.

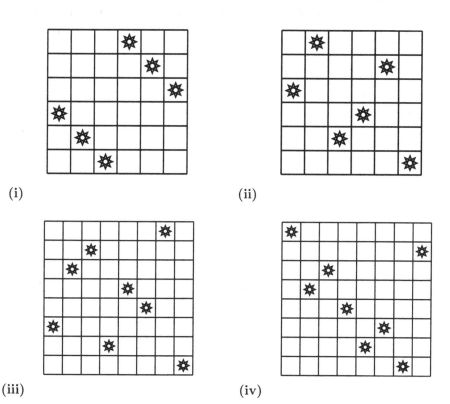

(i)   (ii)

(iii)   (iv)

Problem 283. Prove Theorem 6.

Problem 284. A fixed point in a permutation σ is a number for which $\sigma(i) = i$. Prove the strongest upper bound you can on the number of fixed points, if $\text{VCC}(\sigma) = 0$ (i.e., if M_σ is a Costas array).

Problem 285. Write a program that exhaustively enumerates all Costas arrays of a given order, and, either using your code or working with pencil and paper, compute the largest number of fixed points in Costas arrays of order n, for $n = 1, 2, \ldots, 10$.

Problem 286. Invent and code a deterministic greedy algorithm for placing the successive dots of a permutation matrix so as to minimize the number of violations of the Costas condition. Make your algorithm such that it can find Costas arrays of all orders up to 5.

Problem 287. Compute the number of permutation matrices of order n that are symmetric about their main diagonal. Hint: such permutation matrices correspond to permutations of order 2. Compute also the fraction of all permutation matrices that are symmetric in this fashion.

Problem 288. Essay. Address the following question: would it be a good idea to search for Costas arrays that are symmetric about their main diagonal?

8

Plus-One-Recall-Store

In this chapter, we will do our first computer experiments with *genetic programming*. Genetic programming in its modern form was invented by John Koza and John Rice [38, 39, 40]. An excellent book on the subject is *Genetic Programming: An Introduction* by Wolfgang Banzhaf, Peter Nordin and Robert E. Keller[9]. Genetic programming is the use of evolutionary computation to produce small pieces of computer code. The first attempts at this were done by R. M. Friedberg and his collaborators at IBM in the late 1950s [26, 27]. A fascinating method of evolving self-reproducing computer programs appears in Tom Ray's "An approach to the synthesis of life" [48], which is reprinted in [22].

The first section of the chapter is intended as a general introduction to genetic programming and is less narrowly focused than the rest of the chapter. It also contains no experiments, just preparatory reading and problems intended to build needed mental muscles. The second section introduces the Plus-One-Recall-Store (PORS) problems, previewed in Problems 16 and 17. The technical background of this genetic programming problem appears in [6]. The third section studies in detail the technique of population seeding, using nonrandom initial populations to enhance performance of an evolutionary algorithm. The fourth section applies various generic techniques for improving evolutionary computation from earlier chapters to the PORS problems. The structure of the experiments in this chapter is given in Figure 8.1.

The PORS problems are entirely abstract. One of the two main problems is solvable by theorem and proof techniques, so that the true optimal solutions are already known. (Technical details have been kept to an absolute minimum.) We study PORS not for itself but to learn genetic programming, much as we studied the string evolver in Chapter 2 to learn about evolutionary algorithms. The PORS problems make up a very good test suite for exploring evolutionary computation techniques. The advantages of the PORS problems as a test suite include the following:

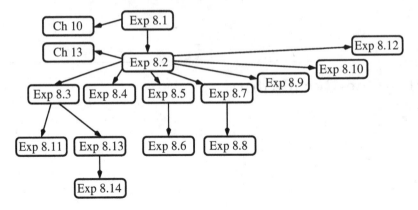

1   Basic routines for PORS trees.
2   The Efficient Node Use problem.
3   The Minimal Description problem.
4   Population seeding for Efficient Node Use problem.
5   Analysis of solutions.
6   Statistical modeling to generate initial populations.
7   Initial trees from $\mathcal{T}_s$.
8   Initial trees from $\mathcal{T}^*, \mathcal{T}_s^*$.
9   Tree-depth lexical product fitness.
10 Range-niche specialization for Efficient Node Use problem.
11 Range-niche specialization for Minimal Description problem.
12 Exploring models of evolution.
13 Initializing the memory to 1.
14 Population seeding for Minimal Description problem.

Fig. 8.1. The topics and dependencies of the experiments in this chapter.

- The PORS problems are exceedingly simple, with only four total operations and terminals. In spite of this simplicity, they have a rich mathematical structure that challenges the power of evolutionary computation by asking the system to learn to use a calculator-style memory.
- The Efficient Node Use problem, introduced in Section 8.2, contains very clear and easily detected building blocks, the "multiply by 2" and "multiply by 3" subtrees or subroutines explored in Section 8.4. The Efficient Node Use problem thus provides a test bed for systems that are supposed to conserve or even recognize building blocks.
- The Efficient Node Use problem is actually three very different problems. The number of nodes in the tree, modulo 3, determines the problem class. These problems have respectively a unique optimal solution, a number that is linear in the number of nodes in the tree, and a number that is a quadratic function of the number of nodes in the tree. The problem thus contains a diverse test suite.

- The Minimal Description problem is one for which there is a large number of possible fitness functions, all of which function less than well. It thus provides a challenging problem in which students may exercise their imagination.

Leaving aside the practical applicability of the PORS problems, these problems have a deep and rich mathematical structure, reflected in the problems given in the chapter. Instructors of a mathematical bent may wish to extend these problems, particularly in a combinatorial way. Engineering classes using this text should assign problems from this chapter with great caution; some require a good deal of familiarity with the various forms of abstract nonsense that so delight mathematicians.

8.1 Overview of Genetic Programming

Genetic programming (abbreviated GP) is defined to be the use of an evolutionary algorithm to evolve computer programs, typically stored as parse trees. A parse tree is shown in Figure 8.2 in tree form as well as in a functional notation used in the computer language LISP. We term the compressed notation *LISP-like*.

A parse tree has interior nodes, called *operations*, and leaves or terminal nodes, called *terminals*. Operations have the usual definition. In the tree shown in Figure 8.2, the operations are sqrt, $/, +, -, *$, and $**$. Terminals can contain external values passed to the program, input/output devices, or constants. The terminals in Figure 8.2 are the external inputs a, b, and c and the constants 2 and 4. The *root* of the tree is the first operation when the tree is written in LISP-like form. In Figure 8.2, the root is the divide operation. The subtrees that are linked to an operation are called the *arguments* of the operation. The *size* of a parse tree is the number of nodes in it. The *depth* of a parse tree is the largest distance, in tree links, from the root to any terminal. The depth of the parse tree in Figure 8.2 is 6.

Genetic Operations on Parse Trees

In genetic programming, crossover is performed by exchanging subtrees as shown in Figure 8.3, with parents above and children below. This crossover operation is called *subtree crossover*. Mutation consists in taking a randomly chosen subtree, deleting it, and replacing it with a new randomly generated subtree. In later chapters, we will add other mutation operators. This mutation operator is called *subtree mutation*. Notice that all notion of a "point mutation" as developed in preceding chapters has gone completely by the board. Mutation at some loci of a parse tree has an excellent chance of changing large numbers of other loci (see Problem 299).

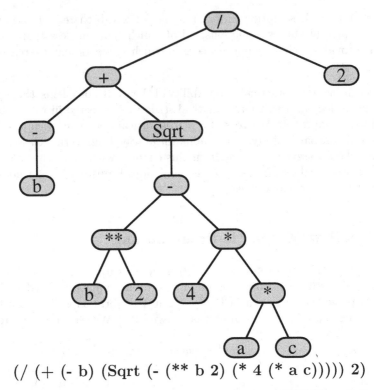

$$(/ \; (+ \; (- \; b) \; (Sqrt \; (- \; (** \; b \; 2) \; (* \; 4 \; (* \; a \; c)))))) \; 2)$$

Fig. 8.2. An example of a parse tree in standard and LISP-like notation.

We will also use a third type of variation operator called *chopping*, which reduces the size of a parse tree. The genomes we have used in previous chapters have had uniform size. A gene did not have a risk of become unmanageably large after several generations. The crossover operation used in genetic programming conserves the total number of nodes in the parents, but one child may get more and another fewer. As a result, the number of nodes in one creature can, theoretically, grow until it swamps the machine being used. The chop operator is used to reduce the size of a parse tree if mutation and crossover have made it "too large." In order to apply the chop operator, you select one argument of the root operation of a program and allow it to become the entire program, deleting the rest of the parse tree.

Genetic programming requires mature use of dynamic memory allocation. A parse tree is not declared as a single data structure but is rather built up recursively. The data structure used to store a node of a parse tree contains the node's type (constant, external input, unary operation, binary operation, etc.), the node's value (e.g., the numerical value of a constant or the identity of an external input), and pointers to the subtrees that form its arguments if it is an operation. Since the node data structure must be able to hold several

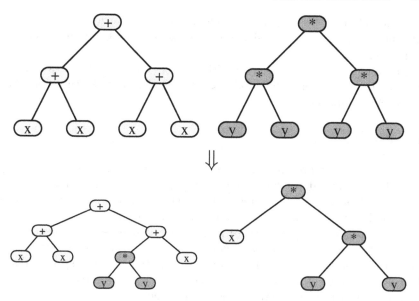

Fig. 8.3. Crossover of two trees by exchanging subtrees.

different types of nodes, parts of its structure are often unused. If the node is a nonconstant terminal, the value field may be empty. If the node is a terminal, it will have all of its pointers empty. An example of a data structure for one node of a parse tree is shown in Figure 8.4.

```
struct node {float val;      //holds value for contant terminals
             type int;       //holds type, e.g., operation code
             node* args[2];  //pointers to argument trees
      };
```

C or C++

Fig. 8.4. Example of a data structure for holding nodes used to build pointer-based parse trees.

Generating Random Parse Trees

The first computational step in an evolutionary algorithm is to generate a random population, so we must learn to generate random parse trees. A random parse tree on n nodes with unary or binary operations is generated recursively as follows. If $n = 1$, then we generate a terminal node. If $n = 2$, we generate a unary operation, from those available, whose argument is a tree of size one. If $n \geq 3$, then we pick an operation from among those available. If that operation is unary, we generate its argument as a tree of size $n - 1$. If that operation is binary, we split $n - 1$ randomly into two nonzero pieces a and b and generate random trees of size a and of size b to be the arguments of the binary operation.

The random tree generation algorithm knows only about 3 sizes of trees: size one (terminal), size two (operation terminal), and size three or more (operation arguments).

Type Checking

Can you do genetic programming with real source code? Many of you may feel a chill running down your spine at the thought of what crossover and mutation might do to real source code. There are many structural reasons why genetic programming evolves limited, small parse trees instead of full computer programs. Most of these reasons are tied to the complexity and fragility of "real" computer code. For the most part we avoid this sort of problem by simply evolving small pieces of code that do very specific tasks. There is one source of fragility that we will avoid by main force: data typing. Our genetic programs will have a single data type.

One of the most common errors a beginning programming student will make is a type check error. The student may read in a number as a character string and then be hurt and surprised when adding one to it causes an error (in any reasonable language) or when adding one mysteriously causes the first digit to vanish (in C or C++). To avoid having our evolving programs make (or spend all sorts of computer power avoiding) type errors, we will have only one type. This means that all terminals must return values of that type, and all operations must take as arguments values of that type and return values of that type.

Requiring that the parse tree have a single type in all its nodes does not completely deprive us of the advantages of having multiple types. Both C and BASIC have numerical Booleans. Zero is considered false and nonzero is considered true. This means that operations like $\leq$, $=$, and $\geq$ can be defined as numerical operations that, in some situations, function as Boolean operations, all within a single data type. In this chapter, our data type will be integers.

Problems

Problem 289. Transform each of the following expressions from functional form into LISP-like notation. Do *not* simplify or modify the expressions algebraically before transforming them.

(i) $f(x) = \frac{x^3}{x^3 - 1}$

(ii) $g(x) = (x - 1)(x - 2)(x - 3)(x - 4)$

(iii) $h(x, y) = \sqrt{(x - a)^2 + (y - b)^2}$

(iv) $d(a, b, c) = \sqrt{b^2 - 4ac}$

(v) $r(\theta) = 2 \cdot \cos^2(\theta)$

(vi) $r(\rho, \theta) = \sqrt[3]{\rho^2 \theta^2 + 1}$

Problem 290. Transform each of the following expressions from LISP-like notation into functional form. Do *not* simplify or modify the expressions algebraically before transforming them. Assume that logical connectives like "$<$" and "$\geq$" return 1 for true and 0 for false. You will need to name the functions. Assume that x, y, and z are variables and that any other letters or names represent constants. Notationally simplify the functions as much as you can.

(i) (+ (Sqr (Tan x)) 1)

(ii) (+ (+ (Sqr x) (Sqr y)) 1)

(iii) (+ (* (b $\geq$ 0) (/ x 2)) (* (b < 0) (/ (- 0 x) 2)))

(iv) (/ (+ (+ x y) z) (+ (* x (* y z)) 1))

(v) (** (+ x (+ y z)) 3)

(vi) (/ 1 (+ 1 (* x x)))

Problem 291. On page 209, the claim is made that the crossover operation for parse trees, if repeated, can cause the size of the tree to grow without bound. Assuming that we start with a population made of trees of the form (+ a b) in generation zero, compute the worst-case size of the biggest tree in generation n. Give examples of worst-case trees for $n = 1, 2, 3$, and 4.

Problem 292. On page 211, there is a description of the algorithm for generating parse trees with operations that are unary and binary. Describe the algorithm for generating parse trees with unary, binary, and ternary operations. Also, show the data structure needed, as in Figure 8.4. (Use your favorite computer language.)

Problem 293. In order to generate parse trees, you must break a remaining number $n - 1$ of nodes ($n \geq 3$) into 2 nonzero pieces a and b. Give pseudocode for doing this.

Problem 294. In order to generate parse trees with ternary operations (see Problem 292), you must break a remaining number $n - 1$ of nodes into 3 nonzero pieces a, b, and c. Give pseudocode for doing this.

Problem 295. Suppose we have a language for parse trees in which there are only terminals and binary operations. Prove that all parse trees in this language have odd size.

Problem 296. The *Catalan numbers* C_n count the number of ways to group terms when you have a nonassociative operation on n variables. The first few values are $C_1 = 1$, $C_2 = 1$, $C_3 = 2$, $C_4 = 5$. The corresponding ways of grouping terms are

$n = 1 : (a),$
$n = 2 : (a \odot b),$
$n = 3 : (a \odot (b \odot c)), ((a \odot b) \odot c),$
$n = 4 : (a \odot (b \odot (c \odot d))), (a \odot ((b \odot c) \odot d)), ((a \odot b) \odot (c \odot d)),$
$\qquad ((a \odot (b \odot c)) \odot d), (((a \odot b) \odot c) \odot d).$

First, find a general formula for the Catalan numbers. Second, explain why C_n is also the number of parse trees with $2n + 1$ nodes, if the language for those parse trees has one binary operation and one terminal.

Problem 297. Suppose we have a GP-language with x terminals and y binary operations. Give a formula for the number of possible trees in terms of the number of nodes n in the tree. Modification of the answer to Problem 296 may serve.

Problem 298. Suppose we have a GP-language with x terminals, y binary operations, and z unary operations. Give a formula for the number of possible trees in terms of the number of nodes n in the tree. If you find this hard, you should first do Problem 297 or other cases in which one of x, y, or z is 0. Your answer should involve a summation over something and may use the Catalan numbers.

Problem 299. The *rate* of a mutation operator for parse trees is the fraction of the nodes it will replace on average. Mathematically or experimentally, estimate the rate of a mutation operator that selects, with uniform probability, a node in a parse tree and replaces the subtree rooted there with a new subtree exactly the same size. If the GP-language in question has only unary operations, then this isn't too hard to compute. A language with binary operations is harder. A language with both unary and binary operations may well be too hard to do other than experimentally. Detail exactly your mathematical or experimental techniques.

Problem 300. Suppose we have a GP-language whose data type is integers. Give a definition of a binary operation that can be used as an If-then and of a ternary operation that can be used as an If-then-else. Discuss briefly the alternatives you considered and why you made the choices you did.

Problem 301. Suppose you were evolving real-valued parse trees that were encoding functions of one variable (stored in a terminal "x") to minimize the sum of squared error at 80 sample points with the fake bell curve (see Equation 2.1). Assume that your parse tree has available the operations of an inexpensive scientific calculator. Show mathematically why

$$\textbf{(Sqr (Cos (Atan x)))} \text{ or } f(x) = \cos^2(\tan^{-1}(x))$$

is not an unreasonable result.

Problem 302. Essay. Reread Problem 301. Clearly, the parse tree
$$\textbf{(/ 1 (+ 1 (* x x)))}$$
would give perfect (zero) sum of squared error with points sampled from the fake bell curve. Why then did this parse tree never come up in 80 runs (done by the author), while the parse tree given in Problem 301 appeared in 5 of the runs? Consider the action of mutation and crossover of the parse trees when answering this question.

8.2 The PORS Language

The Plus-One-Recall-Store (PORS) language has two terminals, one unary operation, and one binary operation. They are summarized in Figure 8.5. The environment in which PORS programs exist contains an external memory to which the program has access. Much like the memory in a cheap pocket calculator, the user may store and recall values to and from the memory. The binary operation of the PORS language is normal integer addition. The unary operation **Sto** does nothing to its argument; it just returns whatever value it is given. It is used for its side effect of storing the number that passes through it into the external memory. The terminal **Rcl** returns the value in the external memory. The terminal **1** is a simple integer constant.

Type	Name	Description
Terminal	1	integer constant, one
	Rcl	recalls an external memory
Unary operation	Sto	stores a value in an external memory and returns the number stored as its value
Binary operation	+	integer addition

Fig. 8.5. Nodes in the Plus-One-Recall-Store language.

Semantics of PORS

Now that we have specified the elements of a PORS parse tree, we can learn how to "run" it as a program. This is called *evaluating* the parse tree. The tree is evaluated from the bottom up. Since we are working with randomly generated parse trees, we will encounter trees that execute a recall instruction before their first store instruction. To prevent this from being undefined and hence a problem, declare the external memory location to be initialized to 0 at the beginning of the evaluation of any parse tree. The execution of the store and recall instructions and the constant 1 are obvious and unambiguous. When executing a plus, execute all the instructions in the left argument before those in the right argument. An example of a PORS parse tree, adorned with the values each node returns when executed, is shown in Example 25.

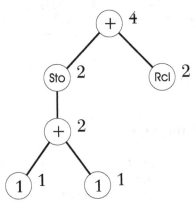

Example 25. The parse tree shown above performs the following steps when evaluated: it adds 1 and 1, stores the resulting 2 in memory (also returning the 2 as the value of the **Sto** operation), and then adds that 2 to the contents of memory to obtain the result of the entire program, 4.

For a PORS parse tree T, we denote by $\epsilon(T)$ the result of evaluating the parse tree (equivalently, running the program described by the tree). The result $\epsilon(T)$ is said to be the *value* of T. To start evolving parse trees, we lack only a problem to solve. We will work on the two tasks implicit in Problems 16 and 17. The *Efficient Node Use problem* asks, "what is the largest number that can be generated by a parse tree with n nodes?" The *Minimal Description problem* asks, "given k, what is the smallest tree T for which $\epsilon(T) = k$?" The first problem is easy, while the second is quite hard, as we shall see. Before attacking either of these problems, we need to get the basic routines for genetic programming built and working.

Experiment 8.1 *Write or obtain a set of routines for manipulating parse trees in the PORS language including all the following, as well as any others you find convenient:*

Name	Argument	Returns	Description
RandTree	integer	pointer	generates a random tree with *int* nodes
SizeTree	pointer	integer	finds the number of nodes in a tree
DeepTree	pointer	integer	finds the depth of a tree
CopyTree	pointer	pointer	makes a copy of a tree, doing all needed dynamic allocation
KillTree	pointer	nothing	disposes of a parse tree
PrinTree	pointer	nothing	prints out a parse tree
SaveTree	pointer	nothing	saves a tree in a file
ReadTree	nothing	pointer	reads a tree from a file
RSubTree	pointer	pointer	returns a pointer to a random subtree of a tree
CrossTree	pointers	nothing	performs the "exchange subtrees" crossover operation used in genetic programming
MuteTree	pointer	nothing	picks a random subtree, deletes it, and replaces it with a random one the same size
ChopTree	pointer, integer	pointer	replaces a tree with one of its subtrees coming off the root, repeatedly, until the tree has *int* or fewer nodes
EvalTree	pointer	integer	computes the value of a tree

When the above routines are ready, test them by computing the following for a population of 1000 *trees of size* 5, 10, *and* 20 *respectively.*

(i) mean and standard deviation of tree depth;
(ii) a histogram of how often each depth occurred;
(iii) a histogram of the values of the trees

Also, print out examples of copying, mutation, crossover, and chopping.

Experiment 8.1 is a baseline for later experiments in this chapter. For the most part, it is included so that you can test and debug the underlying routines. The software in this chapter uses dynamic allocation. Hence, it is more likely to contain subtle bugs than the software you coded in the preceding chapters. In our next experiment, we will put these routines to work attempting to solve the Efficient Node Use problem for small n. The Efficient Node Use problem has a natural fitness function: the evaluation function ϵ. As long as we keep the number of nodes in trees in our population bounded by the number of nodes n for which we are currently trying to solve the Efficient Node Use problem, we can set the tree's fitness to its value.

Experiment 8.2 *Write or obtain software for an evolutionary algorithm to solve the Efficient Node Use problem. Use a population of* 100 *parse trees with* n *nodes. Use tournament selection with tournament size* 4 *as your model of evolution. If a child has more than* n *nodes after crossover, apply the chop operator until it has* n *or fewer nodes. Mutate each child with probability* 0.4. *Fitness of a tree* T *is simply* $\epsilon(T)$.

n	$max\ \epsilon(T)$	n	$max\ \epsilon(T)$	n	$max\ \epsilon(T)$
1	1	10	9	19	72
2	1	11	12	20	96
3	2	12	16	21	128
4	2	13	18	22	144
5	3	14	24	23	192
6	4	15	32	24	256
7	4	16	36	25	288
8	6	17	48	26	384
9	8	18	64	27	512

Fig. 8.6. Evaluation values for solutions to the Efficient Node Use problem for small values of n.

A tree T on n nodes is optimal if $\epsilon(T)$ is as large as possible given n. The table given in Figure 8.6 contains the numbers that optimal trees evaluate to for $1 \le n \le 27$. For $n = 12, 13, 14, 15$, and 16, run the evolutionary algorithm on a population until it finds an optimal tree or for 500 generations. Run 100 populations. Plot the fraction of successful populations as a function of generations of evolution. Also, note the fraction of initial populations that contained an optimal tree and the fraction that failed to find an optimal tree. In your write-up, explain what happened and why. Is the probability of finding an optimal tree within 500 generations a decreasing function of the number of nodes in the tree?

For use with experiments in Section 8.3, write the code for this experiment so that you can optionally save your population of parse trees in a file.

If your version of Experiment 8.2 worked properly, there should be a strange relationship between the number of nodes and the time-to-solution. In this case "strange" means "not monotone." Looking at the table given in Figure 8.6 can give you a clue to what is happening.

Our next experiment will break new ground. The point of this experiment is to give you practice at coming up with fitness functions for a difficult problem. Consider the Minimal Description problem: given an integer k, find the PORS parse tree with the smallest number of nodes that evaluates to k. Problem 309 tells us that we can restrict our population to trees with at most $2k + 1$ nodes. The difficulty comes in figuring out which trees with $2k + 1$ or fewer nodes are "close" to computing k. Examine Figure 8.7. It gives solutions to the Minimal Description problem for $k = 7, 8$. The two trees do not share

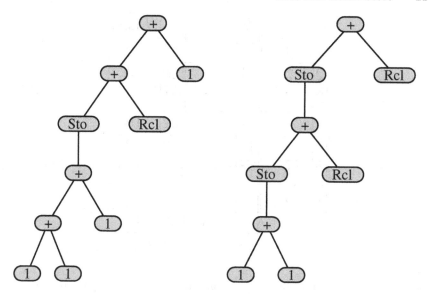

Fig. 8.7. Minimal trees for computing 7 and 8.

much in the way of structure. The Minimal Description trees for $k = 11, 12$ are even less alike.

If minimal-sized trees that compute numbers differing by one are not structurally similar, then minimizing $|k - \epsilon(T)|$ is probably not a good fitness function for the Minimal Description problem. What other choices are there? Are there any other senses in which two numbers a and b can be close? One obvious source of closeness is the divisibility relation. This yields several candidates: the greatest common divisor of a and b; the integer division of a by b or b by a; and the remainder of a divided by b or of b divided by a. These could all profitably appear in a definition of "close." It is possible that lexical products or numerical averages of these divisibility relations might serve as parts of fitness functions for the Minimal Description problem.

In addition, to solve the Minimal Description problem we want to minimize the number of nodes in the tree. This makes the Minimal Description problem an example of *multicriteria optimization* with the multiple criteria being computation of numbers close to k and minimization of the number of nodes used to do so. We will call a PORS parse tree that evaluates to k with the fewest number of nodes a *Minimal Description tree* for k.

For those of you who have not taken a course in discrete mathematics, we will review the method for computing the greatest common divisor (GDC) of two numbers. Suppose we are given integers a and b. Perform the following computations:

k	$|T|_{min}$	k	$|T|_{min}$	k	$|T|_{min}$
1	1	10	11	19	15
2	3	11	13	20	14
3	5	12	11	21	15
4	6	13	13	22	16
5	8	14	13	23	18
6	8	15	13	24	14
7	1	16	12	25	16
8	9	17	14	26	16
9	10	18	13	27	15

Fig. 8.8. The minimal number of nodes in a tree describing k, for small values of k.

$$x_0 \leftarrow a,$$
$$x_1 \leftarrow b,$$
$$x_2 \leftarrow x_0 \mod x_1,$$
$$x_3 \leftarrow x_1 \mod x_2,$$
$$x_4 \leftarrow x_2 \mod x_3,$$

$$\ldots$$

until $x_n = 0$ for the first time. At that point, $x_{n-1} = \gcd(a, b)$.

Experiment 8.3 *Before writing any code, write a short essay on what you think will make a good fitness function for the Minimal Description problem. Using your fitness function, rewrite the code from Experiment 8.2 to evolve parse trees with no more than $2k + 1$ nodes (apply the chop operation to maintain this). Test your code on finding Minimal Description trees for $k = 8, 10, 13,$ and 18 with the same number of runs and reporting techniques as in Experiment 8.2. A table of sizes of answers for the Minimal Description problem appears in Figure 8.8. Once you have done this testing, revise your fitness function and try again. If you find a killer fitness function, try it out on other numbers.*

   Things to think about while doing this experiment include the following: What, if any, is the connection between the Efficient Node Use problem and the Minimal Description problem? Is the factorization of a number an important thing to know when you are trying to improve performance of your algorithm on that number?

If this experiment is being done by a whole class, you should work in teams: share data and ideas, and comment on one another's results in your write-ups.

In this section, we have developed the basic framework of PORS and presented the two main problems. In the next section, we will explore seeding the population and test its effect on performance of evolutionary algorithms for both problems.

Problems

Problem 303. For each of the following parse trees in LISP-like notation, compute $\epsilon(T)$.

(a) **(Sto (+ 1 (+ 1 (Sto 1))))**

(b) **(Sto (+ Rcl (Sto (Sto (+ Rcl (Sto 1))))))**

(c) **(+ (+ (Sto (Sto 1)) 1) (+ Rcl (Sto (+ Rcl Rcl))))**

(d) **(+ (+ (Sto (Sto (Sto Rcl))) (Sto 1)) (+ 1 (+ 1 1)))**

(e) **(Sto (Sto (+ (+ 1 Rcl) 1)))**

(f) **(+ (Sto Rcl) (+ Rcl (Sto Rcl)))**

(g) **(Sto (Sto (+ (Sto (+ (Sto 1) (+ 1 Rcl))) (+ Rcl Rcl))))**

(h) **(+ (+ (Sto 1) (Sto (+ Rcl 1))) (Sto (+ Rcl Rcl)))**

(i) **(+ (Sto (Sto (+ (+ (Sto 1) 1) (Sto 1)))) (Sto Rcl))**

(j) **(+ (+ (Sto (+ (Sto 1) 1)) (Sto (+ 1 Rcl))) (Sto Rcl))**

(k) **(+ (Sto 1) (+ (Sto (+ 1 (+ 1 Rcl))) (+ Rcl Rcl)))**

(l) **(+ (+ (Sto (+ 1 1)) (+ (Sto (+ 1 Rcl)) Rcl)) Rcl)**

Problem 304. Prove that the Minimal Description of 2^n requires no more than $3n$ nodes.

Problem 305. Prove that the Minimal Description of 3^n requires no more than $5n$ nodes.

Problem 306. Let $f(n) = \max\{\epsilon(T) : T$ is a PORS parse tree with n nodes$\}$. Prove that $f(n)$ is strictly increasing save that $f(6) = f(7)$, and $f(1) = f(2)$.

Problem 307. Prove that for $n \geq 6$, all solutions to the Efficient Node Use problem on n nodes contain **Sto** instructions, but for odd n less than 6, solutions to the Efficient Node Use problem do not contain **Sto** instructions. Explain why $n = 2, 4$ are pathological cases.

Problem 308. Essay. Consider the following sets of integers: prime numbers, factorials, powers of two, and all integers. Rank them from most to least difficult, on average, for the Minimal Description problem, and explain your ranking. Assume that the comparison is made between the sets by comparing members of similar magnitude.

Problem 309. We say that a PORS parse tree *describes* a number if it evaluates to that number. Prove that the description of a number k by a PORS parse tree requires at most $2k + 1$ nodes.

In the next several problems we will let $g(k)$ be the number of nodes in a Minimal Description tree for the number k.

Problem 310. Show that $g(k + 1) \leq g(k) + 2$ and give one example of a k where the bound is tight.

Problem 311. Show that $g(2k) \leq g(k) + 3$ and give one example where this bound is tight.

Problem 312. Show that $g(k)$ admits a logarithmic upper bound. This can be done constructively.

Problem 313. Essay. Explain the connections between the Efficient Node Use problem and the Minimal Description problem. What information does one give you about the other?

Problem 314. A *subroutine* in the PORS language is a connected subset of a tree with one link coming in and one going out. Give subroutines that multiply their input by 2, 3, and 5. Advanced students should give such subroutines with a minimal number of nodes.

Problem 315. The chart of answers given in Figure 8.6 strongly suggests a closed form for the maximum number that can be generated by an n-node parse tree. Find that closed form.

Problem 316. Short Essay. Is the evolutionary algorithm in Experiment 8.2 doing an optimization?

8.3 Seeding Populations

In this section, we want to explore the effects of a number of types of population seeding. Population seeding consists in placing creatures in the initial population that you think will help the population to do what you want. The ultimate in population seeding is to place the answer to your problem in the original population. Often this is done in an algorithm using niche specialization in order to find variations on the solution or new types of solutions. We will explore two less-extreme types of seeding. In the first, we will use evolved creatures from one instance of the Efficient Node Use problem as the starting population for another. In the second, we will try to apply common sense to the generation of the initial population.

Experiment 8.4 *Using the software from Experiment 8.2, evolve 10 populations of parse trees that contain solutions for the Efficient Node Use problem for n =8, 9, and 10 respectively. For each of these 10 populations, use that population 10 times (with different random number seeds) as the initial population, as in Experiment 8.2, but with n =25, 26, and 27. First of all, did seeding help? Did the populations find solutions faster than you would have expected based on your experience so far? Second, in addition to the fraction of successes per time graph, make a 3 × 3 matrix showing which populations were helped the most by which seed populations, as measured by fraction of successes after 500 generations.*

At this point, we will try a different population seeding technique. This technique rests on the assumption that a very simple statistical model of PORS parse trees in a successful population of trees with n nodes contains useful information about the trees that generalizes to other values of n.

Let us review how PORS parse trees are generated. A tree with one node is a **1** or an **Rcl** with equal probability. A tree with two nodes is a **Sto** with a one-node tree as its argument. A tree on three or more nodes is, with equal probability, a **+** with the remaining nodes divided into nonzero parts between its arguments or a **Sto** with a tree on the remaining nodes as its argument.

The probability of finding any given parse tree is quite likely different in an evolved population containing optimal trees as compared to in an initial random population. The basic plan is to use a statistical model to generate an initial population with statistics close to those in a successful population and hope this speeds up evolution. In order to gain any advantage, we must apply statistical models derived from populations of trees with a smaller number of nodes to generating initial populations of trees with a larger number of nodes. Our statistical model will look at the probability, for each operation (**+** or **Sto**), of each type of argument (**+**, **1**, **Rcl**, or **Sto**) and bias tree generation with those probabilities. We will divide the experiment into two parts.

Experiment 8.5 *Create or obtain software for a new routine to add to your library of PORS parse tree routines: given a parse tree, compute the number* *of times the left and right arguments of a + and the argument of a* **Sto** *are, respectively, a +,* **Sto**, **Rcl**, *or 1. Using the software from Experiment 8.2, evolve 10 populations of parse trees that contain solutions to the Efficient Node Use problem on n = 12 nodes. Apply your new routine to compute, for the entire population, the fractions of left arguments of +, right arguments of +, and arguments of* **Sto** *that are, respectively, +,* **Sto**, **Rcl**, *or 1. Generate 1000 random trees in the usual manner and compute these same fractions. Are they different? Try to explain why.*

Think about what the results of Experiment 8.5 suggest about useful restrictions on the generation of initial populations of PORS parse trees. Hold that thought until after we have done the second half of the statistical modeling experiment.

Experiment 8.6 *Modify the random parse tree generation routine used to generate initial populations (but not the one used during mutation) so that nodes appear as arguments with probability equal to the corresponding fractions generated in Experiment 8.5.*

This will require a little work. For example, you will need 4 probabilities for the left argument of a +: P_+, P_{Sto}, P_1, and P_{Rcl}. But a one-node tree that is to be the left argument of a + is either a **1** *or an* **Rcl**; P_+ *and* P_{Sto} *are 0 for technical reasons when you are generating a one-node tree. That means, in this situation, that you use the dependent probabilities $P_1^* = \frac{P_1}{P_1 + P_{Rcl}}$ and $P_{Rcl}^* = \frac{P_{Rcl}}{P_1 + P_{Rcl}}$ instead of P_1 and P_{Rcl}. These are the probabilities of* **1** *and* **Rcl** *given that we know that we are choosing only a* **1** *or an* **Rcl**.

The general principle for generating parse trees with a statistical model is as follows. When all the probabilities we have are for events that are technically allowed, we use them in unmodified form. Otherwise, we use the probabilities of things that are currently allowed divided by the sum of things currently allowed. This is exactly the notion of dependent probability that appears in statistics books.

Using your new initial tree generator, rerun Experiment 8.2 for n = 15, 16, and 17. For each of these n, what effect does the new generation method have on speed of solution, fraction of runs with a solution in the initial population, and fraction of runs that fail to converge? Explain your results.

Look again at your results for Experiment 8.5. One clear piece of good sense implicit in the statistics is that a **Sto** should not have another **Sto** as its argument. This is a cumbersome way to get this piece of wisdom, but there it is. (Perhaps you can find other bits of wisdom somehow encoded in the probabilities.) In the next experiment, we will begin using common sense to generate better initial populations by simply requiring that **Sto** operations have as their arguments things other than **Sto**.

Definition 8.1 *Let $\mathcal{T}_s$ be the set of PORS parse trees in which no* **Sto** *operation has a* **Sto** *as an argument.*

Experiment 8.7 *Add to your library of PORS parse tree routines a routine that can generate random trees in* $\mathcal{T}_s$*. Using this routine for generating the initial population (but not for mutation), redo Experiment 8.2 and compare the results with the results obtained in Experiment 8.2.*

If we comb through the populations evolved for the Efficient Node Use problem, another property of optimal trees emerges. When a parse tree is evaluated, the instructions are executed in some order. In particular, if a tree has **Sto** instructions, then it has a *first* **Sto** instruction. This execution order also orders the terminals of a tree, and so we may separate the terminals into those that execute before the first **Sto** and those that execute after the first **Sto**. It is a property of optimal trees that all terminals executed before the first store are **1**'s and all those executed afterwards are **Rcl**'s (see Problem 324).

Definition 8.2 *Let* $\mathcal{T}^*$ *be the set of PORS parse trees with only* **1***'s before the first* **Sto** *and only* **Rcl***'s after. The terms "before" and "after" are relative to the order in which nodes are executed.*

The two conditions that are satisfied by trees in $\mathcal{T}_s$ and $\mathcal{T}^*$ don't interfere with one another, so we will also give a name to their intersection.

Definition 8.3 *Let* $\mathcal{T}_s^* = \mathcal{T}_s \bigcap \mathcal{T}^*$.

This gives us four classes of PORS parse trees: the class of all PORS parse trees and the three classes using common sense to restrict structure. We have already done a version of Experiment 8.2 for all trees and those in $\mathcal{T}_s$. The next experiment will complete the sweep.

Experiment 8.8 *Write or obtain new random tree generation routines to generate initial populations in* $\mathcal{T}^*$ *and* $\mathcal{T}_s^*$*. Run Experiment 8.7 again, doing the full experiment for both methods of generating the initial population. Compare the results with the results of Experiments 8.2 and 8.7.*

There is another possible improvement in the initial population. Notice that the argument of a **Sto** in an optimal tree is never an **Rcl** or a **1**. If we wanted to explore this class of trees, we could have another commonsense class of trees called $\mathcal{T}_r$ and also $\mathcal{T}_r^*$, $\mathcal{T}_{rs}$, and $\mathcal{T}_{rs}^*$. Exploration of the various r-types of trees would require quite a bit of fiddling and wouldn't add much to the development of GP techniques. We will leave exploration of the r classes of trees to those intrigued by the mathematical theory of PORS, and instead dredge out several old friends from earlier chapters and test them in the PORS environment.

Problems

Problem 317. Take the parse trees shown in Figure 8.7, copy them, and number the nodes in the order they are executed.

Problem 318. Essay. Does the answer to Problem 315 suggest an explanation for the 3×3 matrix in Experiment 8.4? Explain your answer.

Problem 319. How many PORS parse trees are there with n nodes?

Problem 320. How many PORS parse trees with n nodes are there in $\mathcal{T}_s$?

Problem 321. How many PORS parse trees are there with n nodes in $\mathcal{T}^*$?

Problem 322. How many PORS parse trees are there with n nodes in $\mathcal{T}_s^*$?

Problem 323. Of the three special classes of parse trees, $\mathcal{T}_s, \mathcal{T}^*$, and $\mathcal{T}_s^*$, which can be generated by simply carefully choosing the probabilities of a statistical model like the one used in Experiment 8.6?

Problem 324. Prove that any optimal tree (a solution to the Efficient Node Use problem) is in $\mathcal{T}_s^*$.

Problem 325. Prove that if an optimal tree with n nodes contains a **Sto** instruction, then there exists an optimal tree with n nodes for which the left argument of the root node is a tree whose root is a **Sto**.

Problem 326. Essay. In Experiments 8.5 and 8.6, we split apart the problem of finding a statistical model for use in generating an initial population and running an algorithm that used that statistical model to generate its initial population. Describe in detail an evolutionary algorithm that would build a statistical model for one n and then start over with a larger n, all in one environment. Conceivably this process could be done in a loop to solve both problems for successive n and refine the statistical model as n grew. In terms of the mathematics of PORS for the Efficient Node Use problem, explain which small n might give good statistics for larger n.

Problem 327. Classify and count solutions to the Efficient Node Use problem for $n \leq 27$. Figure 8.6 will help you to identify such solutions.

Problem 328. Essay. For the Efficient Node Use problem, $n = 15$ yields a fitness function with an odious local optimum. What is it and why?

8.4 Applying Advanced Techniques to PORS

In this section, we will apply various techniques from earlier chapters to the PORS environment and produce a few new ones. It is important to keep in mind that except to certain very odd people like theoretical computer scientists and mathematicians, the various PORS problems are not themselves intrinsically interesting. PORS is intended as a very simple test environment for genetic programming, akin to the string evolver as it was used in Chapter 2.

The form of solutions to the Efficient Node Use problem suggests that optimal trees must have relatively high depth. This gives us a natural place to test an idea from Chapter 5: lexical products of fitness functions.

Experiment 8.9 *Modify the software from Experiment 8.2 so that the fitness function is $\epsilon(t)$ lex $depth(T)$ with ϵ dominant. Do the same data runs and reports. Does using the lexical product fitness function improve performance?*

A technique for improving optimization that we treated in Chapter 3 is niche specialization. For PORS parse trees, range-niche specialization is a more natural choice than domain-niche specialization (see Problem 333). The fitness function of the Efficient Node Use problem produces natural numbers. These natural numbers are spaced out rather strangely, as we saw in Experiment 8.3. This indicates that we should say that two PORS parse trees are similar enough to decrease one another's fitness if they produce exactly the same value. In the terms used in Chapter 3, the similarity radius is zero. To do range-niche specialization, we also need a penalty function. In the next experiment, we will compare a couple of penalty functions as well as assess whether range-niche specialization helps at all.

Experiment 8.10 *If you need to, review Section 3.3 until you remember how range-niche specialization works. Now modify the software from Experiment 8.2 to operate with range-niche specialization for the following penalty functions:*

(i) $q(m) = (m + 3)/4$,

(ii) $q(m) = \begin{cases} 1 & \text{if } m \leq 4, \\ m/4 & \text{otherwise.} \end{cases}$

Do the same runs and make the same report as in Experiment 8.2. Compare the results. Which, if either, penalty function helps more? Do they help at all? Is it worth doing range-niche specialization for the Efficient Node Use problem?

It may be that niche specialization is more helpful for the Minimal Description problem than for the Efficient Node Use problem. The next experiment tests this conjecture.

Experiment 8.11 *Modify the software from Experiment 8.3 to operate with range-niche specialization for the following penalty functions:*

(i) $q(m) = (m + 3)/4$,

(ii) $q(m) = \begin{cases} 1 & \text{if } m \leq 4, \\ m/4 & \text{otherwise,} \end{cases}$

(iii) $q(m) = \sqrt{m}$.

Do the same runs and make the same report as in Experiment 8.3. Compare the results. Which penalty function, if any, helps more? Do they help at all? Is it worth doing range-niche specialization for the Minimal Description problem?

So far each evolutionary algorithm we have run on a PORS problem has used tournament selection with tournament size 4. It might be interesting to explore some other models of evolution. Of the two PORS problems, the Efficient Node Use problem and the Minimal Description problem, we have a much better handle on the Efficient Node Use problem. It should be clear from earlier experiments that there is a large local optimum in the Efficient Node Use fitness function when n is a multiple of 3. Keeping this in mind, do the following experiment.

Experiment 8.12 *Modify the software from Experiment 8.2 so that it can use other models of evolution. For all 8 possible combinations of*

> *Roulette selection*
> *Rank selection*

with

> *Random replacement*
> *Random elite replacement*
> *Absolute fitness replacement*
> *Local elite replacement*

do the same set of runs and report the same data as in Experiment 8.2. Compare the outcome of this experiment with that of Experiment 8.2. Which model of evolution works best for which values of n?

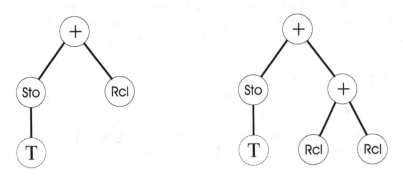

Fig. 8.9. Subroutines for multiplying by 2 and 3.

In Problem 314, the notion of subroutines is casually brought up. Recall that a *subroutine* in the PORS language is a connected subset of a tree with one link coming in and one going out. Figure 8.9 shows you subroutines for multiplying the output of any tree T by 2 or 3. (Alert readers will note that this answers the two easier parts of Problem 314; think of this as a reward for students who read, or at least flip through, the text.)

So far our treatment of the Minimal Description problem has been vague and left all the work to you. The subroutines in Figure 8.9 suggest a way of getting very small trees for numbers of the form $2^x 3^y$. In general, minimal subroutines for multiplying by m may be very good things to have in a population of parse trees being evolved to find a Minimal Description of a multiple of m. The next two experiments will put the Minimal Description problem on a firmer footing. The first, somewhat bizarrely, should be called a PRS experiment.

Experiment 8.13 *This experiment will require a revision of the standard PORS routines. In normal PORS, the memory is initialized with a 0, and the set of terminals is {1,Rcl}. Create a new set of parse tree manipulation routines from the PORS routines in which the external memory is initialized to 1, not 0, and all terminals are* **Rcl***. Rebuild the software from Experiment 8.3 to use the new routines. Use the software to find Minimal Description trees in the PRS language for $k = 2, 3, 5, 7,$ and 11. If you have not already done so, do Problem 330. Report different variations of the Minimal Description tree for each k.*

Once we have these PRS parse trees that form minimal descriptions of various numbers (and have completed Problem 330), we have a source of subroutines that may help with the Minimal Description problem (or cause us to dive into a local optimum, see Problem 332). In any case, the subroutines evolved in Experiment 8.13 may give us an effective way of seeding the population for the standard PORS Minimal Description problem.

Experiment 8.14 *Add to your suite of PORS parse tree routines a routine that takes as arguments a maximum number of nodes and a tree of the type located in Experiment 8.13 for computing k. It should convert the tree into a subroutine for multiplying by k and then concatenate copies of the subroutine together with a one-node tree consisting of the terminal* **1** *so as to obtain a tree that computes k^m in the PORS environment, with m as large as possible. The bound on m is implied by having only n nodes available.*

Rebuild your software from Experiment 8.3 to operate with a seeded population. For each prime factor p of k, incorporate trees of the form p^m (m chosen to yield the largest power of p dividing k) as described above into the population (in close to equal numbers for all prime factors of k).

With the new software do the same type of runs and report the same things as in Experiment 8.3, but for $k = 8, 10, 18, 21, 27, 35,$ and 63. Does the population seeding help at all? You may want to try revising your fitness function.

Problems

Problem 329. Essay. Explain, to the best of your ability, *why* solutions to the Efficient Node Use problem have tree depth much higher than random trees. Would you expect solutions to the Minimal Description problem to share this property?

Problem 330. Read the description of Experiment 8.13. Prove that a Minimal Description tree for k evolved in the PRS environment has the same size as a minimal subroutine in PORS for multiplying by k by constructing one from the other.

Problem 331. Compare and contrast the fitness function used in Experiment 8.2 with that used in Experiment 3.18. In what ways are they similar? Would Alife techniques that enhanced one tend to enhance the other? When or when not?

Problem 332. Prove or disprove: a minimal description tree for k and a minimum-size subroutine for multiplying by k always have the same number of nodes.

Problem 333. Essay. In this section we explored the use of range-niche specialization in the PORS environment. This is not difficult because the range of PORS parse trees is the natural numbers where there is a very natural measure of similarity to use. In order to do domain niche specialization we would need a way of telling whether two PORS parse trees are close together. One possibility is simply to test and see whether they are identical. Give another, with details, and compute the similarities of several example trees.

Problem 334. Essay. Clearly, if we find a tree on n nodes that evaluates to k, then a Minimal Description tree for k has at most n nodes. Defend or attack the following proposition logically: While evolving a population to solve the Minimal Description problem for a number k, we should apply the chop operation to any tree with more nodes than the smallest tree we have found so far that evaluates to k.

9

Fitting to Data

In this chapter we will look at techniques for fitting curves and models to data. We will study three techniques: classical least squares of the sort that appears in Appendix C, least squares with an evolutionary algorithm, and least squares using genetic programming. These three methods, in the order given, go from fast, reliable, and highly restricted to slow, semireliable, and completely unrestricted. Classical least squares restricts the form of the function, e.g., line, parabola, exponential function, etc. Genetic programming has the advantage that you need not select the type of curve you are fitting before you start; it locates the curve and fits its parameters simultaneously. Genetic programming relaxes this restriction at the cost of making the search process for the model enormously harder. Hybrid systems, beyond the scope of this text, stop the genetic programming system occasionally to use standard statistical methods to fit the numerical parameters in the parse trees. These systems can exhibit performance competitive with the best data modeling techniques. The material in this chapter on genetic programming will build on the material in Chapter 8, so a quick review is suggested before beginning this chapter. The dependencies of the experiments in this chapter are given in Figure 9.1.

9.1 Classical Least Squares Fit

In Appendix C, the equations for least squares fit to a line are given in Equations C.1 and C.2. The idea behind least squares fit is simple. Take a model of the data, e.g., "I think that these data come from a linear curve of the form $y = ax + b$," and then use a minimization technique to pick a and b to make the error between the model and data as small as possible. The unknown values a and b are the parameters of the model.

From our experience in Chapters 2 and 3 we know that quadratic curves give unimodal (e.g., easy) optimization problems. Looking at the derivation of the least squares fit to a line in Section C.4, we see that the problem of

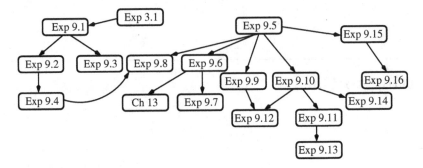

1 Fitting a line to data.
2 Exploring mutation operators.
3 Evolving the model along with the fit.
4 Fitting to a hyperbolic tangent curve.
5 Evolving parse trees to fit data.
6 Exploring mutation operators.
7 Varying sample points.
8 Introducing the if-then-else operation.
9 Introducing automatically defined functions.
10 Generalizing to two data dimensions.
11 Varying the sample points in two dimensions.
12 ADFs in two dimensions.
13 Inducing the formula for the area of an ellipse.
14 Generalizing across dimensions.
15 Exploring parse tree size.
16 Using a designated intron.

Fig. 9.1. The topics and dependencies of the experiments in this chapter.

minimizing the squared error between a data set and a line $y = ax + b$ is exactly a quadratic minimization problem. Because it gives a method of estimating the parameters of our model that is a simple, quadratic minimization, we will use minimizing the square of the differences between our model and the data as our method of finding model parameters (and of finding the whole model when we use genetic programming). Here is an example of a least squares fit of a model to three-dimensional data.

Example 26. Suppose we have 6 data points that we believe, except possibly for modest measurement error, lie in a plane in $\mathbb{R}^3$. The points are given in the table below. Our model will be the general plane

$$z = ax + by + c$$

with parameters a, b, and c. To compute the sum of squared errors (SSE) we take the sum of the z-values minus the model's predicted z-values, squared.

i	x_i	y_i	z_i
1	0	0	1
2	0	1	4
3	1	0	3
4	1	1	6
5	1	2	9
6	2	1	8

So, to compute a, b, and c we want to find the values that minimize

$$\text{SSE}(a, b, c) = \sum_{i=1}^{6} (z_i - a \cdot x_i - b \cdot y_i - c)^2. \tag{9.1}$$

Evaluating the sum, we get

$$\text{SSE}(a, b, c) = 7a^2 + 7b^2 + 6c^2 + 10ab + 10ac + 10bc - 68a - 72b - 62c + 207.$$

Setting the partial derivative with respect to a, b, and c equal to zero, we obtain

$$14a + 10b + 10c = 68,$$
$$10a + 14b + 10c = 72,$$
$$10a + 10b + 12c = 62,$$

which has a unique solution:

$$a = 2,$$
$$b = 3,$$
$$c = 1.$$

And we see that all the data points exactly fit:

$$z = 2x + 3y + 1.$$

In classical least squares fit, the first step is to choose the type of curve to fit to the data. The general term for this curve type is the *model*. The unknown coefficients in the model are called the *parameters* of the model. In Example 26, the model is a plane in 3-space, $f(x, y) = ax + by + c$, with parameters a, b, and c. (The data given in Example 26 were clearly taken from this plane. You can tell by plugging the data into Equation 9.1 and computing the squared error for $a = 2$, $b = 3$, $c = 1$; it is exactly zero.)

The problem we will be ultimately concerned with is finding the model to fit to the data.

Definition 9.1 *The* **sum of squared error,** *or* **SSE,** *of a model with a data set is the sum over the points in the data set of the square of the value of the dependent variable subtracted from the value obtained by plugging the independent variables into the model.*

Examine the data given in Table 9.1. What function best models this data?

x	0.25	0.5	0.75	1	1.25	1.5	1.75	2	2.25	2.5	2.75	3	3.25
y	0.516	0.125	-0.0781	0	0.453	1.38	2.86	5	7.89	11.6	16.3	22	28.8

Table 9.1. A data set from an unknown model.

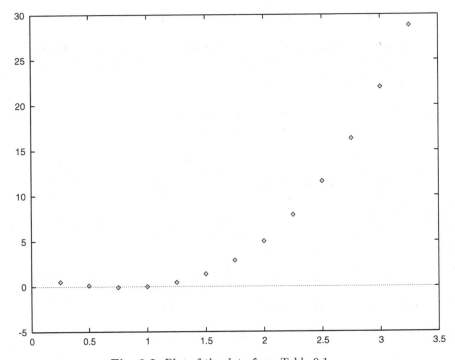

Fig. 9.2. Plot of the data from Table 9.1.

To help you answer the question, you could plot the data. The result of doing this is shown in Figure 9.2. The plot shows that the data are quite unlikely to lie on a line, but gives little other information. The data might lie on a parabola with a minimum near $x = 0.75$, but it is hard to tell by direct observation of a few data points. The Stone–Weierstrass theorem says that any continuous function can be approximated arbitrarily well with polynomials.

The data plotted do look like they come from a continuous function. This means that we need only select the degree of the polynomial to fit. One way to do this is to fit higher- and higher-degree polynomials until the squared error stops decreasing. If we try this for the data given in Table 9.1, then we obtain the results given in Table 9.2.

Model	Least Square Fit	SSE
$y = ax + b$	$y = 8.75x - 7.86$	200.6
$y = ax^2 + bx + c$	$y = 5.25x^2 - 9.62x + 3.62$	5.03
$y = ac^3 + bx^2 + cx + d$	$y = x^3 - 2x + 1$	3×10^{-6}
$y = ax^4 + bx^3 + cx^2 + dx + e$	$y = 0.001x^4 + 0.993x^3 +$ $0.014x^2 - 2.01x + 1$	6×10^{-6}

Table 9.2. Results of fitting successively higher-degree polynomials to the data given in Table 9.1.

Notice that the SSE drops radically from a degree-1 model to a degree-2 model, and again from degree 2 to degree 3. The sum of squared error actually goes up a bit (probably not significantly) when we move to a degree-4 model of the data. The least-squares-curve-fitting software used also found whole number values for the coefficients of the cubic fit as well; this is suggestive. Examining the quartic fit, we see that the coefficients that were zero in the cubic fit remain quite small. (In fact, the data set was generated by plugging into $f(x) = x^3 - 2x + 1$, as can be seen in Figure 9.3.) Notice that the data are plotted as glyphs, while the model is plotted as a continuous line. This is standard practice for displaying data together with a model of that data. In situations other than toy examples, the model will not fit the data exactly, and the minimum SSE will not be so close to zero.

It turns out that fitting polynomial models to bivariate data amounts to inverting matrices (see Problem 338). Least squares fit to nonpolynomial models can be quite daunting, involving very annoying algebraic manipulation. One special class of models, trigonometric series, can be fit using *Fourier series* techniques. Data resulting from time series are an excellent candidate for Fourier analysis, and courses in time series and many courses in mathematical analysis treat Fourier series in depth. A time series is simply a set of data taken across a period of time with the measurement times saved as the independent variable on which the data depend. For example, the time of sunrise each day forms a time series.

Another attack on fitting of data relies on first transforming the data so that a simple model fits, fitting the model, and then inverting the transformation. Example 27 shows a standard example of this kind of transformation.

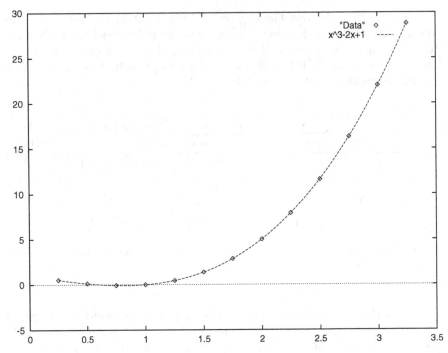

Fig. 9.3. Plot of the data from Table 9.1 together with the best fitting polynomial curve.

Example 27. Assume that we have data drawn from a system undergoing exponential growth:

$$P(t) = A_0 e^{rt} + \epsilon(t), \tag{9.2}$$

with $P(t)$ being the population at time t, A_0 being the population at time $t = 0$, r being the growth-rate parameter, and $\epsilon(t)$ being the random error away from exponential growth. If we assume that $\epsilon(t)$ is small and take the natural log, the model of the data becomes:

$$\ln(P(t)) = rt + \ln(A_0), \tag{9.3}$$

which is a linear function of two variables. Simply taking the natural log of the independent variable permits us to fit a line to the data. The slope r of that line is the growth-rate parameter, while its intercept is the natural log of the initial population.

Problems

Problem 335. In Example 26, we saw that to fit a plane to points in 3-space we needed 3 parameters: the unknown coefficients of x and y, and an unknown constant. A *term* in a multivariate polynomial has degree n if the sum of the exponents of its member variables are n. Thus the third-degree terms in a general 2-variable polynomial, minus their coefficients, would be x^3, $x^2 y$, xy^2, and y^3. In fitting a multivariate polynomial model, each term of each degree requires its own coefficient. Compute the number of unknown coefficients needed to do a least squares fit to a data set of an nth-degree polynomial model with k variables. (This is the number of parameters needed to do the fit.)

Problem 336. Suppose that you wished to fit the model $y = a$, a constant function, to a data set $\{(x_i, y_i) : i = 1, \ldots, n\}$. What type of value would you obtain, relative to the data, for the parameter a? Hint: since we didn't tell you the values of the data, we're not looking for a numerical answer; give a descriptive name.

Problem 337. Suppose you wished to fit a parabola $y = ax^2 + bx + c$ to the data given below. Compute the SSE in terms of a, b, c and, using multivariate calculus, compute the values of a, b, and c that minimize the SSE.

x	1	2	3	4	5	6	7	8
y	2	3	10	15	26	35	49	63

Problem 338. This problem requires familiarity with simple linear algebra. Show, by algebraic derivation, that fitting a polynomial of the form $f(x) = a_0 + a_1 x + \cdots + a_n x^n$ to a data set can be phrased as the solution of a matrix equation

$$A\mathbf{x} = \mathbf{b}.$$

Give the entries of the matrix A and the vector $\mathbf{b}$ as summations over (functions of) the coordinates of the data points.

Problem 339. True or False: The total amount of deviation of the values of the model from the data points is all that matters in minimizing SSE. The distribution among the data points of the deviations from the model is unimportant.

Problem 340. For the population data given below, use the techniques outlined in Example 27 to find the growth rate for the population.

Time:	0	1	2	3	4	5	6	7	8	9	10	11	12
Population:	19	22	26	33	39	47	56	65	80	94	112	140	164

You also obtain an estimate for the initial population, for which you have an observed value. Is there any reason to prefer the observed or estimated value? Explain.

Problem 341. Problem 338 implies that fitting a univariate polynomial requires that we invert a matrix; in essence, the solving of linear equations suffices. Can we do a least squares fit of the model

$$y = \sin(ax + b)$$

to bivariate data without solving nonlinear equations? Justify your answer, assuming an arbitrary number of data points.

Problem 342. The data below were taken from a polynomial of degree no more than 5 with integer coefficients. Either find or write a program that can do a least squares fit of polynomial models to data. Construct a table like Table 9.2 and give the polynomial. Note that the values were rounded off to make the table.

x	0.2	0.4	0.6	0.8	1	1.2	1.4	1.6	1.8	2
y	-2.84	-2.39	-1.69	-0.85	0	0.686	0.998	0.686	-0.538	-3

Problem 343. Essay. Explain why fitting a single continuous curve to an entire data set may not be a good idea. Give at least one example in which splitting the data set into two parts yields a better result than treating the data set as a unified whole. What sorts of processes in the natural world might produce such a data set?

9.2 Simple Evolutionary Fit

In this section, we will adapt the real function optimizer from Chapter 3 to do a least squares fit to various models. We will introduce some evolutionary algorithms (which may be of some interest in their own right) to serve as sources of data for a least squares fit. We will start with a simple example of the art: fitting a line to data.

Experiment 9.1 *Write or obtain software for an evolutionary algorithm operating on genes of the form (a, b), where a and b are real numbers. Initialize a and b uniformly at random in the range $-10 \leq a, b \leq 10$. Let the fitness function be the SSE between a data set of points $\{(x_i, y_i) : i = 1, \ldots, n\}$ and the model $y = ax + b$. This function is to be minimized, of course.*

The evolutionary algorithm should operate on a population of 100 genes using single tournament selection with tournament size 4. Use single-point crossover and single-point real mutation with mutation size $\epsilon = 0.2$. Call a population successful when its SSE drops to within 0.1% of the true minimal SSE. (You must compute the true minimal SSE using calculus; it may be handy to build the ability to do this into your software. See Equations C.1 and C.2 for a formula for the correct values of a and b.)

For each of the data sets below, run the algorithm 50 times, saving the number of generations required for success. Plot the fraction of populations that have succeeded as a function of the number of generations. Report which data sets are the most difficult to fit, which is a separate question from the quality of fit.

Data set 1:

x	0	1	2	3	4	5	6	7	8	9
y	2	2	6	6	10	10	14	14	18	18

Data set 2:

x	0	1	2	3	4	5	6	7	8	9
y	1	2	5	10	17	26	37	50	65	82

Data set 3:

x	0	1	2	3	4	5	6	7	8	9
y	4	0	4	0	4	0	4	0	4	0

Experiment 9.1 is intended mostly to get you started with curve fitting. The data sets are, respectively, a line with periodic errors, a quadratic curve, and a simple periodic function. None of them really fit well with a linear model, although the first data set comes close. An interesting question is whether being less like a line makes a data set easier or harder to fit with a linear model.

The next experiment is intended to make a point about mutation. When doing a least squares fit to data, the sensitivity to changes in one or another dimension in the space of parameters may not be uniform. Likewise, as we approach minimum error, smaller mutations are better at helping the algorithm converge.

Experiment 9.2 *Modify the software from Experiment 9.1 to use different mutation operators as specified below. For each of the mutation operators given, save and plot the time to success on data set 1 from Experiment 9.1 and compare their plots with one another and with the plots of the data for the first data set in Experiment 9.1.*

- *One-point real mutation with mutation size $\epsilon = 0.2$, 50% of the time, and $\epsilon = 0.02$, 50% of the time.*
- *One-point real mutation with mutation size $\epsilon = 0.02$.*
- *One-point real mutation with mutation size $\epsilon = \frac{0.2}{\sqrt{n}}$, where n is the current generation number, starting at 1.*
- *Two-point real mutation with mutation size $\epsilon = 0.02$.*

Which mutation operator gave the best performance? In your write-up, speculate as to the reasons for the differential performance.

One of the things that would be nice for curve-fitting software to do is to automatically figure out the model for us. With the simple evolutionary fit, this requires something like a lexical product of fitness (defined in Section 5.1) in which the degree of the polynomial is used as a tie-breaker when the SSEs of two polynomials are essentially the same. Using the results shown in Figure 9.2, we can get a reasonable notion of "similar" SSE. The next experiment implements an evolutionary curve fitter that tries to find which model to use.

Experiment 9.3 *Modify the software from Experiment 9.1 to operate on a different sort of gene. The genes should have 7 real numbers and an integer. Use one-point crossover with the integer being viewed as being in the gene before the first real number. The real numbers are the coefficients of a polynomial whose degree is specified by the integer. The first real is the constant term, the last is the coefficient of x^6.*

Mutation should be real single-point mutation with mutation size $\epsilon = 0.2$ 7 times in 8, and 1 time in 8 it should modify the integer degree by ± 1, with a lower bound of 1 and an upper bound of 6. Correct impossible mutations by not performing them. The coefficients above the degree given by the integer are unused, but are available for crossover.

As before, we wish to minimize the SSE of the polynomial specified by the gene. When two genes have SSE fitness that differs by less than 0.1 take as more fit the polynomial that is of lower degree.

Test the algorithm on the data sets given in Table 9.1 and Problem 342, running the algorithm 20 times for each polynomial. Report whether the algorithm was able to find the degree correctly and whether there was a consensus degree in the best members of final populations. Repeat the experiment on the first data set with the degree forced always to be 5. The program can reset the degree by zeroing out the higher-degree coefficients: does it do so efficiently?

Now we will analyze a simple artificial life system with evolutionary curve-fitting software. This system is based on a game called the *Public Investment Game*. Suppose that you want to model public spending on something most people need, say roads. Clearly, it would be "fair" for people with similar amounts of money to pay the same amount for roads. However, a person who spends less than the others still benefits from the roads.

A simple model of this type of situation can be constructed as follows. A referee gives each person in the game $100. Each person may secretly put some number of dollars in an envelope. The contents of the envelope are doubled and the result divided among all players evenly. The initial money represents each person's available money. The money placed into the envelope represents the money spent to build the roads: public spending. The doubling of the money in the envelope represents the value of having roads.

The question we will examine is how simple strategies of play evolve under different conditions. As we will see, playing this game as a single-shot game and then selecting for players that have the most money quickly leads to no public spending.

The basic software we will use works as follows. A population of 60 investors is represented by an array of integers initialized to a value of 100. The integer is the amount to put in the envelope. The population is shuffled into random groups of 12 investors that play the Public Investment Game, once.

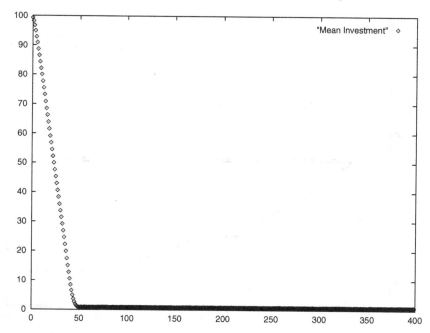

Fig. 9.4. Average investment level of individuals over 50 populations from play 1 to play 400 of the Public Investment Game.

The fitness of each investor is 100 minus the money put in the envelope plus a share of the doubled investment money. The population is then shuffled into random groups of two investors. The investor in each pair with the lower fitness adopts the strategy (integer) of the investor with the higher fitness, plus a uniformly distributed integer in the range −5 to 5. Numbers above 100 or below 0 are mapped to 100 or 0, respectively. Ties are broken uniformly at random.

This is a very simple evolutionary algorithm mimicking a population of inexperienced investors learning from one another by observation. If this algorithm is run 100 times for 400 plays of the Public Investment Game, then the average over all populations and all investors within a population behaves as shown in Figure 9.4. Problem 345 gives insight into why this happens.

An interesting phenomenon occurs if we complicate the model by adding two global features: a minimum investment level called the *law*, and a penalty for failure to invest at least the amount specified by the law called the *fine*. The law could be thought of as the mandated level of taxation and the fine as the penalty for tax evasion. This situation differs from reality in that (i) investors are allowed to overpay tax and (ii) tax cheats are 100% likely to be caught. The software is modified as follows: decrease the fitness of an investor who has an investment amount less than the law by the amount of the fine.

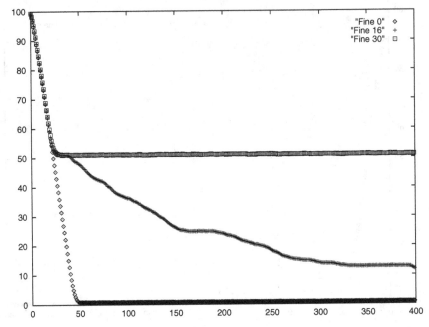

Fig. 9.5. Average investment over 50 populations of 60 investors with a law of 50 and fines of 0, 16, and 30.

We will investigate what happens with a law of 50 and a number of different fine levels. Figure 9.5 shows the average investment with a law (minimum investment) of 50 and fines of 0, 16, and 30 over 100 populations of 60 investors. Figure 9.6 shows the individual population tracks that led to the average results displayed in Figure 9.5.

In the cases of the extreme fines, the results are simple. As fast as the rate of adaption (the ±5 variation in copied strategies) allows, the population either drops to zero or to the law. With the fine of 16, however, the rapid adaption down to the level of the fine is followed by an almost evaporative series of defections to zero. The average track simply descends slowly. For some of the possible combinations of investments near the law, it is possible to be slightly ahead if you don't meet the law. If too many investors arrive at that state at the same time, then the whole population is paying the fine, and it no longer matters.

Logically, the law and fine have two sorts of impact. A small fine is simply a cost of doing business: it is less than the gain from low personal investment. Given that our investors are simply integers, with no sort of moral sense, fines that are less than the profit of bidding low are ignored. On the other hand, high fines will force compliance. A fine of 200, for example, is more than you can possibly make no matter what happens and amounts to instant failure for

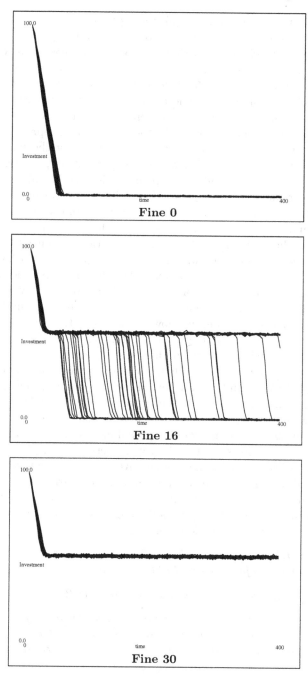

Fig. 9.6. The average investment level for 50 individual populations with the law set to 50 and fines of 0, 16, and 30.

any investor not obeying the law. This means that at some point above zero but below 200 the fines start causing strong compliance.

Examining Figure 9.6, we see that this logic is not complete. With a fine of zero, the investment level moves smoothly to zero (plus a small amount due to the variation of -5 to $+5$ when another investor's strategy is copied). With a fine of 30, a smooth decline to the law of 50 is observed; the curve is similar in character to the decay to zero for no fine. However, when the fine is 16, something odd happens. As the average bid approaches the law, the investors manage to divide between those who pay the fine and those who obey the law. If we examine individual populations, we see that they are either all fine payers or all in compliance with the law after 400 rounds of playing the game and learning from fellow investors. Once a population is made mostly of the exactly lawful (investment levels of the law but not much more) or criminals (investment levels below the law), it tends to stay in that condition. Somehow, the dynamics of the simulation permit either outcome, even though the investors start with generous lawful investment levels of 100.

Investigating more deeply, we perform the following experiment. For fines of 0 to 40, in increments of 2 dollars, we find the average investment after 400 plays of the Public Investment Game. The play and learning from fellow investors takes place as before, but now we save only the average investment level in generation 400. This quantity is plotted versus the fine in Figure 9.7. This is the data we wish to model with a least squares fit.

If the assertion that an individual population must either become uniformly law-abiding or uniformly criminal (within a mutation driven small distance of zero investment), then the expected investment level for a given fine is the law times the probability of a given population becoming lawful. In addition, Figure 9.7 indicates that this probability is zero for small fines, one for large fines, and follows a sigmoid curve in between. This is enough information to choose a model for the average investment as a function of the fine data. We need a sigmoid curve with a minimum of zero, a maximum of the law (50 in this case), and which can have its maximum slope and horizontal position modified. The shifted hyperbolic tangent curve fits our requirements admirably, see Figure 5.7 and Equation 5.4.

Experiment 9.4 *Modify the software from Experiment 9.2, using the best mutation scheme you found there, to fit to the model*

$$y = 50 \times (\tanh(a \cdot (x - b)) + 1)/2.$$

Fit this model to the data given in Figure 9.7. Run for 5000 generations with 10 populations and compare the results. Also, plot the data together with a graph of the fitted curve with the lowest error. Is there a consensus about the values of a and b? Find the average over your 10 runs of a and b, and compute the SSE for those averages. How does the resulting SSE compare with the best result from each individual population? You may want to work Problems 347– 350 concurrently with this experiment.

fine	avg	fine	avg
0	0.734667	22	41.7112
2	0.834667	24	46.1248
4	0.977167	26	50.0627
6	1.9965	28	50.5992
8	3.37967	30	50.8558
10	5.683	32	50.772
12	6.51933	34	50.8998
14	10.0245	36	50.8272
16	16.8492	38	50.8257
18	24.7958	40	50.7795
20	32.6318		

Fig. 9.7. Average investment level after 400 plays over 100 populations of 60 investors as a function of the fine amount with a law of 50.

Problems

Problem 344. In the discussion before Experiment 9.2, the claim was made that "When doing a least squares fit to data, the sensitivity to changes in one or another dimension in the space of parameters may not be uniform." Suppose we fit the data below to a quadratic curve $y = ax^2 + bx + c$. Document the variation in the SSE when varying each of the coefficients a, b, and c by ± 0.1. This will require first performing the quadratic fit, hopefully a *very* simple task.

x	-4	-3	-2	-1	0	1	2	3	4
y	25	16	9	4	1	0	1	4	9

Problem 345. Prove: in the Public Investment Game with $k > 3$ players, if one player's bid is lower than all the others, then that player will make the most money in that round.

Problem 346. Suppose we are having a population of integers learn to play the Public Investment Game as described in this section with 60 players that play in groups of 12. The law is 50, and the fine is 16. If all the players begin with an investment level of 5 or less, show that the population will not become lawful. (A lawful population is one in which a majority have investment levels above the law.)

For the next several problems you need to read Experiment 9.4 and the material leading up to it. Also examine the graph of $(\tanh(x) + 1)/2$ and its derivative, given below.

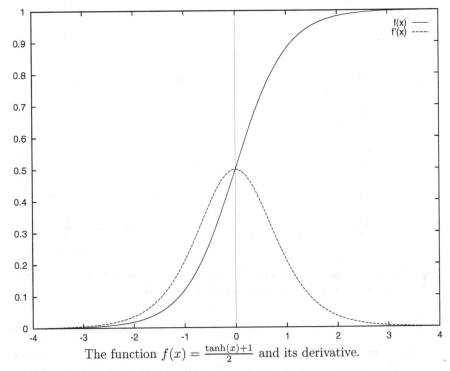

The function $f(x) = \frac{\tanh(x)+1}{2}$ and its derivative.

Problem 347. If $f(x) = 50 \cdot (\tanh(a \cdot (x - b)) + 1)/2$ is fit by the method of least squares to a data set like the one given in Figure 9.7, then the numbers a and b contain information about the data being modeled. Examine the graph above of the function $\tanh(x)$ and its derivative. What do a and b tell you about the location of the sloped part of the data set and about the slope $\frac{dy}{dx}$?

Problem 348. Examine the data in Figure 9.7. The data point for a fine of 10 seems to be high if the curve really is sigmoid. Given the experiment that provided the model, would you expect the value given by the fitted curve or the experimental value of 5.683 to be closer to the value we would obtain by averaging over 1,000,000 populations (10,000 times as many as were used to produce the data)?

Problem 349. Least squares fit will work faster if the gene is initialized close to the correct answer. Give a procedure for initializing the genes of Experiment 9.4 that requires only eyeball estimates of the data, but that will perform better than the random initialization used in the experiment in its current form.

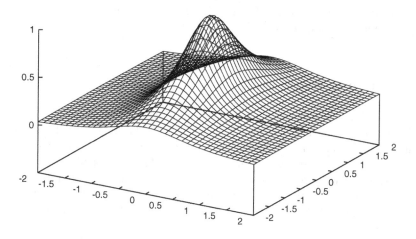

Fig. 9.8. A curve $g(x, y) = \frac{1}{1+2x^2+6y^2}$, where the variation away from the mode is not symmetric.

Problem 350. The current model of Experiment 9.4 presumes that for small values of the fine, the average investment level is 0. Since half of all investors are just mutated, and for 5 out of 11 of those this means that a number between 1 and 5 has been added to their investment, it follows that the average investment must be greater than 0. In addition, some of these mutants survive. Give a generalization of the model used in Experiment 9.4 that models the nonzero nature of the minimum average investment. Also, estimate the value of the minimum investment for a fine of zero.

Problem 351. Suppose we are using least squares to do a high-degree polynomial fit to bivariate data with an evolutionary algorithm. This problem is unimodal and, as Problem 344 shows, subject to different slopes in different directions about the optimum. Would you expect range- or domain-niche specialization to help more in this situation? An example of a unimodal curve with this sort of directional dependence of variation is shown in Figure 9.8.

Problem 352. Essay. We could estimate the current rate of variation of the SSE with each parameter by simply varying each parameter the same amount and directly computing the change in SSE. We could then set mutation size for that parameter to be some global mutation size divided by that variation. Discuss the pros and cons of implementing such as system with (i) the estimate of variation made only for the best gene in each generation and used to calculate mutation size for all genes and (ii) the estimate made individually for each gene. Include the implications of Experiment 9.2 in your discussion (if you performed it) and also remember to consider both evolutionary search advantages and computational cost. Use the curve given in Figure 9.8 as an example of the sort of nonuniform variation in different directions for which we are trying to compensate.

9.3 Symbolic Regression

In this section, we will use genetic programming (GP), a new version of the technique introduced in Chapter 8, to do fits to data. We will be evolving formulas to fit data, i.e., performing *symbolic regression*. Review the basics of genetic programming given in Section 8.1. The fitness function we will be using, as in the preceding section, is minimization of squared error. The difference is that we will not be fitting parameters of a given model, but rather selecting a model (complete with parameters) in the form of a parse tree or parse trees.

The use of parse trees lets us generate random formulas. This means that we are searching an enormously larger space than in the preceding sections. The search is no longer a simple unimodal optimization over the real numbers; it is a mixed search of a discrete space of formulas with incorporated real parameters.

In Chapter 8, the set of operations and terminals was tiny and fixed. In this chapter, we will need to worry about which terminals and operations we need for each experiment. We will also want to be able to enable and disable given terminals and operations. Because of this, you will need an entire class of parse tree routines that let you specify which terminals and operations are available. We will simply assume this capability and leave it to you (or your instructor) to write or download the necessary routines. If only a limited number of GP experiments from this chapter are performed, then much simpler parse tree code can be used.

For our first exploration, we will use the set of terminals and operations given in Table 9.3. The set of terminals and operations used for the plus-one-recall-store problem in Chapter 8 had the advantage that it required no error trapping; each function could accept without error every value that it encountered. The operations in Figure 9.3 share this property, except for division. Division by 0 is not allowed. When doing genetic programming, we modify singularities so as to prevent impossible values. This means that the formulas obtained may not be quite standard, and so a bit of care is required.

Node	Symbol	Semantics
		Terminals
real	varies	A real constant in the range -1 to 1
x	x	Holds the value of the independent variable
		Unary Operations
minus	–	Unary minus, negates its argument
square	sqr	Computes the square of its argument
sine	sin	Computes the sine of its argument
cosine	cos	Computes the cosine of its argument
arctangent	atan	Computes the arctangent of its argument
		Binary Operations
plus	+	Adds its arguments
minus	-	Computes the difference of its arguments
times	*	Multiplies its arguments
divide	/	Computes the quotient of its arguments, returning 0 for a divide by zero

Table 9.3. Terminals and operations for the first set of experiments.

Since error trapping is very system- and language-specific, the problem of detecting impossible divisions is left to the writer of a given piece of GP code. Notice that for the present, real constants are to be in the range $-1 \le c \le 1$. This is less limiting than you might think, because such constants can be used to simulate other constants with a fairly small number of operations. We are now ready to do our experiment.

Experiment 9.5 *Write or obtain software for an evolutionary algorithm to perform genetic programming that operates on a population of 400 parse trees. Initialize the population to have trees with 6 nodes and use the chop operation to contain tree size at no more than 12 nodes. Use single tournament selection with a tournament size of 4 as your model of evolution. When bred, pairs of trees should be crossed over 50% of the time and simply copied 50% of the time. Mutate new trees (crossed or copied) 50% of the time. Let the fitness function be minimization of the SSE with the set of points*

$$\mathcal{P} = \left\{ \left(\frac{i}{40} - 1, \frac{1}{(i/40 - 1)^2 + 1} \right) : i = 0, \ldots, 80 \right\}.$$

Run 30 populations for either 500 generations or until the SSE drops below 10^{-6}.

Report the following:

(i) *One best-fitness formula from each population,*

(ii) *The number of populations that did not need to evolve 500 generations. Create a fraction-of-successes graph where success is defined as finishing early.*

At this point, we pay off on a promise made in Chapter 8 and introduce some new mutation operators. These augment, rather than replace, the subtree mutation introduced in Chapter 8. There are at least three ways to mutate a parse tree without modifying the tree structure. Two of these mutations would have been rather pointless in the PORS environment.

Definition 9.2 *A* **constant mutation** *locates a constant in a parse tree and applies an appropriate mutation to it. A real-valued constant would have a number from some distribution added to it as with real point mutation. An integer or character constant could be replaced with an appropriate new value.*

Definition 9.3 *A* **terminal mutation** *locates a terminal and generates a new terminal. Constant mutations, defined above, are a special case of this. The difference is that this mutation can replace constants with variables and vice versa.*

Definition 9.4 *An* **operation mutation** *locates an operation and changes its identity while preserving its arity (a unary operation is replaced with another unary operation; a binary operation is replaced with another binary operation, etc.). To use this mutation requires that multiple operations of each arity be available.*

Experiment 9.6 *Perform Experiment 9.5 again, but this time instead of using subtree mutation alone, use it 25% of the time and also use constant mutation, terminal mutation, and operation mutation 25% of the time each. Report the same data and compare. Does the new mix of mutation operators help?*

Experiments 9.5 and 9.6 try to approximate a known function on a fixed grid of sample points. It is interesting how many *different* solutions an evolutionary algorithm can find (see Problem 353). Take the problem, for example, of approximating the fake bell curve, $f(x) = \frac{1}{x^2+1}$. What effect does using a fixed set of sample points have on the evolutionary process? It may be the case that using a different random set of sample points for each generation is a good thing. Problem 357 addresses this issue. At present, we will treat the question experimentally.

Experiment 9.7 *Modify the software from Experiment 9.6 so that the SSE used for fitness is computed, in any one generation, on the set of points*

$$\mathcal{P}_r = \left\{ \left(\frac{i}{40} - 1 + r, \frac{1}{(i/40 - 1 + r)^2 + 1} \right) : i = 0, \dots, 80 \right\},$$

where $-0.4 \le r \le 0.4$. *Remember to have a single set of points (value of* r*) for each generation, but change* r *for each new generation's fitness evaluations. Run 30 populations. Take the best creature in the final generation of each population and compute the SSE on the set of points* $\mathcal{P}$ *used in Experiment 9.6. Compare the quality of solutions obtained in each experiment and the number of times the algorithm halted early.*

A traditional operation in genetic programming is the if-then-else operation. It requires that we put a Boolean interpretation on the data type for the genetic programming. For real numbers, one reasonable interpretation is "positive or zero is true, negative is false." We will use this interpretation for the remainder of the chapter. If-then-else, symbolized ITE, is a ternary function with the following definition:

$$\mathrm{ITE}(x, y, z) = \begin{cases} y & \text{if } x \ge 0, \\ z & \text{if } x < 0. \end{cases} \tag{9.4}$$

The ITE operation permits evolution to embed decisions into formulas. If a function has split rules, then ITE permits us to code them in a transparent fashion. The next experiment is intended as a simple demonstration of this.

Experiment 9.8 *First repeat Experiment 9.5, running 30 populations, on the set of points*

$$\mathcal{S}_0 = \left\{ \left(\frac{i}{10} - 1, f\left(\frac{i}{10} - 1 \right) \right) : i = 1, \dots, 20 \right\},$$

where $f(x)$ *is 0 when* x *is negative and 1 otherwise. Now modify your GP software from Experiment 9.5 to use the ITE operation. With the modified parse trees, rerun Experiment 9.5 on the set of points* $\mathcal{S}_0$. *Compare the SSE of the best creatures from the two experiments. How often did the second set of experiments come up with an essentially correct answer?*

Experiment 9.8, in addition to illustrating the use of the ITE operation in dealing with discontinuous functions, does something else entertaining. In the first half, we are using a continuous family of functions to approximate a collection of sample points drawn from a discontinuous function. This is nearly impossible (and gets harder as we increase the number of sample points). The point is this: trying to do an impossible approximation teaches you something about the approximating technique. Examine creatures with low SSE evolved in the first half of Experiment 9.8. How do they do their approximating?

Problems

Problem 353. Review Experiment 9.5. For each of the formulas below, rate that formula as an accurate approximation to $f(x) = \frac{1}{x^2+1}$ in general and on the set of points $\mathcal{P}$.

(i) $f_1(x) = \cos^2(\arctan(x))$,
(ii) $f_2(x) = \cos^2(x)$,
(iii) $f_3(x) = \frac{1}{x^2+0.99975}$, and
(iv) $f_4(x) = \frac{\cos(\arctan(x))\sin(\arctan(x))}{x}$.

Problem 354. For each of the functions in Problem 353, estimate how hard it is for the evolutionary algorithm described in Experiment 9.5 to find that function, given the fitness function and model of evolution used.

Problem 355. Several of the functions used in the experiments in this section are listed below. For each, state and support your opinion as to whether deleting the function from those available to the GP system would decrease or increase time to success.

(i) $\sin(x)$,
(ii) $\cos(x)$,
(iii) x^2,
(iv) $\arctan(x)$.

Problem 356. Take 3 of the best functions evolved in Experiment 9.5 that did *not* come from populations that finished early. Graph them on the interval $-2 \leq x \leq 2$. Recalling that the fitness function used to produce these functions operated in the interval $-1 \leq x \leq 1$, comment on their behavior outside the fitness relevant region.

Problem 357. Essay. Review Experiment 9.7. When we sample the SSE of the function on different sets of points each time, the fitness values of different parse trees jump about. First: prove that they do not do so in a manner that preserves relative fitness. By that we mean that if on one set of sample points parse tree 1 has higher fitness than parse tree 2, then the situation may be reversed on another set of sample points. Second: comment on the effect that this stochastic method of computing fitness has on the population's tendency to get stuck in local optima. Reason in terms of fitness landscapes, if possible.

Problem 358. Review Experiment 9.8. The set of points $\mathcal{S}_0$ is drawn from the Heaviside function,

$$H_0(x) = \begin{cases} 1 & \text{if } x \geq 0, \\ 0 & \text{if } x < 0. \end{cases}$$

Assume that we measure the true SSE of a function $f(x)$ on an interval $I = [a, b]$ with an integral:

$$\mathrm{SSE}_{[a,b]} = \int_a^b \left(f(x) - H_0(x)\right)^2 \cdot dx.$$

Discuss the minimum SSE possible if $f(x) = a \cdot \tanh(b \cdot x)$.

Problem 359. In Problem 18, you were asked to find a minimal GP representation for a given polynomial. Supposing you can use arbitrary constants, the symbol x, and the operations $+$ and $*$, give the minimum number of operations needed for a single-tree GP representation of a polynomial

$$f(x) = a_0 + a_1 x + \cdots + a_n x^n$$

when you have no ability to factor and when each a_i is nonzero.

Problem 360. Review the definition (Definition 3.4) of fitness landscapes. Describe the domain space for Experiment 9.5. If we are to graph a function, we must have some way of representing the domain as a space where distances between points are apparent. What definition of distance works for this space?

Problem 361. What extensions to the GP system described in this section would you need to solve differential equations? In addition to describing how to automatically extract the derivative, describe the modifications to the SSE fitness function required to attempt to solve $y' = x + 2y$.

Problem 362. Essay. Examine the following function in LISP-like notation, an evolved formula from a run of Experiment 9.5.

```
(Cos (Div (Sin (Div(Sin X1) -0.898361))
  (Sub 0.138413 0.919129)))
```

The subtree (Sub 0.138413 0.919129) is a verbose way of coding the constant -0.780716. It would be possible to shrink parse trees by detecting subtrees that compute a constant value and replacing them with a real constant terminal. The question is this: is this helpful? Consider the effects on evolution of such compaction.

9.4 Automatically Defined Functions

A standard technology used in genetic programming is the *automatically defined function*, or *ADF*. In standard programming, an ADF would be called a subroutine. In order to use ADFs, we need to modify our parse tree routines to accommodate them. We will restrict ourself to one-variable ADFs for the present.

Our parse tree routines should be modified to allow or disallow the use of a unary operation called ADF. Our basic creature will now be made of two parse trees. The first is a "main" parse tree that can use the ADF operation, and the second is an ADF that does not use that operation. In the main tree, when the ADF operation appears, its argument is sent to the ADF parse tree as the terminal x, and the ADF parse tree is evaluated. (Both parse trees use the terminal x, but in the ADF the value of x is the value of the argument of the call to the ADF, while in the main parse tree, it is the input variable.)

Again: the ADF parse tree is not allowed to use the ADF operation. This would amount to recursion, which would open a substantial can of worms. An example of a parse tree and accompanying ADF appear in Figure 9.9.

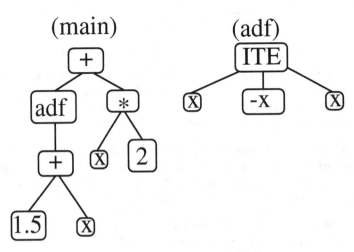

Fig. 9.9. A parse tree and ADF.

ADFs fulfill the same role in genetic programming that subroutines do in normal programming. They provide code fragments that can be used multiple times during the execution of the program. This implies that some problems will benefit more from ADFs than others. We will test this notion in the next experiment.

Experiment 9.9 *In this experiment, we will approximate two polynomials from a fixed set of points. Use 41 points with x-coordinates equally spaced in the range $-2 \leq x \leq 2$. Modify your parse tree routines to enable the use of ADFs. Also, disable the trigonometric functions used in previous experiments. Use the software of Experiment 9.5.*

Let the main parse trees have a maximum size of 36 and an initial size of 12. When ADFs are used, let the ADFs have an initial size of 4 and a maximum size of 8. Reproduction of parse trees with ADFs copies both the main tree and the ADF.

When crossover is required in a population with ADFs, do one of the following, selecting uniformly at random: cross the main trees in the usual fashion, cross the ADFs in the usual fashion, or take the main tree and ADF from distinct parents. Mutation should pick the main tree or the ADF to mutate with equal probability. A mutation should be used 50% of the time.

Approximate the following functions:

(i) $f_1(x) = x^3 - 2x^2 - x + 2$,
(ii) $f_2(x) = x^6 - 4x^5 + 2x^4 + 8x^3 - 7x^2 - 4x + 4$.

Treat success as an SSE of 0.001 and run 50 populations for each of the two functions, both with and without ADFs. Report and graph the time to success for all 4 sets of runs. (The first function is much easier to approximate, so concentrate on the difference the ADF made.)

In addition, for the best parse trees, state the fraction that used an ADF for each polynomial. Finally, examine the best parse trees from each population and state whether factorization of f_1 or f_2 was used. How did the parse tree represent the functions? (The best parse trees are the ones with the lowest SSEs and the fewest nodes.)

There is no reason, beyond simplicity, to limit ourselves to a single ADF. It is possible, for example, to design structures that contain a large set of parse trees that refer to one another. Problem 363 looks at possible generalizations of the notion of ADF given here. In Chapter 10, we will look at other ways of fragmenting parse trees to form something more like a computer program and less like a single statement.

Many other such methods exist. You could brainstorm and come up with some good ones yourself. Remember, though, that being a neat idea is not enough. If you are going to modify the basic model of genetic programming, there is a series of questions you should ask yourself.

What does it cost in software? A different way to do genetic programming must be implemented if it is to be of any use. If you did the modifications to permit ADFs, you have some idea that moving beyond basic parse trees can be an annoying and bug-ridden task. When vetting a new idea, think carefully through what will be needed to build the software that supports it. You can do ADFs with variable numbers of arguments, libraries of useful ADFs, libraries of parse tree fragments, and other similar things. The question is, can you get them running reliably?

What is the cost in instruction cycles? If you modify the representation used by GP software, you will make it faster (small chance) or slower (large chance). If the change of representation reduces the evolution's time-to-solution (as ADFs should have done in at least one part of Experiment 9.9), then a modest decrease in the speed of evaluating the fitness of individual creatures is acceptable. Be sure to do back-of-the-envelope estimates of the increased time your modifications will take.

What is the cost in readability? As well as coding your new representation for use on a computer, you will need some method of representing it to people. (There are some cases in which a black box that nobody ever looks under the hood of is acceptable, but these are rare.) Ask yourself whether you have a method of displaying the results of your new representation that is no worse than the LISP-like display of parse trees. If you've got one that is substantially better, good; this even pays for some added computational or programming cost. Always keep in mind at design time that a little work now on your I/O and display routines can pay off substantially at data analysis time.

Problems

Problem 363. Essay. It is stated in the discussion of ADFs on page 254 that "The ADF parse tree is not allowed to use the ADF operation." Explain why not and discuss the difficulties and potentialities of relaxing this restriction.

Problem 364. Design a set of operations and terminals, including 3 terminals a, b, and c, that permit you to write a parse tree that computes the roots of the quadratic $f(x) = ax^2 + bx + c$. The first root should be returned the first time the parse tree is evaluated, the second the second time the parse tree is evaluated. You may either define a value NAN (not a number) or, for advanced students, return the complex roots. Since the parse tree must return two values, it must have some sort of memory that lets it know whether it has been called before. If the root is unique, return it twice. Question: would an ADF help?

Problem 365. Find a minimal GP representation, as a single parse tree, of the polynomial $f_2(x)$ from Experiment 9.9 using the operations used in that experiment. Give the answer both with and without ADFs. Does using an ADF result in fewer total nodes?

Problem 366. In the spirit of Chapter 8, we give the following question. Suppose you are allowed to build a parse tree and ADF using only one real constant a, the variable x, and the operations ADF, +, and *. If the sum of the nodes used in the main parse tree and ADF is at most n, what is the highest-degree polynomial you can manage for $n = 8$, 12, or 16 nodes? Beginning students can simply present their best construction; advanced students should prove that they have a correct answer.

Problem 367. Reread the three boldfaced questions at the end of Section 9.3. Keeping them in mind, consider the following structure. We generate 6 parse trees, T_1–T_6, with at most 8 nodes over some otherwise reasonable set of operations and terminals, including only one variable terminal x. Each parse tree can use the parse trees with smaller index as ADFs. We use an evolutionary algorithm (leave the details vague) to fit (by minimizing SSE) to a system of two equations $f(x)$ and $g(x)$, where T_5 is used to approximate $f(x)$ and T_6 is used to approximate $g(x)$. Answer all three questions for this system.

9.5 Working in Several Dimensions

Thus far, we have only done one-dimensional data fitting with genetic programming. We will now look at issues that arise in multiple dimensions. Our old friend, the fake bell curve, will again be the function of choice for our experiments. The first issue we choose to deal with is the curse of dimensionality.

In Experiment 9.5, we fitted a fake bell curve using a collection of 81 sample points with x-coordinates uniformly spaced on the interval $-1 \leq x \leq 1$. Imagine that we wish to fit the fake bell curve $\mathcal{B}_2(x, y) = \frac{1}{x^2+y^2+1}$ in two dimensions. If we used the same spacing of points, we would need to use $81^2 = 6561$ points. Clearly, as the number of dimensions increases, the number of points needed for a given spacing skyrockets. Before attempting anything too clever, we will check how much harder the problem gets when we hold the number of points constant.

Experiment 9.10 *Rerun Experiment 9.5 with a new termination condition: SSE at most 0.001 or 2000 generations. Also increase the initial and maximum tree size to 8 and 16 respectively. Save the time-to-solution data. Now modify the software from Experiment 9.5 to permit two variable terminals and use the evolutionary algorithm to fit parse trees representing functions $f(x, y)$ to the fake bell curve in two dimensions using points:*

$$\mathcal{P}_{B2} = \left\{ \left(\frac{i}{4} - 1, \frac{j}{4} - 1, \frac{1}{(i/4 - 1)^2 + (j/4 - 1)^2 + 1} \right) : i = 0, \ldots, 8, j = 0, \ldots, 8 \right\}.$$

Save time-to-solution and estimate the increase in difficulty of adding a dimension while holding the number of points constant. Does the two-dimensional version of the experiment find any essentially correct solutions?

This curse of dimensionality is not unique to us; engineers and physicists have struggled with it for a long time (and found a way to cope). In engineering and physics, it is often necessary to integrate a function of many variables. If we suppose that error estimates require a rectilinear grid with a spacing of 0.05, then 9 dimensions will require, at 21 points per dimension, some 7.94×10^{11} sample points. Present-day computers are not up to working with so many points, and future, more efficient, computers would have trouble with just slightly higher dimensions.

The solution engineers and physicists use for integrating multidimensional functions is to pick points at random, termed *Monte Carlo* integration. An advanced text on numerical analysis will go over the techniques of Monte Carlo integration, including the critical error estimates used to decide how many random points are enough. We will simply use the idea of sampling points to learn to approximate functions in many dimensions.

Experiment 9.11 *Modify the software from the second half of Experiment 9.10 to operate on n randomly selected points $(x, y, \mathcal{B}_2(x, y))$ with $-1 \leq x, y, \leq$*

1. *An array of points should be generated in each generation and used to evaluate the fitness of all the parse trees in that generation. Use this fitness for selection, but also compute the fitness as in Experiment 9.10 and use it to determine success. Compare the time to success of the previous experiment (measured in generations) with that of this experiment using data taken for $n = 20, 81,$ and 200 points. Do random points work any better than regularly spaced ones? How does the number of points used affect the time to success? Remember to compensate for the effect of computing more sample points.*

The fake bell curve, in any number of dimensions, could be defined as $\mathcal{B}_b(p) = \frac{1}{d(p)^2+1}$, where p is a point in $\mathbb{R}^n$ and $d(p)$ is the distance from the point to the origin. Given that this distance function enormously simplifies computation of the fake bell curve, it is possible that the fake bell curve would benefit from the use of ADFs.

Experiment 9.12 *Modify the parse tree routines from Experiment 9.9 and/or 9.10 to permit two-variable ADFs and then reperform the 3 sets of runs from Experiment 9.11 using ADFs. Use the variation operators given in Experiment 9.9. Permit the ADF trees an initial 6 and a maximum of 10 nodes. Compare with the results of Experiment 9.11: does using ADFs help?*

One problem with genetic programming, as performed so far in this chapter, is the mutation of real constant terminals. Real constant terminals are called *ephemeral constants*. Subtree mutation never makes small changes to the value of such real constants: it either leaves them alone or completely changes them. There are a number of ways to deal with this. The simplest is to add a mutation operator that locates a constant within a tree and performs a real mutation, as in Chapter 3.

The fake bell curve uses only the constant value 1, so we will switch to a different function with more constants. Recall that the area of an ellipse with major axis $2a$ and minor axis $2b$ is πab.

Experiment 9.13 *Implement a mutation operator that can locate a real constant (equal chance among all constants in a parse tree) and do a real mutation with mutation size ϵ to it. Add this new mutation operator to the software used in Experiment 9.11.*

In this experiment we will evolve parse trees to find the area of an ellipse of diameters $2a$ and $2b$. Use a population of 400 parse trees with an initial size of 6 and a maximum size of 12 nodes. Let your model of evolution be single tournament selection with tournament size 4. Use crossover in 80% of all mating events and subtree mutation in 50% of mating events. Minimize the SSE of your parse trees with the formula $Area = \pi ab$ for 80 randomly chosen values of a and b in the range $0.1 \leq a, b \leq 4$, choosing new random values in each generation. Test for success by looking for an SSE of 0.001 or less on the ellipses with a, b valued at any multiple of 0.5 in the range $0.5 \leq a, b \leq 4$.

Do 30 runs, saving time-to-success data and giving up after 2000 generations. Now, run the 30 populations again with each new tree mutated 50%

of the time with the new mutation operator (in addition to and with probability independent of the subtree mutation) with $\epsilon = 0.2$. Did the new mutation operator help?

Now let's do an experiment to explore the problems caused by increasing the number of dimensions.

Experiment 9.14 *Modify the software used in Experiment 9.10 to use* 200 *randomly selected sample points for fitness evaluation. Perform* 30 *runs each, attempting to approximate the fake bell curve in* 1, 2, 3, *and* 4 *dimensions. Compare the results obtained in each case. Estimate the relative difficulty of the problems.*

As is always the case, there is a large number of things we could do to further explore these issues, comparing random sample points with a fixed grid that moves (as in Experiment 9.7), for example. This chapter has dealt only with artificial, manufactured data. You are encouraged to locate real data of interest and use the techniques you have learned to try to fit to the data. In later chapters, we will explore using genetic programming to model discrete data and heterogeneous data. In real-life applications, mixed real values and discrete data are the rule rather than the exception.

Problems

Problem 368. Experiment 9.10 uses the 81 points of Experiment 9.5 to form a regular two-dimensional grid with 81 points. Suppose that we wish to do a similar comparison across 1, 2, and 3 dimensions with the number of points the same for each dimension. What numbers of points less than 1000 permit this? Hint: this is almost trivial.

Problem 369. Examine the operations given in Table 9.3. When we move to multiple dimensions, there is more potential utility to having operations that take many arguments. With some operations, it is obvious what modifications are needed to allow more arguments. For example, it is obvious what an n-ary addition operator does: it adds up all of its arguments. For each of the operations in Table 9.3, give your best shot at defining how that operation would act on an arbitrary list of arguments. Also, rate your construction as completely natural, as reasonable but different from the original, or as contrived.

Problem 370. Essay. Read the three boldfaced questions at the end of Section 9.3. Keeping them in mind, consider the following idea. Modify all operations so that they operate on a list of arguments of arbitrary size. Answer all three of the questions with respect to this system.

Problem 371. Essay. Read the three boldfaced questions at the end of Section 9.3. Keeping them in mind, consider the following idea. Our experience in approximating the fake bell curve has shown that one weakness of genetic programming is the poor job it does at discovering constants. Consider a system with exactly 10 available constant terminals (instead of the generic real number constant terminal). Each parse tree is augmented with an array of 10 real numbers. These array genes undergo two-point crossover and single-point real mutation with an appropriate mutation size during breeding. Answer all three of the questions with respect to this system.

Problem 372. Review Experiment 9.13. We observed in Experiment 9.5 that it was easier for the evolutionary algorithm to find the approximation $f(x) = \cos^2(\tan^{-1}(x))$ for the fake bell curve in one dimension than for it to find the correct formula. For each of the following features, comment in a sentence or two on the degree to which it might help the software from Experiment 9.5 find the formula $\frac{1}{x^2+1}$.

(i) A terminal **one** that returns the value 1.
(ii) The mutation operator introduced in Experiment 9.13.
(iii) A reciprocal operator.

Problem 373. Review. Using calculus, verify the formula for the area of an ellipse given on page 258. Recall that an ellipse is defined by the relation

$$\frac{x^2}{a^2} + \frac{y^2}{b^2} = 1.$$

Refer to the graph below for inspiration.

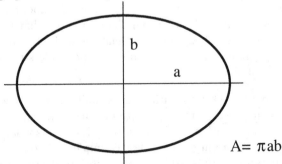

$$A = \pi ab$$

Problem 374. Suppose that in the course of more than 100 populations run working out Experiments 9.5 and 9.7, only one formula appeared as the best-of-run that approximated the functional form of the fake bell curve. This was

$$f(x) = \frac{0.993126}{0.991796 + x^2}.$$

A large number of variations on the solution

$$g(x) = \cos^2(\arctan(x))$$

appeared as best-of-runs. Why? Consider the shape of the cosine function and the number of ephemeral constants required.

9.6 Introns and Bloat

Genetic programming is the first instance of a variable-sized representation we have studied. In Chapter 8, we did genetic programming, but in a fashion that minimized the impact of size variation. The PORS trees were used to explore the behavior of a particular size of parse tree. This limited their ability to use their variable size for much of anything.

The *depth* of a parse tree is the maximum number of links from the root to any terminal. The main function shown in Figure 9.9 has depth 3, while the ADF has depth 1. In standard genetic programming, as opposed to genetic programming described thus far in this text, the depth of trees is controlled, but not their size. The decision to control size in the genetic programming in this text was motivated by research on the subject of bloat. In working with a variable-sized genome, *bloat* is defined to be the use of space to gain advantages not directly related to fitness. In this section, we will do a few experiments to explore the phenomenon of bloat.

Bloat came as a surprise to the first researchers to encounter it. They observed that the average size of parse trees grows to near the maximum possible, whatever restrictions on size are present. In Problem 291, we saw that parse trees can grow without bound as a result of subtree crossover. The question remains, however, as to *why* they grow in this manner. Since there are more large trees than small trees, bloat could be the result of simple random drift. However, providing secondary fitness pressure for "parsimony" (small tree size) does not have much effect. There seems to be a positive selection pressure for large size.

One explanation of parsimony-resistant bloat makes an analogy to natural genetics. Genes, in living creatures, often contain DNA that does not code for anything and that is spliced out at the RNA stage, during translation of DNA into protein. These chunks of DNA are called *introns*. Introns permit DNA to undergo crossover that does not disrupt the protein-coding portions. Since subtree crossover of the sort used in genetic programming is highly disruptive, there is a selection pressure to develop substructures in the parse trees that allow the parse tree to resist crossover-based disruption. The question then becomes, "what do these substructures look like?" It is intuitive that they will depend on numerical identities, but their exact form is not clear. Let us experiment.

Experiment 9.15 *Rewrite the software used in Experiment 9.5 so that it saves the average size of parse trees in the final population. Rerun the experiment in its original form and with the upper bound on the size of parse trees changed from 12 to 60. Make the following analysis.*

(i) Plot the generation of solution versus the average size of tree in the population. Is there a correlation in either of the two sets of runs?

(ii) Which set of runs had the better average number of generations to solution?

(iii) Which set of runs had the better fitness among those runs that ran out of time?

(iv) Examine the best-of-run formulas. Are there any structures that might be introns? Are there any structures that look like they were created by repeated subtree crossover?

One way to see whether bloat is useful is to make it easier. A *designated intron* is a function that returns its argument or, if it is not a unary function, one of its arguments. If bloat is useful for avoiding crossover-based disruption, making it easy by including a designated intron should enhance bloat. It is an open question whether it will enhance the performance of a GP system.

Experiment 9.16 *Rewrite the software used in Experiment 9.5 so that it saves the average size of parse trees in the final population. Also, modify the parse tree software to include a unary operation, say, that simply returns its argument. Rerun the experiment twice with the maximum size of parse trees set to 12 and to 60. Compare with the results from Experiment 9.15. Is the say operation used oftener than it would be by random chance? Do the runs with the say operation exhibit superior performance?*

Problems

Problem 375. The following formulas are results that occurred in our version of Experiment 9.5. Verify that each equals $f(x) = \frac{1}{x^2+1}$, and then spot the potential bloat in each. Some of the functions have more than one instance of something that could be bloat.

(i) $q(x) = \cos(-\arctan(x))\cos(\arctan(x))$,

(ii) $g(x) = \cos(\arctan(x))\cos(-(\arctan(x) - (x - x)))$,

(iii) $h(x) = \cos^2(\arctan(x)(\cos(x - x))^2)$,

(iv) $a(x) = \cos^2(x\frac{\arctan(x)}{x})$,

(v) $b(x) = \frac{x}{\left(\left(\frac{x}{\frac{\cos(\arctan(x))}{\cos(\arctan(x))}}\right)\right)}$,

(vi) $c(x) = \cos^2((x - x) + \arctan(x))$.

Problem 376. Give a method of using introns for string genes of the sort used in Sunburn or VIP, from Chapter 4. Explain how the introns might help. Give also the design of an experiment to tell whether the introns help.

Problem 377. A *recessive gene* in biology is a gene that has no effect on the phenotype of the creature carrying it. Call a subtree recessive if it can be changed without changing the fitness of the tree above it. Construct a tree that has a recessive subtree and that is an exact answer for Experiment 9.5.

Problem 378. Reread Problem 362. Are such constant subtrees a possible source of bloat? What other purpose could they serve?

10

Tartarus: Discrete Robotics

In this chapter, we will deal with an environment called *Tartarus* in which a virtual robot (thought of as a bulldozer) shoves boxes around trying to get them up against the walls. The bulldozer lives on a grid and is supplied with information about adjacent grids via its sensors. The Tartarus environment was proposed by Astro Teller in his paper "The Evolution of Mental Models" [55]. In the first section we will use a string evolver to test the support routines that maintain the Tartarus environment. In the second section we will use genetic programming (GP) with a parse tree representation to run the bulldozer. In the third we will modify the GP language to allow the programs a form of memory like the one in the PORS environment. In the fourth section, we will use a different sort of memory, that of Chapter 6. This last modification has many applications that will be useful later.

GP automata, described in Section 10.4, is our first example of a complex, hybrid representation. Keep in mind that complexity should *never* be sought for its own sake. In the initial Tartarus paper the idea of adding calculator-style memory to parse trees, the subject of Section 10.3, was explored. These memories permit parse trees to have internal state information that is critical for good performance in the Tartarus environment. Entangling the memories that enable internal state information with the operations and terminals that form the processing power of the parse tree poses a formidable challenge for evolution (or any search technique). GP automata divide the parse trees from the state information by moving the storage of state information to a finite state machine. This decomposition of the method of storing state information from the parse trees that process sensory information makes searching for a good Tartarus controller a far easier task for evolution to perform.

Another important notion this chapter introduces is the use of a simple string evolver as a baseline for the difficult Tartarus task. A standard string evolver performs poorly on Tartarus, getting at most one-third of the fitness of more complex representations. Using variable-length strings and exotic mutation operators permits a Tartarus string evolver to exhibit performance superior to the initial published Tartarus experiments. The meaning of this

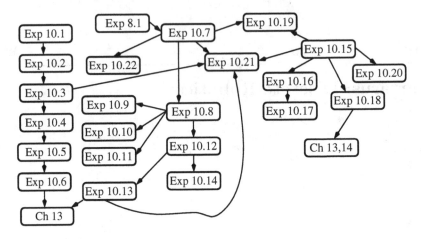

1   Implementing the Tartarus environment.
2   Applying a string evolver to Tartarus.
3   Exploring a periodic string representation.
4   Gene doubling and halving neutral mutations.
5   Another crossover operator for variable length genes.
6   Nonaligned crossover.
7   Genetic programming: a parse tree representation.
8   Adding a random number terminal.
9   Analysis of evolved Tartarus controllers.
10 Adding automatically defined functions.
11 Exploring less subtree crossover.
12 Adding calculator-style memories to the parse trees.
13 Adding more memory.
14 Using indexed memory.
15 Tartarus with GP automata, a complex hybrid representation.
16 Exploring mutation operator usage frequency.
17 More Tartarus controller analysis.
18 Adding think actions, λ-transitions, in the automata.
19 Population seeding.
20 Exploring fitness validation.
21 Exploring variable numbers of fitness trials.
22 Evolving Tartarus boards instead of controllers.

Fig. 10.1. The topics and dependencies of the experiments in this chapter.

result is explored in the Problems; it highlights the importance of state information. A string controller specifies a fixed dance of moves with no use of sensory information. That a string can get high fitness by storing only a sequence of moves and its current position in that sequence of moves (a form of state information) implies that state information is more valuable than sensory information. As we will see, the best performance comes from using both sensory and state information. The relative importance of these two resources would not have been understood without the string evolver baseline.

This chapter requires familiarity with genetic programming as introduced in Chapter 8 and would benefit from familiarity with Chapter 9. This chapter uses a slightly more complex GP language than PORS, but with simpler data types: the integers or the integers (mod 3). The use of integer operations makes implementation of the parse tree evaluation routines much easier than in Chapter 9. The dependencies of the experiments in this chapter are given in Figure 10.1.

10.1 The Tartarus Environment

A Tartarus board is a $k \times k$ grid, like a checkerboard, with impenetrable walls at the boundary. Each square on a Tartarus board contains nothing, a box, or the robot (henceforth called the *dozer*). A valid starting configuration in Tartarus consists of m boxes together with a placement and heading of the dozer. In the starting configuration, no boxes are in a block of 4 covering a 2×2 area of the board, and no box is adjacent to the wall. The dozer starts away from the wall and can be heading up, down, left, or right. An example of a valid Tartarus starting configuration is given in Figure 10.2.

The goal of Tartarus is for the dozer to shove the boxes up against the walls. On each move, the dozer may go forward, turn left, or turn right. If a single box with space to move into is ahead of it, then, when it moves forward, the box moves as well. If a dozer is facing a wall or a box already against a wall or a box with another box in front of it, a go forward move does nothing.

We will use a fitness function called the *box-wall* function for our work in this chapter. Given a valid initial configuration with $k = m = 6$, the dozer is allowed 80 (or more) moves. Each side of a box against a wall, after all the moves are completed, is worth 1 point to the dozer. This means that a box in the corner is worth 2 points; a box against the wall, but not in the corner, is worth 1. Add the scores from a large number of boards to get the fitness value used for evolution. Figure 10.3 shows 4 boards with scores. The maximum possible score for any Tartarus board with 6 boxes is 10: boxes in all 4 corners and the other 2 boxes against the wall.

Definition 10.1 *A starting configuration of a Tartarus board is said to be* **impossible** *if its score is 4 or less no matter what actions the dozer takes.*

The starting configurations with 2×2 blocks of 4 are impossible, which is why they are excluded.

In the later sections, we will be evolving and testing various sorts of representations of dozer controllers in the Tartarus environment. We will need working routines to support and display the Tartarus environment as well as a baseline for measuring performance. In this section, we will explore the Tartarus environment *without* genetic programming. These non-GP experiments

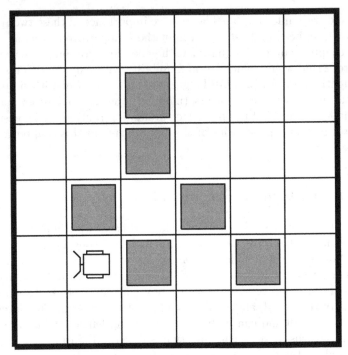

Fig. 10.2. A 6 × 6 Tartarus board with $m = 6$, dozer heading left.

will serve as a baseline for comparison of the various genetic programming techniques.

Experiment 10.1 *Build or obtain routines for maintaining and displaying the Tartarus environment. Develop a data structure for holding Tartarus boards that saves the positions of the boxes and the dozer's position and heading. Include the following routines in your software.* **MakeBoard** *should create a $k \times k$ Tartarus board with m boxes and an initial dozer position and heading in a valid starting configuration.* **CopyBoard** *should copy Tartarus boards.* **Move** *should take a board and a move (turn left, turn right, or go forward) and update the box positions and dozer position and heading.* **Score** *should compute the box-wall fitness function of a board, given a set of moves.* **DisplayBoard** *should print out or display a Tartarus board.*

For $k = m = 6$ (a 6 × 6 world with 6 boxes), generate 40 valid starting Tartarus configurations and save them so they can be reused. Use each board 20 times. Randomly generate 320 moves (turn right, turn left, and go forward). Record the fitness at $80, 160, 240,$ and 320 moves, and compute the average score. Do this experiment with the following 3 random number generators:

(i) Equal chance of turning left, turning right, and going forward,
(ii) Going forward 60%, turning left 20%, turning right 20%,

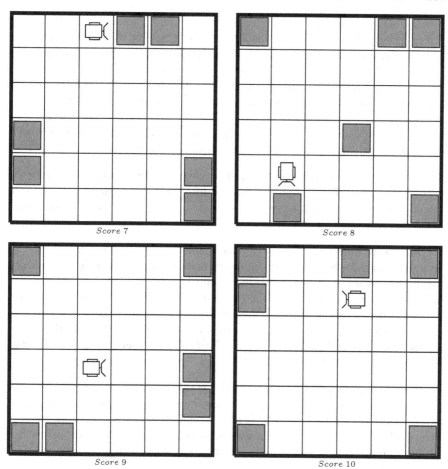

Fig. 10.3. Boards after 80 moves with scores.

(iii) A left turn never follows a right turn, a right turn never follows a left turn, but everything equally likely otherwise.

Report the average scores in a table indexed by the number of moves and types of random number generators. Explain why this forms a baseline experiment for Tartarus; would you expect nonrandom dozers to do better or worse than random moving points?

Experiment 10.1 gives us the basic routines for maintaining Tartarus. We now move on to a minimal Alife technology for Tartarus, the string evolver.

Experiment 10.2 *This experiment fuses string-evolver technology with the Tartarus routines from Experiment 10.1. Write or obtain software for an evolutionary algorithm for evolving strings over the alphabet $\{L, R, F\}$. Use a population of 400 strings of length 80 evolving under tournament selection* *with two-point crossover and one of one-, two-, or three-point mutation. Each string specifies 80 moves for the dozer to make. Evaluate the fitness of a string by testing it on 12 Tartarus boards and summing the scores on the boards to get a fitness value. All strings in the population should be evaluated on the same 12 boards in a single generation. A new set of 12 boards should be created for each generation. Run the algorithm for 100 generations.*

Graph the average fitness divided by 12 (the average per-board score) and the fraction of L, R, and F in the population, averaged over 20 runs, for each of the 3 mutation operators. Summarize and explain the results.

At this point, we digress a bit and do two experiments that illustrate an odd feature of evolution and use a very interesting mutation operator first introduced by Kristian Lindgren in a paper on Iterated Prisoner's Dilemma [42].

Experiment 10.3 *A string with fewer than 80 moves can still be used to control a dozer for 80 moves by cycling through the string. Modify the software from Experiment 10.2 to use strings of any length up to 80. Using only one-point mutation, rerun Experiment 10.2 for strings of length 5, 10, 20, and 40. Compare the fitness and average final fitness with those for the one-point-mutation 80-character strings in Experiment 10.2. Keeping in mind that every shorter string gives behavior exactly duplicated by a longer string, explain what happened.*

For the next experiment, we need the notion of *neutral mutations*. A neutral mutation is one that does not change a creature's fitness. The null mutation, mentioned in Chapter 3, is an example of a neutral mutation, but it does nothing. In his work on Prisoner's Dilemma, Lindgren came up with a neutral mutation that changes the creature's genotype substantially without changing its phenotype.

Lindgren stored a Prisoner's Dilemma strategy in a lookup table indexed by the last several plays. He doubled the length of the lookup table by increasing the index set by one play and then making a new lookup table from two copies of the old lookup table, one for each possible value of the new index. The additional index was vacuous in the sense that no matter what that index said, cooperate or defect, the creature's response was the same.

"Why bother?" I hear you cry. Imagine that we add Lindgren's doubling mutation to a population of Prisoner's Dilemma players evolving under an evolutionary algorithm. Occasionally, one of the creatures doubles the size of its gene. Its behavior does not change, but the strategies that can be derived from it under a point mutation change radically.

Look back at the results of Experiment 10.3. Evolution finds it much easier to optimize a short string than a long one. If we optimize a short string, double its length, and then continue the optimization in a completely new space (with *constructively* the same fitness), we may be able to enormously speed the process of evolution.

If we include a gene-doubling mutation operator in the mix of genetic operations, then we will need another that cuts genes in half. The reason for this has to do with the behavior of one-sided random walks. Imagine that you flip a coin and stand still for heads, while moving to the right for tails. The net effect will be substantial movement to the right. If we occasionally double a gene and never shorten one, there is danger that all genes will attain some maximum length *before* the search of the smaller gene lengths is complete.

Definition 10.2 *A* **gene-doubling mutation** *for a string-type gene is a mutation that doubles the length of the string by concatenating two copies of the string.*

Definition 10.3 *A* **gene-halving mutation** *for a string type gene is a mutation that halves the length of the string by splitting the string into its first and second halves and saving one at random.*

Not that we have defined gene-doubling and gene-halving mutations, we must confront the problem of crossing over pairs of genes of different lengths. There are at least three ways to do this, given in the experiments below.

Experiment 10.4 *Modify the software from Experiment 10.3 in the following fashion. First, modify the string data structure so that it includes the length of the string. Second, modify the crossover operator so that when two strings of different lengths cross over, the crossover points are chosen inside the smaller string. Perform mutation so that 90% of the time the algorithm does a single point mutation, 5% of the time the algorithm does a gene-doubling mutation, and 5% of the time the algorithm does a gene-halving mutation. Generate the initial population with a 50% chance each of being length 5 or 10. Ignore gene-halving mutations on strings of length 5 and gene-doubling mutations on strings of length 80.*

In addition to average fitness and fraction of types of moves, save the average fitness of each length of string and plot the average of averages within length classes over the course of evolution. Also, give a histogram of the string lengths in the final generation, summed over all runs. In your write-up, state whether better strings were located at all, whether better strings of length 80 were located, and whether evolution was faster.

There is an obvious bias in the crossover operator above. It favors, to some degree, genetic material in the latter part of the strings for preservation as an intact chunk. This is offset to a modest degree by the ability of the gene-halving mutation to "bring the back half forward," but this may not help. If the moves in the front part of the genes are "more evolved" (better), then exhumed back halves may have trouble competing. A crossover operator that is more difficult to code, but perhaps more fair, is explored in the next experiment.

Experiment 10.5 *Repeat Experiment 10.4 with a new crossover operator as follows. When two strings of differing lengths are crossed over, copy the shorter gene into a scratch variable and double its length, until it is the same length as the longer gene. Cross over the two resulting genes, now the same length, and then truncate one of the resulting strings so that it is the length of the shorter parent gene. This operator produces children of lengths matching the parents. Present the same data as in Experiment 10.4 and discuss the impact of changing the crossover operator.*

Definition 10.4 *A crossover operator is said to be* **conservative** *if two identical parents cross over to create identical children. Such crossover operators are also sometimes called* **pure**.

One thing that distinguishes evolutionary algorithms that use string genes from genetic programming is the conservative nature of the crossover operator. Crossing over strings, even in our last two experiments, lines up equivalent positions in the string. Information doesn't migrate around the gene, and identical parents clone themselves when they are reproduced and crossed over. Subtree crossover in genetic programming, on the other hand, does not preserve positions. Identical parents produce radically different children. In the next experiment, we will strike off into the middle ground between conservative string crossover and subtree crossover by using *nonaligned* crossover.

Nonaligned crossover exchanges substrings of the same length that may not be lined up in the two strings. This permits genetic information to move about the collective genome of the population over time.

Experiment 10.6 *Repeat Experiment 10.4 using nonaligned crossover. Present the same data as in Experiment 10.4 and discuss the impact of changing the crossover operator. Compare also with Experiment 10.5 if you performed it.*

Problems

Problem 379. Essay. In Experiments 10.2–10.4 we saved data on the fraction of moves of each type. Explain a way of using that data to modify and improve performance in Experiment 10.1.

Problem 380. For $k = 6$ and $m = 4, 5$, and 6, compute the probability that placement of the boxes away from the wall, but otherwise at random, will produce an invalid starting configuration (4 boxes in a 2×2 group).

Problem 381. Essay. Given the answer to Problem 380, is it worth excluding the impossible configurations? Consider both the cost in computer time and the effects on evolution in your answer.

Problem 382. Give counting formulas for the number of strings of length k over the alphabet $\{L, R, F\}$ in which:

(i) L and R are never adjacent,
(ii) At least half the characters are F,
(iii) The sequences LLL and RRR never appear,
(iv) (i) and (iii) both hold.

Hint: find a recursion in terms of the string length.

Problem 383. Explain why each of the classes of strings given in Problem 382 might be a good restricted class from which to draw dozer genes.

Problem 384. Compute the fitness of the following strings when run on the Tartarus board shown in Figure 10.2. Place the numbers 1–80 on an empty board to show where the dozer moves and give the final configuration of boxes. Warning: this is a very time-consuming piece of busy work; the purpose is to build mental machinery for debugging Tartarus code.

(i) $FFFLF$,
(ii) $FFLFR$,
(iii) $FFLFFRFLFR$.

Problem 385. Essay. Refer to the material in Appendix B on Markov chains. Describe a genetic algorithm that evolves the matrix of a Markov chain controlling a dozer in the Tartarus environment. What is a reasonable set of states for such a Markov chain? Can a chain be designed that outperforms the 80-move version of Experiment 10.1?

Problem 386. Essay. In Chapter 8, we did some work with seeding populations. First, explain why the special strings in Problem 382 would be good for seeding a population for Experiments 10.2–10.4. Next, give a description of such an experiment, including pseudocode for generating strings in classes (i)–(iv).

Problem 387. Suppose we are looking at string genes for the Tartarus problem studied in this section with $k = m = 6$. Define a set G of genes to have all of the following properties:

(a) The longest substring of pure L's or pure R's is of length 2.
(b) The longest substring of pure F's is of length 5.
(c) The substrings LR and RL never occur.

Which types of genes from Problem 382 appear in G? Prove that if we have a collection of boards B that we are using to test fitness of genes of length n, then, for every gene h of length n outside of G, there is a gene $g \in G$ of length n such that g scores as much as h on the collection B.

Problem 388. Short Essay. In what sense is the set G of genes in Problem 387 like the set T_s^* from Definition 8.3?

Problem 389. Programming Problem. Write a short program that efficiently enumerates the members of the set G of genes from Problem 387. Using this program, report the number of such genes of length $n = 1, 2, \ldots, 16$. For debugging ease, note that there are 23,522 such genes of length 12.

Problem 390. Essay. The advantage of using aligned crossover operators is that the population can tacitly agree on the good values for various locations and even agree on the "meaning" of each location. Nonaligned crossover (see Definition 7.21) disrupts the position-specificity of the locations in the string. Consider the following three problems: the string evolver from Chapter 2 on the reference string

01101001001101100101101,

real function optimization on a unimodal 8-variable function with a mode at $(1, 2, 3, 4, 5, 6, 7, 8)$, and the string evolver in Experiment 10.6. Discuss the pros and cons of using nonaligned crossover to solve these problems.

Problem 391. Essay. Consider the three techniques discussed in this section: gene doubling, nonaligned crossover, and imposing restrictions (e.g., "no LR or RL") on the strings used in the initial population. Discuss how these techniques might help or interfere with one another.

10.2 Tartarus with Genetic Programming

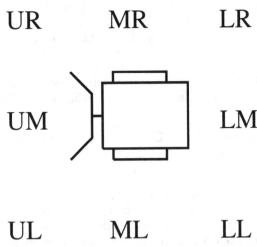

Fig. 10.4. Dozer sensor terminal placement.

In this section, we will develop a system for evolving parse trees to control the dozer. The data type for our parse trees will be the integers (mod 3) with the translation to dozer actions $0 = L$, $1 = R$, $2 = F$. The terminals include the constants 0, 1, and 2. In most of the experiments, we will use 8 "sensor" terminals that will report what is in the squares adjacent to the dozer. These terminals are UM (upper middle), UR (upper right), MR (middle right), LR (lower right), LM (lower middle), LL (lower left), ML (middle left), and UL (upper left). The positions are relative to the dozer, not absolute. Figure 10.4 shows on which square, relative to the dozer, each sensor terminal reports. A sensor terminal returns 0 for an empty square, 1 for a box, and 2 for a wall. We also may use a terminal called RND that returns a uniformly distributed random number. The terminals are summarized in Table 10.1.

Name	Type	Description
0, 1, 2	constants	The integers (mod 3)
$UM, UR, MR, LR,$ LM, LL, ML, UL	sensors	Reports on a square adjacent to the dozer
RND	special	Returns a uniformly distributed constant (mod 3)

Table 10.1. Dozer GP language terminals.

We will be using a number of operations in the dozer GP language, changing which are available in different experiments. All the experiments in this section will use unary increment and decrement (mod 3), addition and subtraction (mod 3), a binary maximum and minimum operation that imposes the usual order on the numbers 0, 1, and 2, and a ternary if-then-else operator that takes zero as false and nonzero as true. These operations are summarized in Table 10.2. This list will be substantially extended in the next section, so be sure to make your code able to handle the addition of new operations and terminals.

If you wrote and documented the parse tree manipulation code used in Chapter 8, you will be able to modify it for use in this chapter. We will need the same parse tree routines as those used in Experiment 8.1, but adapted to

Name	Type	Description
INC	unary	Adds one (mod 3) to its argument
DEC	unary	Subtracts one (mod 3) from its argument
ADD	binary	Adds its arguments (mod 3)
SUB	binary	Subtracts its arguments (mod 3)
MAX	binary	Returns the largest of its two arguments, $0 < 1 < 2$
MIN	binary	Returns the smallest of its two arguments, $0 < 1 < 2$
ITE	ternary	If first argument is nonzero, returns second argument; otherwise returns third argument

Table 10.2. Dozer GP language operations.

the terminals and operations given above. The first experiment will test the technique of genetic programming in the Tartarus environment using the 8 environmental sensors. The next experiment will test the effects of adding the RND random number terminal. The parse tree manipulation routines should be able to allow and disallow the use of the RND terminal.

There is some entirely new code needed: given a board including dozer position and heading, compute the values of the 8 terminals used to sense the dozer's environment.

Experiment 10.7 *Look at the list of tree manipulation routines given in Experiment 8.1. Create and debug or obtain software for the same set of routines for the terminals and operations given in Tables 10.1 and 10.2, and also terminal and operation mutations (see Definitions 9.3 and 9.4). In this experiment, do not enable or use the RND terminal.*

Using these parse tree routines, set up a GP system that allows you to evolve parse trees for controlling dozers. Recall that the three possible outputs are interpreted as 0, turn left; 1, turn right; and 2, go forward. You should use a population of 120 trees under tournament selection with a probability of 0.4 of mutation. Your initial population should be made of parse trees with 20 program nodes, and you should chop any parse tree with more than 60 nodes.

Sum the box-wall fitness function over 40 boards to get fitness values, remembering to test each dozer controller in a generation on the same boards, but generate new boards for each new generation. Evolve your populations for 100 generations, saving the maximum and average per-board fitness values in each generation as well as the best parse tree in the final generation of each run. Do 30 runs and plot the average of averages and average of best fitness.

Answer the following questions:

(i) *How do the parse trees compare with the strings from Section 10.1?*
(ii) *Do the "best of run" parse trees differ from one another in their techniques?*
(iii) *Are there any qualitative differences between the strings and the parse trees?*

Now we again ask the question, can evolution make use of random numbers? In this case, we do so by activating the RND terminal in our GP software.

Experiment 10.8 *Modify your parse tree routines from Experiment 10.7 to include the RND terminal. Do the same evolution runs, answer question (ii) from Experiment 10.7, and compare the results of this experiment with those obtained without the RND terminal.*

It is of interest to know to what degree the parse trees in the "best of run" file from Experiment 10.8 are simply generating biased random numbers as opposed to reacting to their environment.

Experiment 10.9 *Build or obtain a piece of software that can read in and evaluate the parse trees saved from Experiment 10.8 and determine the fraction of moves of each type (turn left, turn right, or go forward) and detect the use of RND and of the sensor terminals. Do 100 evaluations of each parse tree confronted with no adjacent boxes, one adjacent box in each of the 8 possible locations, a wall in each of the 4 possible positions, and each of the 20 possible combinations of a box and a wall. Answer the following questions:*

(i) Do the parse trees act differently in the presence of boxes and walls?
(ii) What is the fraction of the parse trees that use the RND terminal?
(iii) What is the fraction of the parse trees that use (some of) the sensor terminals?
(iv) Do any parse trees use only sensors or only the RND terminal?

In a paragraph or two, write a logically supported conjecture about the degree to which the parse trees under consideration are only biasing mechanisms for the RND terminal.

Now, we will revisit the automatically defined function. The ADF is the GP structure acts like a subroutine. This is accomplished by adding a second parse tree to the structure of the dozer controller. As in Section 9.4, the second parse tree contains the "code" for the ADF. In the "main program" parse tree we add a new operation called ADF. An example of a parse tree and its accompanying ADF are shown in Figure 10.5. In our implementation, ADF will be a binary operation: there will be two new terminals called x and y in the ADF parse tree that have the values of the two arguments passed to the ADF. This will require fairly extensive modifications of our GP software. These include the following:

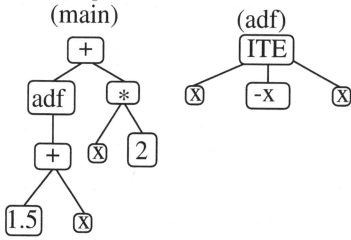

Fig. 10.5. Main parse tree and ADF parse tree.

- We must be able to maintain two separate (but closely related) GP languages for the main and ADF parse trees. In the main parse tree, there is a new binary operation called ADF. In the parse tree describing the ADF, there are two new terminals x and y.
- The parse tree evaluator must be able to evaluate the two parse trees appropriately. Whenever an ADF operation is encountered in the main parse tree, the evaluator must compute the values of its arguments and then evaluate the ADF parse tree with those values used for x and y.
- Randomly generated dozer controllers now contain two parse trees, and all the utility routines must work with both trees.
- Our mutation operator must now choose whether to mutate the main or the ADF parse tree.
- Crossover should do one of three things:
  - Cross over the main parse trees of the parents.
  - Cross over the ADF parse trees of the parents.
  - Take the main and ADF trees from distinct parents in both possible ways.

The use of ADFs in genetic programming gives many of the benefits of subroutines in standard programming. They are small pieces of code that can be called from several locations in the main parse tree. A good ADF can evolve and spread to multiple different members of the population, roughly a form of code reuse. A program with an ADF is more nearly modular and may therefore be easier to mine for good ideas than a single large parse tree. Let us see to what degree using ADFs helps our dozer controllers.

Experiment 10.10 *Make the modifications needed to allow the software from Experiment 10.8 to use ADFs. Use a starting size of 20 nodes for the main parse trees and 10 nodes for the ADF trees. Chop the main parse trees when their size exceeds 40 nodes, and the ADF trees when their size exceeds 20.*

Redo the runs for Experiment 10.8 using ADFs both with and without the RND terminal. Set the mutation rate to 0.4. Half the time mutate the ADF; half the time mutate the main parse tree. For crossover, cross over the ADF with probability 0.4, the main tree with probability 0.4, and simply trade the main tree and ADF tree with probability 0.2.

In your write-up, compare the performance of the 4 possible types of evolved controllers, with and without RND and with and without ADF.

There is another issue to treat before we conclude this section on plain genetic programming for Tartarus. Crossover is a very disruptive operation in genetic programming. In a normal evolutionary algorithm, it is usually the case that if we cross over two identical creatures, then the resulting children will be identical to the parents. This is manifestly not so for genetic programming, as we saw in Problem 291.

The standard crossover operation in genetic programming, subtree crossover, is an example of a *nonconservative* crossover operator.

In the theory of evolutionary computation as presented in the earlier chapters, the role of crossover was to mix and match structures already in the population of creatures rather than to generate new ones. The GP operator is far more capable of creating new structures than a standard (conservative) crossover operator. It is thus probably a good idea to use null crossover, the do-nothing crossover defined in Chapter 2, a good deal of the time. When applying null crossover, the children are simply verbatim copies of the parents.

Experiment 10.11 *Rebuild the software from Experiment 10.8 to use standard crossover with probability p and null crossover with probability $(1 - p)$. Redo the data acquisition runs for $p = 0.1, 0.5$ and also use the runs done in Experiment 10.8, which can be viewed as $p = 1.0$ runs. Compare performance and redo the sort of analysis done in Experiment 10.9 on the best-of-run files from all 3 sets of runs.*

Problems

Problem 392. Which subsets of the set of the 7 operations given in Table 10.2 are complete sets of operations for $\mathbb{Z}_3$? A set S of operations is said to be *complete* if any operation with any number of arguments can be built up from the members of S.

Problem 393. We can view constants as being operations that take zero arguments. Taking this view, there are 3 operations named 0, 1, and 2 in our GP language. If we add these 3 operations to those in Table 10.2, then which subsets of the resulting set of 10 operations are complete? (See Problem 392 for a definition of complete.)

Problem 394. Essay. Why would it be nice to have a complete set of operations in a GP language? Is it necessary to have such a complete set? Is there any reason to think having more than a minimal complete set of operations is nice?

Problem 395. Give a clear description of (or pseudocode for) a conservative crossover operation for use on parse trees. Compare its computational cost with the standard crossover operator.

Problem 396. Suppose we have a GP language with only binary operations and terminals, including a binary ADF. If the ADF and the main parse tree between them possess a total of exactly n nodes, give a formula for the maximum number of operations executed in evaluating a parse tree pair. Assume that every instruction of the ADF must be reexecuted each time it is called by the main parse tree.

Problem 397. Prove that the total number of nodes n in the main parse tree and ADF in Problem 396 is even.

Problem 398. Evaluate the parse tree with ADF shown in Figure 10.5 for:

   (i) $ML = MR = 0$,
  (ii) $ML = 1$, $MR = 0$,
 (iii) $ML = 0$, $MR = 1$,
  (iv) $ML = MR = 1$,
   (v) $ML = 2$, $MR = 0$,
  (vi) $ML = 0$, $MR = 2$,
 (vii) $ML = 1$, $MR = 2$, and
(viii) $ML = 2$, $MR = 1$.

Problem 399. Write by hand a parse tree in the GP language used in this section that will exhibit the following behaviors in the presence of isolated blocks:

- If a block is in the square to the left or right of the dozer, the dozer will turn toward it.
- If a block is behind the dozer, the dozer will turn.
- If a block is ahead of the dozer, the dozer will go forward.

You may use an ADF if you wish.

Problem 400. Take the parse tree you wrote for Problem 399 and diagram its behavior for 80 time steps on the board shown in Figure 10.2.

Problem 401. Essay. In Problem 399, you were asked to create a parse tree that exhibited three behaviors. This is an ambiguous task; each behavior responds to a particular environmental event, but these events are not always distinct. First, explain how many events the three behaviors requested are responding to and detail carefully which combinations of the events can happen at the same time. Second, establish and defend a precedence structure on the behaviors that will yield a good score for Problem 400. Advanced students should instead defend their precedence structure against the entire space of possible boards.

Problem 402. When you did Problem 399, you probably noticed that the dozer can push a single box over to the wall and get stuck. Assume that the answer to Problem 399 was done as a single parse tree, and that that parse tree is being used as a terminal ADF (an ADF that takes no arguments). Write a main parse tree that returns the output of its ADF with probability p and otherwise does something random (i.e., returns an evaluation of RND) for at least 5 different values of p, using no more than 20 nodes in the main parse tree.

10.3 Adding Memory to the GP language

The setup of the Tartarus world makes it impossible for the dozer controller, in many circumstances, to tell the difference between pushing a box forward and pushing it in a futile fashion against a wall. How long should a dozer push a box before giving up? How can a dozer controller know how long it has been pushing a block? There are two broad approaches to this problem. One is for the dozer controller to learn to count to the width of the world minus one; the other is for the controller to avoid instructing the dozer to push a box more than once without turning. The latter strategy may require more time steps to accomplish its goal, but is much simpler for evolution to discover. To see this, try to come up with a method of counting in any of the GP technologies used so far. Tricky, isn't it?

In his paper "The Evolution of Mental Models" [55], Astro Teller added to his dozer control language a new type of structure called *indexed memories*. He had 20 memories that were accessed by a binary store operation (arguments were the number to store and an index into the memories) and a unary recall operation (the argument was an index into the memories). We will use a much simpler form of memory, mimicking that used for the PORS problem in Chapter 8. We will add a varying number of new unary operations and terminals for using one or more memory locations with our GP language.

Experiment 10.12 *Add to the GP language we have been using so far an STO and RCL instruction, as in Chapter 8. The store operation is unary. It places a number (mod 3) into an external memory location and also returns the number stored as its value. The RCL operation is a terminal returning the contents of the external memory. Initialize the external memory to zero.*

With this modified GP language, use the same evolution as in Experiment 10.8 to test the utility of having a memory. We had 4 sorts of GP languages for Tartarus available before we added the memory: with and without ADF and with and without the RND terminal. Adding in a memory gives a total of 8 possible GP languages, 4 of which we have already tested. Do evolutionary runs for one of these 4 possibilities. (Later experiments will refer to the memory-with-ADFs version of this experiment.)

- *memory alone,*
- *memory with ADFs,*
- *memory with RND terminals available, or*
- *memory with ADFs and RND terminals available.*

For those runs you perform, compare performance with other available data. If you have runs using and not using the RND operator, discuss the degree to which the RND terminal can interfere with use of the memory. Place your discussion in an evolutionary context: in your opinion, does use of the RND terminal create local optima that arrest the development of memory-using strategies?

Having a single memory gives the dozer controller, in some sense, three internal states, 0, 1, and 2, stored in the memory. Since the version of Tartarus we are using is 6 squares across, this may not be enough. It may well be worth trying more memories. We will try two distinct techniques: indexing 3 memories and making 3 separate pairs of memory manipulation terminals available.

Experiment 10.13 *Modify the software from Experiment 10.12 to have 3 unary operations, STO1, STO2, and STO3, and 3 terminals, RCL1, RCL2, and RCL3, that access 3 separate memories. Do the evolutionary runs with ADFs but no RND terminals and compare the results with the single-memory version.*

The next experiment will be more like the indexed memories used by Teller. We will transform the unary store instruction into a binary store instruction and the recall terminal into a unary recall operation.

Experiment 10.14 *Rebuild the GP language used in Experiment 10.12 to use 3 memory locations as follows. Instead of a unary store operation STO, have a binary store operation so that (STO x y) places the value x into memory location y. Instead of a recall terminal RCL, have a unary recall operation so that (RCL x) returns the contents of memory x. Do the same evolutionary runs with ADFs but no RND terminals and compare the results with the single-memory version and with Experiment 10.13.*

So far, we have evolved dozer controllers for the Tartarus problem that have access to two types of resources: sensory information and memory. If you look carefully, you can find controllers that use only sensory information, that use only memory, and that use both. To our surprise, memory is more valuable, by itself, than sensory information, at least as they appear in the implementations of this chapter. This in turn raises the questions, "What other types of memory are available?" and "Is there another collection of sensory information we could present to the controllers?" We will address this issue in the Problems and attempt an exploratory answer to the question, "How much memory is enough memory?"

Problems

Problem 403. Using the version of the GP language from Experiment 10.14, write by hand a dozer controller to attack the board shown in Figure 10.2. We recommend that this problem be worked in small groups. Cruel instructors may wish to assign grades competitively.

Problem 404. Identify and explain your identification of dozer controllers from the chapter so far that:

(i) use memory only,

(ii) use sensory information only, and
(iii) use both.

Problem 405. How many internal states (internal means not depending on the values returned by sensor terminals) can a dozer controller in Experiment 10.13 or 10.14 have?

Problem 406. Define a *state* of the Tartarus board to be a set of possible positions for the boxes together with a position and heading of the dozer. Compute the number of possible states of Tartarus for $k = m = 6$ (6×6 board with 6 boxes). How many memories, storing a value (mod 3) would a GP language require to give the dozer controller that many internal states?

Problem 407. Essay. Reread Problem 406. Discuss the value of having at least one internal state per external state. There are three things you may wish to consider: the problem of recognizing a state of the world, the problem of managing the information involved, and the effect symmetries of the board may have. Based on your discussion, try to estimate a useful number of states.

Problem 408. In his experiments, Teller had 20 indexed memories, which could store a value in the range $0 \le n \le 19$. Based on your answer to Problem 406, how many internal states did Teller's dozer controllers have per possible external state of the world?

Problem 409. Refer to Problem 385 if you did it. A Markov chain controller (Markov chains are discussed in Appendix B) for a dozer is a 3-state Markov chain with the states corresponding to the 3 actions the dozer uses: turn right, turn left, and go forward. Suppose we were using a language with a single memory and the RND terminal, as in Experiment 10.12; then which of the following Markov chain dozer controllers could we simulate in that language? The Markov chains are represented by their transition matrices indexed in the order L(eft), (R)ight, and (F)orward. Give a constructive proof of your answer when possible.

$$(i) \begin{bmatrix} 0 & 1 & 0 \\ 0 & 0 & 1 \\ 1 & 0 & 0 \end{bmatrix}, \quad (ii) \begin{bmatrix} 0 & 0 & 1 \\ 0 & 0 & 1 \\ 1/2 & 1/2 & 0 \end{bmatrix}, \quad (iii) \begin{bmatrix} 1/3 & 0 & 2/3 \\ 0 & 1/3 & 2/3 \\ 1/3 & 1/3 & 1/3 \end{bmatrix}.$$

Problem 410. For each of the Markov chains given in Problem 409, compute which strings of moves they can produce. You answer should be a specification of a subset of the set of all strings over the alphabet $\{L, R, F\}$.

Problem 411. Give the transition matrix of a Markov chain controller that can produce sequences of moves where the following rules hold:

- no more than 2 left turns in a row,
- no more than 1 right turn in a row,
- no more than 5 moves ahead in a row,

- no right turns follow left turns,
- no left turns follow right turns.

Hint: how many states do you need at a minimum?

Problem 412. Programming Problem. Write a program that takes as input a parse tree and produces a C, C++, Pascal, or LISP (your choice of one) routine that takes a sensor state of Tartarus as input and returns a dozer move as output, in a fashion exactly that of the parse tree used as input. This is a parse tree compiler.

10.4 Tartarus with GP Automata

In this section, we will fuse the finite state automata we learned to evolve in Chapter 6 with parse trees to produce an artificial life technology called a GP automaton. A number of experiments with GP automata appear in the literature [1, 2, 28]. Given the obvious desirability of being able to count to small numbers, a very easy and natural function for a finite state automaton, it would be nice if we could adapt finite state automata to Tartarus.

In Chapter 6, we used finite state automata with a small input and output alphabet. Consider a finite state automaton used to control a dozer. The output alphabet would be simply $\{L, R, F\}$. What about the input alphabet? Since there are 8 terminals that sense adjacent squares, each of which can take on 3 values (0-empty, 1-box, 2-wall), there are naively $3^8 = 6561$ possible inputs. The true number is smaller, as you are asked to compute in Problem 413, but there are still many hundreds of possible inputs to the dozer in the Tartarus environment, a dauntingly large "input alphabet" for our putative finite state automaton.

If we were to implement a finite state automaton in the tabular fashion of Chapter 6, then we would require several hundred columns in the transition table, each specifying the next state and action for a possible state of the sensor terminals. This means that each data structure would have a number of genetic loci in the thousands. This structure is too large to use with an evolutionary algorithm that finishes in a reasonable amount of computer time.

The natural solution to this dilemma is to filter the input data down to a manageable number of objects *before* presenting it to the finite state controller. In effect, we want to compress the input bandwidth of Tartarus down to something manageable. If we were working with real-valued input data, for example, we could divide the data into a few ranges and use those as the input alphabet for our finite state automaton. With Tartarus, it is not obvious what sort of data compression to use, and so we will leave the matter to evolution.

In Sections 10.2 and 10.3, we have already created a system for building and testing potential bandwidth compression devices: parse trees! Our parse trees took a set of sensor terminal values and distilled from them a move: turn left, turn right, or go forward. Those parse trees with a fitness of 1 or more were able to make at least one fairly sensible decision for at least one configuration. If we fuse those decision-makers with a finite state automaton that can do exactly the sort of counting that is natural for the Tartarus environment, we may reap great performance improvements.

Name	Type	Description
0, 1, 2	terminal	constants equal to possible input values
UM,UR,MR,LR, LM,LL,ML,UL	terminal	Report on squares adjacent to the dozer
ODD	unary	Return 1 if the argument is odd, 0 otherwise
COM	unary	Compute one minus the argument (complement)
~	unary	Unary minus
+, −	binary	The usual addition and subtraction
=, <>, >=, <=, >, <	binary	Comparisons that return 0 for false and 1 for true
MAX,MIN	binary	maximum and minimum
ITE	trinary	if-then-else, $0 =$ false and $1, 2 =$ true

Table 10.3. Decider language, terminals, and operations.

Definition 10.5 *A* **GP automaton** *is a finite state automaton that has a parse tree associated with each state.*

You should review Section 6.1. The GP automata we use for Tartarus are defined as follows. Each GP automaton has n states, one of which is distinguished as the initial state. The initial response will always be to do nothing. As with standard finite state automata, we will have a transition function and a response function. The response function will produce moves for the dozer from the alphabet {**L**, **R**, **F**}. The transition function will produce a number representing one of the states. Both will use input from parse trees called *deciders*. They will operate on the parity (odd or even) of the output of the deciders. Each state will have a decider associated with it. These deciders will be integer-valued (any integer, not integers (mod 3)) parse trees using operations and terminals as given in Table 10.3. Their job is to "look at" the current Tartarus board and send back a small amount of information to the finite state automaton.

The data structure used for the GP automaton is an integer specifying the initial state together with an array of states. Each state is a record containing a decider, an array of the responses that a state makes if the decider returns an odd or even number, and the next state to go to if the decider returns an odd or even number. This data structure is selected with an eye toward the genetic operators we will be using. An example of a GP automaton is given in Figure 10.6.

Start: 1→7			
State	If Even	If Odd	Deciders
0	F→6	F→3	(+ (< LM 2) (~ LR))
1	F→4	L→4	(<= 2 (min LL (> -1 MR)))
2	R→3	T→0	(ITE (~ LR) LR (Com (Com -1)))
3	F→4	R→1	(ITE UR (Odd (ITE (~ LR) LR (Com 2))) 2)
4	F→5	R→0	(max UR (- 2 (Com UM)))
5	L→6	R→0	(< (Com ML) (<= 2 LM))
6	F→3	T→1	(ITE (~ LR) LR -1)
7	T→5	R→3	(ITE (~ UR) (Odd (~ LR)) 2)

Fig. 10.6. An 8-state GP automaton for controlling a dozer in Tartarus. (Responses and transitions are given in the form *response* → *transition*. The responses L, R, and F are the standard moves for Tartarus; T is the new think action unique to GP-Automata.)

One interesting property of GP automata, in contrast to parse trees, is the divorce of input and output types. This divorce is inherited from finite state automata and represents a potentially valuable difference from pure parse tree systems. While it is always possible to interpret the output of a parse tree to change its type, e.g., $0 = L, 1 = R, 2 = F$, the process can rest on quite arbitrary decisions.

Genetic Operations on GP Automata

One of the bugaboos of genetic programming is the crossover operator. With the parse trees appearing in several separate parts of the GP automaton, we need no longer mix and match our entire data structure when doing crossover. The crossover operator we will use for GP automata treats the array of states as a string and simply copies the initial state of each parent to the corresponding child. We thus may use any of the string crossover operators from Chapter 2 on GP automata: one-point, two-point, as well as the more exotic operators.

As crossover becomes more straightforward, mutation becomes more vexed. There is a large number of fairly natural mutation operators that make sense for GP automata. There are two classes of mutations: mutation of the nondecider parts of the finite state controller and mutation of the deciders.

Definition 10.6 *To do a* **finite state point mutation** *of a GP automaton, choose uniformly at random one of: the initial state, a single transition, and a single response. Replace it with a valid random value.*

Definition 10.7 *To do an* **exchange mutation**, *exchange two uniformly chosen deciders.*

Definition 10.8 *To do a* **replacement mutation,** *copy one uniformly selected decider over another.*

Definition 10.9 *To do a* **cross mutation,** *perform a subtree crossover operation on a pair of uniformly selected deciders.*

Definition 10.10 *To do a* **decider point mutation,** *perform a normal parse tree point mutation (subtree replacement) on a decider.*

In our first experiment in this section, we will simply get some GP automata evolving. In the second experiment, we will explore the utility of the various mutation operators. In the third experiment, we attempt to characterize the behavior of the various controllers evolved. In the fourth experiment, we will make a substantial modification to the GP automata paradigm to allow the possibility of extended computation. In the fifth, we will take another look at population seeding, but with a new twist.

Experiment 10.15 *Build or obtain routines for handling GP automata. This should include all 5 of the mutation operators given above, as well as two-point crossover on the array of states. The deciders should be parse trees using the operations and terminals given in Table 10.3. Build an evolutionary algorithm that operates on a population of 60 GP-automata-controlled dozers with 8 states. Use tournament selection with tournament size 4 as your model of evolution. Do one mutation on each new dozer controller, split evenly among the 5 mutation operators given and null mutation (doing nothing). Chop any deciders that grow to have more than 20 nodes. For fitness, evaluate each controller on 40 Tartarus boards in the usual fashion.*

Run the algorithm for 200 generations saving the average and best fitness of each population in each generation and the fraction of actions of each type. Do 20 runs and save the best controller in the final generation. Compare the per-board fitness with other experiments you have done. For comparison, write a program that tests each of the best-of-run controllers on 5000 Tartarus boards. The average over 5000 boards can be used to rate controllers in a stable fashion. In your write-up, graph the average, over populations and runs, of the average and best fitness in each generation.

With GP automata software in hand, we are ready to explore the various mutation operators.

Experiment 10.16 *Take the software from Experiment 10.15 and rebuild it to allow you to set the probability of using each mutation operator. Redo the experimental runs but with the following mixes of mutation operators (use all the mutation operators in a run equally often):*

(i) Finite state point mutation and decider point mutation.
(ii) Finite state point mutation and cross mutation.

(iii) Finite state point mutation, cross mutation, and exchange mutation.

Report the same data as in Experiment 10.15 and compare the performances of the two experiments.

The next experiment is intended to help you understand the results of the preceding two experiments. If things are going as they did for us, then using GP automata caused a substantial increase in performance. It would be nice to try to achieve some degree of qualitative understanding of this increase.

Experiment 10.17 *To help understand and analyze your results, write or obtain a lab program that can read in the best-of-run GP automata from Experiments 10.15 and 10.16. This program should allow you to test the GP automata on a number of boards and also should allow you to watch the dozer move, time step by time step, on boards that you specify. The details are left to you, and the project should be graded in part on the design of the lab. Use the lab to develop a taxonomy of dozer behaviors. Write up an explanation of how (and how well) each type of behavior you found works.*

In many ways, a GP automaton is more like a normal computer program than a standard GP parse tree. Instead of a single statement (parse tree) living in a trivial or iterative control structure, the program represented by a GP automaton has many statements (deciders and actions) that live inside a nontrivial control structure (the finite state automaton). One important difference between a standard computer program and a GP automaton is that a computer program may execute many statements before it produces output, while a GP automaton produces one piece of output per statement (decider) executed.

It is not too hard to extend the GP automata paradigm to include the execution of multiple deciders before producing output. In standard finite state automata, there are sometimes transitions that produce no response. These are called λ-*transitions* after λ the empty string. We will add a new action, the *null action*, to the responses that the GP automata can make. When a GP automaton produces a null action, it immediately makes the transition to the next specified state and executes it normally, without waiting for the next time step, a form of λ-transition. This means that null actions can allow the execution of multiple deciders per time step in the GP automata. There is one potential pitfall: infinite loops. A GP automaton that uses null actions has the potential to get stuck in a loop in which one null action leads to another forever. To finesse this potential problem, we will simply cut off a GP automaton if it evaluates more than some fixed number of deciders in a time step.

Experiment 10.18 *Modify the software from Experiment 10.15 to allow null actions and extended computation in the fashion described above. Cut off the GP automaton if it uses 8 null actions in a row and have the dozer sit in place for a time step in this event. Do the same evolution runs as in Experiment 10.15 and compare performance.*

There is an interesting point to be made by our attempt to improve performance by using null actions. When a null action is executed, the automata is, in effect, saying, "I don't have enough information; I need to think about this some more." Then it moves to another state and uses the decider in that state to determine its action. It is using two parse trees instead of one to make its decision. In fact, it can use up to eight parse trees. Adding the null action increases the complexity of the decisions being made while making the data structure only marginally more complex.

There are many outputs possible from each decider (any integer). In a GP automaton, it is possible to divorce actions (**L**, **R**, **F**) from computations (done using integers) from inputs (sensors). This means that without type checking or the attendant computational complexity, it is possible to use three distinct data types. In dealing with discrete events in a real-parameterized situation that requires discrete responses, the advantage of this separation of the three classes of data (input, computational, and output) will become more pronounced.

At the beginning of this section, while setting the stage for GP automata, we noted that using finite state automata in the Tartarus environment is quite tricky due to the large size of the input bandwidth, and then cheerfully noted that the parse tree technology from the preceding sections gave us a very natural bandwidth-compression technology. We then used random parse trees in our GP automata, even though we have software that could give us parse trees. In the next experiment, we will take the natural step of creating our initial populations of GP automata from preevolved parse trees. In order to do this next experiment, you will need the software from Experiment 10.7.

Experiment 10.19 *First change the software from Experiment 10.7 to have a chop limit of 20 nodes and to use the language used by the deciders in this section. Interpret the integer output (mod 3) to produce actions. Perform 10 runs, saving the entire final populations of all the runs into a single file. Now modify the software from Experiment 10.15 to use parse trees chosen randomly from this file, instead of randomly generated parse trees. Perform the same evolutionary runs as in Experiment 10.15 and compare the performance. Be sure to account for, or at least mention, the additional computation done before the GP automata in this experiment started evolving.*

There are, as usual, a large number of other experiments one could perform with the technologies in this chapter. Students looking for projects might try exploring the mutation operators more completely, changing the board size, changing the fitness function, or changing the task being attempted by the dozers. Variation of task will be done in a large way in Chapter 12. It also might be of interest to see how performance varies as the number of states in the automata are varied. Finally, we have made no effort to test GP automata with access to randomness or memories in their deciders. We will come back to the idea of GP automata in different environments in later chapters.

Problems

Problem 413. Using any of the GP languages from Sections 10.2 and 10.3, the dozer controller may "see" a large different number of configurations of adjacent squares. Compute how many such configurations there are.

Problem 414. Is the crossover operator given in this section for GP automata conservative? See Definition 10.4.

Problem 415. Give an example of a GP automata application in which it would be valuable to exploit the divorcing of input, computation, and output data types, so as to use 3 separate data types.

Problem 416. Essay. Consider a parse tree dozer controller and a GP automaton dozer controller. Assuming that both have the same number of nodes of parse trees somewhere inside, which executes more code per time step on average? Treat both models of GP automaton computation given in this section (with and without null actions).

Problem 417. Programming Problem. Write a program that takes as input a GP automaton and produces a C, C++, Pascal, or LISP (your choice of one) routine that takes a sensor state of Tartarus as input and returns a dozer move as output, in a fashion exactly like that of the GP automaton used as input. This is a GP automaton compiler.

Problem 418. Essay. Explain the possible evolutionary utility of the exchange and replacement mutation operators. If available, support your ideas with data from the experiments.

Problem 419. Using 4-state GP automata similar to those used in Experiment 10.18, write by hand a dozer controller to attack the board shown in Figure 10.2. We recommend that this problem be worked in small groups. Cruel instructors may wish to assign grades competitively. (Compare with Problem 403.)

Problem 420. Essay. In this chapter, we enhanced standard genetic programs with random number terminals, ADFs (subroutines), memory, indexed memory, and finite state automata. Describe techniques for and discuss the possible advantages and disadvantages of adding stack manipulation operations and terminals to the GP languages used in the Tartarus environment. Assume an operation PUSH that puts x on the top of the stack and returns the value of x, and terminals POP (pops the stack), TOP (reports the value on the top of the stack), and EMP (returns 0 if the stack is not empty, 1 if it is).

Problem 421. Explain the sense in which the string genes are finite state automata. Where is the state information stored? How many transitions are there out of each state when a string gene is viewed as a finite state automaton?

Problem 422. Give a procedure for constructing a GP automaton with n states that has the property that its transition and response functions duplicate the behavior of a string of length n.

Problem 423. Essay. Suppose we have a source of excellent string genes of length 8. Would you expect a population generated by the technique suggested in Problem 422 to do better compared to the random populations used in Experiment 10.15: (i) at first, and (ii) by the end of the experiment?

Problem 424. Essay. Review Problem 408. In "The Evolution of Mental Models" Teller reports fitnesses in the ballpark of 4.5, averaged over a large number of boards. If all went as it should, the best GP automata you evolved in this section got fitnesses up in the area of 5 to 6.5. In light of the various string baseline experiments in Section 10.1, what is a good number of states to use?

10.5 Allocation of Fitness Trials

How many Tartarus boards should we use to evaluate the fitness of a dozer controller? We dealt with a similar problem in Chapter 5 in evaluating the fitness of symbols (see Problem 178). While we want to have a controller that does well on all 300,000+ possible Tartarus boards, we cannot afford to test it on them all. If we do not use a large number of boards, there is a danger that an inferior controller will outscore a superior controller on the Tartarus boards we happen to pick. The conflict then is between computer time needed to test controllers and quality of fitness evaluation needed to save the superior controllers. Let us verify that this problem exists.

Experiment 10.20 *Take the best-of-run GP automata that you saved in Experiment 10.15. Rank order them by the average per-board fitness they achieved in the evolutionary algorithm that produced them. After that, fix a set of 5000 Tartarus boards selected uniformly at random, and evaluate the GP automata's per-board average fitness on those boards. Rank order them according to the new fitness numbers. Did the order change?*

If your experiment went as ours did, a 40-board test does not produce the same rank ordering that a 5000-board test does. It is not unreasonable to imagine that this problem is more acute for better controllers. With a population of random controllers that have not undergone evaluation and selection, it is likely that most of them will be quite unfit. Testing on only a few boards is enough to sort out the slightly-more-fit minority. At the other end of evolution, we have many controllers that are quite good, and we are likely to be making fine distinctions among them. Since the best controller in the population may have to survive for dozens of fitness evaluations before a better controller arrives, it has dozens of chances to die from an unlucky selection of test boards. With this thought in mind, we propose our first experiment in wisely allocating fitness trials.

Experiment 10.21 *Ask your instructor which evolutionary algorithms from this chapter are to be modified to perform this experiment. Good ones would be from among Experiments 10.3, 10.7, 10.13, and 10.15. Replace the fitness evaluation in the evolutionary algorithms you are going to modify with a fitness evaluation that uses 10 boards for the first generation and a number of boards that is the larger of 10 or 10 times the maximum average per-board fitness in the preceding generation, rounded to the nearest integer thereafter. Report the effect, if any, on the average and standard deviation (over runs performed for a given setup) of the maximum fitness in the final generation.*

Let us turn our attention now from the issue of how many boards to the issue of *which* boards. The ideal situation would be to locate a small collection of boards with the property that if a dozer controller does well on those boards, then it will do well on the full set of boards. The problem is that there is no a priori way to know which those boards are. We can turn to nature for inspiration in this regard, by transforming our noisy optimization problem of finding a good dozer controller into a coevolution of boards and controllers.

The basic idea is this: Any 6×6 Tartarus board has 10 points of fitness "at risk" (the maximum possible score for a dozer on that board). Whenever a dozer meets a board, those 10 points are divided between the dozer and the board: the dozer gets points as usual, and the board gets those points not earned by the dozer. This gives us a fitness function for boards; boards with high fitness are those on which it is difficult for the dozers to score points. An *impossible board* is one on which it is impossible for *any* dozer to score 10. Now we can evolve the boards. Recall that a "board" is a placement of 6 boxes with no close group of 4 boxes, together with a placement and heading of the dozer.

Experiment 10.22 *Rebuild the software from Experiment 10.7 to use an evolving population of boards in the following manner. Have 40 slots for boards and fill them with 40 distinct boards (allow no repetition). Make sure that no board with a close group of 4 is included in the group. For fitness evaluation, test all dozer controllers on all 40 of the boards, awarding fitness to the boards in the manner described in the text. After the fitness evaluation, sort the boards by fitness and delete the k worst boards, replacing them with new ones. Save the best-of-run dozer controllers for each run and compare them, using the software developed in Experiment 10.20, with the best-of-run dozer controllers from Experiment 10.7. Perform this experiment for $k = 2, 10$, and 20. In addition to saving the best dozer controller in each run, save the highest-scoring board in the final generation.*

Problems

Problem 425. The evolutionary algorithm used to find difficult boards in Experiment 10.22 works by replacing the worst members of the population of boards with new random boards. Give a design for an evolutionary algorithm that would function more efficiently, in your opinion, at locating difficult boards.

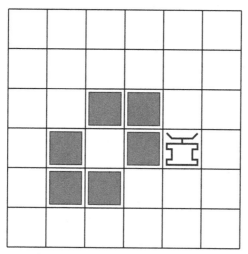

Problem 426. Examine the board above. Prove that the maximum score possible on the above board is 4. This board was discovered by Stephen Willson and is an example of an impossible board.

Problem 427. Essay. When we excluded boards with close groups of 4 boxes, it was because such configurations are impossible. If we are evolving difficult boards as in Experiment 10.22, then the system may locate impossible ones. Comment on whether an evolutionary algorithm like the one you designed in Problem 425 or the minimal system used in Experiment 10.22 is more efficient at finding impossible boards.

Problem 428. In this chapter, we have used random number generators, strings, parse trees, parse trees with memories, and GP automata as dozer controllers. If you were to build an evolutionary algorithm whose express purpose was to locate impossible boards, which of these would you pick as the controller? Would it be better to evolve the controller or to pick a set of really good controllers you had already evolved?

Problem 429. Construct another impossible configuration besides the one given in Problem 426 and anything involving a close-packed group of 4. You may make the board larger and use more boxes.

11

Evolving Logic Functions

In this chapter, we will be evolving logic functions. We will use neural nets like those used for the symbots in Chapter 5. We will explore different ways of coding them and different mutation and crossover operators. We will also compare neural nets with both basic and one advanced type of genetic programming as data structures for evolving logic functions. We assume you are familiar with the logic of NOT, AND, OR, NAND, NOR, XOR, parity, and majority functions. When writing out logic functions, we will use the convention that NOT is the unary operation $\neg$, and AND and OR are the binary operations $\wedge$ and $\vee$, respectively. In the genetic programming section we will switch to printable characters ($\sim$, *, and +) for these logic functions. The dependencies of the experiments in this chapter are given in Figure 11.1.

Evolving logic functions is a traditional test problem for both neural nets and genetic programming. This chapter presents basic techniques for evolving neural nets, and many others appear in the literature. Section 11.4 not only introduces the application of genetic programming to evolving logic functions but also introduces a rather clever idea of Peter Angeline's called a *Multiple Interacting Programs system*, or *MIPs, net*. Dr. Angeline once described MIPs nets in the following fashion: "Take a neural net. Put a parse tree on each neuron. Now throw away the neural net." MIPs nets are an example of a representation that permits a number of really interesting crossover and mutation operators as well as a nonstandard operator similar to gene-doubling in which a parse tree is duplicated. This operator incorporates a powerful technique from biology into evolutionary computation. Duplicating a gene and then varying one copy is an important source of new genetic diversity.

11.1 Artificial Neural Nets

Artificial neural nets are structures inspired by biological nervous systems. In a living creature, neurons are connected to one another at *synapses*. The neuron is both excited and inhibited by impulses coming in from different

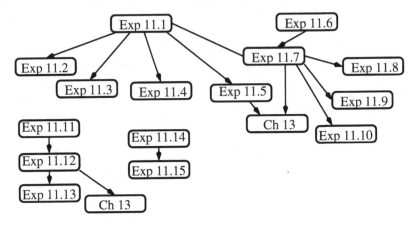

1   Evolving a 2-input logical exclusive OR.
2   Evolving a matrix with row-peeling crossover.
3   Evolving threshold weights.
4   Extending to NOR, AND, and "echo" functions.
5   Evolving 3-input functions.
6   Using the connection list representation.
7   Evolving functions with the connection list representation.
8   Lexical fitness with the number of neurons.
9   Evolving a binary adder with carry.
10 Adding neutral mutations.
11 A parse tree representation for logic functions.
12 Evolving logic functions with parse trees.
13 Exploring leaf mutation.
14 Extending the parse tree representation to MIPs nets.
15 Gene duplication with MIPs nets.

Fig. 11.1. The topics and dependencies of the experiments in this chapter.

synapses. When the total level of excitation passes a threshold, the neuron *fires*, or *turns on*. A biological nervous system functions through the effects of many such firings in a net of interconnected neurons. It changes by changing its physical connections and their properties. This is accomplished both by changing the strength of synaptic connections and by changing the chemical environment. This chemical modification happens directly when neurons release biochemicals and indirectly when the neurons cause various organs to release biochemicals. The bandwidth and operation of a biological neural net often defies understanding or even adequate description.

An artificial neural net is simpler. It has neurons with inputs and outputs and intraneuron connections that are analogous to biological synapses. Each connection has a weight, usually a real number, and a direction. The direction establishes which neuron is sending the message along the connection and which neuron is receiving. The character of a neuron is determined by its *transfer function.* This function computes the neuron's output from the sum of its weighted inputs. When functioning, a neuron follows these steps: first, the inputs from connected neurons are multiplied by their connection weights and summed; the neuron then passes that sum through its transfer function; then, the result is sent along all the neuron's output connections. The neurons in a neural net can update their output values synchronously or asynchronously. The choice of updating method has a large effect on the behavior of the net. Graphs of several common transfer functions appear in Figure 11.1.

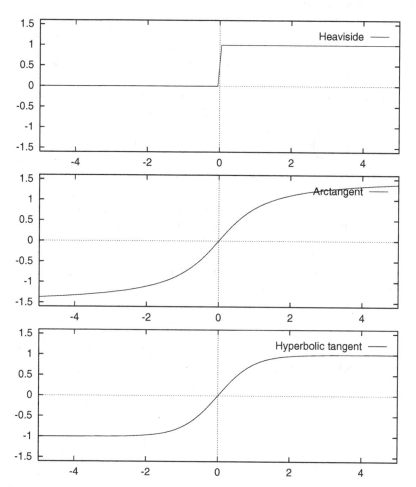

Fig. 11.2. Heaviside, arctangent, and hyperbolic tangent transfer functions for neurons.

When one is building a neural net, there are many choices to be made. What type of transfer function are you going to use? Should you allow the transfer functions to take parameters? What sort of connection topology are you going to use? How are you going to program your neural net?

Definition 11.1 *The* **connection topology** *of a neural net is a specification of which neurons are connected to which other other neurons. The connection topology is a directed graph. If this graph contains no directed cycles, then the net is said to be* **feed-forward**; *otherwise, it is* **recurrent**.

There are many ways to program neural nets. We will use evolutionary algorithms. For large net sizes and for some problems, evolutionary-algorithm-based programming of neural nets appears to be inferior to other methods. Where the evolutionary method shines is in the programming of small neural nets that have to do an ill-defined task and in programming systems of small neural nets that are later integrated together to do a complex task as we did with the symbots of Chapter 5.

The simplest transfer function used in neural nets is a step function. A neuron that uses this function is called a *Heaviside* neuron. It sums its weighted inputs, and it outputs 1 if the result is greater than a threshold value t, and it outputs 0 if the result is t or less. The first graph in Figure 11.1 is a Heaviside transfer function. These exceedingly simple neurons are more than adequate to the job of building neural nets that implement logic functions. Logic functions take n binary inputs and produce a single binary output. Examples of logic functions are *AND* functions (which output 1 only if all their inputs are 1), *OR* functions (which output 1 if any of their inputs are 1, and 0 otherwise), and *NOT* functions (which has a single input and whose output is 1 minus its input). Logic functions are described with truth tables, examples of which appear in Figure 11.3. Truth tables list all possible combinations of inputs with their corresponding outputs.

2-input **XOR** Function		
Input 1	Input 2	Output
0	0	0
0	1	1
1	0	1
1	1	0

3-input **AND** Function			
Input 1	Input 2	Input 3	Output
0	0	0	0
0	0	1	0
0	1	0	0
0	1	1	0
1	0	0	0
1	0	1	0
1	1	0	0
1	1	1	1

Fig. 11.3. Examples of truth tables for logic functions.

There are two broad classes of neural nets: *feed-forward* and *recurrent*. *Recurrent* nets have loops; *feed-forward* nets don't have loops. What is a loop? Recall that all connections in a neural net have a direction. A *loop* is a sequence of connections, followed in the positive direction, that returns to its starting point. Other things being equal, a recurrent net is much more complex in its behavior than a feed-forward net. Recurrent nets can, for example, develop long-term memory; the ability of feed-forward nets to remember previous inputs is limited to the longest sequence of connections following an input.

Problems

In the following problems, assume that the neurons are Heaviside neurons. You select the threshold value t. Use the same t for all the neurons in a single net.

Problem 430. Implement each of the following logic functions as a neural net:

(i) NOT,
(ii) 2-input AND,
(iii) 2-input OR,
(iv) 2-input XOR.

Problem 431. A *majority function* is a logic function with an odd number of inputs whose output is equal to the majority of its inputs (majority vote). Construct neural nets that implement 3-input and 5-input majority functions.

Problem 432. Compute the number of different n-input logic functions. Hint: count the number of ways to fill in the output column in a truth table.

Problem 433. Show that you can implement any n-input logic function using multiple instances of a single 2-input logic function. Give a neural net implementation of such a function with the fewest possible neurons.

Problem 434. Give a bound on the number of neurons needed to implement an n-input AND function, trying to make the bound as small as possible.

Problem 435. Suppose we are building an n-input logic function that outputs 1 for exactly k of the possible inputs. Prove that we need at most $k + 1$ neurons to build the function. This bound is not tight. Give an example of a 4-input function that outputs 1 for exactly 6 of its possible inputs and that uses fewer than 7 neurons.

Problem 436. What is the smallest number of 3-input majority functions and NOT functions needed to implement a 4-input AND function? Assume that you have a input constant of 0 or 1 available. Advanced students: prove that the constant is necessary.

Problem 437. Essay. Given that evolution is the primary method we will use to find, design, or program the structures we are studying in this book, why is question 436 relevant?

Problem 438. Essay. The connections in a neural net hold the knowledge (or functionality) of the neural net. The pattern of connections and the weights are both relevant. How much information can be packed into a single abstract real number? Be sure to *state* the definition of the term "information" you are using in your essay. Discuss the relevance of your answer to a neural net that might actually get built.

Problem 439. Construct a recurrent neural net with two inputs r and t that acts as follows. If t is 1, then the output of the net should be equal to r. If t is 0, then the output should be whatever it was the last time t was 1. This net is a very simple form of memory device: r is what is to be remembered, and t tells the device when to remember it.

Problem 440. Suppose you want to store in a computer unambiguous specifications of logic functions. How many bits will you need to specify an n-input logic function? Notice that the two examples in Figure 11.3 use 12 and 32 bits, respectively. (There are 12 or 32 1's and 0's in the truth tables.) These are *not* minimal representations.

11.2 Evolving Logic Functions

In this section, we will be evolving logic functions that use the truth values $0 = $ false and $1 = $ true. The neurons in our neural nets will be Heaviside neurons, making this assignment of truth values a natural one, since it is already the output type of the neurons.

An n-input logic function can be thought of as a string over the alphabet $\{0, 1\}$ of length 2^n that is the output column of its truth table when inputs are listed in the standard binary order, the one that corresponds to counting in binary. The truth table of the AND function, shown in Table 11.1, would correspond to the string 0001, for example. Table 11.2 gives the output strings for several standard logic functions.

In order to implement a logic function as a neural net, we will evolve the output strings of logic functions to match a reference string that is the output string of the desired function. How many neurons should we use? What connection topology should we use? In this section, we will implement an n-input function as a feed-forward network with 3 layers of neurons with all possible connections between one layer and the next and no connections between neurons in the same layer. There will be m neurons in the first layer, k neurons in the second layer, and 1 in the last. The real parameters of each neural net will be an $n \times m$ matrix F that holds the connection weights of

2-input AND function		
Input1	Input2	Output
0	0	0
0	1	0
1	0	0
1	1	1

Table 11.1. AND function truth table.

Inputs	Function	Output String
1	NOT	10
2	OR	0111
2	AND	0001
2	XOR	0110
2	NOR	1000
2	NAND	1110
3	Majority	00010111
3	Parity	01101001
3	AND	00000001
3	OR	01111111

Table 11.2. Output strings for various logic functions.

the inputs to the first layer of neurons, an $m \times k$ matrix S that holds the connection weights of the first layer to the second, and a $k \times 1$ matrix T that holds the connection weights of the second layer to the output neuron, together with $(m+k+1)$ different threshold values for the Heaviside functions of the neurons. This involves $nm + mk + m + 2k + 1$ total real values. For 2-input functions, we will find that $m = k = 2$ will suffice, which means that we will need 15 real parameters.

In addition to a model of evolution and some mutation operators taken from real function optimization, we need a fitness function. Our initial fitness function will be the number of agreements of the output string of a neural net representing a logic function (see Problem 443) with a reference string that is the output string of the desired logic function. We also need a way of evaluating the neural net efficiently. If $\mathbf{v}$ is the vector of inputs to the net, then $\mathbf{v}F$ is the set of inputs to the first layer. Running this vector through the thresholds of the first layer gives us a new $0, 1$ vector $\mathbf{u}$, and $\mathbf{u}S$ is the vector of summed inputs to the second layer. Running this through the thresholds of the second layer produces a $0, 1$ vector $\mathbf{t}$ that is the input to the third layer, and so $\mathbf{t}T$ is the number that the final neuron thresholds to give an output. From this discussion, one can see that running the neural net amounts to multiplying vectors by matrices and thresholding the resulting vectors. Thresholding in this case amounts to mapping values above the neuron's individual threshold value to 1 and those below to 0. The last piece needed to evolve neural nets is a crossover operator.

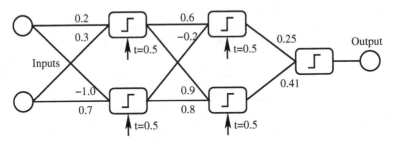

Fig. 11.4. The neural net logic function based on matrices F, S, and T.

The structure of the neural net as 3 matrices and a list of individual neural thresholds is not a string of real numbers. Each of the matrices could be peeled either by rows or by columns to make a string of real numbers. The 6 strings, 3 from the matrices and 3 from the thresholds of each layer, could be concatenated in any order. For simplicity, we will always use one of two ways of transforming the neural net into a string: row peeling of all 3 matrices or column peeling of all 3 matrices. Suppose we have a 2-input neural net with $m = k = 2$. If all the neurons have threshold values of 0.5, and if we have

$$F = \begin{bmatrix} 0.2 & 0.3 \\ -1 & 0.7 \end{bmatrix},$$

$$S = \begin{bmatrix} 0.6 & -0.2 \\ 0.9 & 0.8 \end{bmatrix},$$

$$T = \begin{bmatrix} 0.25 \\ 0.41 \end{bmatrix},$$

then the row peeling would yield the string of reals

$$(0.2, 0.3, -1, 0.7, 0.5, 0.5, 0.6, -0.2, 0.9, 0.8, 0.5, 0.5, 0.25, 0.41, 0.5),$$

while the column peeling would yield

$$(0.2, -1, 0.3, 0.7, 0.5, 0.5, 0.6, 0.9, -0.2, 0.8, 0.5, 0.5, 0.25, 0.41, 0.5).$$

In both row and column peeling, we put the matrix entries and the threshold values together in the order in which we use them in traversing the net. Once we have the neural net parameters peeled into a string of reals, we can use any of the crossover operators we have used in the past for real function optimization.

Notice that in this experiment we are simplifying matters by leaving the neuron's internal parameter (the threshold value) constant, rather than permitting it to evolve. The neural net specified by F, S, and T is given in Figure 11.4.

Experiment 11.1 *Create or obtain software for an evolutionary algorithm to operate on a gene containing an $n \times m$ matrix, an $m \times k$ matrix, and a $k \times 1$ matrix, coding a logic function as described in the text. Let all the neurons have the same threshold value α. Take, as your fitness function, the number of positions on which the output string of the evolving neural net agrees with the output string of the desired function. Use single tournament selection with tournament size 4 and single-point uniform real mutation with mutation size $\epsilon = 0.2$. Use column peeling to generate the strings for crossover. Set $n = k = m = 2$, take $\alpha = 0.5$, and perform exactly one mutation on each gene generated by crossover.*

With a population size of 200, run an evolutionary algorithm 100 times for at most 400 generations on the task of locating weights for an XOR function. Let the initial connection weights be in the range $-1 < x < 1$, and do not permit the threshold numbers to change. Save the number of generations to solution. Repeat the experiment with $\alpha = -0.5$. In your write-up, use the normal approximation to the binomial at the time half of one set of runs are completed to decide whether the choice of α makes a significant difference.

The next experiment is for the suspenders-and-belt student. It is intuitive that the choice between column peeling and row peeling isn't all that important. The only way to tell *for sure* is to do the experiment.

Experiment 11.2 *Modify the software from Experiment 11.1 to use row peeling instead of column peeling. Do the same runs and compare the time-to-solution results in the same fashion, both within this experiment and between experiments.*

Definition 11.2 *Call a logic function* **irreducible** *if for each input i, there is some combination of values for the other inputs for which changing the value of i will change the value of the output.*

What distinguishes interesting logic functions from boring ones? The definition of irreducibility is an attempt to formally define that property. (For more about irreducibility, see Problems 444, 445, and 446.) One quite simple and interesting function is the *parity* function on n inputs. If an odd number of its inputs are 1, the parity function returns 1; otherwise, it returns 0. Notice that XOR is 2-input parity. Because the n-input parity function is so interesting, it is a standard test case for systems that are supposed to induce logic functions.

A *cascade* of (2-input) functions is shown in Figure 11.5. Very often, a cascade of functions extends the functionality of the function to a larger number

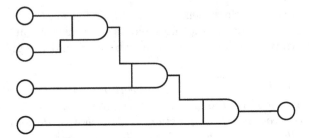

Fig. 11.5. A cascade of 3 AND functions.

of inputs. A cascade of n 2-input AND or OR functions yields an $(n + 1)$-input AND or OR function, for example. Likewise, a cascade of $(n - 1)$ XOR functions yields n-input parity. This implies that for an evolutionary algorithm that works on large logic functions, some sort of variation operator that creates cascades would be a good thing.

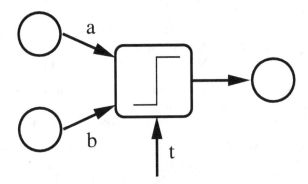

Fig. 11.6. A minimal fragment of a neural net representing a logic function.

At the moment, we want to look at a small, meaningful subunit of a neural net. One neuron connected to two inputs can implement a 2-input function. Such a net fragment is depicted in Figure 11.6. For a fixed threshold value t, we can determine the neural net's behavior as a function of its two input weights, a and b. There are 16 possible behaviors, corresponding to the 16 2-input logic functions, only some of which are possible. We find which logic functions we have by applying the 4 possible combinations of zero-one inputs (00, 11, 01, and 10) and examining the output behavior. For input 00, the neural net outputs $0 \geq t$; for 11, it outputs $a + b \geq t$; for 01, it outputs $b \geq t$, and for 10, it outputs $a \geq t$. The result is a "phase diagram," indexed by the values of a and b. A pair of such phase diagrams for threshold values $t = 0.5$ and $t = -0.5$ are given in Figure 11.7.

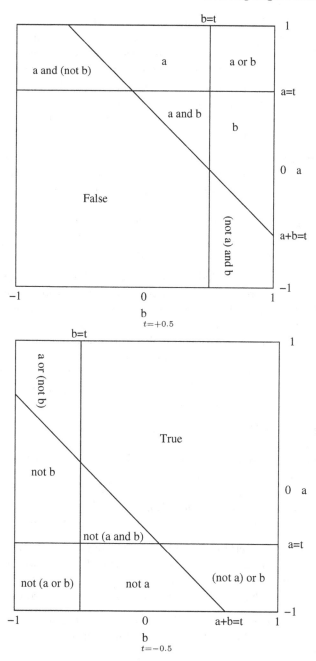

Fig. 11.7. Phase diagram of logic in the neural net fragment from Figure 11.6 in terms of connection weights a and b. Note what logic functions are possible for positive and negative thresholds, respectively.

Make sure that Figure 11.7 makes sense to you. It seems that the two-connection-weights-and-one-neuron fragment shown in Figure 11.6 can produce very different sets of logic functions depending on whether the threshold is positive or negative. In particular, the two functions from which any other function can be built, $\neg(a \wedge b)$ (NAND) and $\neg(a \vee b)$ (NOR), appear only when the threshold is negative. This sheds light on the outcome of Experiment 11.1. It also suggests a second experiment to see what thresholds evolution will select, left to its own devices.

Experiment 11.3 *Modify the software from Experiment 11.1 to make the threshold weight of each neuron evolvable and initialized in the range $(-1, 1)$. Rerun the experiment, documenting whether solution time with floating thresholds is significantly different from that with fixed positive or negative threshold values. Also, plot the fraction of neuron thresholds that are positive and negative in the final generation of each population. (The instructor may wish to assign Problem 452 with or before this experiment.)*

Experiment 11.3 is intended to check whether the theoretical suggestions of Figure 11.7 are correct. If they are, then most populations should have at least one negative threshold weight. This question is explored in more detail in Problem 452, but be sure at least to consider the question of what outcomes in terms of distribution of negative thresholds would be interesting or suggestive.

At this point, we suggest an experiment designed to verify something hinted at in other places: XOR is hard. In earlier experiments, we used an evolutionary algorithm to locate XOR functions. Now we will locate some other functions as well for comparison.

Experiment 11.4 *Modify the software from Experiment 11.1 to search for a NOR (output string 1000), an AND (output string 0001), and an "a" function 0101 (it just echoes one of its inputs). Save the time-to-solution data and decide which of these functions is significantly easier to locate than (i) the others and (ii) the XOR function located in Experiment 11.1. Notice that all 3 of the logic functions in this experiment can be implemented as a neural net fragment of the type shown in Figure 11.6.*

To conclude this section, we will increase the size of the neural nets for which we are searching. All the experiments thus far have been on 5-neuron nets with 2 inputs. In the remainder of the chapter, we will work with larger neural nets, and so this experiment will provide a basis for comparison.

Experiment 11.5 *Modify the software from Experiment 11.1 to search for 3-input AND and parity functions. This means that $n = 3$, and we will take $m = k = 3$ as well. Set the neuron thresholds to -0.3 and do 100 runs for each function, leaving the parameters the same. Save the times-to-solution and compare them for these two functions.*

In this section, we have built code for evolving 3-layer feed-forward neural nets with all possible connections between layers. In the next section, we will retool our representation of neural nets to permit us to evolve connections as well as connection weights.

Problems

Problem 441. Compute the output string for the following logic functions: (Assume that the inputs are listed in the usual binary counting or lexicographic order.)

(i) 5-input parity,
(ii) 3-out-of-5 majority,
(iii) a cascade of 4 NAND functions,
(iv) a cascade of 4 NOR functions.

Problem 442. Give a logical function that will yield the following output strings. Call your inputs $x_1, x_2, \ldots, x_n$ and use only AND, OR, and NOT to build your logical expressions. Assume that x_1 is the lowest-order bit of the input (the one that changes the most often).

(i) **0110**,
(ii) **01101001**,
(iii) **00000001**,
(iv) **11111110**,
(v) **01110000**.

Problem 443. Suppose you have an n-input neural net called NET that somehow implements a logic function L. Give pseudocode that produces the output string of L. Do *not* nest n loops; this is inelegant.

Problem 444. Give an algorithm for detecting irreducibility (see Definition 11.2).

Problem 445. Explain why irreducible logic functions are interesting.

Problem 446. Prove that n-input parity, n-input AND, and n-input OR functions are all irreducible.

Problem 447. Suppose that the inputs of a neural net are on the vertices of a regular pentagon. Give the output string for a neural net that outputs 1 when no symmetry of the pentagon (rotation or flip) takes the pattern of inputs to itself, and outputs 0 otherwise. Hint: if we had said square instead of pentagon, the answer would be 0110100110010110.

Problem 448. Suppose we generalize Problem 447 to an n-gon. We call such a neural net an n-input asymmetry detector. Prove that the output string of the neural net must be a palindrome.

Problem 449. Examine Figure 11.7. The choice of the neuron threshold in the 1-neuron 2-input neural net fragment creates a good deal of variation in which logic functions are possible. Each of the negative threshold functions is a logical inversion of one of the positive threshold functions, and 7 functions appear in each diagram. Since there are 16 functions altogether, two are missing. Which two?

Problem 450. Examine Figure 11.7. If the two weights a and b are generated uniformly at random between 1 and -1, then what is the probability of each of the functions (i) for $\alpha = +0.5$ and (ii) for $\alpha = -0.5$?

Problem 451. Examine Figure 11.7. For what value of the negative threshold would the NAND ($\neg(a \wedge b)$) and NOR ($\neg(a \vee b)$) have equal probability of appearing? Assume that the selection of a and b is made uniformly at random on the interval $-1 \leq x \leq 1$. Reading (or doing) Problem 450 may clarify this question.

Problem 452. Essay. A fact of digital logic is that the only 2-input functions from which you can construct any other are NAND and NOR. Because of this, the information in Figure 11.7 suggests that a negative threshold is probably a good thing. Experiment 11.3 tests how often negative thresholds evolve. Here is the essay question: what evidence from Experiment 11.3 demonstrates the value of negative weights and why? Possible answers include overabundance and appearance at least once in each neural net. Be sure to state *and defend* your conclusions.

11.3 Selecting the Net Topology

Even with a fixed number of inputs, some logic functions require more neurons than others. Compare, for example, minimal neural-net implementations of a 3-input AND and a 3-input parity function. In addition, a net done by layers with all possible connections present will often have connections it does not need. When we do not know ahead of time what the good number of neurons or connections is, it would be nice to let evolution select not only the connection weights, but the topology (wiring diagram, layout) of the neural net. One way to do this is to fix the number of neurons and evolve a list of weighted connections between neurons. Neuron number can, to some degree, be evolved by noticing when evolution has chosen to connect a neuron in a fashion that makes it irrelevant to the output of the neural net.

In order to ensure that we get a sensible feed-forward net, we need to impose some restrictions on the possible connections (read Section 11.1 for terminology). Our evolutionary algorithm will operate on a list of connections. Each connection will be of the form (a_i, b_i, ω_i), meaning that the output of neuron a_i is connected to the input of neuron b_i with connection weight ω_i.

An example of such a neural net is given in Figure 11.8. From that example, it is clear that we need the following restrictions. First, the inputs and neurons are numbered consecutively and all connections go from smaller numbers to larger numbers ($a_i < b_i$). This ensures that the net is feed-forward. Second, all b_i must be large enough *not* to be inputs. Third, the output is taken from the highest numbered neuron. We will call this scheme for specifying a neural net a *connection-list-specified neural net*.

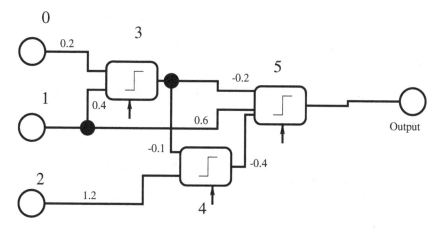

Fig. 11.8. Neural net derived from the connection list: $(0, 3, 0.2)$, $(1, 3, 0.4)$, $(1, 5, 0.6)$, $(2, 4, 1.2)$, $(3, 4, -0.1)$, $(3, 5, -0.2)$, $(4, 5, -0.4)$. (Notice that inputs are numbered and treated as neurons.)

There are some restrictions that seem natural, but are unnecessary. For example, it would be natural to demand that when $i \neq j$, $a_i \neq a_j$ or $b_i \neq b_j$. In other words, no double connections. Instead, we will simply interpret a double connection as a connection whose strength is the sum of the individual connections. This gives evolution the freedom to reduce the number of connections in a net by superimposing them. Implicit in this is a need to simplify an evolved net once we have it in hand.

Evaluating a Connection-List-Specified Neural Net

Once we have a list of connections that specify a neural net, we have the problem of actually evaluating the list and producing a net. What follows is one possible method. Suppose our neurons are Heaviside neurons with a fixed threshold, and that including inputs we have n neurons. From the list, we will derive a *weight matrix* as follows. First, initialize an $n \times n$ matrix M to all zeros. Traversing the connection list, add w_i to $M[a_i, b_i]$. Now all the connections are stored in M.

Evaluating the net is done recursively. Initialize a Boolean n-vector $\mathbf{v}$ to [input]; that is, put true in $\mathbf{v}[i]$ for i equal to the index of an input. Keep also

```
//global definitions
int v[n];       //is neuron output i known?
double m[n];    //value of neuron output i, if known
int i;

[...]
    //prepare to call recursive net evaluator
    for(i=0;i<n;i++){
        v[i]=(i<inputs);
        if(v[i])m[i]=INPUT[i];else m[i]=-1;  //load known input values
    }
    return(receval(n-1));
[...]

int receval(int index){  //return the output value of neuron index

double scratch;
int i;

    if(v[index])return(m[index]);  //if value is known, return it
    //otherwise compute it
    scratch=0;
    for(i=0;i<index-1;i++)if(M[i,index]!=0)scratch+=M[i,index]*recval(i)
    v[index]=1;  //value now known
    if(scratch<alpha)m[index]=0;else m[index]=1;  //Heaviside computatio
    return(m[index]);
}
```

Fig. 11.9. Code for recursive evaluation of a neural net, given the connection matrix M. ("Alpha" is the Heaviside threshold. "Inputs" is the number of input neurons.)

an output n-vector **m** of reals that holds the output values of each neuron. It is initialized to hold the known input values. These vectors are used by a recursive routine that returns the output value of any desired neuron, given values for the input neurons, by summing the appropriate row of M times the outputs of relevant neurons connected to the desired neuron. When the output of a given neuron is known, it is in **m** and the corresponding entry of **v** is true. Otherwise, a recursive call is used. Code for this evaluation routine is given in Figure 11.9.

Notice that the algorithm given in Figure 11.9 does not compute a neuron's output value until it needs it. This is called *lazy evaluation* and is a standard method of reducing computational work. Before we define evolutionary algorithms for connection-list neural nets, let's do a sampling experiment that will permit us to implement and debug our evaluator.

Experiment 11.6 *Implement or obtain software for generating and evaluating connection-list-specified neural nets. The number of inputs and neurons should be easy to change. Generate 1000 different 3-input nets with 5 neurons other than the inputs and 15 connections. Use Heaviside neurons with a threshold value of $\alpha = -0.5$. Evaluate them, determining their output string, and tally how many times each output string appears. Report the empirical probability of each of the 256 possible functions. Which was more likely, AND or parity?*

To expand the software from Experiment 11.6 into an evolutionary algorithm for evolving logic function specifications, we need variation operators for connection-list-specified neural nets. Crossover is not difficult. If we treat the triples (a_i, b_i, ω_i) that specify connections as atomic objects, then we can do crossover on the list as if it were a string, treating connections as characters.

There are two different types of objects inside a connection that should be modifiable by mutation. Thus, we need two distinct notions of point mutation. A *topological mutation* replaces the values of a_i and b_i with new, semantically correct values, e.g., $a_i < b_i$ and b_i is not an input neuron. A *weight mutation* is a real point mutation of the parameter ω_i. We will use Gaussian real point mutations with mean zero and specified variance. These add a zero-mean Gaussian to the weight ω_i. The formula for Gaussian random numbers is given in Equation 3.1.

Experiment 11.7 *Using the connection-list-specified neural net routines from Experiment 11.6 and the variation operators described above, build or obtain software for an evolutionary algorithm to evolve logic functions. Use the fitness function, population size, and model of evolution from Experiment 11.1. Perform two-point crossover of the list of connections and mutate each new gene 0 to 2 times with the number of mutations selected uniformly at random. Do a topological mutation one time in four and a weight mutation with a variance of 0.2 three times in four. Use 3 input and 6 noninput neurons with 20 connections.*

Run 100 populations for at most 400 generations, attempting to locate a 3-input AND and a 3-input parity function. Compare time-to-solution data. Compute and record the average number of noninput neurons used in correct solutions (computing the number of neurons used should be built into your evaluator).

It may be that logic functions implemented by neural nets with fewer parameters are easier to locate. Rather than trying several different numbers of neurons, the next experiment employs lexical fitness to encourage the evolutionary algorithm not to use some of its neurons.

Experiment 11.8 *Modify the software from Experiment 11.7 to put "number of neurons used" in a lexical fitness function. For two neural nets that get the same number of positions correct in the output string, break ties by judging* the neural net that used fewer of its neurons more fit. Do the same runs as in Experiment 11.7 and compare the results.

Binary Adder with Carry In and Out				
Inputs			Outputs	
Bit1	Bit2	Carry	Carry	Sum
0	0	0	0	0
0	0	1	0	1
0	1	0	0	1
0	1	1	1	0
1	0	0	0	1
1	0	1	1	0
1	1	0	1	0
1	1	1	1	1

Fig. 11.10. Two-bit adder with carry, input, and output.

It is possible that we will want a neural net with more than one output. It's not hard to code such an object, but we need to modify (a little) the coding of our connection-list-specified neural nets and (somewhat more) our fitness function. A common 2-output object is the 2-bit adder, which, as its name suggests, adds 2 bits. The outputs are the sum and carry of the 2 bits. To make an adder for binary natural numbers, we also need to accept carry bits into the adder. The adders can then be cascaded to add any number of bits. The truth table for a binary adder is given in Figure 11.10.

The structure of the connection list is modified as follows. The value of a_i may never be that of an output neuron. In the past, we had only one output neuron. The binary adder shown in Figure 11.10 has two output neurons. This means that the two largest neuron indices are off limits for values of a_i, where before only the largest was.

The fitness function is modified by changing the definition of the output string. The binary adder has 3 inputs, and so has an 8-character output string, but this character string is over the alphabet $\{00, 01, 10, 11\}$. We will turn this into a 16-character output string by running through the sum bits, then the carry bits. That makes the output string for the binary adder a 16-character string, to wit, 0110100100010111. To generate the output string for an evolving neural net, we generate, for the combinations of 3 inputs in lexicographical order, the string of sum outputs followed by the string of carry outputs. This is compared to the correct output string, and fitness is the number of bits in agreement.

Experiment 11.9 Modify the software from Experiment 11.7 as outlined to permit the evolutionary algorithm to search for a binary adder with carry. Use the same parameters as Experiment 11.7, save that there should be 8 noninput *neurons and* 28 *connections. Do* 100 *runs. Save solution times and compare to the difficulty of locating a 3-input parity function.*

A *neutral mutation operator* is one that does not modify the fitness of an organism, but that does modify its gene. After a creature has undergone a neutral mutation, it may have different children than it otherwise would have had. For a connection-list-specified neural net, there is a natural choice of neutral mutation. The connections appear in the list in some order, and that order is irrelevant to the functioning of the net. Our neutral mutation is some means of reordering the list of connections. In the tradition of keeping mutations small, we will implement neutral mutation by swapping a pair of list elements selected uniformly at random.

Experiment 11.10 *Modify the software from Experiment 11.7 to use neutral mutations half the time on new genes, in addition to and separate from other mutations. Using the same parameters, do* 100 *runs. Save solution times and compare with the results of not using the neutral mutation. Do a second set of runs in which the rate of neutral mutations drops smoothly from* 1.0 *after the first generation to* 0.0 *in generation 200. Compare with the other runs. Does neutral mutation help? Does it disrupt convergence near the end of evolution? Option: add to the evaluator the ability to compute the number of nonzero connections. Does neutral mutation affect this statistic?*

In the next section, we will compare neural nets to genetic programming as a method of performing logic function induction.

Problems

Problem 453. Suppose we have n input and k noninput neurons. If there are no multiple or repeated connections, then what is the maximum number of connections possible in the connection-list scheme used in this section?

Problem 454. Suppose we have n input and k noninput neurons. A *connection topology* is a list of connections with unknown weights. If there are no multiple or repeated connections, then what is the maximum number of net connection topologies possible?

Problem 455. Reread Problem 454. Find two connection topologies that are different, in the sense of Problem 454, but neither of which can be used to implement a logic function the other cannot.

Problem 456. Short Essay. Reread Problem 455. Call two topologies *different* if one can, by selecting weights, implement a logic function in one that no selection of weights will implement in the other. Is it difficult to count the number of different connection topologies?

Problem 457. Compute the minimal number of Heaviside neurons, with threshold $\alpha = 0.5$, needed to implement a 3-input AND and a 3-input parity function. Prove that your solutions are minimal.

Problem 458. On page 307, it is asserted that the restriction $a_i < b_i$ in a connection list forces the neural net to be feed-forward. Prove this assertion.

Problem 459. Examine the code given in Figure 11.9. When computing the output value of the neural net for a given set of inputs, does it evaluate the output value of every neuron? If so, explain why it needs to; if not, explain how the list of neurons it does use might be useful.

Problem 460. Compute the truth table of the logic function implemented in Figure 11.8. Assume that the neurons are Heaviside neurons with a threshold of $\alpha = -0.5$.

Problem 461. For the following list of connections, for 3 input and 3 noninput neurons, give the neural net that is coded and compute its truth table. Assume that the neurons are Heaviside neurons with a threshold of $\alpha = -0.5$. Remember that the output of the neural net is computed from the highest-numbered neuron. The input neurons are numbered 0–2, the others 3–5.

$(a,b,\quad \omega\quad)$
$(0,3,-0.41)$
$(3,5,-0.43)$
$(1,4,-0.32)$
$(2,4,-0.39)$
$(4,5,-0.44)$
$(1,3,-0.34)$

Problem 462. Give an algorithm that takes as its input a list of connections for specifying a neural net and outputs a reduced list of connections and the numbers of neurons actually used in the output. This reduced list should be minimal; explain why yours is. This type of software is called a target compiler: it takes a specification for a digital object, trims out the unneeded parts, and gives a minimal functional object as output.

Problem 463. Design, by hand, a feed-forward neural net to perform the binary-adder-with-carry task given in Figure 11.10. Use as few neurons and connections as you can.

Problem 464. Short Essay. In the first two sections of this chapter we insist that our neural nets be feed-forward. Explain why.

Problem 465. Essay. In Experiment 11.7, we use a somewhat arbitrary ratio, 1:3, of topological to weight mutations. Explain the functions of these mutations. The evolutionary algorithm is searching both for a neural net connection topology and its weights. Of these two sorts of mutations, which one makes more changes in the output string, on average? Justify your answer mathematically or experimentally. Would varying the ratio of the two mutation types over evolutionary time be a good idea? Why?

11.4 GP Logics

In the last two sections, we have tried various neural network technologies to search for logic functions. Another evolutionary computation technique that is used to search for logic functions is genetic programming. There are pros and cons. An advantage of genetic programming is that it starts with a complete set of logic operations. A disadvantage is that since genetic programs are organized as trees, their operations have a fan-out of just one. Fan-out is an electrical engineering term. In a circuit, the *fan-out* of a device is the number of inputs of other devices to which the output of the device is connected. Neuron 3 in Figure 11.8 has a fan-out of two, for example. Inputs to a parse tree, in contrast to operations, have an arbitrarily large fan-out since they are terminals.

Before proceeding with this section, you should review Sections 8.1 and 8.2. We will need a genetic programming language that operates on logical variables with logical operators. The tree manipulation routines given in Section 8.2 will be needed, respecialized to a logical language. We will begin by specifying a logical GP language and testing it. The logical language we will use is given in Table 11.3.

Language Element	Type	Symbol	Semantics
True	terminal	**T**	constant: logical true
False	terminal	**F**	constant: logical false
x_i	terminal	x_i	input variable
NOT	unary operation	$\sim$	inverts argument
OR	binary operation	$+$	true if either argument is true
AND	binary operation	$*$	true if both arguments are true

Table 11.3. Logical language for genetic programming.

Experiment 11.11 *Write or obtain software for a parse tree language that implements the logical GP language given in Table 11.3. Make sure that the number of input variables in a given tree is easy to specify on the fly. Once the parse tree routines are written and debugged, compute the output strings of 1000 random parse trees with 12 nodes and 3 input variables. Compute the empirical probability of AND and of 3-input parity, comparing the result with the results of Experiment 11.6.*

Notice that parse trees form a drop-in replacement for the neural nets we used in the earlier sections of this chapter. The definition of "output string" and with it the fitness functions we used in the last chapter remain the same. Since we have radically changed the coding used for the logic functions, the time to convergence for various logic functions, all other things being as close to equal as possible, may well be different. The next experiment checks this.

Experiment 11.12 *Using the parse tree routines from Experiment 11.11 and the variation operators described in Section 8.1, build an evolutionary algorithm to evolve parse-tree logic functions. Use the fitness function, population size, and model of evolution from Experiment 11.1. Use 3 input variables and trees with at most 16 nodes. Let initial trees have from 6 to 16 nodes with the number of nodes selected uniformly at random. Run 100 populations for at most 400 generations, attempting to locate a 3-input AND function and a 3-input parity function. Compare the time-to-solution of these two tasks, and also compare to the results obtained in Experiment 11.7. Option: try performing these experiments again without subtree crossover and see what the effect is on solution time.*

For neural nets topological mutations and weight mutations had different effects on evolution. The next experiment tries using a different mutation operator for parse trees. Subtree mutation consists in deleting a subtree and replacing it with another. Another possible mutation operator is a *leaf mutation*, which picks a terminal of the tree uniformly at random and replaces it with another terminal. In Problem 469, we ask you to suggest why leaf mutation might be helpful in getting a population to use cascading.

Experiment 11.13 *Modify the software from Experiment 11.12 to incorporate leaf mutation. Do 100 runs, modified as follows. Start trying to evolve a 2-input parity function. Ten generations after a 2-input parity function is found, modify the rate of the mutation operators from 100% subtree replacement to half-and-half subtree replacement and leaf mutation. At the same time, change the fitness function to search for 3-input parity. At this time the number of variables must be increased from 2 to 3. Save time-to-solution and compare with the results of Experiment 11.12.*

Now we will introduce a multitree variation on the parse tree approach that reduces the fan-out problem. It is based on an idea called MIPs nets, or *Multiple Interacting Program systems nets*, invented by Peter Angeline. An example of such a system is shown in Figure 11.11. A MIPs net contains multiple parse trees. The outputs of some trees are made available to other trees as input terminals. This is similar to the automatically defined functions (ADFs) used in other genetic programming systems, but there is no master–slave relationship between the trees.

We still want to have a feed-forward structure for searching. In a feed-forward MIPs net we number the trees, and higher-numbered trees have access

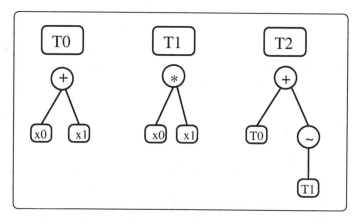

Fig. 11.11. A 3-tree logic function. (The output of trees with smaller number is available as input terminals to trees with larger numbers.)

to the input of lower-numbered trees. To do this, we will use "other tree" terminals which are numbered 0–255. In a given tree, number i, we take the index of the "other tree" terminal, mod i. This ensures that we will always access the output value of a tree with a smaller number and does not presume any particular number of trees. Tree0 will get a result of false from its other tree terminals. During evaluation, we evaluate Tree0 first. After evaluation, we know its output value, and this is then made available to the other trees. Tree1 is evaluated next, and so on, until the highest-numbered tree is evaluated, and its output is the output value of the net.

A MIPs net does not have any particular number of trees; rather, it has a limit on its total number of nodes. The modifications of our parse tree software needed to use MIPs nets are as follows. First, add a new type of terminal with 256 flavors, T_0–T_{255}. Second, change the basic structure to be a vector of pointers, large enough to permit a generous number of trees. We will require new mutation operators: add a new random tree (*enlargement*), and delete one tree selected at random (*contraction*). MIPs nets do not use crossover; the system uses four types of mutations: subtree mutation, leaf mutation, enlargement, and contraction.

Experiment 11.14 *Write or obtain software for an evolutionary algorithm to evolve MIPs nets as described. Assume that a net has at most 24 nodes and at most 8 trees. When generating the initial population, generate from 1 to 4 trees in each MIPs net with from 3 to 6 nodes in each tree. Use a steady-state evolutionary algorithm in which a pair of nets are selected and compared. A mutant of the better net is copied over the worse one. The output of a MIPs net is the output of its highest-numbered tree; use agreement with the desired output string as the fitness function.*

Use 1 to 2 mutations on each new tree, of which 25% pick a tree at random and do a subtree mutation, 25% pick a tree at random and do a leaf mutation,

25% are enlargement (add a tree with 3 to 6 nodes), and 25% are contraction. When a MIPs net has 8 trees, perform contraction in place of enlargement. When a MIPs net has 1 tree, perform enlargement in place of contraction. When a MIPs net exceeds the total node boundary, perform contractions unless it has 1 tree with too many nodes, in which case use a chop operation.

Do 100 runs on a population of 200 MIPs nets for at most 400 generations, looking for a 3-input parity function. Compare with other experiments searching for a 3-input parity function.

One of the mechanisms that biological evolution uses is to duplicate a gene, freeing one copy of that gene to change without disrupting critical biological function. Following some ideas from Chapter 10 on GP automata deciders, we note that MIPS nets have the potential to exploit this type of evolutionary mechanism. Call this a *gene duplication* and implement it by picking a tree uniformly at random and copying it over another picked uniformly at random. Let us experimentally check the effect of such a mutation operator.

Experiment 11.15 *Modify the software from Experiment 11.14 to include gene duplication. Use the 4 mutation operators from Experiment 11.14 80% of the time, and use gene duplication the remaining 20% of the time. Perform the same runs and compare the time-to-solution with and without gene duplication.*

Problems

Problem 466. Compute the number of 12-node parse trees in the language given in Table 11.3. Assume 3 input variables. You may want to look at Problem 296.

Problem 467. Prove or disprove: all 3-input logic functions can be realized with 12-node parse trees in the language given in Table 11.3. There are 256 of these, so you must use symmetries to represent entire classes of functions with single examples.

Problem 468. Compute the output string for each of the following parse trees:

(i) $(+ (\sim x_0) (+ (\sim x_1) (\sim x_2)))$,
(ii) $(+ (* (\sim x_0) (* x_1 (\sim x_2))) (* x_0 (* (\sim x_1) (\sim x_2))))$,
(iii) $(+ (+ x_1 x_0) (\sim (+ x_0 x_1)))$,
(iv) $(+ (* (\sim x_2) (* x_0 (\sim x_1))) (* x_1 (* (\sim x_2) (\sim x_0))))$.

Problem 469. Show, pictorially, how subtree crossover makes it easier to discover cascades of functions. Discuss, in a few sentences, the problem with getting the terminals right. A cascade of AND functions is given in Figure 11.5.

Problem 470. Construct, by hand, a parse tree that computes the logic function with output string 0110100110010110.

Problem 471. What logic function does the MIPs net given in Figure 11.11 compute? Give the truth table and name it if it has a name.

Problem 472. In Figure 11.10, a 2-output logic function (a binary adder with carry in and out) is described. In a few sentences, explain how to use a MIPs net to code for multiple-output logic functions, and illustrate your technique by constructing a MIPs net that computes the binary adder function.

Problem 473. In doing genetic programming, there is no need to limit ourselves to binary operations. Both the AND and OR functions are defined for any number of inputs. Describe how to modify the GP software used in this section to accommodate arbitrary-input AND and OR operations. Be sure to explain how mutation (or crossover) can change the number of arguments a given operator uses.

Problem 474. If we use AND, OR, and NOT, then what is the minimum number of operations needed to compute (i) 3-input parity, (ii) 4-input parity, and (iii) 5-input parity.

Problem 475. We can change the function computed by a parse tree by modifying its terminals. True or false: if two parse trees have the same operations (possibly connected differently), then each has the same number of functions it could code for under various modifications of its terminals. In essence, you are asked to establish whether some trees are inherently more diverse than others. Prove your answer.

Problem 476. Essay. Give the design of an experiment used to compute the optimum ratio of leaf mutations to subtree mutations in evolving parse trees as in Experiment 11.13.

Problem 477. In Chapter 10, we discovered that storing internal state information was key to performing well on the Tartarus task. Suppose we modify MIPs net evaluation so that the output value of each tree in the last time step is stored and is the value of the terminal referring to that tree in the current time step. In essence, this is a one-time-step delay. Can a MIPs net evaluated in this manner store internal state information? Justify your answer in a few sentences.

12

ISAc List: Alternative Genetic Programming

In this chapter, we will look at ISAc lists, another method of producing programs by evolution. ISAc lists are far simpler than the genetic programming techniques we studied in previous chapters and can be stored as a simple array of records. This data structure is also, for several test problems, better adapted to grid robots than the GP automata from Chapter 10. In addition to introducing ISAc lists, this chapter will also explore several new grid robot tasks. A version of the Vacuum Cleaner task, whose goal is to visit every square in the grid, appears in [38]. The Herbivore task, in which we add the ability to eat boxes to the grid robots' palette of actions, is a simple model of foraging. The North Wall Builder task asks the grid robots to construct a simple structure.

While the main point of this chapter is to introduce ISAc lists as a new representation for evolvable code, there are other important points. We explore population seeding with the new representation, both within and between grid robot tasks. The thread of using a simple string evolver as a baseline for more complex representations continues. Some effort is made to explore the scaling of grid robot tasks by increasing the size of the boards.

12.1 ISAc Lists: Basic Definitions

The structure we will now study, the If-Skip-Action list (ISAc list), was invented by Mark Joenks [3] during a fit of boredom in a machine-language programming class. Joenks observed that machine-language instructions come in a linear order, followed during execution, except when interrupted by jumps. It struck him that if we wrote a very simple machine language that tested conditions and, based on the result, either specified an action or made a jump, then we would have an evolvable programming language. An example of an ISAc list is given in Figure 12.2.

An *ISAc list* is a vector, or array, of ISAc nodes. An *ISAC node* is a quadruple $(a, b, \text{act}, \text{jmp})$ where a and b are indices into a data vector, act is

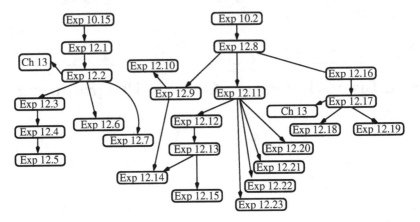

1   ISAc lists for Tartarus.
2   Exploring board and population size.
3   Population seeding.
4   Population seeding.
5   Adding a random action.
6   Shorter ISAc lists.
7   Larger boards.
8   String evolver baseline for the Vacuum Cleaner task.
9   ISAc lists for the Vacuum Cleaner task.
10 Population seeding.
11 String evolver baseline for the Herbivore task.
12 Herbivore with ISAc lists.
13 Population seeding.
14 Seeding Herbivore with Vacuum Cleaner genes.
15 Seeding Herbivore with string baseline genes.
16 String baseline with North Wall Builder.
17 North Wall Builder with ISAc lists.
18 Alternative fitness function.
19 Larger boards with population seeding.
20 Use of null actions.
21 Adaptive strings.
22 Stochastic alphabet strings.
23 Adaptive stochastic strings.

Fig. 12.1. The topics and dependencies of the experiments in this chapter.

an action that the ISAc node may take, and jmp is a specification of where
to jump, if the action happens to be a jump action. The *data vector* holds
inputs, scratch variables, and constants, in other words, everything we might
put into the terminals of a genetic programming setup. An ISAc list comes
equipped with a fixed Boolean test used by every node. Execution in an ISAc
list is controlled by the *instruction pointer*.

Node	a	b	act	jmp	comment
0	3	2	2	6	if $1 - 0 > 0$ set register to zero
1	0	2	1	2	If $x - 0 > 0$ jump to 3
2	3	2	1	5	If $1 - 0 > 0$ jump to 6
3	3	2	3	9	If $1 - 0 > 0$ increment register
4	3	2	4	4	If $1 - 0 > 0$ decrement x
5	3	2	1	0	If $1 - 0 > 0$ jump to 1
6	1	2	1	7	If $y - 0 > 0$ jump to 8
7	3	2	1	10	If $1 - 0 > 0$ jump to 11
8	3	2	3	7	If $1 - 0 > 0$ increment register
9	3	2	5	2	If $1 - 0 > 0$ decrement y
10	3	2	1	5	If $1 - 0 > 0$ jump to six
11	3	2	0	4	If $1 - 0 > 0$ NOP

Fig. 12.2. An example of an ISAc list that operates on 2 variables x, y and a scratch register. (The data vector $\mathbf{v}$ is of the form $[x, y, 0, 1]$. The ISAc actions are 0-NOP, 1-jump, 2-zero register, 3-increment register, 4-decrement x, and 5-decrement y. The Boolean test for this ISAC list is if $(\mathbf{v}[a] - \mathbf{v}[b] > 0)$, i.e., is item "$a$" larger than item "$b$"?)

The operation of an ISAc list is as follows. We start with the instruction pointer at the beginning of the ISAc list, indexing the zeroth node. Using the entries a, b, act, jmp of that node, we look up data item a and data item b in the data vector and apply the Boolean test to them. If the test is true, we perform the action in the act field of the node; otherwise, we do nothing. If that action is "jump," we load the contents of the jmp field into the instruction pointer. We then increment the instruction pointer. Pseudocode for the basic ISAc list execution loop is shown in Figure 12.3.

```
IP← 0                                        //Set Instruction Pointer to 0.
LoadDataVector(v);                           //Put initial values in data vector.
Repeat                                       //ISAc evaluation loop
    With ISAc[IP] do                         //with the current ISAc node,
    If v[a] − v[b] > 0 then PerformAction(act); //Conditionally perform action
    UpdateDataVector(v);                     //Revise the data vector
    IP ← IP + 1                              //Increment instruction pointer
Until Done;
```

Fig. 12.3. Algorithm for executing an ISAc list.

There are 3 types of actions used in the act field of an ISAc node. The first is the *NOP* action, which does nothing. The inclusion of the NOP action is inspired by experience in machine-language programming. An ISAc list that has been evolving for a while will have its performance tied to the pattern of jumps it has chosen. If we insert new instructions, the target addresses of many of the jumps change. We could tie our "insert an instruction" mutation operator to a "renumber all the jumps" routine, but this is computationally complex. Instead, we have a "do nothing" instruction that serves as a placeholder. Instructions can be inserted by mutating a NOP instruction, and they can be deleted by mutating into a NOP instruction *without* the annoyance of renumbering everything.

The second type of action used in an ISAc list is the *jump* instruction. The jump instructions are *goto* instructions. For those of you who have been brainwashed by the structured programming police, any kind of control structure, "for," "repeat-until," "do-while," is really an "if (condition) then goto (label)" structure, carefully hidden from the delicate eyes of the software engineer by his compiler or codegenerator. In a high-level language, this "goto hiding" aids in producing sensible, easy-to-read code. ISAc lists are low-level programming and are rich in jump instructions. These instructions simply load a new value into the instruction pointer when the Boolean test in their node is true. Notice that to goto node n we issue a jump to node $n-1$ instruction. This is because even after a jump instruction, the instruction pointer is incremented.

The third type of action is the one of interest to the environment outside the ISAc list. We call these *external* actions. Both NOP and jump instructions are related to the internal bookkeeping of the ISAC list. External actions are reported to the simulator running the ISAc list. In the ISAc list shown in Figure 12.2, the external actions are "zero register," "increment register," "decrement x," and "decrement y." Notice that an ISAc list lives in an environment. It sees the environment through its data vector and may modify the environment through its actions.

Done?

As usual, we will generate random objects and randomly modify them during the course of evolution. Since the objects we are dealing with in this chapter have jump statements in them, they will often have infinite loops. This is similar to the situation we faced in Chapter 10 with the null action. We will adopt a similar solution, but one that is more forgiving of long finite loops or non-action-generating code segments. We will place a limit on the total number of instructions that can be executed before fitness evaluation is terminated. Typically, this limit will be a small integer multiple of the total number of external instructions we expect the ISAc list to need to execute to do its job.

Even an ISAc list that does not get into an infinite loop (or a long finite loop) needs a termination condition. For some applications, having the instruction pointer fall off of the end of the list is an adequate termination condition. The example ISAc list in Figure 12.2 terminates in this fashion. We will call this type of ISAc list a *linear* ISAc list. Another option is to make the instruction pointer function modulo the number of instructions in the ISAc list. In this variation, the first instruction in the list immediately follows the last. We call this type of ISAc list a *circular* ISAc list. With circular ISAc lists, we either have explicit "done" actions, or we stop when the ISAc list has produced as many external actions as the simulator requires.

Generating ISAc Lists, Variation Operators

Generating an ISAc list is easy. You must choose a data structure, either an array of records $(a, b, \text{act}, \text{jmp})$ or 4 arrays $a[]$, $b[]$, $\text{act}[]$, $\text{jmp}[]$ with records formed implicitly by common index. Simply fill the array with appropriately sized integers chosen uniformly at random. The values of a and b are in the range $0, \ldots, n_v - 1$, where n_v is the number of items in the data vector. The act field is typically in the range $0, \ldots, n_a + 1$, where n_a is the number of external actions. The two added actions leave space for the jump and NOP actions. Since NOP and jump are always present, it is a good idea to let action 0 be NOP, action 1 be jump, and then for any other action, subtract 2 from the action's number and return it to the simulator. This will make using ISAc lists evolved on one task for another easier, since they will agree on the syntax of purely ISAc list internal instructions. The jmp field is in the range $0, \ldots,$ listsize, where listsize is the length of the ISAc list.

The variation operators we will use on ISAc lists should seem comfortably familiar. If we treat individual ISAc nodes as atomic objects, then we can use the string-based crossover operators from Chapter 2. One-point, two-point, multipoint, and uniform crossover take on their familiar meanings with ISAc lists.

Point mutation of an ISAc list is a little more complex. There are three sorts of fields in an ISAc node: the data pointer fields a and b, the action field act, and the jump field jmp. A *point mutation* of an ISAc list selects an ISAc node uniformly at random, selects one of its 4 fields uniformly at random, and then replaces it with a new, valid value. For finer resolution, we also define the *pointer mutation*, which selects a node uniformly at random and then replaces its a or b field, an *action mutation* that selects a node uniformly at random and then replaces its act field, and a *jump mutation* that selects a node uniformly at random and replaces its jmp field.

Data Vectors and External Objects

In Chapter 10, we augmented our basic parse trees with operations that could affect external objects: calculator-style memories. In Figure 12.2, various actions available to the ISAc list were able to modify an external scratch register and two registers holding variables. Much as we custom-designed the parse tree language to the problems in Chapters 8, 9, and 10, we must custom-design the environment of an ISAc list.

The primary environmental feature is the data vector, which holds inputs and constants. Figure 12.2 suggests that modifiable registers are another possible feature of the ISAc environment. To use memory-mapped I/O, we could permit the ISAc list to directly modify elements of its data vector, taking these as the output. We could give the ISAc list instructions that modify which data set or part of a data set is being reported to it in its data vector. We will deal more with these issues in later chapters, but you should keep them in mind.

With the basic definitions of ISAc lists in hand, we are now ready to take on a programming task. The next section starts us off in familiar territory, the Tartarus environment.

Problems

Problem 478. What does the ISAc list given in Figure 12.2 do?

Problem 479. Using the notation from Figure 12.2, give a sequence of ISAc nodes that implements the structure

```
while(v[1]>v[2])do Action(5);
```

Problem 480. Using the notation from Figure 12.2, give a sequence of ISAc nodes that implements the structure

```
while(v[1]==v[2])do Action(5);
```

Problem 481. Using the notation from Figure 12.2, give a sequence of ISAc nodes that implements the structure

```
while(v[1]!=v[2])do Action(5);
```

Problem 482. Using the notation from Figure 12.2, give a sequence of ISAc nodes that implements the structure

```
while(v[1]>=v[2])do Action(5);
```

Problem 483. Take the commands given for the ISAc list in Figure 12.2 and add commands for incrementing x and y and decrementing the register. Write a linear ISAc list that places $x - y$ into the register.

Problem 484. Essay. The code fragments from Problems 479, 480, 481, and 482 show that any comparison of two data vector items can be simulated (less-than comparisons simply require that we reverse a and b). It may, however, be the case that the cost in space and complexity of simulating the test you need from the one test you are allowed will impede discovery of the desired code. Describe how to modify ISAc lists to have multiple different Boolean tests available as primitive operations in an ISAc node.

Problem 485. Essay. Those readers familiar with Beginners All-purpose Symbolic Instruction Code (BASIC) will recognize that BASIC's method of handling subroutines is easily adapted to the ISAc list environment. For our BASIC-noncompliant readers, a BASIC program has a number associated with each line. The command "GOSUB < *linenumber* >" transfers control to the line number named. When a "RETURN" command is encountered, control is returned to the line after the most recent "GOSUB" command. Several GOSUB commands can be executed followed by several returns, with a stack of return locations needed to decide where to return. Describe a modification of the ISAc list environment to include jump-like instructions similar to the BASIC GOSUB and RETURN commands. Does the current method for disallowing infinite loops suffice, or do we need to worry about growth of the return stack? What do we do if the ISAc list terminates with a nonempty return stack? Should this be discouraged, and if so, how?

Problem 486. Describe the data vector and external commands needed to specialize ISAc lists to work on the Plus-One-Recall-Store Efficient Node Use problem, described in Chapter 8. For the Efficient Node Use problem, we were worried about the total number of nodes in the parse tree. Be sure to state what the equivalent of "size" is for an ISAc list and carefully restate the Efficient Node Use problem. Give a solution for size 12.

Problem 487. Reread Problem 486. Is it possible to come up with an ISAc list that solves a whole class of Efficient Node Use problems?

Problem 488. Essay. In Chapter 9, we used evolutionary algorithms and genetic programming to encode formulas that were being fit to data. Using an analogy between external ISAc actions and keys on a calculator, explain how to use an evolutionary algorithm to fit a formula embedded in an ISAc list to data. What, if any, real constants go in the data vector? Give an ISAc list that can compute the fake bell curve

$$f(x) = \frac{1}{1+x^2}.$$

Be sure to explain your external commands, choice of Boolean test, and data vector.

Problem 489. Following the setup in Problem 488, give an ISAc list that can compute the function

$$f(x) = \begin{cases} x^2 & \text{if } x \geq 0, \\ -x^2 & \text{if } x < 0. \end{cases}$$

Problem 490. Following the setup in Problem 488, give an ISAc list that can compute the function

$$f(x,y) = \begin{cases} 1 & \text{if } x^2 + y^2 \leq 1, \\ \frac{1}{x^2+y^2} & \text{if } x^2 + y^2 > 1. \end{cases}$$

Problem 491. Describe the data vector and external commands needed to specialize ISAc lists to play Iterated Prisoner's Dilemma (see Section 6.2 for details). Having done so, give ISAc lists that play each of the following strategies. Use the commented style of Figure 12.2.

 (i) Always cooperate,
 (ii) Always defect,
 (iii) Tit-for-tat,
 (iv) Tit-for-two-tats,
 (v) Pavlov,
 (vi) Ripoff.

Problem 492. Are ISAc lists able to simulate finite state automata in general? Either give an example of a finite state automaton that cannot, for some reason, be simulated by an ISAc list, or give the general procedure for performing such a simulation, i.e., coding a finite state automaton as an ISAc list.

12.2 Tartarus Revisited

The first thing we will do with ISAc lists is revisit the Tartarus task from Chapter 10. You should reread the description of the Tartarus problem on page 265. We will specialize ISAc lists for the Tartarus problem as follows. We will use a data vector that holds the 8 sensors (see Figure 10.4) and 3 constants, $\mathbf{v} = [UM, UR, MR, LR, LM, LL, ML, UL, 0, 1, 2]$. The ISAc actions will be 0, NOP; 1, Jump; 2, Turn Left; 3, Turn Right; 4, Go Forward. This specification suffices for our first experiment.

Experiment 12.1 *Implement or obtain software to create and run circular ISAc lists, as well as the variation operators described in Section 12.1. Be sure to include routines for saving ISAc lists to a file and reading them from a file. With these routines in hand, as well as the Tartarus board routines from Chapter 10, build an evolutionary algorithm that tests ISAc lists specialized for Tartarus on 40 boards. Use 80 moves (external actions) on each board with a limit of 500 ISAc nodes evaluated per board. Use a population of 60 ISAc lists of length 60. Use point mutation and two-point crossover for your variation operators and single tournament selection with tournament size 4 for your model of evolution. Do 20 runs for 200 generations each and compare your results with Experiment 10.15.*

One thing you may notice, comparing this experiment with Experiment 10.15, is that ISAc lists run much faster than GP automata. Let's see whether we can squeeze any advantage out of this.

Experiment 12.2 *Redo Experiment 12.1, testing ISAc lists on* 100 *Tartarus boards with population size* 400. *Do* 100 *runs and compare the results with Experiment 12.1. Compare both the first* 20 *runs and the full set of* 100. *Also save the best ISAc list from each run in a file. We will use this file later as a "gene pool."*

In animal breeding, the owner of a high-quality animal can make serious money selling the animal's offspring or stud services. In our attempts to use evolution to locate good structures, we have, for the most part, started over every time with random structures. In the next couple of experiments, we will see whether using superior stock as a starting point can give us some benefit in finding good ISAc list controllers for Tartarus dozers. There is an enormous literature on animal breeding. Reading this literature might inspire you with a project idea.

Experiment 12.3 *Modify the code from Experiment 12.2 so that instead of generating a random initial population, it reads in the* 100 *best-of-run genes from Experiment 12.2, making 4 copies of each gene as its initial population. Do 25 runs and see whether any of them produce Tartarus controllers superior to the best in the gene pool.*

In Experiment 12.3, we started with only superior genes. There is a danger in this; the very best gene may quickly take over, causing us simply to search for variations of that gene. This is especially likely, since each member of a population of superior genes has pretty complex structure that does not admit much disruption; crossover of two different superior genes will often result in an inferior structure. To try to work around this potential limitation in the use of superior stock, we will seed a few superior genes into a population of random genes and compare the result with that of using only superior genes. In future chapters we will develop other techniques for limiting the spread of good genes.

Experiment 12.4 *Modify the code from Experiment 12.3 so that instead of generating a random initial population, it reads in the* 100 *best-of-run genes from Experiment 12.2. The software should then select 10 of these superior genes at random and combine them with 390 random ISAc lists to form an initial population. Do 25 runs, each with a different random selection of the initial superior and random genes, and see whether any of them produce Tartarus controllers superior to the best in the gene pool. Also, compare results with those obtained in Experiment 12.3.*

We have, in the past, checked to see whether the use of random numbers helped a Tartarus controller (see Experiment 10.8). In that experiment, access to random numbers created a local optimum with relatively low fitness. Using gene pools gives us another way to check whether randomness can help with the Tartarus problem.

Experiment 12.5 *Modify your ISAc list software from Experiment 12.4 to have a fourth external action, one that generates a random action. Generate the random action so that it is Turn Left 20% of the time, Turn Right 20% of the time, and Go Forward 60% of the time. Redo Experiment 12.4, permitting this random action in the randomly generated parts of the initial population, but still reading in random-number-free superior genes. Do 100 runs and compare the scores of the final best-of-run creatures with those obtained in past experiments. Do the superior creatures use the random action? Did the maximum fitness increase, decline, or remain about the same?*

The choice of length-60 ISAc lists in Experiment 12.1 was pretty arbitrary. In our experience with string baselines for Tartarus in Chapter 10 (Experiment 10.1), string length was a fairly critical parameter. Let us see whether very short ISAc lists can still obtain decent fitness scores on the Tartarus problem.

Experiment 12.6 *Modify your ISAc list software from Experiment 12.2 to operate on length-10 and length-20 ISAc lists. Do 100 runs for each length and compare the results, both with one another and with those obtained in Experiment 12.2. Do these results meet with your expectations?*

We conclude this section with another Tartarus generalization. In the past, we played Tartarus on a 6×6 board with 6 boxes. We now try it on a larger board.

Experiment 12.7 *Modify your ISAc list software from Experiment 12.2, and the Tartarus board routines, to work on an 8×8 board with 10 boxes. Do 100 runs, saving the best genes in a new gene pool. Verify that fitness, on average, increases over time and give a histogram of your best-of-run creatures.*

The brevity of this section, composed mostly of experiments, is the result of having already investigated the original Tartarus problem in some detail. The Tartarus task is just one of a large number of tasks we could study even in the limited environment of a virtual agent that can turn or go forward on a grid with some boxes. In the next section, we will take a look at several other tasks of this sort that will require only modest variations in software. We leave for later chapters the much harder problem of getting multiple virtual agents to work together.

Problems

Problem 493. In Chapter 10, we baselined the Tartarus problem with fixed sets of moves, and used a simple string evolver (Experiment 10.2) to locate good fixed sets. Give a method for changing such a string of fixed moves into an ISAc list that (i) exhibits exactly the same behavior, but (ii) is easy to revise with mutation.

Problem 494. In Experiment 12.1, we permitted up to 500 ISAc list nodes to be evaluated in the process of generating 80 moves on the Tartarus board. This may be an overgenerous allotment. Design a software tool that plots the fitness of an ISAc list for different limits on ISAc nodes. It should perform its tests on a large number of boards. Is there a way to avoid evaluating the ISAc list on one board several times? This is a tool for postevolution analysis of a fixed ISAc list.

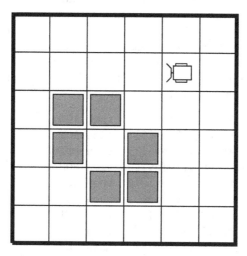

Fig. 12.4. An example of an impossible Tartarus starting configuration of the type discovered by Steve Willson.

Problem 495. When studying the Tartarus problem, Steve Willson noticed that the close grouping of 4 blocks wasn't the only impossible Tartarus configuration (see Definition 10.1). Shown in Figure 12.4 is another such impossible configuration. Explain why the Willson configuration is impossible. Is it impossible for all initial positions and headings of the dozer?

Problem 496. Compute the number of starting 6×6 Tartarus boards for which there is a close grouping of 4 boxes and the number that are Willson configurations. Which sort of impossible board is more common? Be sure to include dozer starting positions in your count.

Problem 497. Reread Problem 495. Describe an evolutionary algorithm that locates impossible boards. You will probably need to coevolve boards and dozer controllers. Be careful to choose a dozer controller that evolves cheaply and easily. Be sure, following the example from Problem 495, that dozer position is part of your board specification.

Problem 498. Essay. In the experiment in which we inject a small number of superior genes into a large population of random genes, there is a danger we will get only variations of the best gene in the original population. Discuss a way to use superior genes that decreases the probability of getting only variations of those superior genes. Do not neglect nonelite selection, use of partial genes, and insertion of superior genes other than in the initial population. Try also to move beyond these suggestions.

Problem 499. Given below is a length-16 circular ISAc list. It uses the data vector described in this section, and the NOP, jump, and external actions are given explicitly; i.e., ACT 0 is turn left, ACT 1 is turn right, and ACT 2 is go forward. For the Tartarus initial configuration shown in Figure 10.2, trace the action of a dozer controlled by this ISAc list and give the score after 80 moves. Put the numbers 1–80 on a blank Tartarus board, together with heading arrows, to show how the dozer moves.

```
 0: If v[9]>v[8]    then ACT 2
 1: If v[7]>v[9]    then ACT 1
 2: If v[4]>v[1]    then ACT 2
 3: If v[10]>v[10]  then NOP
 4: If v[4]>v[9]    then ACT 2
 5: If v[10]>v[9]   then ACT 1
 6: If v[4]>v[0]    then ACT 2
 7: If v[9]>v[2]    then ACT 2
 8: If v[2]>v[8]    then JMP 3
 9: If v[3]>v[6]    then ACT 0
10: If v[10]>v[7]   then ACT 2
11: If v[7]>v[10]   then NOP
12: If v[6]>v[7]    then ACT 0
13: If v[0]>v[8]    then ACT 2
14: If v[10]>v[7]   then ACT 0
15: If v[3]>v[6]    then ACT 1
```

Problem 500. If we increase board size, are there new variations of the Willson configuration given in Figure 12.4? Please supply either pictures (if your answer is yes) or a mathematical proof (if your answer is no). In the latter case, give a definition of "variations on a Willson configuration."

Problem 501. Recompute the answer to Problem 496 for 6 boxes on an $n \times n$ board.

Problem 502. Give a neutral mutation operator for ISAc lists. It must modify the list without changing its behavior. Ideally, it should create variation in the children the ISAc list can have.

Problem 503. Essay. Suppose we use the following operator as a variation operator that modifies a single ISAc list: Generate a random ISAc list, perform two-point crossover between the random ISAc list and the one being modified, and pick one of the two children at random as the new version of the list. Is this properly a mutation operator? Is it an operator that might help evolution? Give a quick sketch of an experiment designed to support your answer to the latter question.

Problem 504. In Experiment 12.5 we incorporate a left–right symmetric random action into the mix. Would an asymmetric random action have been better? Why or why not?

Problem 505. In Section 10.1 (string baseline), we used gene-doubling and gene-halving mutations (see Definitions 10.2 and 10.3). Give definitions of gene-doubling and gene-halving mutations for ISAc lists. Your mutations should not cause jumps to favor one part of the structure.

12.3 More Virtual Robotics

In this section, we will study several virtual robotics problems that can be derived easily from Tartarus. They will incorporate modest modifications of the rules for handling boxes and substantial modifications of the fitness function. We will make additional studies using initial populations that have already undergone some evolution. As a starting point, they will, we hope, perform better than random structures. In addition to rebreeding ISAc lists for the same task, we will study crossing task boundaries.

The first task we will study is the *Vacuum Cleaner* task. The Vacuum Cleaner task does not use boxes at all. We call the agent the *vacuum* rather than the dozer. The vacuum moves on an $n \times n$ board and is permitted $n^2 + 2n$ moves: turn left, turn right, go forward, or stand still. When the *vacuum* enters a square, that square is marked. At the end of a trial, the fitness of the vacuum is $+1$ for the first mark in each square, -1 for each mark after the first in each square. We call this the *efficient cleaning* fitness function. The object is to encourage the vacuum to visit each square on the board once. We will need to add the fourth action, *stand still*, to the external actions of the ISAc list software.

The only variation between boards is the starting position and heading of the vacuum. In addition, the heading is irrelevant in the sense that the average fitness over the set of all initial placements and headings and the average fitness over all initial placements with a single heading are the same. Because of this, we will always start the vacuum with the same heading and either exhaustively test or sample the possible placements. We now perform a string baseline experiment for the Vacuum Cleaner task.

Experiment 12.8 *Modify the Tartarus software to use a world with walls but no boxes and to compute the efficient cleaning fitness function. Use a*

9×9 *board and test fitness on a single board starting the vacuum facing north against the center of the south wall. Use a string evolver to evolve a string of* 99 *moves over the alphabet {left, right, forward, stand still } in three ways. Evolve* 11-*character strings,* 33-*character strings, and* 99-*character strings, using the string cyclically to generate* 99 *moves in any case.*

Have your string evolver use two-point crossover and 0- *to n-point mutation (where n is the length of the string divided by* 11*), with the number of point mutations selected uniformly at random. Take the model of evolution to be single tournament selection with tournament size* 4*. Do* 100 *runs for up to* 400 *generations on a population of* 100 *strings, reporting time-to-solution and average and maximum fitness for each run. Compare different string lengths, and if any are available, trace one maximum-fitness string.*

With a baseline in hand, we now will evolve ISAc lists for the Vacuum Cleaner task. Notice we are sampling from 16 of the 81 possible initial placements, rather than computing total fitness.

Experiment 12.9 *Modify the evolutionary algorithm from Experiment 12.8 to operate on ISAc lists of length* 60*. Evolve a population of* 400 *ISAc lists, testing fitness on* 16 *initial placements in each generation. Report fitness as the average score per board. Use two-point crossover and from* 0- *to 3-point mutation, with the number of point mutations selected uniformly at random. Do* 100 *runs lasting at most* 400 *generations, saving the best ISAc list in each run for later use as a gene pool. Plot the fitness as a function of the number of generations and report the maximum fitnesses obtained. How did the ISAc lists compare with the string baseline?*

We will now jump to testing on total fitness (all 81 possible placements), using the gene pool from the last experiment. The hope is that evolving for a while on a sampled fitness function and then evolving on the total fitness function will save time.

Experiment 12.10 *Modify the evolutionary algorithm from Experiment 12.9 to read in the gene pool generated in that experiment. For initial populations, choose* 10 *genes at random from the gene pool and* 390 *random structures. Evolve these populations, testing fitness on all* 81 *possible initial placements in each generation. Report fitness as the average score per board. Do* 100 *runs lasting at most* 400 *generations. Plot the fitness as a function of the number of generations and report the maximum fitnesses obtained.*

The Vacuum Cleaner task is not, in itself, difficult. Unlike Tartarus, where a perfect solution is difficult to specify, it is possible to simply write down a perfect solution to the Vacuum Cleaner task. You are asked to do this in Problem 512. The Vacuum Cleaner task does, however, require that the ISAc list build some sort of model of its environment and learn to search the space efficiently. This efficient space-searching is a useful skill and makes the Vacuum Cleaner gene pool from Experiment 12.9 a potentially valuable commodity.

In the next task, we will use this efficient space sweeping as a starting point for learning a new skill, eating.

The *Herbivore* task will add a new agent name to our roster. Tartarus has the dozer; the Vacuum Cleaner task has the vacuum. The agent used in the Herbivore task is called the *cowdozer*. The Herbivore task takes place in an $n \times n$ world and uses boxes. The rules for boxes, pushing boxes, and walls are the same as in Tartarus, save that an additional action is added: the *eat* action. If the cowdozer is sitting with a box directly ahead of it and it executes an *eat* action, then the box vanishes. Our long-term goal is to create agents that can later be used in an ecological simulation. For now, we merely wish to get them to eat efficiently.

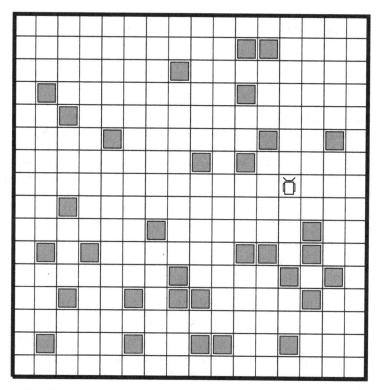

Fig. 12.5. A valid starting configuration for the Herbivore task.

A single Herbivore board is prepared by scattering k boxes at random on the board. These boxes may be anywhere, since the eat action causes a complete lack of "impossible" boards. A valid starting configuration is shown in Figure 12.5. Be sure to keep the numbering of ISAc actions (NOP, jump, turn left, turn right, and go forward) consistent in the Vacuum Cleaner and Herbivore tasks. Give *stand still* and *eat* the same action index. This facilitates

334 Evolutionary Computation for Modeling and Optimization

the use of a Vacuum Cleaner gene pool as a starting point for Herbivore. For the Herbivore task, our fitness function will be the number of boxes eaten.

With the Vacuum Cleaner task, it wasn't too hard to select an appropriate number of moves. The task of justifying that choice is left to you (Problem 507). For the Herbivore task, this is a harder problem. The cowdozer must search the board, and so would seem to require at least as many moves as the vacuum. Notice, however, that the cowdozer does not need to go to every square; rather, it must go beside every square. This is all that is required to find all the boxes on the board. On the other hand, the dozer needs to turn toward and eat the boxes. This means, for an $n \times n$ board with k boxes, that we need some fraction of $n^2 + 2n$ moves plus about $2k$ moves. We will err on the side of generosity and compute moves according to Equation 12.1:

$$\text{moves}(n, k) = \frac{2}{3}n^2 + 2(n + k), \ n \times n \text{ board}, k \text{ boxes.} \qquad (12.1)$$

We will now do a series of 5 experiments that will give us an initial understanding of the Herbivore task and its relation to the Vacuum Cleaner task.

Experiment 12.11 *Modify your board maintenance software to permit the eat action and the generation of random Herbivore boards. Be sure to be able to return the number of boxes eaten; this is the fitness function. With the new board routines debugged, use the parameters from Experiment 12.8, except as stated below, to perform a string baseline for the Herbivore task. Use board size $n = 9$ with $k = 27$ boxes, and do 126 moves per board. Test fitness on 20 random boards. Use strings of length 14, 42, and 126. Do 0- to q-point mutation, where q is the string length divided by 14, and the number of point mutations is chosen uniformly at random. Do 100 runs and save the best strings from each of the 42-character runs in a gene pool file. Plot average and maximum fitness as a function of time.*

Now that we have a string baseline, we can do the first experiment with adaptive structures.

Experiment 12.12 *Modify the evolutionary algorithm from Experiment 12.11 to operate on ISAc lists of length 60. Evolve a population of 400 ISAc lists, testing fitness on 100 Herbivore boards. Report fitness as the average score per board. Use two-point crossover and from 0- to 3-point mutations, with the number of point mutations selected uniformly at random. Do 100 runs lasting at most 400 generations, saving the best ISAc list in each run for later use as a gene pool. Plot the fitness as a function of the number of generations and report the maximum fitnesses obtained. How did the ISAc lists compare with the string baseline?*

And now let's check to see how the breeding-with-superior-genes experiment works with Herbivore genes.

Experiment 12.13 *Modify the evolutionary algorithm from Experiment 12.12 to read in the gene pool generated in that experiment. For initial populations, choose 10 genes at random from the gene pool and 390 random structures. Evolve these populations, using the same fitness evaluation as in Experiment 12.12. Report fitness as the average score per board. Do 100 runs, lasting at most 400 generations. Plot the fitness as a function of the number of generations and report the maximum fitnesses obtained.*

At this point, we will try something completely different: starting a set of Herbivore simulations with Vacuum Cleaner genes. There are three general outcomes for such an experiment: vacuum genes are incompetent for the Herbivore task and will never achieve even the fitness seen in Experiment 12.12; vacuum genes are no worse than random, and fitness increase will follow the same sort of average track it did in Experiment 12.12; Vacuum Cleaner competence is useful in performing the Herbivore task, so some part of the fitness track will be ahead of Experiment 12.12 and compare well with that from Experiment 12.13.

Experiment 12.14 *Redo Experiment 12.13 generating your initial population as follows. Read in the gene pool generated in Experiment 12.9 and use 4 copies of each gene. Do not use any random genes. For each of these genes, scan the gene for NOPs and, with probability 0.25 for each NOP, replace them with eat actions. Stand-still actions should already use the same code as eat actions. Do 100 runs, plotting average fitness and maximum fitness over time. Compare with the fitness tracks from Experiments 12.12 and 12.13.*

Now we look at another possible source of superior genes for the Herbivore task: our string baseline of that task. In Problem 493, we asked you to outline a way of turning strings into ISAc lists. We will now test one way to do that.

Experiment 12.15 *Redo Experiment 12.13 with an initial population of 84-node ISAc lists generated as follows. Read in the gene pool of length-42 strings generated in Experiment 12.11, transforming them into ISAc lists by taking each entry of the string and making it into an ISAc node of the form If$(1 > 0)$Act(string-entry), i.e., an always-true test followed by an action corresponding to the string's action. After each of these string-character actions, put a random test together with a NOP action. All JMP fields should be random. Transform each string 4 times with different random tests together with the NOPs. Do not use any random genes. Do 100 runs, plotting average fitness and maximum fitness over time. Compare with the fitness tracks from Experiments 12.12, 12.13, and 12.14.*

We have tried a large number of different techniques to build good cowdozer controllers. In a later chapter, these cowdozers will become a resource as the starting point for ecological simulations. We now move on in our exploration of ISAc list robotics and work on a task that is different from Tartarus, Vacuum Cleaner, and Herbivore in several ways. The *North Wall Builder* task

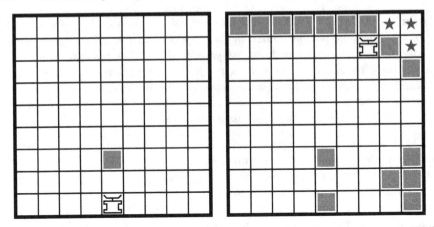

Fig. 12.6. The starting configuration and a final configuration for the North Wall Builder task. (The stars denote uncovered squares, on which the constructor lost fitness.)

tries to get the agent, called the *constructor*, to build a wall across the north end of the trial grid.

The differences from the grid robotic tasks we have studied so far are as follows. First, there will be a single fitness case and so no need to decide how to sample the fitness cases. Second, the blocks are delivered to the board in response to the actions of the constructor. Third, we remove the walls at the edges of the world. The constructor will have the same 8 sensors that the other grid robots had and will still detect a "2" at the edge of the world. What changes is the result of pushing a box against the edge of the board and of a go forward action when facing the edge of the board. A box pushed against the edge of the board vanishes. If the constructor attempts to go forward over the edge of the board, it vanishes, and its fitness evaluation ends early. For a given constructor, the *survival time* of the constructor is the number of moves it makes without falling off of the board.

The starting configuration and a final configuration are shown in Figure 12.6. North Wall Builder uses a square board with odd side lengths. In order to evolve constructors that build a north wall, we compute fitness as follows. Starting at the north edge of the board in each column, we count the number of squares before the first box. These are called *uncovered squares*. The *North Wall Builder fitness function*(NWB fitness function) is the number of squares on the board minus the number of uncovered squares. The final configuration shown in Figure 12.6 gets a fitness of $9 \times 9 - 3 = 78$.

The box delivery system is placed as shown in Figure 12.6, two squares from the south wall in the center of the board along the east-west axis. It starts with a box in place and the constructor pushes boxes as in Tartarus, save that boxes and the constructor may fall off the edge of the board. If the position of the box delivery system is empty (no box or constructor on it), then a new box appears. The constructor always starts centered against the south wall facing north toward the box delivery system. We permit the constructor $6n^2$ moves on an $n \times n$ board. Having defined the North Wall Builder task, we can begin experiments with a string baseline.

Experiment 12.16 *Modify your board software to implement the rules for the North Wall Builder task, including pushing boxes off the edge of the world, returning a value that means that the constructor has fallen off the edge of the world, and computing the NWB fitness function. In addition to fitness, be sure to write the routines so that survival time is computed. Having done this, run a string baseline experiment like Experiments 12.8 and 12.11. Use a population of 100 strings of length 100, cycling through the strings to generate the required number of moves. Work on a 9×9 board, allowing each constructor 486 steps. Report maximum and average fitness and survival time as a function of generations of evolution, doing 30 runs. Add survival time as a tiebreaker in a lexical fitness function and do 30 additional runs. Does it help?*

If your string baseline for the North Wall Builder task is working the way ours did, then you will have noticed that long survival times are somewhat rare. In North Wall Builder, reactivity (the ability to *see* the edge of the world) turns out to be quite valuable. We now can proceed to an ISAc experiment on North Wall Builder.

Experiment 12.17 *Modify the evolutionary algorithm from Experiment 12.16 to operate on ISAc lists of length 100. Evolve a population of 400 ISAc lists, using the NWB fitness function by itself and then the lexical product of NWB with survival time, in different sets of runs. Report mean and maximum fitness and survival time. Use two-point crossover and 0- to 3-point mutation, with the number of point mutations selected uniformly at random. Do 100 runs for each fitness function, lasting at most 400 generations. Save the best ISAc list in each run for later use as a gene pool. How did the ISAc lists compare with the string baseline for North Wall Builder? Did the lexical fitness help? Did it help as much as it did in the string baseline? Less? More?*

What is the maximum speed at which the constructor can complete the NWB task? In the next experiment, we will see whether we can improve the efficiency of the constructors we are evolving by modifying the fitness function. Suppose that we have two constructors that get the same fitness, but one puts its blocks into their final configuration several moves sooner than the other. The faster constructor will have more "post fitness" moves for evolution to modify to shove more boxes into a higher fitness configuration. This suggests that we should place some emphasis on brevity of performance.

One way to approach this is to create a *time-averaged* fitness function. At several points during the constructor's allotment of time, we stop and compute the fitness. The fitness used for selection is the average of these values. The result is that fitness gained as the result of block configurations that occur early in time contributes more than such fitness gained later. Notice that we presume that there are reasonable construction paths for the constructor for which a low intermediate fitness is not required.

Experiment 12.18 *Modify the software from Experiment 12.17 to use an alternative fitness function. This function is the average of the old NWB fitness function sampled at time steps* $81, 162, 243, 324, 405,$ *and* $486(n^2, 2n^2, \ldots, 6n^2)$. *Retain the use of survival time in a lexical fitness function. Use the new fitness function for selection.*

Report the mean and maximum of the new fitness function, the old fitness function at each of the 6 sampling points, and the survival time. Do 100 runs. Does the new fitness function aid in increasing performance as measured by the original fitness function? Do the data suggest an answer to Problem 525?

Problem 529 asks you to speculate on the value of a gene pool evolved for one size of board for another size of board. We will now look at the answer, at least for the North Wall Builder task.

Experiment 12.19 *Modify the software from Experiment 12.17 to use* 11×11 *and* 13×13 *boards. Use the gene pool generated in Experiment 12.17 for population seeding. Create initial populations either by uniting 10 of the genes from the gene pool, selected uniformly at random, with 390 random genes, or by generating 400 random genes. Using the NWB fitness function lex survival time, do 100 runs for at most 400 generations for each of the two new board sizes and for each of the initial populations. Report mean, deviation, and maximum of fitness and survival time. Did the genes from the gene pool help?*

As has happened before, the combinatorial closure of the experiments we *could* have performed is enormously larger than those we did perform. You are encouraged to try other experiments (and please write us if you find a good one). Of especial interest is more study of the effect that skill at one task has on gene-pool quality for another task. This chapter is not the last we will see of ISAc lists; we will look at them in the context of epidemiology and ecological modeling in future chapters.

Problems

Problem 506. On page 331 it is asserted that for the Vacuum Cleaner task, the average fitness over the set of all initial placements and headings and the average fitness over all initial placements with a single heading are the same. Explain why this is so.

Problem 507. For an $n \times n$ board is $n^2 + 2n$ a reasonable number of moves for the Vacuum Cleaner task? Why or why not?

Problem 508. Short Essay. Given the way fitness is computed for the Vacuum Cleaner task, what use is the stand-still action? If it were eliminated, would solutions from a evolutionary algorithm tend to get better or worse? Explain.

Problem 509. Are the vacuum's sensors of any use? Why or why not?

Problem 510. Would removing the walls and permitting the board to wrap around at the edges make the Vacuum Cleaner task harder or easier? Justify your answer in a few sentences.

Problem 511. Is the length of an ISAc list a critical parameter, i.e., do small changes in the lengths of the ISAc lists used in an evolutionary algorithm create large changes in the behavior of the evolutionary algorithm, on average? Justify your answer in a few sentences.

Problem 512. For a 9×9 board, give an algorithm for a perfect solution to the Vacuum Cleaner task. You may write pseudocode or simply give a clear statement of the steps in English. Prove, probably with mathematical induction, that your solution is correct.

Problem 513. In other chapters, we have used neutral mutations, mutations that change the gene without changing fitness, as a way of creating population diversity. Suppose we change all turn lefts to turn rights and turn rights to turn lefts in an agent. If we are considering fitness computed over all possible starting configurations, then is this a neutral mutation for (i) Tartarus, (ii) Vacuum Cleaner, (iii) Herbivore, or (iv) North Wall Builder? Justify your answers in a few sentences.

Problem 514. Experiment 12.11 samples 20 boards to estimate fitness. How many boards are in the set from which this sample is being drawn?

Problem 515. Suppose instead of having walls at the edge of the board in the Herbivore task we have the Herbivore board wrap around, left to right and top to bottom, creating a toroidal world. Would this make the task easier or harder? Would the resulting genes be better or worse as models of foraging herbivores? Explain.

Problem 516. Suppose that we were to use parse trees to code cowdozers for the Herbivore task. Let the data type for the parse trees be the integers (mod 4) with output interpreted as 0=turn left, 1=turn right, 2=go forward, 3=eat. Recalling that 0=empty, 1=box, 2=wall, explain in plain English the behavior of the parse tree $(+ \; x_0 \; 2)$, where x_0 is the front middle sensor.

Problem 517. Short Essay. How bad a local optimum does the parse tree described in Problem 516 represent? What measures could be taken to avoid that optimum?

Problem 518. Compute the expected score of the parse tree described in Problem 516 on a 16×16 board with 32 boxes.

Problem 519. Give the operations and terminals of a parse tree language on the integers (mod 4) in which the parse tree described in Problem 516 could appear. Now write a parse tree in that language that will score better than $(+\ x_0\ 2)$. Show on a couple of example boards *why* your tree will outscore $(+\ x_0\ 2)$. Advanced students should compute and compare the expected scores.

Problem 520. For the four grid robotics tasks we've looked at in this chapter (Tartarus, Vacuum Cleaner, Herbivore, and North Wall Builder), rate the tasks for difficulty (i) for a person writing a controller, and (ii) for an evolutionary algorithm. Justify your answer. Is the number of possible boards relevant? The board size?

Problem 521. Invent and describe a new grid robotics task with at least one action not used in the grid robotics tasks studied in this chapter. Make the task interesting and explain why you think it is interesting.

Problem 522. In Experiment 12.14, we used 4 copies each of the Vacuum Cleaner gene pool members to create an initial population. Given the three possible classes of outcomes (listed on page 335) among which the experiment was attempting to distinguish, explain why we did *not* include random genes in the initial population.

Problem 523. In Experiment 12.14, we tried using a Vacuum Cleaner gene pool as a starting point for an evolutionary algorithm generating Herbivore controllers. For each pair of the four tasks we study in this chapter, predict whether using a gene pool from one task would be better or worse than starting with a random population for another. Give one or two sentences of justification for each of your predictions.

Problem 524. Does using survival time in a lexical fitness for the North Wall Builder create the potential for a bad local optimum? If so, how hard do you estimate it is to escape and why; if not, explain why not.

Problem 525. Is the allotment of $6n^n$ moves for a constructor to complete the North Wall Builder task generous or tight-fisted? Why? (see Experiment 12.18)

Problem 526. Short Essay. In Experiment 12.18, we examine the use of time-averaged fitness to encourage the constructor to act efficiently. On page 338, it is asserted that for this to help, it must be possible to reach high-fitness configurations without intermediate low-fitness ones. In other words, fitness should pretty much climb as a function of time. First, is the assertion correct? Explain. Second, are there solutions to the North Wall Builder task in which the fitness does not increase as a function of time?

Problem 527. Hand code, in the language of your choice or in pseudocode, a perfect solution to the North Wall Builder task. Beginning students work on a 9×9 board; advanced students provide a general solution.

Problem 528. A *macromutation* is a map that takes a gene to a single other gene making potentially large changes. Give, in a few sentences, a method for using gene pool members in a macromutation operator.

Problem 529. Essay. For the Tartarus, Vacuum Cleaner, Herbivore, and North Wall Builder tasks, we can vary the board size. With this in mind, cogitate on the following question: would a gene pool created from experiments on one board size be a good starting point for experiments on another board size? Feel free to cite evidence.

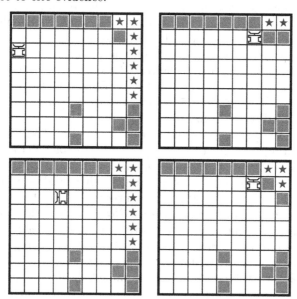

Problem 530. Short Essay. The four boards above are final states for the North Wall Builder task. They came from simulations initialized with gene pools, in the manner of Experiment 12.14. What can you deduce from these boards about the process of evolution from a gene-pool-derived initial population containing no random genes? Is anything potentially troublesome happening?

12.4 Return of the String Evolver

This section includes some ideas not in the main stream of the ISAc list material, but suggested by it and valuable. Thus far, we have done several string baseline experiments: 10.2, 10.3, 10.4, 10.5, 12.8, 12.11, and 12.16. The

4 experiments from Chapter 10 developed the following methodology for a string baseline of a virtual robotics experiment.

The simplest string baseline evolves fixed-length strings of a length sufficient to supply a character for each move desired in the simulation (Experiment 10.2). The next step is to use shorter strings, cyclically, as in Experiment 10.3. The advantage of this is not that better solutions are available—it is elementary that they are not—but rather that it is much easier for evolution to *find* tolerably good solutions this way. Experiments 10.4 and 10.5 combine the two approaches, permitting evolution to operate on variable-length strings. This permits discovery at short lengths (where it is easier), followed by revision at longer lengths, reaping the benefits of small and large search spaces, serially.

In the baseline experiments in the current chapter, we simply mimicked Experiment 10.3 to get some sort of baseline, rather than pulling out all the stops and getting the best possible baseline. If your experiments went as ours did, this string baseline produced some surprising results. The difference between the string baseline performance and the adaptive agent performance is modest, but significant, for the North Wall Builder task. In the Herbivore task, the string baseline substantially outperforms the parse tree derived local optimum described in Problem 516.

The situation for string baselines is even more grim than one might suppose. In preparing to write this chapter, we first explored the North Wall Builder task with a wall at the edge of the world, using GP automata and ISAc lists. Much to our surprise, the string baseline studies, while worse on average, produced the only perfect gene (fitness 81 on a 9×9 board). The string baseline showed that with walls the North Wall Builder task was too easy. With the walls removed, the adaptive agents, with their ability to not walk off the edge of the board, outperformed our string baseline population. This is why we present the wall-free version of the North Wall Builder task.

At this point, we will introduce some terminology and a point of view. An agent is *reactive* if it changes its behavior based on sensor input. The parse trees evolved in Experiment 10.7 were purely reactive agents. An agent is *state conditioned* if it has some sort of internal state information. The test for having internal state information is, does the same input result in different actions at different times (neglect the effect of random numbers)? An agent is *stochastic* if it uses random numbers.

Agents can have any or all of these three qualities, and all of the qualities can contribute to fitness. Our best agents thus far have been reactive, state conditioned, nonstochastic agents. When used individually, we find that the fitness contributions for Tartarus have the order

$$\text{reactive} < \text{stochastic} < \text{state conditioned}.$$

In other words, purely reactive agents (e.g., parse trees with no memory of any sort) perform less well than tuned random number generators (e.g., Markov chains), which in turn achieve lower fitness than purely state conditioned agents (e.g., string controllers).

Keeping all this in mind, we have clear evidence that the string baselines are important. Given that they are also a natural debugging environment for the simulator of a given virtual robotic world, it makes no sense not to blood a new problem on a string baseline. In the remainder of this section, we will suggest new and different ways to perform string baseline studies of virtual robotics tasks.

Our first attempt to extend string controllers involves exploiting a feature like the NOP instruction in ISAc lists. In Experiments 10.3, 10.4, and 10.5, we tinkered with varying the length of the string with a fairly narrow list of possibilities. The next experiment improves the granularity of these attempts.

Experiment 12.20 *Start with either Experiment 12.11 (Herbivore) or Experiment 12.16 (North Wall Builder). Modify the string evolver to use a string of length 30 over the alphabet consisting of the actions for the virtual robotics task in question, together with the null character "*". Start with a population of 400 strings, using the string cyclically to generate actions, ignoring the null character. Do 100 runs and compare with the results for the original string baseline. Plot the fraction of null actions as a function of time: are there particular numbers of null actions that seem to be desirable? Now redo the experiment, but with a 50% chance of a character being null, rather than a uniform distribution. What effect does this have?*

The use of null characters permits insertion and deletion, by mutation, into existing string controllers. It is a different way of varying the length of strings. We now will create reactive strings for the Herbivore environment and subject them to evolution. Examine Figure 12.7. This is an alphabet in which some characters stand for one of two actions, depending on information available to the cowdozer's sensors. We call a character *adaptive* if it codes for an action dependent on sensory information. Nonadaptive characters include the traditional actions and the null character from Experiment 12.20.

Experiment 12.21 *Rebuild your Herbivore board routines to work with the adaptive alphabet described in Figure 12.7, except for the null character. Run an evolutionary algorithm operating on a population of 400 adaptive strings of length 30, used cyclically during fitness evaluation. For fitness evaluation, use a sample of 20 boards to approximate fitness. Use 9×9 Herbivore boards. Use single tournament selection with tournament size 4, two-point crossover, and 0- to 3-point mutation, with the number of mutations selected uniformly at random.*

Do 100 runs and compare with other experiments for the Herbivore task. Save the fraction of adaptive characters in the population and plot this in your write-up. Now perform these runs again with the null character enabled. Was the effect different from that in Experiment 12.20 (assuming comparability)?

Character	Meaning	Adaptive
L	Turn left	No
R	Turn right	No
F	Move forward	No
E	Eat	No
A	If box left, turn left; otherwise, go forward	Yes
B	If box right, turn right; otherwise, go forward	Yes
C	If wall ahead, turn left; otherwise, go forward	Yes
D	If wall ahead, turn right; otherwise, go forward	Yes
Q	If box ahead, eat; otherwise, go forward	Yes
*	Null character	No

Fig. 12.7. Alphabet for the adaptive Herbivore string controller.

The adaptive characters used in Experiment 12.21 are not the only ones we could have chosen. If we include failure to act, there are $\binom{5}{2} = 10$ possible pairs of actions. A choice could be made from each pair, based on information from any of 8 sensors with 3 return values. One is tempted to use a metaevolutionary algorithm to decide which adaptive characters are the most valuable (but one refrains). Rather, we look at the preceding experiment's ability to test the utility of adaptive characters and note that stochastic characters can also be defined. A stochastic character is one that codes for an action based on a random number. In Figure 12.8, we give a stochastic alphabet.

Character	Meaning	Stochastic
L	Turn left	No
R	Turn right	No
F	Move forward	No
E	Eat	No
G	Turn right, turn left, or go forward with equal probability	Yes
H	Turn right 20%, turn left 20%, go forward 60%	Yes
I	Turn left or go forward with equal probability	Yes
J	Turn right or go forward with equal probability	Yes
K	Turn left 30%, go forward 70%	Yes
M	Turn right 30%, go forward 70%	Yes
*	Null character	No

Fig. 12.8. Alphabet for the stochastic Herbivore string controller.

Experiment 12.22 *Rebuild your Herbivore board routines to work with the stochastic alphabet described in Figure 12.8, except for the null character. Run an evolutionary algorithm operating on a population of 400 adaptive strings of length 30, used cyclically during fitness evaluation. For fitness evaluation, use a sample of 20 boards to approximate fitness. Use 9 × 9 Herbivore boards.*

Use single tournament selection with tournament size 4, two-point crossover, and 0- to 3-point mutation, with the number of mutations selected uniformly at random.

Do 100 runs and compare with other experiments for the Herbivore task. Save the fraction of stochastic and of each type of stochastic characters in the population and plot this in your write-up. Now perform these runs again with the null character enabled. Compare with other experiments and comment on the distribution of the stochastic characters within a given run.

We conclude with a possibly excessive experiment with a very general sort of string controller. We have neglected using string doubling and halving mutations on our adaptive and stochastic alphabets; these might make nice term projects for students interested in low-level design of evolutionary algorithm systems. Other more bizarre possibilities are suggested in the Problems.

Experiment 12.23 *Rebuild your Herbivore board routines to work with the union of the adaptive and stochastic alphabets described in Figures 12.7 and 12.8, except for the null character. Run an evolutionary algorithm operating on a population of 400 adaptive strings of length 30, used cyclically during fitness evaluation. For fitness evaluation, use a sample of 20 boards to approximate fitness. Use 9×9 Herbivore boards. Use single tournament selection with tournament size 4, two-point crossover, and 0- to 3-point mutation, with the number of mutations selected uniformly at random.*

Do 100 runs and compare with other experiments for the Herbivore task. Save the fraction of stochastic, of adaptive, and of each type of character in the population and plot these in your write-up. Comment on the distribution of the types of characters within a given run.

Problems

Problem 531. On page 343, it is asserted that string controllers, like those from Experiments 10.2, 10.3, 10.4, 10.5, 12.8, 12.11, and 12.16, are purely state conditioned agents. Explain why they are not reactive or stochastic and identify the mechanism for storage of state information.

Problem 532. Give pseudocode for transforming adaptive string controllers, à la Experiment 12.21, into ISAc lists with the same behavior. Hint: write code fragments for the adaptive characters and then use them.

Problem 533. Give a segment of an ISAc list that cannot be simulated by adaptive string controllers of the sort used in Experiment 12.21.

Problem 534. How many different adaptive characters of the type used in Experiment 12.21 are there, given choice of actions and test conditions?

Problem 535. Give and defend an adaptive alphabet for the Tartarus problem. Include an example string controller.

Problem 536. Give and defend an adaptive alphabet for the Vacuum Cleaner task. Include an example string controller.

Problem 537. Give and defend an adaptive alphabet for the North Wall Builder task. Include an example string controller.

Problem 538. Examine the adaptive alphabet given in Figure 12.7. Given that the Q character is available, what use is the E character? Do not limit your thinking to its use in finished solutions; can the E character tell us anything about the evolutionary algorithm?

Problem 539. Essay. Stipulate that it is easier to search adaptive alphabets for good solutions, even though they code for a more limited collection of solutions. Explain how to glean adaptive characters from evolved ISAc lists and do so from some evolved ISAc lists for one of the virtual robotics tasks studied in this chapter.

Problem 540. Either code the Herbivore strategy from Problem 516 as an adaptive string controller (you may choose the length) or explain why this is impossible.

Problem 541. Short Essay. Does the lack of stochastic actions involving eating represent a design flaw in Experiment 12.22?

Problem 542. Give and defend a stochastic alphabet for the Tartarus problem or explain why any stochasticity would be contraindicated. Include an example string controller if you think stochasticity could be used profitably.

Problem 543. Give and defend a stochastic alphabet for the Vacuum Cleaner task or explain why any stochasticity would be contraindicated. Include an example string controller if you think stochasticity could be used profitably.

Problem 544. Give and defend a stochastic alphabet for the North Wall Builder task or explain why any stochasticity would be contraindicated. Include an example string controller if you think stochasticity could be used profitably.

Problem 545. Reread Problems 385, 409, and 411. Now reread Experiment 12.15. The thought in Experiment 12.15 was to transform a string baseline gene into an ISAc list. In the three problems from Chapter 10, we were using Markov chains as controllers for the Tartarus problem. Explain why a string controller is a type of (deterministic) Markov chain. Explain how to transform a string gene into a deterministic Markov chain that can lose its determinism by mutation, sketching an evolutionary algorithm for starting with string genes and evolving good Markov controllers.

Problem 546. Short Essay. Reread Experiment 12.22 and then answer the following question: does a Markov chain controller ever benefit from having more states than the number of types of actions it needs to produce? Explain.

Problem 547. Give an evolutionary algorithm that locates good adaptive or stochastic characters. It should operate on a population of characters *and* a population of strings using those characters, simultaneously, so as to avoid using a metaevolutionary (multilevel) algorithm.

13

Graph-Based Evolutionary Algorithms

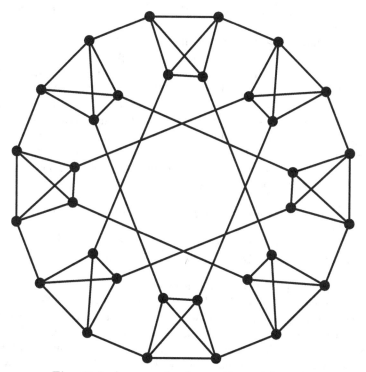

Fig. 13.1. An example of a combinatorial graph.

This chapter is in some ways a capstone chapter, pulling together problems and representations from many other chapters. In this chapter we will use combinatorial graphs to add geographic structure to the populations being evolved. Many experiments from previous chapters will be taken up and expanded. If you are unfamiliar with combinatorial graphs, you should read Appendix D. Refer to [58] for a comprehensive treatment of graph theory. An example of a combinatorial graph is given in Figure 13.1. The first several experiments develop tools for assessing the geographic character of graphs. We will place individuals on the vertices of the graph and permit mating to take place only between individuals on vertices connected by an edge.

This is a generalization that gives a feature present in the biological world to evolutionary algorithms. Consider a natural population of rabbits. No matter how awesome the genetics of a given rabbit, it can breed only with other rabbits nearby. Placing the structures of our evolving population into a geography and permitting breeding only with those nearby limits the spread of information in the form of superior genes. Single tournament selection is already available as a method of limiting the number of children of a high-fitness individual, but as we will see, using a graphical population structure gives us far more flexibility than varying tournament size does. Even tournament selection has a positive probability of any good gene breeding and replacing another.

You may wonder why we wish to limit the spread of good genes. The answer lies in our eternal quest to avoid local optima or premature convergence to a suboptimal solution. Limiting the spread of a good structure without utterly destroying it permits the parts of the population "far" from the good structure to explore independently selected parts of the search space. Where a standard evolutionary algorithm loses population diversity fairly rapidly and ends up exploring a single sector of the fitness landscape of a problem quite soon, an evolutionary algorithm with an imposed geography can continue to explore different parts of the fitness space in different geographic regions.

Our primary questions in this chapter are these:

1 *Does placing a combinatorial graph structure on a population ever change performance?*

2 *If so, what sorts of graph structures help which problems?*

3 *How do we document the degree to which a given graph structure helps?*

Throughout the chapter, you should think about the character of the example problems in relation to the diversity preservation (or other effects) induced by the use of a graph as a geographic population structure. Recall the broad classes of problems that exist: unimodal and multimodal, optimization as opposed to coevolution. The long-term goal that underlies this chapter is to obtain a theory, or at least a sense, of how the fitness landscapes of problems interact with graphical population structures. The relationship of the experiments in the chapter is given in Figure 13.2. Note the very large number of unconnected components.

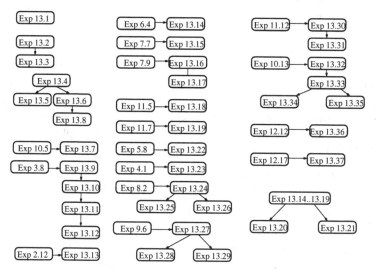

1   Neutral selection on the hypercube.
2   Analysis of data from neutral-selection experiments.
3   Neutral selection on random regular graphs.
4   Graph-based evolutionary algorithms for one-max.
5   For 2-max, 4-max.
6   The Royal Road problem.
7   Tartarus using a string representation.
8   Changing population size.
9   Fake bell curve.
10   Starting with a bad initial population.
11   Functions with two optima.
12   Changing the local mating rule.
13   Self-avoiding walks.
14   String prediction with finite state machines.
15   Permutations.
16   Traveling Salesman problem.
17   Population seeding.
18   The 3-parity problem with neural nets.
19   Logic functions with connection lists.
20   Change of local mating rule.
21   Changing graphs.
22   Symbots.
23   Sunburn.
24   PORS.
25   Elite fitness.
26   Absolute fitness.
27   Fake bell curve.
28   Standard tournament selection baseline.
29   Random regular graphs.
30   The 3-parity problem with genetic programming.
31   The 4-multiplex and 6-parity problems.
32   Tartarus with genetic programming.
33   Tartarus with GP automata.
34   Using a complex vertex/fitness map.
35   Varying the fitness function.
36   Herbivore.
37   North Wall Builder.

Fig. 13.2. The topics and dependencies of the experiments in this chapter.

13.1 Basic Definitions and Tools

We will impose a graphical structure on an evolutionary algorithm by placing a single population member at each vertex of the graph and permitting reproduction and mating only between neighbors in the graph. (Note that this is not the only way one could use a graph to impose a graphical geography on an evolutionary algorithm.) Our model of evolution will need a way of selecting a gene from the population to be a parent, a way of selecting one of its neighbors to be a coparent, and a way of placing the children.

Definition 13.1 *The* **local mating rule** *is the technique for picking a neighbor in the graph with which to undergo crossover and for placing children on the graph. It is the graph-based evolutionary algorithm's version of a model of evolution.*

There is a large number of possible models of evolution. Not only are there many possible local mating rules, but also, there are many methods for picking the vertex that defines the local neighborhood. We will explore only a few of these models. Following Chapter 2, we will define the methods for picking the parent, locally picking the coparent, and of placing children. A local mating rule will consist in matching up three such methods.

Definition 13.2 *The second parent, picked from among the neighbors of the first parent, is termed the* **coparent.**

The parent may be picked by roulette, rank, random, or systematic selection operating on the entire population. The first three of these methods have the exact same meaning as in Chapter 2. *Systematic* selection orders the vertices in the graph and then traverses them in order, applying a local mating rule at each vertex. Any selection method may have *deferred* or *immediate* replacement. Deferred replacement does not place any children until coparents have been picked for each vertex in the graph, matings performed, and children generated. The children are held in a buffer until a population updating takes place. Deferred replacement yields a generational graph-based algorithm. *Immediate* replacement places children after each application of the local mating rule and is akin to a steady-state version of the algorithm.

Coparents may be selected (from the neighbors of the parent) by roulette, rank, random, or absolute fitness. The first three terms again have the exact same meaning as in Chapter 2. Absolute fitness selects the best neighbor of the parent as the coparent.

Replacement will involve one of: the parent, parent and coparent, the neighbors of the parent including or not including the parent. *Absolute replacement* replaces both parent and coparent with the children. *Absolute parental replacement* replaces only the parent with one of the children selected at random. *Elite parental replacement* replaces the parent with one of the children selected at random *if* the child is at least as fit as the parent. *Elite replacement* places the best two of parent, coparent, and children into the slots formerly occupied by the parent and coparent. *Random neighborhood replacement* picks a vertex in the neighborhood of the parent (including the parent) and places one child selected at random there. *Elite neighborhood replacement* picks a vertex in the neighborhood of the parent (including the parent) and places one child selected at random there if it is at least as good as the current occupant of the vertex. *Elite double neighborhood replacement* picks two neighbors at random and replaces each with a child selected at random if the child is better.

Neutral Behavior of Graphs

Before we run evolutionary algorithms on graphs, we will develop some diagnostics to study the way different graphs interact with them. These diagnostics will approximate the amount of biodiversity and of potentially useful mating enabled by a given graph structure. (By useful mating we mean mating between creatures who are substantially different from each other; crossover between similar creatures is wasted effort.) In order to study the effects of the graph independently from the effects of the evolutionary algorithm, we will use random selection and no variation operators. First, we need some way to measure biodiversity and the potential for useful mating.

Definition 13.3 *If we have identifiable types* $\{1, \ldots, k\}$ *in a population of N creatures with n_i creatures of type i, then the* **entropy** *of the population is*

$$E = -\sum_{i=0}^{k} \frac{n_i}{N} \cdot \log_2 \left(\frac{n_i}{N} \right).$$

If you have studied information theory, you will recognize the entropy defined above as the Shannon entropy of the "probability" of encountering a creature of a given type when sampling from the population. Entropy will be our surrogate for biodiversity. It has a number of properties that make it a good choice as a diversity measure. First of all, it increases as the number of types of creatures increases. This makes sense: more types, more biodiversity. The second good property is that if the number of types is fixed, then entropy increases as the population is more evenly divided among the types. Imagine a cornfield with one foxtail and one dandelion. Now imagine a field evenly divided between foxtails, dandelions, and mustard plants. Both fields have 3 types of plants in them, but the second field is far more diverse. The third desirable property of entropy is that it is independent of the total number of creatures, and so permits comparison between populations of different sizes.

Definition 13.4 *The* **edge** *of a population on a graph is the fraction of edges with different types of creatures at their ends.*

Evolutionary algorithms generate new solutions to a problem in three ways. The initialization of the population is a source of new solutions, though if initialization is random, a crummy one. Mutation generates variations of existing solutions. Crossover blends solutions. When it works well, crossover is a source of large innovations. An unavoidable problem with standard evolutionary algorithms is that they lose diversity rapidly; soon, crossover is mostly between creatures of the same approximate "type." The result is that most crossover is wasted effort. The edge of a graph is the fraction of potential crossovers that could be innovative. Figure 13.3 shows the edge and entropy for 5 graphs over the course of 1,000,000 mating events in a neutral selection experiment.

Definition 13.5 *A* **neutral selection** *experiment for a graph G with k vertices is performed as follows. The vertices of the graph are labeled with the numbers 1 through k in some order. Each label represents a different type of creature. A large number of mating events are performed in which a vertex is chosen at random, and then the label on one of its neighbors, chosen at random, is copied over its own label. At fixed intervals, the labels are treated as a population and the entropy and edge are computed.*

Since neutral selection experiments are stochastic, typically one must average over a large number of them to get a smooth result. We will explore this stochasticity in the first experiment of this chapter.

Experiment 13.1 *For the 9-hypercube, H_9, perform 5 neutral selection experiments with 1,000,000 mating events and a sampling interval of 1000 mating events. The graph H_9 is described in Appendix D. Graph the edge and entropy for each of the 1000 samples taken and for each of the 5 experiments separately. Report the number of the sampling event on which the entropy drops to zero (one label remaining) if it does. Compare your plots with the average, over 100 experiments, given in Figure 13.3. Comment on the degree to which the plots vary in your write-up and compare in class with the results of other students.*

Looking at the tracks for entropy and edge from Experiment 13.1, we see that there is a good deal of variability in the behavior of the evolution of individual populations on a graph. Looking at Figure 13.3, we also see that there is substantial variation in the behavior using different graphs.

Definition 13.6 *An* **invariant** *of a graph is a feature of the graph that does not change when the way the graph is presented changes, without changing the fundamental nature of the graph.*

Experiment 13.2 *Write or obtain software that can gather and graph data as in Figure 13.3. The graphs ($K_{512}, H_9, T_{4,128}, P_{256,1}$, and C_{512}) used in this experiment are described in Appendix D. Perform a neutral selection experiment using 1,000,000 mating events with the edge and entropy values sampled*

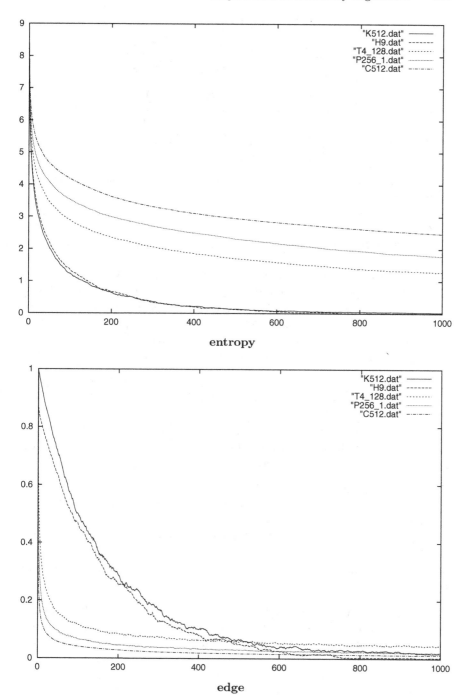

Fig. 13.3. Graphs showing the entropy and edge in a neutral selection experiment for the complete graph K_{512}, the 9-dimensional hypercube H_9, the 4×128 torus $T_{4,128}$, the generalized Petersen graph $P_{256,1}$, and the 512-cycle C_{512}. (The plots are averaged over 100 experiments for each graph. Each experiment performs 1,000,000 mating events, sampling the entropy and edge each 1000 mating events.)

Name	degree	diameter	edges
C_{512}	2	256	512
$P_{256,1}$	3	129	768
$T_{4,128}$	4	66	1024
H_9	9	9	2304
K_{512}	511	1	130,816

Table 13.1. Some graph invariants for the graphs whose neutral selection results appear in Figure 13.3.

every 1000 mating events and averaged over 100 runs of the experiment. Test the software by reproducing Figure 13.3.

Now rerun the software on the graphs $P_{256,1}$, $P_{256,3}$, $P_{256,7}$, and $P_{256,15}$. Graph the entropy and edge for these 4 Petersen graphs together with C_{512} and K_{512}. This experiment checks to see whether edge and entropy vary when degree is held constant (3) and also compares them to graphs with extreme behaviors. In your write-up, comment on the degree to which the edge and entropy vary and on their dependence on the degree of the graph.

Experiment 13.2 leaves the number of edges the same while changing their connectivity to provide a wide range of diameters (excluding C_{512} and K_{512} which serve as controls). The next experiment will generate graphs with the same degree and a relatively small range of diameters. This will permit us to check whether there is variability in neutral selection behavior that arises from sources other than diameter.

Experiment 13.3 *Write or obtain software that can generate random regular graphs of the sort described in Definition D.22. Starting with the Petersen graph $P_{256,1}$ and making 3500 edge moves for each instance, generate at least 4 (consult your instructor) random 3-regular graphs. Repeat Experiment 13.2 for these graphs, including the C_{512} and K_{512} controls. In addition, compute the diameter of these graphs and compare diameters across all the experiments performed.*

For the remainder of the chapter, you should keep in mind what you have learned about neutral selection. Try to answer the following question: what, if any, value does it have for predicting the behavior of graph-based evolutionary algorithms?

Problems

Problem 548. List all the local mating rules that can be constructed from the parts given in this section. Put a check mark by any that don't make sense and give a one-sentence explanation of why they don't make sense.

Problem 549. For a single population in a neutral selection experiment, are entropy and edge, as functions of time measured in mating events, monotone decreasing? Prove your answer.

Problem 550. Examine the neutral selection data presented in Figure 13.3. Say which of the invariants in Table 13.1 is most predictive of entropy behavior and of edge behavior and why you think so.

Problem 551. Essay. Describe the interaction between edge and entropy. If edge is high, is entropy decreasing faster on average than if edge is low? At what level of entropy, on average, is it likely for edge to go up temporarily? Can you say anything else?

Problem 552. Find a sequence of mate choices for a neutral selection experiment on K_5 for which the entropy stays above 1.5 forever. Do you think this will ever happen? Why?

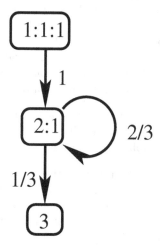

Problem 553. Consider a neutral selection experiment on K_3. It starts with distinct labels on all 3 vertices. After the first mating event, there are 2 labels that are the same and 1 that is different. The next mating event has $\frac{1}{3}$ chance of making all the labels the same and $\frac{2}{3}$ chance of leaving two labels the same and one different. Once all the labels are the same, the system is stuck there. This behavior, encompassing all possible histories, can be summarized in the Markov chain diagram shown above. First convince yourself that the probabilities given are correct for K_3. Draw the equivalent diagram for K_4, using the states [1:1:1:1], [1:1:2], [1:3], [2:2], and [4].

Problem 554. Suppose we are performing neutral selection on C_n, the n-cycle. Describe all possible collections of labels that can occur during the course of the experiment.

Problem 555. Suppose that n is even and that we are running a neutral selection experiment. For K_n, $P_{\frac{n}{2},2}$, and C_n, consider the sets of all vertices that have the same label. Do these collections of vertices have to be connected in the graph? If the answer isn't obvious, you could run simulations. The answer should be quite obvious for K_n.

Problem 556. Extend Table 13.1 by copying the current entries and adding the information for the Petersen graphs used in Experiment 13.2.

Problem 557. Two graphs are *isomorphic* if you can exactly match up their vertices in a fashion that happens also to exactly match up their edges. Notice that, unless two graphs have the same number of vertices and edges, they cannot be isomorphic. For the graphs $P_{16,n}$ with $1 \le n \le 15$, find out how many "different" graphs there are. (Two graphs are "the same" if they are isomorphic.)

If you are unfamiliar with graph theory, this is a very difficult problem; so here are four hints. First, consider equivalence of numbers (mod 16). If you go around the inner circle of the Petersen graph by jumps of size one to the left or right, you still draw the same edges, and so pick up some obvious isomorphisms with which to start. Second, remember that isomorphism is transitive. Third, isomorphism preserves all substructures. If we have a closed cycle of length 4 in one graph and fail to have one in another, then they cannot be isomorphic. Fourth, notice that isomorphism preserves diameter.

Problem 558. Compute the diameters of $P_{256,k}$ for $1 \le k \le 255$. What are the maximum and the minimum, and how do these numbers reflect on the choice of the 4 Petersen graphs in Experiment 13.2?

Problem 559. Generate 1000 graphs of the sort used in Experiment 13.3. For the diameters of these graphs, make a histogram of the distribution and compute the mean, standard deviation, minimum value, and maximum value. Finally, compare these with the results of Problem 558. Are the Petersen graphs representative of cubic graphs?

Problem 560. Suppose that we adopt an extremely simple notion of connectivity: the fraction of edges a graph has relative to the complete graph. Thus, C_5 has a connectivity of 0.5, while K_5 has a connectivity of 1.0. Answer the following questions:

(i) What is the connectivity, to 3 significant figures, of the 5 graphs used to generate Figure 13.3?
(ii) Approximately how many graphs are there with connectivity intermediate between H_9 and K_{512}?

(iii) In what part of the range of connectivities from 0 to 1 is most of the variation in neutral selection behavior concentrated?

Problem 561. Suppose you have a random graph with edge probability $\alpha = 0.5$ (for all possible edges, flip a fair coin to see whether the edge is present). Compute, as a function of the number n of vertices in the graph, the probability that it will have diameter 1.

Problem 562. Essay. Modify the code you used in Problem 559 to report whether the graph is connected (see Definition D.7). What fraction of graphs were connected? Now perform the experiment again starting with C_{512} instead of $P_{256,1}$ and find what fraction of the graphs are connected. Explain the results.

13.2 Simple Representations

In this section, we will examine the effects of imposing a geography using graphs on the simplest of evolutionary algorithms, those with data structures of fixed-sized strings or vectors of real numbers. The initial work will be a comparison of the 1-max problem and one of its variations, called the k-max problem.

Definition 13.7 *The k-max fitness function $F_{k\text{-}max}$ maps strings of length l over a k-member alphabet to the count of their most common character. Thus, $F_{2\text{-}max}(0110100100110) = 7$, while $F_{5\text{-}max}(ABBCCCDDDDEEE) = 4$.*

The advantage of the k-max problem is that it is constructively polymodal. This permits us to compare the 1-max problem, a completely unimodal problem, with a number of constructively polynomial problems with a very similar character. Let us start by doing an experiment that baselines the behavior of graph-based evolutionary algorithms on the 1-max problem.

Experiment 13.4 *For the 5 graphs K_{512}, H_9, $T_{4,128}$, $P_{256,3}$, and C_{512}, using random selection of the parent, roulette selection of the coparent, and immediate elite replacement of the parent by the better child, run a graph-based evolutionary algorithm on the 1-max problem over the binary alphabet with length 16. Use two-point crossover and single-point mutation. In addition, run a baseline evolutionary algorithm on the 1-max problem using single tournament selection with size 4.*

For each of the 6 evolutionary algorithms, save time-to-solution (cutting off algorithms at 1,000,000 mating events) for 100 runs of each algorithm. Give a 95% confidence interval for the mean time-to-solution for each algorithm. Discuss which graphs are superior or inferior for the 1-max problem and compare with the single tournament selection baseline.

The 1-max problem has two distinct evolutionary phases. In the first, crossover mixes and matches blocks of 1's and can help quite a lot. In the second, a superior genotype has taken over, and we have 0's in some positions throughout the population, forcing progress to rely on mutation. In this latter, longer phase, the best thing you can do is to copy the current best structure as fast as possible. The dynamics of this second phase of the 1-max problem suggest that more connected graphs should be superior.

Experiment 13.5 *Repeat Experiment 13.4 for the 2-max problem and for the 4-max problem. The fitness function changes to report the largest number of any one type of character, first over the binary alphabet and then over the quaternary alphabet. Compare results for 1-max, 2-max, and 4-max.*

The Royal Road function (see Section 2.5) is the standard "hard" string evolver problem. Let's use it to explore the effects of varying the mutation rate (which we already know is important) with the effects of changing the graph used.

Experiment 13.6 *Repeat Experiment 13.4 for the classic Royal Road problem (l = 64 and b = 8) or for l = 36, b = 6, if the run time is too long on the classic problem. Do 3 groups of 6 collections of 100 runs, using each of the following mutation operators: one-point mutation, probabilistic mutation with rate one, and probabilistic mutation with rate two. Compare the effects of changing graphs with the effects of changing mutation operators.*

There is no problem with running Experiments 13.4, 13.5, and 13.6 as steady-state algorithms. A Tartarus-type problem in which we are sampling from among many available fitness cases, requires a generational algorithm. The following experiment thus uses deferred replacement.

Experiment 13.7 *Create a generational graph-based evolutionary algorithm (one using deferred replacement) to evolve string controllers for the Tartarus problem. We will use variable-length strings and the gene-doubling and gene-halving operators described in Section 10.1. The initial population and variation operators are as in Experiment 10.5. Test fitness on 100 random 6 × 6 Tartarus boards with 6 boxes. Use the same graphs that were used in Experiment 13.4. The algorithm will visit each vertex systematically as a parent. Select the coparent by roulette selection and use absolute replacement.*

Baseline the experiment with an evolutionary algorithm using size-4 tournament selection. Perform 100 runs of length 200 generations and compare average and best results of all 6 sets of runs. If possible, apply knowledge about what strings are good from Chapter 10 to perform a qualitative assessment of the algorithm.

Another issue that is worth examining is that of population size.

Experiment 13.8 *Redo Experiment 13.6 with the following list of graphs:* $P_{2^n,3}$, $n = 4, 6, 8$, *and* 10. *Run a steady-state algorithm and measure time-to-solution in mating events. Use the data you already have from* $n = 8$ *to establish a time after which it would be reasonable to give up. Compare across population sizes and, if time permits, fill in more points to document a trend.*

We now turn to the problem of real function optimization with graph-based algorithms, beginning with the numerical equivalent of the 1-max problem, the fake bell curve.

Experiment 13.9 *For the 5 graphs* K_{512}, H_9, $T_{4,128}$, $P_{256,3}$, *and* C_{512}, *using roulette selection of the parent, rank selection of the coparent, and elite replacement of the parent by the better child, run a graph-based evolutionary algorithm to maximize the function*

$$f(x_1, x_2, \ldots, x_8) = \frac{1}{1 + \sum_{i=1}^{8}(x - i)^2}.$$

Use one-point crossover and Gaussian mutation with standard deviation $\sigma = 0.1$. *Generate the initial population by selecting each coordinate uniformly at random from the integers 0 to 9. The function given is a shifted version of the fake bell curve in 8 dimensions.*

Run a baseline evolutionary algorithm using single tournament selection with tournament size 4 as well. For each of the 6 evolutionary algorithms, save time-to-solution (cutting off algorithms at 1,000,000 mating events) for 100 runs of each algorithm. Take a functional value of 0.999 *or more to be a correct solution. Give a 95% confidence interval for the mean time-to-solution for each algorithm. Discuss which graphs are superior or inferior and compare with the single tournament selection baseline.*

Even in 8 dimensions, random initialization of 512 structures will yield some fairly good solutions in the initial population. It would be nice to document whether mutational diversity can build up good solutions where none existed before and then later recombine good pieces from different parts of a diverse population. We will perform an experiment that attempts to do this by starting with a uniformly awful population.

Experiment 13.10 *Perform Experiment 13.9 again, but this time initializing all creatures to the point* $(0, 0, 0, 0, 0, 0, 0, 0)$. *Compare with the previous experiment. Did the ordering of the graphs' performances change?*

For the next experiment, we will use the function,

$$f(x, y) = \frac{3.2}{1 + (40x - 44)^2 + (40y - 44)^2} + \frac{3.0}{1 + (3x - 5.4)^4 + (3y - 5.4)^4},$$
$$(13.1)$$

constructed to have two optima, one very near $(1.1, 1.1)$ and the other very near $(1.8, 1.8)$. The former is the global optimum, while the latter is broader and hence much easier to find.

Experiment 13.11 *Using the graphs K_{32}, H_5, $T_{4,8}$, $P_{16,3}$, and C_{32}, write or obtain software for a graph-based evolutionary algorithm to maximize Equation 13.1. Use roulette selection of the parent and rank selection of the coparent, absolutely replacing the parent with the first child. Use exchanging x and y coordinates as the crossover operator and use crossover 50% of the time. Use single-point mutation with a Gaussian mutation with variance $\sigma = 0.1$ 100% of the time. For each graph, run a steady-state algorithm until a population member first comes within $d = 0.001$ of either $(1.1, 1.1)$ or $(1.8, 1.8)$. Discuss which graphs are better at finding the true optimum at $(1.1, 1.1)$. If this experiment is done by an entire class, compare or pool the results for random graphs.*

Experiment 13.12 *Repeat Experiment 13.11, but change the local mating rule to (i) random selection of the parent and rank selection of the coparent, and then to (ii) systematic selection of the parent and rank selection of the coparent. Compare the local mating rules and document their impact (or lack of impact).*

Experiment 13.11 tests the relative ability of graphs to enable evolutionary search for optima. Function 13.1, graphed in Figure 13.4, has two optima. The local optimum is broad, flat, and has a large area about it. The global optimum is much sharper and smaller. Let's move on to a complex and very highly polymodal fitness function: the self-avoiding walks from Section 2.6.

Experiment 13.13 *Modify the graph-based string evolver to use the coverage fitness function for walks on a 5×5 grid, given in Definition 2.16. Use two-point crossover and two-point mutation. Compute the number of failures to find an answer in less than 250,000 mating events for each of the following graphs: K_{256}, H_8, $T_{4,64}$, $T_{8,32}$, $T_{16,16}$, $P_{128,1}$, $P_{128,3}$, $P_{128,5}$, and C_{256}. Give a 95% confidence interval on the probability of failure for each of the graphs. Also run a size-4 tournament selection algorithm without a graph as a baseline. Are there significant differences?*

The k-max problem has several optima with large basins of attraction and no local optima which are not global optima. The coverage fitness for walks has thousands of global optima and tens of thousands of local optima. The basins of attraction are quite small. The value of diversity preservation should be much greater for the coverage fitness function than for the k-max fitness function. Be sure to address this issue when writing up your experiments.

Problems

Problem 563. Clearly, the optima of the k-max problem are the strings with all characters the same, and so the problem has k optima. Suppose, for each optimum, we define the *basin of attraction* for that optimum to be strings that go to that optimum under repeated application of helpful mutation (mutating

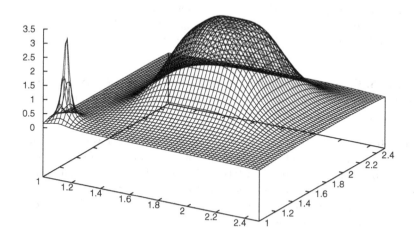

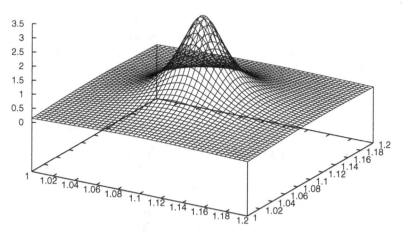

Fig. 13.4. Function 13.1 showing both optima and a closeup of the true optimum.

a character not contributing to fitness into one that does). For $k = 2, 5$ and $l = 12, 13$, compute the size of the basin of attraction for each optimum (they are all the same) and the number of strings that are not in any basin of attraction.

Problem 564. The optima of the k-max problem are all global optima; they all have the same fitness value, and it is the highest possible. Come up with a simple modification of the k-max problem that makes the optima have different heights.

Problem 565. Explain why running a graph-based evolutionary algorithm for Tartarus with evaluation on 100 randomly chosen boards should not be done as a steady-state algorithm.

Problem 566. Suppose that you have a problem for which you know that repeated mutation of a single structure is not as good as running an evolutionary algorithm with some population size $n > 1$. If you are running a steady-state algorithm and measuring time-to-solution (or some other goal) in mating events, then prove that there is a population size such that increasing the population beyond that size is not valuable.

Problem 567. In Experiment 13.11, the graphs use a much smaller population than in other experiments in the section. Why? Explain in a few sentences.

Problem 568. For Function 13.1, perform the following computations. Place a 200×200 grid on the square area with corners at $(0, 0)$ and $(3, 3)$. For each grid cell start at a point in the center of the grid cell. Use a gradient follower with a small step size to go uphill until you reach a point near one of the two optima. This means that you repeatedly compute the gradient and then move a small distance in that direction, e.g., 0.003 (1/1000 the side length of the search grid). The grids that move to a given optimum are in its *gradient basin of attraction*. What is the relative size of the gradient basins of attraction of the two optima? For a discussion of the gradient, see Appendix C.

Problem 569. Explain how to generalize Function 13.1 to n dimensions.

Problem 570. Essay. Would you expect the population diversity of a graph-based evolutionary algorithm to be greater or smaller than the population diversity of a standard evolutionary algorithm.

Problem 571. Short Essay. A thought that occurs quite naturally to readers of this book is to evolve graphs based on their ability to help solve a problem. Discuss this idea with attention to (i) how to represent graphs and (ii) time complexity of the fitness evaluation for graphs.

Problem 572. Short Essay. Reread Experiments 13.9 and 13.10. Now suppose you are attacking a problem you do not understand with graph-based algorithms. One danger is that you will place your initial population in a very small portion of the search space. Does Experiment 13.10 give us a tool for estimating the cost of such misplacement? Answer the question for both string and real number representations (remember the reach of Gaussian mutation).

Problem 573. Essay. Assume the setup used in Experiment 13.6, but with far more graphs. Suppose that you order the graphs according to mean time-to-solution. Changing the mutation operator changes mean time-to-solution. Must it preserve the order of the graphs? In your essay, try to use the notion of fitness landscape and the interaction of that fitness landscape with the gene flow on the graph.

Problem 574. Essay. In this section, we have been testing the effect of changing graphs on the behavior of a graph-based evolutionary algorithm. Can graphs themselves be used as probes for that type of problem? If not, why? If so, how?

Problem 575. For the experiments you have performed in this section, order the graphs within each experiment by the rule $G > H$ if the performance of G on the problem is significantly better than the performance of H. Give these partial orders; if you have performed more than one experiment, comment on the differences between the orders.

13.3 More Complex Representations

Simple representations with linear chromosomes, such as those used in Section 13.2, have different evolutionary dynamics from more complex representations, like those used in Chapters 6–10 and Chapter 12. In this section, we will examine the behavior of some of these systems in the context of GBEAs (graph-based evolutionary algorithms).

In the remainder of this chapter, we will try to control for degree versus topology in the graphs we use by using random regular graphs. We used these graphs in Experiment 13.3. They are described in Appendix D, but we will briefly describe them again here. The technique for producing random regular graphs is not a difficult one, and it generalizes to many other classes of random object generation problems.

The task is to generate a random member of a class of objects. The technique is to find a random transformation that makes a small modification in the object (so that the modified object is still in the class), very like a mutation. As with mutations, we need the transformation to have the property that any object can be turned into any other eventually. We start with any member of the class and make a very large number of transformations, in effect randomly walking it through the object configuration space. This results in a "random" object. Since we want to generate random regular graphs with the same degree as the ones in our other experiments, we use the graphs from the other experiments as starting points.

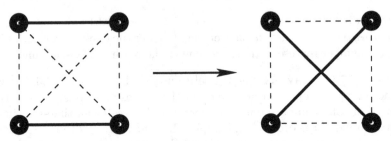

Fig. 13.5. The edge swap operation. (Solid lines denote present edges; dotted lines denote absent edges.)

The transformation used to generate random regular graphs is the edge swap, illustrated in Figure 13.5 and performed in the following manner. Two edges of the graph are located that have the property that they are the only two edges with both ends in the set of 4 vertices that constitute their ends. Those two edges are deleted, and two other edges between the 4 vertices are added. This transformation preserves the degree of the graph while modifying its connectivity.

One point should be made about using random regular graphs. There are an incredible number of different random regular graphs for each degree and number of vertices, as long as there are at least a few tens of vertices. This means that the random regular graph generation procedure is sampling from some distribution on a space of graphs. So what? So, it's important to generate the random regular graphs *before* performing multiple experimental runs and to remember that you are performing experiments on *instances* of a family of graphs. When pooling class results, you may notice large variations in behavior based on which random graphs were used. There are some really good and some really bad random regular graphs out there.

Experiment 13.14 *Review Experiment 6.4, in which we used a lexical partner to enhance performance of a finite state automaton on a string prediction task. Redo this experiment as a baseline, and then, using the same fitness function and representation for finite state automata, perform the experiment as a graph-based algorithm. Use the following graphs: K_{120}, $P_{60,1}$, $P_{60,7}$, $T_{8,15}$, C_{120}, and 3 instances of random regular graphs derived from $P_{60,1}$ and $T_{8,15}$ using twice as many edge swaps as there are edges in a given graph. Use random selection for the parent and roulette selection for the coparent, and use automatic, immediate replacement of the parent with the better of the two children produced. Comment on the impact of the graphs used. Is there much variation between graphs of the same degree? Does the use of graphs increase or decrease the impact of the lexical partner function?*

Experiment 13.14 covers a lot of territory. The graphs may enhance the performance of the lexical partner function or retard it; this effect may not be uniform across graphs. The choice of graph in a GBEA is a very complex sort of "knob," more complex than the mutation rate or population size, but still a parameter of the algorithm that can be tuned. More complex issues, such as the impact of graphs on coevolution, we defer to the future.

Definition 13.8 *A* **partial order** *is a binary relation $\leq$ on a set S with the following 3 properties:*

(i) for all $a \in S$, $a \leq a$,
(ii) for all $a, b \in S$, $a \leq b$ and $b \leq a$ implies $a = b$, and
(iii) for all $a, b, c \in S$, $a \leq b$ and $b \leq c$ implies $a \leq c$.

These 3 properties are called the **reflexive, antisymmetric,** *and* **transitive** *properties, respectively. Divisibility is a partial ordering of the positive integers.*

Definition 13.9 *The* **performance partial order** *of a set of graphs on a problem for a given GBEA is a partial ordering of graphs in which $G \leq H$ if the time-to-solution using G is significantly less than H. In this case, "significantly" implies a statistical test such as disjoint confidence intervals for time-to-solution.*

In theory, crossover is putting pieces together, while mutation is tweaking existing pieces and, at some rate, generating new pieces. In evolving ordered structures (Chapter 7), the nature of the pieces is less clear than it is in a problem with a simple linear gene. Let's check the impact of GBEAs on a couple of ordered gene problems.

Experiment 13.15 *Modify the software used in Experiment 7.7 to run as a GBEA and also compare the standard and random key encodings. Use the same graphs and graph algorithm settings as in Experiment 13.14. For each representation, give the performance partial order for 95% confidence intervals on time-to-solution. What is the impact of the graphs on maximizing the order of permutations?*

And now on to the Traveling Salesman problem. This problem has the potential for segments of a given tour to be "building blocks" that are mixed and matched. This, in turn, creates room for graphs to usefully restrict information flow as good segments are located.

Experiment 13.16 *Redo Experiment 7.9 as a GBEA; use only the random key encoding. Use the same graphs and graph algorithm settings as in Experiment 13.14. If your instructor thinks it's a good idea, run more cases of the Traveling Salesman problem from those given in Chapter 7. Give the performance partial order on 95% confidence intervals for time-to-solution. What is the impact of the graphs on the given examples of the Traveling Salesman problem?*

Since graphs restrict information flow, population seeding may interact with GBEAs to yield novel behavior. Review Algorithms 7.1 and 7.2 in Chapter 7.

Experiment 13.17 *Redo Experiment 13.16 but with population seeding. Do 3 sets of runs. In the first, put a tour generated with Algorithm 7.1 on a vertex 5% of the time. In the second, put a tour generated with Algorithm 7.2 on a vertex 5% of the time. In the third, use both heuristics, each on 5% of the vertices. Give the performance partial order on 95% confidence intervals for time-to-solution. What is the impact of the 3 population seeding methods?*

In Chapter 11, we studied a number of representations for evolving logic functions. Let's check the impact of adding graphs to the experiments using a couple of these representations (the direct representation and connection lists). First, let's add graphs to the experiment using a direct representation.

Experiment 13.18 *Review Experiment 11.5. For the 3-input parity problem only, redo the experiment as a GBEA, using the graphs and graph algorithm parameters from Experiment 13.14. Give the performance partial order for 95% confidence intervals on time-to-solution. What impact does the use of graphs have on this logic function evolution problem?*

Now let's add graphs to the experiment using a connection list representation. Since this is a substantially different representation, it may behave differently.

Experiment 13.19 *Review Experiment 11.7. For the 3-input parity problem only, redo the experiment as a GBEA, using the graphs and graph algorithm parameters from Experiment 13.14. What impact does the use of graphs have on this logic function evolution problem? Compare the results with those from Experiment 13.18.*

This section draws on material from many previous chapters. Prior to this, we have studied interactions of the underlying problem we are trying to solve with the choice of variation operators. In Chapter 3, we argued that these variation operators create the connectivity between values of the independent variable, while the fitness function computes the dependent variable: the fitness of a point in the search space. With GBEAs we add another element of complexity to the system: interaction with the graph-based geography. That geography controls the spread of information both by the nominal connectivity of the graph and by the choice of local mating rule.

Given the number of design features available (representation, fitness function(s), choice of variation operators, rate of application of those operators, model of evolution, and now choice of graph), a coherent, predictive theory of behavior for a GBEA system seems distant. Until someone has a clever idea, we must be guided by rules of thumb and previous experience with similar experiments. Thus far in this chapter we have simply reprised various experiments from previous chapters to assess the impact of using graphs. We have not yet explored the effect of the local mating rule.

Experiment 13.20 *Pick one or more experiments in this section that you have already performed. Modify the local mating rule to use elite rather than absolute replacement and perform the experiment(s) again. What is the impact? Does it change the relative impact of the graphs?*

Experiment 13.20 tests what happens when a GBEA changes its local mating rule. Experiment 13.8 tested what happened when we changed population size while leaving the graph as close to the same as possible.

Experiment 13.21 *Pick one or more experiments in this section that you have already performed. Perform it again using the graphs $T_{n,m}$ for the following values of n and m: 5, 24; 6, 20; 8, 15; 10, 24; 12, 20; 16, 15; 10, 48; 12, 40; and 16, 30. What has more impact, shape or population size? Time should be measured in mating events, so as to fairly compare the amount of effort expended.*

Chapter 5 demonstrated that we could obtain fairly complex behavior from very simple structures (symbots). Review Experiment 5.8, in which we tested the effect of various types of walls on symbots trying to capture multiple sources. In the next experiment, we will test the impact of different graphs on the ability to adapt to those walls.

Experiment 13.22 *Rebuild Experiment 5.8 as a GBEA using random selection of the parent and roulette selection of the coparent with absolute replacement of the parent by a randomly selected child. Test the original algorithm against the graph-based algorithm with the graphs C_{256}, $T_{16,16}$, and H_8. Are different graphs better for different types of walls?*

So far, we have not experimented with competing populations on a graph. Let's draw on the Sunburn model from Chapter 4 for a foray in this direction.

Experiment 13.23 *Implement or obtain software for the following version of the Sunburn evolutionary simulator. Use a graph topology to control choice of opponent. Choose a first ship at random and one of its neighbors at random. Permit these two ship designs to fight. If there is no victor, repeat the random selection until a victor arises. Now pick a neighbor of the losing ship and one of its neighbors. Permit these ships to fight. If there is not a victory, then pick a neighbor of the first ship picked in the ambiguous combat and one of its neighbors and try again until a victory is achieved. The victors breed to replace the losers, as before. Notice that the graph is controlling choice of opponent and, to a lesser degree, choice of mating partner.*

Perform 100 standard Sunburn runs and 100 graph-based Sunburn runs for the graphs C_{256}, $T_{16,16}$, and H_8. Randomly sampling from final populations, compare opponents drawn from all 6 possible pairs of simulations. Is there any competitive edge created by constraining the topology of evolution with graphs?

The material covered in this section gives a few hints about the richness of interactions between graphs and evolutionary computation. Students looking for final projects will ideas here. In Experiment 13.17, we introduced yet another parameter for population seeding, the rate for each heuristic used. Further exploration of that is not a bad idea. Experiment 13.23 opens a very small crack in the door to a huge number of experiments on the impact of graphs on competing populations.

Problems

Problem 576. The graphs used in the experiments thus far have had degree 2, 3, 4, $\log_2(n)$, and $n - 1$, where n is the population size. Give constructions for an infinite family of graphs of degrees 5, 6, and 7.

Problem 577. Suppose that we have a graph on n vertices created by flipping a coin for each pair of vertices and putting an edge between them if the coin shows heads. Compute the probabilities, as a function of n, that such a graph has diameter 1, 2, and more than 2.

Problem 578. Suppose we are generating random regular graphs of degrees 2 and 3 starting with C_{400} and $P_{200,1}$, respectively. Experimentally or logically, estimate the probability that a given random regular graph will be connected.

Problem 579. If we generate a random regular graph, and by accident, it is not a connected graph, does this cause a problem? Why? Is the answer different for different problems?

Problem 580. In the definition of partial order, divisibility of the positive integers was given as an example. Prove that divisibility on the positive integers is a partial order (by checking properties (i)–(iii)) and also show that divisibility does *not* partially order the nonzero integers.

Problem 581. Does the relationship "s is a prefix of t" on strings form a partial order? Prove your answer.

Problem 582. A *total order* is a partial order with the added property that every pair of elements can be compared, e.g., the traditional operation $<$ on the real numbers. What prevents the performance partial order from being a total order?

Problem 583. Reread Problem 582 and give 3 examples of total orders, including a total order on the set of complex numbers.

Problem 584. Verify that the performance partial order is, in fact, a partial order. This is done by checking properties (i)–(iii).

Problem 585. For Experiments 13.14–13.19, decide whether it possible to compute the edge and entropy of the graphs as we did in the neutral graph behavior experiments in Section 13.1. What is required to be able to make these computations?

Problem 586. Compute the diameter of $T_{n,m}$, the $n \times m$ torus.

Problem 587. Compute the diameter of H_n, the n-hypercube.

Problem 588. What is the smallest number of edges that can be deleted from the 5-hypercube to drive the diameter to exactly 6?

Problem 589. The operation *simplexification* is described in Appendix D. We can create graphs with degree n by starting with K_{n+1} and simplexifying vertices. For $n = 3, 4, 5$, determine what population sizes are available, by starting with a complete graph and repeatedly simplexifying vertices.

Problem 590. Reread Problem 589. Would graphs created by simplexification behave differently from other graphs used in this section in a GBEA?

Problem 591. Essay. The list of graphs used in this chapter is modest. Pick and defend a choice of graph for use with the Traveling Salesman problem. An experimental defense is time-consuming, but superior to a purely rhetorical one.

Problem 592. Essay. The list of graphs used in this chapter is modest. Pick and defend a choice of graph for use with the 3-input parity problem. An experimental defense is time-consuming, but superior to a purely rhetorical one.

Problem 593. Essay. A lexical fitness function seeks to smooth the landscape of a difficult fitness function by adding a tie-breaker function that points evolution in helpful directions. A graph-based algorithm breaks up a population, preventing an early good gene from taking over. Do these effects interfere, reinforce, or act independently?

Problem 594. Essay. The ordered sequence of degrees, the number of vertices, and number of edges in the graph are all examples of invariants. Choose the invariant that you think most affects performance in an experiment you have performed. Defend your choice.

Problem 595. Essay. The use of crossover is controversial. At one extreme, people claim that the ability to mix and match building blocks is the key one; at the other extreme, people claim that crossover is unnecessary and even counterproductive. Since both sides have experimental evidence in favor of their assertions, the truth is almost certainly that crossover is only helpful when there are building blocks to be mixed and matched and is potentially very helpful then. Question: given what you have learned from the experiments in this chapter, can the behavior of a problem for graphs of different connectivities be used as a probe for the presence of building blocks? Good luck; this is a hard question.

13.4 Genetic Programming on Graphs

The most complex representations we have examined have been various different genetic programming representations including parse trees, GP automata, and ISAc lists. In this section, we will check the impact of graphs on solving problems using these representations. The simplest genetic programming problem available is the PORS problem from Chapter 8. Review the PORS problem and Experiments 8.2–8.4. Let's check the impact of graphs on the three classes of PORS trees.

Experiment 13.24 *Build or obtain software for a graph-based evolutionary algorithm to work with the PORS problem. Use random selection of the parent and roulette selection of the coparent, with elite replacement of the parent with the better of the two children. Use subtree mutation 50% of the time and subtree crossover 50% of the time, with the 50% chances being independent. Use the graphs C_{720}, $P_{360,1}$, $P_{360,17}$, $T_{4,180}$, $T_{24,30}$, H_9, modified by simplexifying 26 randomly selected vertices, and K_{512}. Simplexification is described in Appendix D. Be sure to create these graphs once and save them, so that the same graph is used in each case.*

Do 400 runs per graph for the Efficient Node Use problem on $n = 14, 15$, and 16 nodes. Document the impact of the graphs on time-to-solution with 95% confidence intervals. Do any of the graphs change the relative difficulty of the three cases of the PORS problem?

We have not explored, to any great extent, the impact of local mating rules on the behavior of the system. Experiment 13.24 uses a very extreme form of local mating rule that insists on improvement before permitting change and which refuses to destroy a creature currently being selected with a fitness bias (the coparent). Let's check the impact of protecting the coparent in this fashion.

Experiment 13.25 *Modify Experiment 13.24 to use elite replacement of parent and coparent by both children. Of the 4 structures, the best two take the slots occupied by the parent and coparent. Compare the results to those obtained in Experiment 13.24.*

Elitism amounts to enforced hill climbing when used in the context of local mating rules. Different regions of the graph may be working on different hills. If a given problem has local optima or other traps, then this hill climbing may cause problems. On the other hand, the boundary between subpopulations on distinct hills may supply a source of innovation. Let's do the experiment.

Experiment 13.26 *Modify Experiment 13.24 to use absolute replacement of parent and coparent by both children. Compare the results to those obtained in Experiments 13.24 and 13.25. Is the impact comparable on the different PORS problems?*

The PORS problem is very simple and highly abstract. Fitting to data, i.e., using genetic programming to perform symbolic regression, is less simple and a good deal more abstract. In Experiment 9.5, we found that it is not difficult to perform symbolic regression to obtain formulas that accurately interpolate points drawn from the fake bell curve

$$f(x) = \frac{1}{x^2 + 1}.$$

Let's see whether time-to-solution or the rate of accurate solutions can be increased with a GBEA.

Experiment 13.27 *Rebuild Experiment 9.6, symbolic regression to samples taken from the fake bell curve, as a GBEA. Use random selection for the parent and roulette selection for the coparent and absolute replacement of the parent by the better child. Also, perform baseline studies that use tournament selection. For each graph, perform tournament selection with tournament size equal to the graph's degree plus one. Don't use normal tournament selection; rather, replace the second-most-fit member of the tournament with the best child. This makes the tournament selection as similar as possible to the local mating rule, so that we are comparing graphs to mixing at the same rate without the graph topology. Use the graphs C_{512}, $P_{256,7}$, $T_{16,32}$, and H_9.*

Perform 400 runs per graph. Determine the impact both by examining the number of runs that find a correct solution (squared error less than 10^{-6} over the entire training set) and by examining the time-to-solution on those runs that achieve a correct solution.

The somewhat nonstandard tournament selection used in Experiment 13.27 is meant to create amorphous graph-like structures that have the same degree but constantly moving edges. This controls for the effect of degree as opposed to topology in another way than using random regular graphs. It does not exactly compare normal tournament selection to a GBEA. It's also not clear that it's the "right" control.

Experiment 13.28 *Perform Experiment 13.27 again, but this time use standard tournament selection of the appropriate degree (reuse your graph results). Compare the results with both the graphs and the nonstandard tournaments from the previous experiment.*

To complete the sweep of the controls for degree versus connectivity in the graphs, let's perform the experiment again with random regular graphs.

Experiment 13.29 *Perform an extension of Experiment 13.27 as follows. Pick the best- and worst-performing graphs and generate 5 random regular graphs of the same degree using twice as many edge swaps as there are edges in the graphs. Run the GBEA again with these graphs. Does this control for degree versus topology have a different effect than the one used in Experiment 13.27?*

We have already reprised the neural net 3-input parity problem from Chapter 11 in Section 13.3. Evolving a parity function is one of the standard test problems in evolutionary computation. Let's take a look at the problem using genetic programming techniques.

Experiment 13.30 *Rebuild Experiment 11.12 to run as a GBEA. Use two local mating rules: roulette selection of the parent and coparent, with absolute replacement of both parent and coparent; random selection of the parent and rank selection of the coparent with absolute replacement of both parent and coparent. Use the graphs C_{720}, $P_{360,1}$, $P_{360,17}$, $T_{4,180}$, $T_{24,30}$, and K_{720}. For each graph and local mating rule, perform 400 evolutionary runs. Compare the performance of the different graphs. Were different graphs better for the AND and parity problems?*

Another common target problem for evolutionary computation is the *multiplexing* problem. The 2^n multiplexing problem takes 2^n *data inputs* and n *encoding inputs*. An encoding input is interpreted as a binary integer that selects one of the data inputs. The output is set to the value of the selected data input. The truth table for a 4-data-input/2-encoding-input 4-multiplexer is given in Figure 13.6.

Data Inputs				Encoding Inputs		Output
0	1	2	3	low	high	
0	*	*	*	0	0	0
1	*	*	*	0	0	1
*	0	*	*	1	0	0
*	1	*	*	1	0	1
*	*	0	*	0	1	0
*	*	1	*	0	1	1
*	*	*	0	1	1	0
*	*	*	1	1	1	1

Fig. 13.6. Truth table for the 4-multiplexer. (* entries may take on either value without affecting the output.)

The truth table given in Figure 13.6 nominally has 64 entries, one for each of the 2^6 possible inputs. The use of the symbol * for "either value" compresses the table to one with 8 inputs. The number of times you can use a * in this fashion in writing a truth table of minimal length is the *degeneracy* of a logic function. On average, logic functions are easier to create via evolutionary computation if they have a higher degeneracy. Let's check this assertion experimentally.

Experiment 13.31 *Modify the software for Experiment 13.30 to work with the 4-multiplexing function and the 6-parity function. Use whichever local mating rule worked best for the 3-input parity problem. Which of these two problems is harder?*

Let us now turn to the various grid robot tasks in Chapters 10 and 12. Experiment 13.7 has already touched on the Tartarus problem and the need to use generational rather than steady-state GBEAs on these problems. This requirement for a generational algorithm comes from the need to compare oranges to oranges in any problem where sampled fitness is used as a surrogate for the actual fitness. Recall that in the 6×6 Tartarus problem, there are in excess of 300,000 boards, and we can typically afford no more than a few hundred boards for each fitness evaluation. This means that rather than computing the true fitness in each generation (the average score over all boards), we use a sample of the boards to compute an estimated fitness.

Experiment 13.32 *Use the GP language from Experiment 10.13, without the RND terminal, to run a generational GBEA, i.e., one with deferred updating. Use 100 Tartarus boards rather than 40 for the fitness function, selecting the 100 boards at random in each generation. For each vertex in the graph, roulette-select a coparent and create a pair of children using subtree crossover 25% of the time and subtree mutation 50% of the time, independently. Use absolute replacement. Use 20-node random initial trees and chop trees that exceed 60 nodes. Run the algorithm on C_{256}, $T_{16,16}$, and H_8, as well as on 2 random regular graphs of degrees 4 and 8.*

For each graph, perform 100 runs. Compare the graphs with two statistics: the time for a run to first exhibit fitness of 3.0 and the mean final fitness after 500 generations. If results are available for Experiment 10.13, also compare with those results. Compute the performance partial order for this experiment.

As we know, the GP trees with 3 memories were not our best data structures for Tartarus. Both GP automata and ISAc lists exhibit superior performance.

Experiment 13.33 *Repeat Experiment 13.32 but use GP automata this time. Use the GP automata with 8 states and null actions (λ-transitions) of the same kind as were used in Experiment 10.18. Be sure to use the same random regular graphs as in Experiment 13.32. Does the identity of the best graph change at all? Compute the performance partial order for this experiment and compare it with the one from Experiment 13.32.*

When we have a sampled fitness, as with Tartarus, there is room to experiment with the allocation of fitness trials. The graph topology gives us another tool for allocating fitness trials. Recall that the hypercube graph can be thought of as having a vertex set consisting of all binary words of a given length. Its edges connect pairs of binary words that differ in one position. The *weight* of a vertex is the number of 1's in its binary word.

Experiment 13.34 *Modify the software from Experiment 13.33 as follows. First, run only on the graph H_8. The possible vertex weights are $0, 1, 2, \ldots, 8$. For vertices of weight $0, 1, 7,$ or 8, evaluate fitness on 500 boards. For words of weight 2 or 6, evaluate fitness on 100 boards. For words of weight 3 or 5, evaluate fitness on 40 boards. For words of weight 4, evaluate on 20 boards.*

Use a fixed set of 500 boards in each generation, giving GP automata that require fewer evaluations boards from the initial segment of the 500. In a given evolutionary run of 500 generations, this will result in 20,480 instances of a dozer being tested on a board. Also, rerun the unmodified software so that each dozer on the H_9 graph uses 80 fitness evaluations. This results in the exact same number of fitness evaluations being used.

Perform 100 runs for both methods of allocating fitness. Using a fixed 5000-board test set, as in Experiment 10.20, make histograms of the 100 best-of-run dozers from each set of runs. Do the methods have different results? Which was better?

There is a possibility implicit in the use of the distributed geography of a GBEA that we have not yet considered. Suppose that we have different fitness functions in different parts of the graph. If the tasks are related, then the easier instance of the problem may prime progress on the harder instance.

Experiment 13.35 *Modify the software from Experiment 13.33 as follows. First, run only on the graph H_8. Do two sets of runs with 100 Tartarus boards used for each fitness evaluation in the usual fashion. In one set of runs, use only the 8×8 Tartarus problem with 10 boxes. In the other, use the 8×8 10-box Tartarus problem on those vertices with odd weight and the 6×6 6-box Tartarus problem on those vertices with even weight. For both sets of runs, use a fixed 5000-board test set, as in Experiment 10.20, to make histograms of the 100 best-of-run dozers for the 8×8 task from each set of runs. How different are the histograms?*

Let's shift both the virtual robotics task and the representation in the next experiment. The Herbivore task was an easier problem when judged by the rate of early progress in enhancing fitness.

Experiment 13.36 *Rebuild the software from Experiment 12.12 to work as a generational GBEA. Use the same graphs and local mating rule as in Experiment 13.32 and check the impact of the graphs on performance in the Herbivore task. Compute the performance partial order for this experiment and compare it with the ones from Experiments 13.32 and 13.33 if they are available.*

The North Wall Builder task from Chapter 12 has the advantage that it has only one fitness case (board) and so runs much faster than Tartarus or Herbivore. Let's do a bi-factorial study of graph and board size.

Experiment 13.37 *Rebuild the software from Experiment 12.17 to work as a GBEA. Use the same graphs as in Experiment 13.32, but change the local mating rule to be random selection of the parent, roulette selection of the coparent, and elite replacement of the parent by the better of the two children. Check the impact of the graphs on performance of the North Wall Builder task for board sizes 5×5, 7×7, and 9×9. Compute the performance partial order for each board size and compare these orders with one another and with all available experiments using the same graphs.*

We have not experimented much with the impact of graphs on competitive tasks (other than Sunburn). We leave this topic for the future, but invite you to design and perform your own experiments. One thing to consider is that it may be hard to compare two populations of competitive agents meaningfully.

Problems

Problem 596. In the PORS system, the two "subroutines" (+ **(Sto T) Rcl**) (multiply by 2) and (+ **(Sto T) (+ Rcl Rcl)**) or (+ (+ **(Sto T) Rcl) Rcl**) (multiply by 3, which has two forms) together with the very similar trees, (+ 1 1), (+ (+ 1 1) 1), and (+ 1 (+ 1 1)), which encode the constants 2 and 3, can be used to build up all optimal PORS trees. For the PORS $n = 15$ Efficient Node Use problem, either compute or experimentally estimate the probability that a random initial tree will contain a subtree that encodes 3 or multiplication by 3.

Problem 597. Reread Problem 596 and either compute or experimentally estimate the probability that a random initial tree will contain a subtree that encodes 2 or multiplication by 2.

Problem 598. Why do we use large numbers of edge swaps when we generate random regular graphs? What would the effects of using a small number be?

Problem 599. In Experiment 13.29 we checked for the difference in behavior of algorithms using the standard graphs for this chapter and those using random graphs with the same degree. How does the diameter of the standard graphs compare with the diameter of random regular graphs of the same degree? Why?

Problem 600. In order to generate random regular graphs of degree d with n vertices, we need a starting graph with the given degree and vertex count. Give a scheme for creating starting graphs of degree d with n vertices for as many degrees and vertex sizes as you can. Remember that the number of vertices of odd degree must be even.

Problem 601. The notion of degeneracy of a truth table is explained on page 375. Compute the degeneracy for each of the following families of logic functions and give the resulting shortest possible truth tables.

(i) n-input OR,
(ii) n-input AND,
(iii) 2^n-multiplexing,
(iv) n-bit parity.

Problem 602. Prove that a logic function and its negation have the same degeneracy.

Problem 603. Prove that the parity function and its negation are the only logic functions whose truth tables have zero degeneracy.

Problem 604. Carefully verify the assertion in Experiment 13.34 that there will be 20,480 evaluations of a dozer on a board in each generation.

Problem 605. Reread Experiment 13.34. Come up with compatible methods of varying the number of fitness trials for (i) C_{256}, (ii) $T_{16,16}$, (iii) $P_{128,1}$, and (iv) $P_{128,7}$. Make sure that your total fitness evaluations in a generation are a multiple of the number of vertices, to permit evaluation on a fixed number of boards as a baseline.

Problem 606. Experiment 13.35 mixed the 8×8 and 6×6 Tartarus problems. Is there a problem in this experiment with having to compare dozers evaluated with different fitness functions?

Problem 607. Reread Experiment 13.35. Suppose that instead of dividing the 8×8 and 6×6 problems by odd- and even-weight vertices, we had divided the hypercube so that the 6×6 fitness function was on the vertices with most-significant bit "0" and the 8×8 fitness function was on the vertices with most-significant bit "1." In this case, there would be vertices that had neighbors evaluated with each of these fitness functions. Give a means of finding a scaling factor that permits comparison of these two fitness functions and defend it. Consider: while the 8×8 function can return higher values, initial progress is probably more rapid on the 6×6 function.

Problem 608. In some sense, Experiment 13.35 mixes the 8×8 and 6×6 Tartarus problems as much as possible. Is this good or bad?

Problem 609. Reread Experiment 13.35. Would you expect this sort of fitness mixing to work better or worse on the PORS problem with $n = 12$ and $n = 15$? Assume that trees are chopped to fit their node.

Problem 610. Essay. For the most part, we have used regular graphs as a way of controlling for one important graph parameter. Is there any reason to think the performance with graphs that are *not* regular, but have similar average degree, would be different? Explain.

Problem 611. Essay. The correct answer to the PORS $n = 15$ Efficient Node Use problem is EVAL(T)=32. Given the answers you found to Problems 597 and 596, discuss why the cycle is the best graph, of those used, for the PORS $n = 15$ Efficient Node Use problem.

Problem 612. Essay. Is the tournament selection used in Experiment 13.27 better or worse than the standard type of tournament selection? For which problems?

Problem 613. Essay. Create a system for evolving graphs. Give a representation including data structure and variation operators. Do not worry about the fitness function.

Problem 614. Essay. A persistent theme in this chapter is the comparison of graphs to see which graph helps the most on a given problem. Discuss the practicality of searching for good graphs by using performance of a GBEA as a fitness function.

Problem 615. Essay. Explain why high degeneracy in a logic function yields an easier evolutionary search problem.

Problem 616. Essay. Suppose that we use a graph with several thousand vertices and place 5 Tartarus boards as well as a dozer on each vertex. We then run a GBEA in which parents are selected at random and coparents are selected by roulette selection after they are evaluated on the boards sitting on the parent's node. Will this scheme find effective dozers? Explain.

Problem 617. Essay. Invent and describe a system for using GBEAs to locate hard Tartarus boards for 8×8 or larger boards.

14

Cellular Encoding

Cellular encoding [31, 32] is a technique for representing an object as a set of directions for constructing it, rather than as a direct specification. Often, this kind of representation is easier to work with in an evolutionary algorithm than a direct coding of the object. There are several examples of cellular encodings in this chapter. The examples, while they are all cellular encodings, have little else in common. Because of this, the experiments in this chapter are organized section by section. The name "cellular encoding" comes from an analogy between the developmental rules governing construction of the desired objects and the biology governing construction of complex tissues from cells. The analogy is at best weak; don't hope for much inspiration from it. A more helpful way to think of the cellular encoding process is as a form of *developmental biology* for the structure described (as with Sunburn in Chapter 4).

Suppose we have a complex object: a molecule, a finite state automaton, a neural net, or a parse tree. A sequence of rules or productions that transform a starting object into an object ready to have its fitness evaluated can be used as a linear gene.

Instead of having complex crossover operators (which may require repair operators) for complex objects, we can use standard crossover operators for linear genes. The behavior of those crossover operators in the search space is often difficult to understand, but this is also often true of crossover operators used with direct encodings.

The idea of evolving a set of directions for constructing an object is an excellent one with vast scope. We will start by building 2-dimensional shapes using instructions from a linear gene. There are a number of possible fitness functions for such shapes; we will explore two. In the second section of this chapter, we will create a cellular encoding for finite state automata and compare it with the direct encodings used in Chapter 6. In the third section, we will give a cellular encoding method for combinatorial graphs. In the fourth section, we will give a method for using context free grammars to control the production of the parse trees used in genetic programming. This permits the evolution of simple linear genes rather than parse trees and allows the user to include domain-specific knowledge in the evolutionary algorithm.

14.1 Shape Evolution

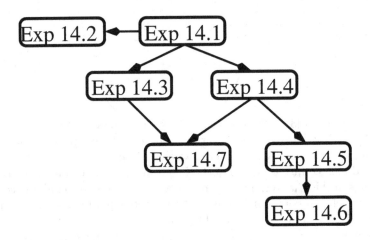

1   Cellular polyomino encoding.
2   Polyomino creator.
3   Maximize bounding box.
4   Space-filling, sliding fitness function.
5   Space-filling fitness.
6   Obstructed space-filling fitness.
7   Population seeding.

Fig. 14.1. The topics and dependencies of the experiments in this section.

A *polyomino* is a shape that can be made by starting with a square and gluing other squares onto the shape by matching up sides. A production of a 3-square polyomino is shown in Figure 14.2. A polyomino with n squares is called an n-omino. Our first cellular encoding is a scheme for encoding n-ominos.

We will use an array of integers as our encoding for n-ominos. The key is interpretation of the integers in the array. Divide each integer by 4. The integer part of the quotient is the number of the square in the n-omino; the remainder encodes a direction: up, down, left, or right.

Algorithm 14.1 Polyomino Development Algorithm
Input: *An array of integers G[] of length k*
Output: *A labeled n-omino and a number F of failures*

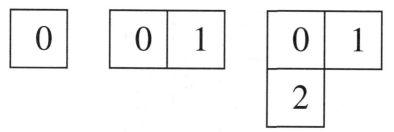

Fig. 14.2. Start with initial Square 1; add Square 2; then, add Square 3 to make a 3-square polyomino.

Details:

> *Initialize a $(2k + 1) \times (2k + 1)$ array A with zeros;*
> *Place a 1 in the center of the array;*
> *Initialize a list of squares in the n-omino with the first square;*
> *Initialize C=1, the number of squares so far;*
> *Initialize F=0, the failure counter;*
> *For(i= 0; i < k; i + +)*
> > *Interpret G[k] (mod 4) as a direction X in (U,L,D,R);*
> > *Find square S of index (G[k]/4) (mod C) in the growing structure;*
> > *If(square in direction X from S in A is 0)*
> > > *C ← C+1;*
> > > *Put C in square in direction X from S in A;*
> > *Else F ← F+1;*
> *End For;*
> *Return A,F;*

Let's do a small example. We will use one-byte integers in this example ($0 \leq x \leq 255$), which limits us to at most 65 squares in the n-omino. This should suffice for the examples in this section; shifting to two-byte integers permits the encoding of up to 16,385-ominos, more than we need.

Example 28. Examine the gene $G = (126, 40, 172, 207, 15, 16, 142)$. Interpret the gene as follows:

Locus   Interpretation
126=4*31+2 Direction 2(down) from Square 0(31(mod 1)); add Square 1
40=4*10+0 Direction 0(up)from Square 0(10(mod 2)); add Square 2
172=4*43+0 Direction 0(up)from Square 1(43(mod 3)); wasted
207=4*51+3 Direction 3(left)from Square 0(51(mod 3)); add Square 3
15=4*3+3   Direction 3(left)from Square 3(3(mod 4)); add Square 4
16=4*4+0   Direction 0(up)from Square 4(4(mod 5)); add Square 5
142=4*35+2 Direction 2(down)from Square 5(35(mod 6)); wasted

The result of this interpretation is the 6-omino

with $F = 2$ failures (wasted loci). We label the n-omino to track the order in which the squares formed. Notice that not all of the 15×15 array A is shown, in order to save space.

Now that we have an array-based encoding for polyominos, the next step is to write some fitness functions. Our first fitness function is already available: the number of failures. A failure means that a gene specified a growth move in which the polyomino tried to grow where it already had a square. If our goal is to grow large polyominos, then failures are wasted moves.

Experiment 14.1 *Create or obtain software for an evolutionary algorithm that uses the array encoding for polyominos. Treat the array of integers as a string-type gene. Initialize the arrays with numbers selected uniformly at random in the range 0–255. Use arrays of length 12 with two-point crossover and single point mutation. The single-point mutation should replace one location in the array with a new number in the range 0–255. Use a population size of 400 with a steady-state algorithm using single tournament selection of size 7.*

Record the number of tournament selections required to obtain a gene that exhibits zero failures for each of 100 runs of the evolutionary algorithm and save a 0-failure gene from each run. Report the time-to-solution and the shapes of the resulting polyominos. Runs that require more than 100,000 tournaments should be cut off, and the number of such tournaments should also be reported.

Experiment 14.1 suffers from a problem common in evolutionary computation; it's hard to tell what the results mean. The only certain thing is that it is possible to evolve length-12 arrays that code for 13-ominos. One interesting question is, are these "typical" 13-ominos? It seems intuitive that some shapes will be better at avoiding failure than others. Let's develop some measures of dispersion for polyominos.

Definition 14.1 *The **bounding box** of a polyomino is the smallest rectangle that can contain the polyomino. For the polyomino in Example 28, the bounding box is a 3×3 rectangle. The **bounding box size** of a polyomino is the area of the polyomino's bounding box.*

Definition 14.2 *The* **emptiness** *of a polyomino is the number of squares in its bounding box not occupied by squares of the polyomino. The emptiness of the polyomino given in Example 28 is 3.*

Experiment 14.2 *Create or obtain software for a random n-omino creator that works in the following fashion. Start with a central square, as in the initialization of Algorithm 14.1. Repeatedly pick a random square in the array holding the polyomino until you find an empty square adjacent to a square of the polyomino; add that square to the polyomino. Repeat this square-adding procedure until the polyomino has n squares.*

The random polyominos will serve as our reference set of polyominos. Generate 100 random 13-ominos. For these 13-ominos and the ones found in Experiment 14.1, compute the bounding box sizes and emptinesses. If some runs in Experiment 14.1 did not generate 13-ominos, then perform additional runs. Compare histograms of the bounding box sizes and emptinesses for the two groups of shapes. If you know how, perform a test to see whether the distributions of the two statistics are different.

The bounding box size is a measure of dispersion, but it can also be used as a fitness function. Remind yourself of the notion of lexical fitness function from Chapter 5 (page 128).

Experiment 14.3 *Modify the software from Experiment 14.1 to maximize the bounding box size for polyominos. For length-12 genes (size-13 polyominos), the maximum bounding box has size 49.*

Do two collections of 900 runs. In the first, simply use bounding box size as the fitness. In the second set of runs, use a lexical product of bounding box size and number of failures in which bounding box size is dominant and being maximized, and the number of failures is being minimized. In other words, a polyomino with a larger bounding box size is superior, and ties are broken in favor of a polyomino with fewer failures.

Compare the time to find an optimal bounding box for the two fitness functions, and explain the results as well as you can. Save the best genes from each run in this experiment for use in a later experiment.

The shape of polyominos that maximize bounding box size is pretty constrained. They appear somewhere along the spectrum from a cross to a Feynman diagram. Our next fitness function will induce a different shape of polyomino.

Experiment 14.4 *Modify the software from Experiment 14.1 to be a generational algorithm that works with the following fitness function on a population of 60 polyominos with genes of length 19. Fitness evaluation requires an empty 200×200 array that wraps in both directions.*

Repeatedly perform the following steps. First, put the polyominos in the population into a random order. Taking each in order, generate a random point in the 200×200 array. If the upper left corner of the current polyomino's *bounding box is placed in that location and all squares of the polyomino can be placed, then place the polyomino there, marking those squares as full and adding the number of squares in the polyomino to its fitness. If the polyomino does not fit in the current location, try other locations by scanning first in the horizontal direction, until either you have tried all locations or a location is found where the polyomino fits.*

Once all the polyominos have had one try, a new random order is generated. Perform fitness evaluation until at least 75% of the 200 × 200 array is occupied or until all shapes have had a chance to find a place in the array and failed. Do 100 runs of 1000 generations length and, comparing expressed shapes rather than genes, show and explain the most common shapes in each run. Are some shapes more common than others? Why?

The fitness function used in Experiment 14.4 lets the shapes compete for space. There are two forces at work here: the need to occupy space and the need to fit into the remaining space. The former pressure should make large shapes, while the latter one will make small shapes. Consider how these pressures balance out when writing up your experiment.

Experiment 14.5 *Modify the fitness function from Experiment 14.4. If a shape does not fit at the randomly chosen location, do not try other locations. Go until the array is 50% full (rather than 75% full). Are the resulting shapes different from those found in Experiment 14.4?*

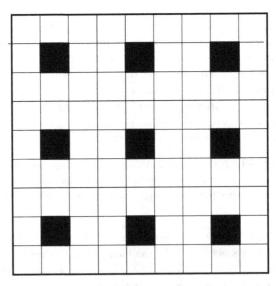

Fig. 14.3. A 9 × 9 grid with all squares that have both coordinates congruent to 1(mod 3) initially filled.

The shapes obtained in our versions of Experiments 14.4 and 14.5 were not too different. Let's see whether we can cause the experiment to produce a different sort of shape by modifying the fitness function again.

Experiment 14.6 *Modify the software from Experiment 14.5 so that there is a 201×201 array used for fitness evaluation in which the squares with both coordinates congruent to $1(\bmod 3)$ start already occupied. Are the resulting shapes different from those found before?*

The outcomes of Experiments 14.4 and 14.5 suggest that compact shapes are favored. Let's try initializing Experiment 14.4 with genes that are not at all compact and see whether we end up with a different sort of solution.

Experiment 14.7 *Modify the software from Experiment 14.4 to read in randomly selected genes chosen from those created during Experiment 14.3 instead of initializing with random genes. Are the resulting shapes any different from those obtained in Experiment 14.4?*

The shape gene is a simple example of cellular encoding, and the experiments in this section are interesting mostly because of their coevolutionary character. The competitive exclusion game the shapes are playing when competing for space is a fairly complex game. You could generalize this system in other directions. Suppose, for example, that we scatter shape "seeds" at the beginning of fitness evaluation and then grow shapes by executing one genetic locus per time step of the development simulation. The partial shapes would need to guard space for additional development. This would put an entirely new dynamic into the shape's growth.

Problems

Problem 618. Run the Polyomino Development Algorithm on the following length-7 polyomino genes:

(i)   G=(146, 155, 226, 57, 9, 84, 25),
(ii) G=(180, 158, 146, 173, 187, 85, 200),
(iii) G=(83, 251, 97, 241, 48, 92, 217),
(iv) G=(43, 241, 236, 162, 250, 194, 204),
(v)   G=(100, 139, 229, 184, 111, 46, 180).

Problem 619. For each of the following polyominos, find a gene of length 12 that will generate that polyomino. The numbers on the squares of the polyomino give the order in which the squares were added to the polyomino during development. Your gene must duplicate the order in which the squares were added.

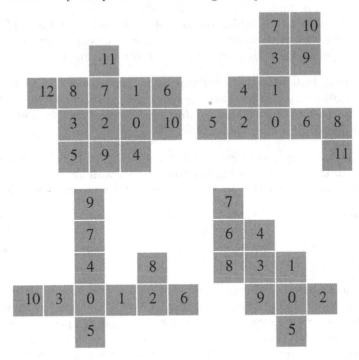

Problem 620. The point of cellular encoding is to specify a complex structure as a linear sequence of construction rules. Suppose that we instead stored
polyominos in a 2-dimensional array. Create a crossover operator for polyominos stored in this fashion.

Problem 621. Consider a 2×2 square polyomino. Disregarding the gene
and considering only the order in which the squares were added, how many
different representations are there?

Problem 622. Enumerate all length-5 polyomino genes that code for a 2×2
square polyomino.

Problem 623. Give an example of a gene of length k that creates a polyomino
of size 2 (for every positive integer k).

Problem 624. Prove that the maximum bounding box size for a polyomino
with n squares is smaller than the maximum bounding box size for a polyomino with $n + 1$ squares.

Problem 625. For as many n as you can, compute the maximum bounding
box size for an n-omino.

Problem 626. Reread Experiment 14.4. If a shape fails to find space once,
is there any point in checking to see whether it fits again? Would a flag array
that marks shapes as having failed once speed up fitness evaluation?

Problem 627. The encoding given for shapes in this section is one possible choice. Try to invent an encoding for shapes (or an alternative algorithm for expressing the shapes) that eliminates wasted moves.

Problem 628. Does Experiment 14.4 need to be generational? If not, how would you modify it to be steady-state?

Problem 629. In Experiments 14.4–14.6, why leave 25%–50% of the board unfilled?

Problem 630. Essay. In Experiments 14.4 and 14.5 we are placing shapes by two different methods and then evaluating them based on their success at filling space. Which strategy is better: fitting well with yourself or blocking others?

Problem 631. Essay. Does Experiment 14.4 or Experiment 14.5 favor compact shapes, like rectangles, more?

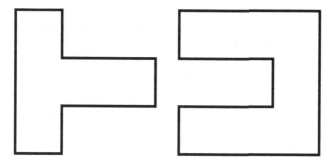

Problem 632. Essay. In Experiment 14.4, shapes are allowed to search for a place they will fit. It's not too hard to come up with complementary shapes that fit together, e.g., the two shown above. Would you expect populations of coexisting shapes that fit together but have quite dissimilar genes to arise often, seldom, or almost never?

Problem 633. Essay. Since different shapes are evaluated competitively in Experiments 14.4–14.7, the algorithms are clearly coevolutionary rather than optimizing. If most of the genes in a population code for the same shape, does the algorithm behave like a converged optimizer?

14.2 Cellular Encoding of Finite State Automata

The evolution of finite state automata was studied in Chapter 6. We evolved finite state automata to recognize a periodic string of characters and then used finite state automata as game-playing agents. In this section, we will examine what happens when we use a cellular encoding for finite state automata. With

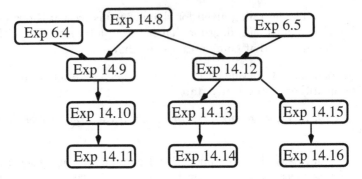

Fig. 14.4. The topics and dependencies of the experiments in this section.

polyominos we started with a single square and added additional squares. In order to "grow" a finite state automaton, we will start with a single-state finite state automaton and modify it to make a larger automaton. In order to do this, we will need editing commands. We will work with automata with k possible inputs and outputs, named $0, 1, \ldots, k - 1$. When we need a specific input or output alphabet (like $\{C, D\}$ for Prisoner's Dilemma), we will rename these integer inputs and outputs to match the required alphabet.

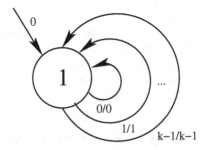

Fig. 14.5. The Echo machine (initial action is zero; last input is its current output).

The starting automaton we will use is the *Echo* machine shown in Figure 14.5. While editing a finite state automaton, we will keep track of the *current state* being edited. The current state will be denoted by a double circle in the state diagram. The current state specifies where editing is to happen, the position of a virtual editing head. The cellular representation will consist of a sequence of editing commands that either modify the automaton or move the current state. Most of the editing commands take a member of the input alphabet of the automaton as an argument and are applied to or act along the transition associated with that input. This specifies unambiguous actions, because there are exactly k transitions out of the current state, one for each possible input. (The exception, B, is an editing command that modifies the initial response, no matter which state is the current state.)

Command	Effect
B (Begin)	Increment the initial action.
F_n (Flip)	Increment the response associated with the transition for input n out of the current state.
M_n (Move)	Move the current state to the destination of the transition for input n out of the current state.
D_n (Duplicate)	Create a new state that duplicates the current state as the new destination of the transition for input n out of the current state.
P_n (Pin)	Pin the transition arrow from the current state for input n to the current state. It will move with the current state until another pin command is executed.
R (Release)	Release the pinned transition arrow if there is one.
I_n	Move the transition for input n out of the current state to point to the state you would reach if you made two transitions associated with n from the current state.

Table 14.1. Commands for editing finite state automata. (Incrementing is always modulo the number of possible responses.)

The commands we will use to edit finite state automata are given in Table 14.1. They are only one possible set of editing commands for finite state automata. We chose a small set of commands with little redundancy that permit the encoding of a wide variety of finite state automata.

The pin command (P_n) requires some additional explanation. This command chooses one of the transitions out of the state currently being edited and "pins" it to the current state. That means that if the current state is moved with an M_n command, then the transition arrow moves with it. This state of affairs continues until either the transition arrow is specifically released with an R command, or until another pin command is executed. (The definition permits only one arrow to be pinned at a time, though it would be possible to pin one arrow of each type unambiguously if the release command also took arguments.) If a transition arrow is still pinned when the editing process ends, then the arrow is left where it is; it is implicitly released when the current state ceases to have meaning, because the editing process ends.

The I_n command is the only command other than the pin command that can be used to move transition arrows. The command I_n moves a transition arrow to the state that the automaton would reach from the current state if two transitions were made in response to inputs of n. (This edit is difficult to perform with the pin command for some configurations of automaton.)

We reserve for the Problems the question of completeness of this set of editing commands. A set of editing commands is *complete* if any finite state automaton can be made with those commands. Even with a complete set of commands, the "random" finite state automata we can generate are very different from those we obtain by filling in random valid values on blank automata with a fixed number of states. When filling in a table at random, it is quite easy to create states that cannot be reached from the initial state. An automaton created with editing commands is much less likely to have many isolated states.

Let's look at an example of several edits applied to the version of the Echo machine that plays Prisoner's Dilemma. Let action 0 be cooperate and action 1 be defect. Then Echo becomes Tit-for-Tat.

Example 29. Let's look at the results of starting with Echo (Tit-for-Tat in Prisoner's Dilemma) and applying the following sequence of editing commands: D_1, M_1, P_0, F_1, F_0, or, if we issue the commands using the inputs and outputs of Prisoner's Dilemma: $D_D, M_D, P_C, F_D. F_C$. The current state is denoted by a double circle on the state.

Tit-for-Tat is the starting point.

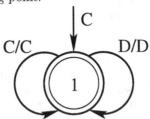

D_D (duplicate) inserts a copy of 1 as the new destination of 1's D-transition.

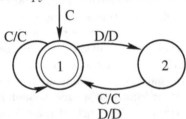

M_D (move) moves the active state to state 2.

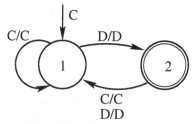

P_C (pin) pins the C-transition from the current state to the current state.

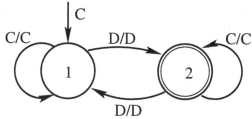

F_D (flip) increments the response on the D-transition from the current state.

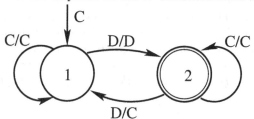

F_C (flip) increments the response on the C-transition from the current state.

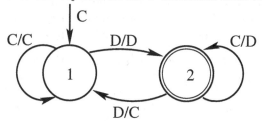

So, this sequence of editing commands transforms our starting automaton, Tit-for-Tat, into a version of Pavlov.

Let's start characterizing the behavior of the system. The following experiment examines the sizes of automata produced by genes of length 20.

Experiment 14.8 *Use an input/output alphabet of size 2. This gives us a total of 12 editing commands. Implement or obtain software for an automaton editor that builds an automaton from a sequence of editing commands. Generate 1,000,000 strings of 20 random editing commands and express them as automata. Compute the number of states in each of these automata and the fraction that have at least one state not connected to the initial state. Make a* *histogram of the lengths of the self-play strings of the automata. (The self-play string is defined in Chapter 6, page 158.)*

Generate 1,000,000 additional automata and collect the same numbers, but this time make the two commands D_0 and D_1 three times as likely as the other commands. What effect does this have on the statistics collected?

One of the issues that must be dealt with in using this cellular encoding for finite state automata is that of state number. Review Experiment 6.4. The number of states used was pretty critical to performance in that experiment. In the next experiment, we will perform Experiment 6.4 again, attempting to find the "right" length for the encoding.

Experiment 14.9 *Modify the software from Experiment 6.4 to use the cellular encoding scheme described in this section. Use the string prediction fitness function alone and the lexical product of string prediction with self-driving length, with string prediction dominant. Use strings of $20, 40, 60,$ and 80 editing commands. Use two-point crossover and one-point mutation that replaces a randomly selected editing command with a new one selected at random. Attempt to match the reference string 111110 using 12 bits.*

Report mean and standard deviation of the number of generations to solution. In your write-up, compare the difference between the plain and lexical fitness functions. What is the effect of changing the length of the strings of editing commands? Is the impact of the lexical fitness partner different for different lengths of strings?

In Experiment 14.8, we tried tinkering with the statistics governing the random generation of editing commands. The relative probability of choosing various commands can be optimized for any experiment. Which probability distributions are "good" depends on the choice of problem.

Experiment 14.10 *Perform Experiment 14.9 again with the D_n commands first twice as likely as the others and then three times as likely. Use whichever fitness function worked best in the previous experiment. Also, choose a set of probabilities of your own for the editing commands, trying to get better performance.*

The idea of placing a nonuniform distribution on the set of editing commands used in a cellular representation can be generalized a good deal. See Problem 654 for one such generalization. The effect of increasing the probability of the D_n commands is to increase the number of states produced relative to the length of the string of editing commands. There are other ways we could control this.

Experiment 14.11 *Perform Experiment 14.10 with the following technique for generating the initial population of strings of editing commands. Use only 40-character strings. Place exactly 7 D_n commands and 33 other commands in the edit strings in a random order. This will cause all the automata to have*

8 states. Compare the impact of this initialization method with the results obtained in Experiment 14.10.

At this point, we will leave the bit-grinding optimization tasks and return to the world of game theory. The basic Iterated Prisoner's Dilemma experiment was performed as Experiment 6.5. Let's revisit a version of this and compare the standard and cellular encodings.

Experiment 14.12 *Rebuild the software from Experiment 6.5 to optionally use cellular encodings. Also, write a tournament program that permits saved files of Prisoner's Dilemma players to play one another. Run the original software with 8-state automata and also run a cellular encoding with a gene length of 48 (yielding an average of 8 states). Perform 30 evolutionary runs for each encoding.*

Compare the resulting behaviors in the form of fitness tracks. Save the final populations as well. For each pair of populations, one evolved with standard encoding and the other evolved with cellular encoding, play each population against the other for 150 rounds in a between-population round robin tournament. Record which population obtained the highest total score. Did either encoding yield substantially superior competitors?

Prisoner's Dilemma has the property that there are no "best" strategies in round robin tournaments. There are pretty good strategies, however, and a population can stay pretty stable for a long time.

You may already be familiar with another game called Rock Paper Scissors. This game is also a simultaneous two-player game, but unlike Prisoner's Dilemma, there are three possible moves: rock (**R**), paper (**P**), and scissors (**S**). Two players choose moves at the same time. If they choose the same move, then the game is a tie. If the players choose different moves, then the victor is established by the following rules: rock smashes scissors; scissors cut paper; paper covers rock. We will turn these results into numbers by awarding 1 point each for a tie, 0 points for a loss, and 3 points for a victory. Table 14.2 enumerates the possible scoring configurations.

Rock Paper Scissors is a game with 3 possible moves, and so we will have 17 editing commands instead of the 12 we had with Prisoner's Dilemma. We have compared the standard and cellular encodings of finite state automata for playing Prisoner's Dilemma already. Let's repeat the experiment for Rock Paper Scissors.

Experiment 14.13 *Rebuild the software from Experiment 14.12 to play Rock Paper Scissors using the scoring system given above. Do agents encoded with the standard or cellular representation compete more effectively, or is there little difference? Add the ability to compute the number of states in a finite state automaton that cannot be reached from the starting state and track the mean of this statistic in the population over the course of evolution. Does one representation manage to connect more of its states to the starting state? Is*

Move		Score	
Player1	Player2	Player1	Player2
R	R	1	1
R	P	0	3
R	S	3	0
P	R	3	0
P	P	1	1
P	S	0	3
S	R	0	3
S	P	3	0
S	S	1	1

Table 14.2. Scoring for Rock Paper Scissors.

the answer to the preceding question different at the beginning and end of the evolutionary runs?

Now let's look at a strategy for Rock Paper Scissors that has a fairly good record for beating human beings.

Definition 14.3 *The strategy* **LOA** *(law-of-averages) for playing Rock Paper Scissors works as follows. If one move has been made most often by its opponent, then it makes the move that will beat that move. If there is a tie for move used most often, then LOA will make the move rock if the tie involves scissors, and the move paper otherwise.*

Experiment 14.14 *Rebuild the software from Experiment 14.13 to play Rock Paper Scissors against the player LOA. In other words, we are now optimizing finite state automata to beat LOA rather than coevolving them to play one another. You must write or obtain from your instructor the code for LOA. Evolve both standard and cellular encodings against LOA playing 120 rounds. Do 30 runs each for 8- and 16-state finite state automata and cellular encodings of lengths 68 and 136. Which encoding works better? Do more states (or editing commands) help more?*

We have done several comparisons of the standard and cellular encodings of finite state automata. The most recent test the ability of the two encodings to adapt to a strategy that cannot be implemented on a finite state automaton (see Problem 648). The ability of a representation to adapt to a strategy written using technology unavailable to it is an interesting one, and you can invent other non-finite-state methods of playing games if you want to try other variations of Experiment 14.14.

One thing we have not done so far is to test two representations directly in a competitive environment. In the next two experiments, we will modify the tournament software used to assess the relative merits of strategies evolved with the standard and cellular encodings into a fitness function. This will permit a form of direct comparison of the two representations.

Experiment 14.15 *Write or obtain software for an evolutionary algorithm that operates on two distinct populations of finite state automata that encode Prisoner's Dilemma strategies. The first should use standard encoding and have 16 states. Use the variation operators from Experiment 6.5. The second population should use cellular encoding with editing strings of length 96, two-point crossover, and two-point mutation that replaces two editing commands with new ones in a given 96-command editing string.*

Evaluate fitness by having each member of one population play each member of the other population for 150 rounds of Iterated Prisoner's Dilemma. As in Experiment 6.5, pick parents from the top $\frac{2}{3}$ of the population by roulette selection and let them breed to replace the bottom $\frac{1}{3}$ of the population. Perform 100 evolutionary runs.

Record the mean fitness and standard deviation of fitness for both populations in a run separately. Record the number of generations in which the mean fitness of one population is ahead of the other. Report the total generations across all populations in which one population outscored the other.

The character of the game may have an impact on the comparison between representations. We have already demonstrated that Iterated Prisoner's Dilemma and Rock Paper Scissors have very different dynamic characters. Let's see whether the last experiment changes much if we change the game.

Experiment 14.16 *Repeat Experiment 14.15 for Rock Paper Scissors. Compare and contrast.*

The material presented in this section opens so many doors that you will probably have thought of dozens of new projects and experiments while working through it. We leave the topic for now.

Problems

Problem 634. Is there a single string of editing commands that produces a given automaton A?

Problem 635. Using the set of editing commands given in Table 14.1, find a derivation of the strategy Tit-for-Two-Tats. This strategy is defined in Chapter 6.

Problem 636. Using the set of editing commands given in Table 14.1, find a derivation of the strategy Ripoff. This strategy is defined in Chapter 6.

Problem 637. What is the expected number of states in an automaton created by a string of n editing commands if all the commands are equally likely to be chosen and we are using a k-character input and output alphabet.

Problem 638. Reread Experiment 14.9. Find a minimum-length string of editing commands to create an automaton that would receive maximum fitness in this experiment.

Problem 639. A *connection topology* for an FSA is a state transition diagram with the response values blank. Assuming any version of a topology can be created with the set of editing commands given in Table 14.1, show that the responses can be filled in any way you want.

Problem 640. Is the representation used for polyominos in Section 14.1 complete? Prove that your answer is correct. Hint: this isn't a difficult question.

Problem 641. A polyomino is *simply connected* if it does not have an empty square surrounded on all sides by full squares. Give an example of a gene for a polyomino that is not simply connected. Then, write out a cellular encoding that can create only simply connected polyominos.

Problem 642. Is the set of editing commands given in Table 14.1 complete? Either prove that it is or find an automaton that cannot be made with the commands. You may find it helpful to do Problem 639 first.

Problem 643. Essay. The Echo strategy, used as the starting point for editing finite state automata, turns out to be Tit-for-Tat when used in the context of Prisoner's Dilemma. In Iterated Prisoner's Dilemma, Tit-for-Tat is a pretty good strategy. In Rock Paper Scissors, is Echo (effectively, rock first, and then repeat your opponent's last action) an effective strategy?

Problem 644. Prove that the population average score in a population playing Rock Paper Scissors with the scoring system given in this chapter is in the range, $1 \leq$ average ≤ 1.5. Prove that if a population consists of a single strategy, then the population gets an average score of exactly 1.

Problem 645. Give a pair of strategies for Rock Paper Scissors that get an average score of 1.5 if they play one another an even number of times.

Problem 646. Is the population average score for a population equally divided between two strategies that are correct answers to Problem 645 completely predictable? If so, what is it? If not, explain why not.

Problem 647. Is it possible for a 60-member population playing 120 rounds of Rock Paper Scissors to achieve the upper bound of 1.5 on population average fitness? Explain.

Problem 648. Prove that the strategy LOA, given in Definition 14.3, cannot be implemented with a finite state automaton.

Problem 649. Is the Graduate School Game (defined in Section 6.3) more like Prisoner's Dilemma or Rock Paper Scissors?

Problem 650. In the single-shot Prisoner's Dilemma, there is a clear best strategy: defect. Does Rock Paper Scissors have this property? Prove your answer.

Problem 651. Essay. The claim is made on page 396 that the strategy LOA for playing Rock Paper Scissors does well against humans. Verify this fact by playing with a few friends. What sort of strategies does LOA do well against?

Problem 652. Essay. Either examining the experimental evidence from Experiment 14.14 or working by pure reason, answer the following question. Will the strategy LOA be one that performs well against finite state automata, or will it perform poorly?

Problem 653. Essay. The number of states in a finite state automaton is not explicitly given in cellular encoding. Suppose you want a certain number of states. You could simply go back to the beginning of the string of edit commands and keep editing until you had as many states as desired. Your assignment: figure out what could go wrong. Will this method always generate as many states as you want? Will the type of automata be different than it would be if instead you used a very long string of edit commands and stopped when you had enough states?

Problem 654. Essay. Discuss the following scheme for improving performance in Experiments 14.9 and 14.10. Do a number of preliminary runs. Looking at the genes for FSAs that achieve maximal fitness, tabulate the empirical probability of seeing each command after each other command in these genes. Also, compute the probability of seeing each editing command as the first command in these successful genes. Now generate random initial genes as follows. The first command is chosen according to the distribution of first commands you just computed. Generate the rest of the string by getting the next command from the empirical distribution of next commands you computed for the current command. Do you think this empirical knowledge reuse will help enough to make it worth the trouble? What is the cost? Can this scheme cause worse performance than generating initial populations at random?

Problem 655. Essay. A *most common strategy* is one that occupies the largest part of the population among those strategies present in a given population. If we look at the average time for one most common strategy to be displaced by another, we have a measure of the volatility of an evolving system. If you surveyed many populations, would you expect to see higher volatility in populations evolving to play Prisoner's Dilemma or Rock Paper Scissors?

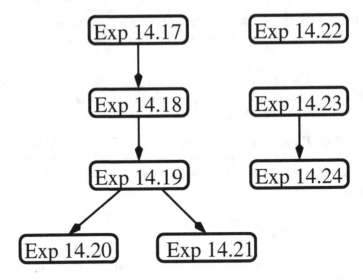

17 Evolving combinatorial graphs.
18 Controlling graph size a different way.
19 Maximizing graph diameter.
20 Cyclic use of the gene.
21 Trying a different fitness function.
22 Sampling graph invariants within the representation.
23 Maximizing girth.
24 Population seeding.

Fig. 14.6. The topics and dependencies of the experiments in this section.

14.3 Cellular Encoding of Graphs

In this section, we venture into the realm of combinatorial graph theory to give a fairly general encoding for 3-connected cubic graphs.

Definition 14.4 *A graph is* **k-connected** *if there is no set of fewer than k edges that we could delete and thereby disconnect the graph.*

Definition 14.5 *A graph is* **cubic** *if each vertex is of degree 3.*

We will start with a very simple graph and use editing rules to build up more complex graphs. In some ways, the cellular encoding we will use for 3-connected cubic graphs is very similar to the one we used for finite state automata. The transition diagrams of finite state automata are directed graphs with regular out degree (always exactly k output arrows). In other ways, the encoding will be quite different; there will be two editing "agents," or bots, at work rather than a single current state.

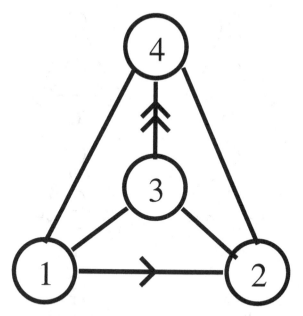

Fig. 14.7. The initial configuration for the graph-editing bots.

The starting configuration for our cellular encoding for graphs is shown in Figure 14.7. The single and double arrows denote the edges that will be the focus of our editing commands. We will refer to these arrows as graph bots with the single arrow denoting the first graph bot and the double arrow denoting the second. During editing, the vertices will be numbered. There are two sorts of editing commands that will be used in the cellular encoding. The first group will be used to move the the graph bots; the second will be used to add vertices and edges to the graph.

The movement commands will use the fact that the vertices are numbered. The commands **R1** and **R2** cause the first and second graph bots, respectively, to reverse their directions. These commands are spoken "reverse one" and "reverse two." The command **AS1** causes the first graph bot to advance past the vertex at which it is pointing so that that vertex is now at its tail. There are two ways to do this, since each vertex has degree 3. **AS1** causes the bot to point to the vertex with the smaller number of the two available. The command **AL1** also advances the first graph bot, but moves it toward the larger of the two available vertices. The commands **AS2** and **AL2** have the same effect as **AS1** and **AL1** for the second graph bot. These commands are spoken "advance small one," "advance large one," "advance small two," and "advance large two," respectively. The effect of the movement commands on the starting graph are shown in Figure 14.8. One important point: We never permit the graph bots to occupy the same edge. If a command causes the two graph bots to occupy the same edge, then ignore that command.

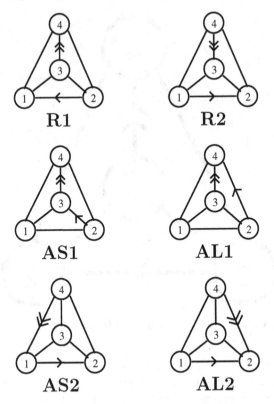

Fig. 14.8. The result of applying each of the editing commands to the initial configuration.

The commands that modify the graph are **I1** and **I2**, spoken "insertion type one" and "insertion type two." Both of these commands insert two new vertices into the middle of the edges with graph bots and join them with a new edge. The new vertices are given the next two available numbers with the smaller number given to the vertex inserted into the edge containing the first graph bot. The two insertion commands are depicted pictorially in Figure 14.9. **I1** differs from **I2** in the way the graph bots are placed after the insertion. **I1** reverses the direction of the second graph bot; **I2** reverses the direction of the first graph bot. In all cases, the bots are placed so that they are pointing away from the new vertices.

To prove that the 8 commands given are sufficient to make any 3-connected cubic graph requires graph theory beyond the scope of this text. In general, however, any cellular encoding requires the designer to deal with the issue of completeness or at least the issue "can the encoding I've dreamed up find the objects that solve my problem?" We next give a derivation of the cube using the set of editing commands just described.

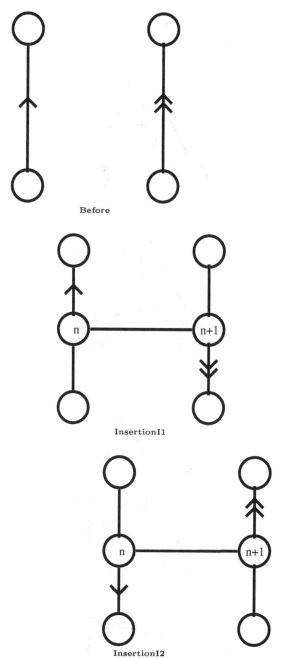

Fig. 14.9. A generic positioning of the graph bots and the results of executing the two insertion commands. (The commands differ only in their placement of the graph bots after the insertion.)

Example 30. The sequence of commands **AL2**, **I2**, **AL1**, **AS1**, **I1** yields the cube. Let's look at the commands one at a time:

Start:

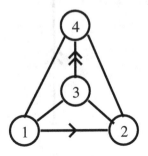

Apply **AL2**:

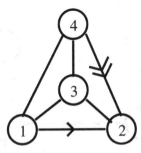

Apply **I2**:

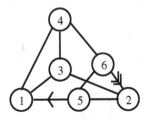

Apply **AL1**:

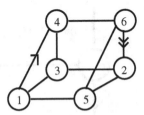

Apply **AS1**:

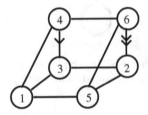

Apply **I1**:

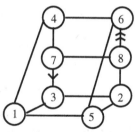

The resulting graph is a somewhat bent version of the cube. Redrawn as a standard cube, we get

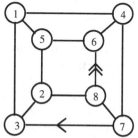

This derivation leaves the graph bots on the graph at the end. When the graph is passed on to another routine, e.g., a fitness evaluation, the graph bots are discarded.

We now have an encoding for graphs as a string of editing commands. We can use an evolutionary algorithm to evolve graphs by just dusting off a string evolver over an 8-character alphabet. There is still a very important piece missing, however: the fitness function.

What do people want in a graph, anyhow? Cast your mind back to the graphs used in Chapter 13. We used various cubic Petersen graphs that were quite symmetric and had fairly high diameter, and we used random cubic graphs, obtained with edge moves, that were not at all symmetric and had pretty low diameter, given their size. What about graphs with intermediate diameters? Our first task will be to search for these.

Instead of just using the diameter as a fitness function, we are going to break up the notion of diameter into smaller pieces with a few additional definitions. Read Section D.3 in Appendix D on distances in graphs.

Definition 14.6 *The **eccentricity** of a vertex in a connected graph is the largest distance between it and any other vertex in the graph. For a vertex v, this quantity is denoted by $\mathrm{Ecc}(v)$.*

The diameter of a graph is simply the maximum eccentricity of any of its vertices. To get a graph with an intermediate diameter, we will minimize the sum of the squared deviations of the eccentricities of all of the graph's vertices from the desired diameter. This will push the graph toward the desired diameter.

Definition 14.7 *The* **eccentricity deviation fitness function** *for eccentricity E for a graph G with vertex set $V(G)$ is defined to be*

$$\mathrm{ED}_E(G) = \sum_{v \in V(G)} (E - \mathrm{Ecc}(v))^2.$$

Notice that this fitness function is to be minimized.

When we were working with cellular encodings of finite state automata, we found that controlling the number of states in the automaton required a bit of care. In contrast to the standard encoding, the number of states was not directly specified as a parameter of the experiment. It was, however, one more than the number of D_n commands. Since the insertion commands for graph editing insert two vertices, the number of vertices in a graph is four plus twice the number of I commands. If we generate genes at random, we will not have good control over the graph size.

In order to get graphs of a specific size, we will execute edit commands until the graph is the desired size. This means that the genes need to be long enough to have enough insertion commands. On average, one command in four is an edit. We will test two methods of getting graphs that are big enough.

Experiment 14.17 *Implement or obtain software for a string evolver over the alphabet of graph-editing commands defined in this section. Use genes of length 130 with two-point crossover and three-point mutation. When creating a graph from a string of editing commands, continue editing until either there are 256 vertices in the graph or you reach the end of the edit string. Use lexical fitness, in which fitness is the number of vertices in the graph with ties broken by the function*

$$\mathrm{ED}_{12}(G),$$

given in Definition 14.7. Evolve populations of 200 graphs using a steady-state algorithm with size-7 single tournament selection.

Report the mean and deviation of both fitness functions and the best value of

$$\mathrm{ED}_{12}(G)$$

from each run. Permit evolution to continue for 500 generations. Also, save the diameter of the most fit graph in each run. Report a histogram of the diameters of the most fit graphs in each run. How fast do the genes converge to size-256 graphs? Was the process efficient at minimizing $E_{12}(G)$?

This is a new way of using a lexical fitness function. Instead of putting the fitness of most interest as the dominant partner, Experiment 14.17 puts a detail that has to be gotten right as the dominant partner. This forces a sufficient number of insertion commands into the genes. Once this has happened, we can get down to the business of trying to match a mean eccentricity of 12. Now let's try another approach to the size-control problem.

Experiment 14.18 *Repeat Experiment 14.17 using a different method for managing the size of the graphs. Instead of lexical fitness with size of the graph dominant, use only the fitness function* $ED_{12}(G)$. *Use genes of length 60 cycling through until the graph is of sufficient size. Explicitly check each gene to make sure it has at least one insertion command and award it a fitness of zero if it does not. (This is unlikely to happen in the initial population but may arise under evolution.)*

Sampling from a region of eccentricity space that is difficult to reach with the explicit constructions and random algorithms given in Appendix D elicits graphs that might be interesting to use in the kind of experiments given in Chapter 13. The reason for thinking such graphs might behave differently is that they have one parameter, average eccentricity, that is different.

Looking at the diameters of the various cubic graphs we used in Chapter 13, we also see that the large-diameter cubic graphs were all generalized Petersen graphs and hence highly symmetric. The random graphs are all at the very low end of the diameter distribution. An evolved population of graphs created with our editing commands is unlikely to contain a highly symmetric graph. Let's see how it can do at sampling the extremes of the diameter distribution.

Experiment 14.19 *Modify the software from Experiment 14.18 to maximize the diameter of graphs. Use the diameter as the fitness function. Use genes of length 60. Since vertices are needed to build diameter, no lexical products will be needed to encourage the production of diameter. Run 10 populations for 5000 generations and save a best gene from generation 50, 100, 500, and 5000 in each run.*

Examine the genes and report the fraction of insertion commands in the best genes from each epoch. Also, save and graph the mean and variance of population fitness, the best fitness, the mean and variance of the vertex set sizes for the graphs, and the fraction of insertion commands in each generation.

It may be that the only imperative of evolution in the preceding experiment is to have all insertion commands. Let's perform a second experiment that speaks to this issue.

Experiment 14.20 *Repeat Experiment 14.19 with genes of length 30 and cycle through them twice. Compare with the results of Experiment 14.19.*

The last two experiments attempted to make high-diameter graphs. Such graphs are "long" and may resemble sausages when drawn. We will now try to do the opposite. Since having few vertices always yields very low diameter, we will write a more complex fitness function that encourages many vertices and low diameter (compactness).

Definition 14.8 *For a graph G with vertex set* $V(G)$, *let*

$$\text{CP}(G) = \frac{|V(G)|}{\sum_{v \in V(G)} \text{Ecc}(v)}.$$

*This function is called the **large compact graph** function. It divides the number of vertices by the sum of their eccentricities. This function is to be maximized.*

Experiment 14.21 *Repeat Experiment 14.19 with the large compact graph function as the fitness function. Compare the resulting statistics and explain. Did the fitness function in fact encourage large compact graphs?*

So far, we have used the graph-editing representation to sample the space of cubic graphs for rare diameters and eccentricities. The resulting graphs are amorphous and probably not of any great interest to graph theorists. They may have application to the kind of work done in Chapter 13. These problems were mostly intended to help us to understand and work with the graph-editing system. At this point, we will go on to a much more difficult mathematical problem.

Definition 14.9 *The **girth** of a graph is the length of the shortest closed cycle in the graph. The **girth at** v, for a vertex v of a graph, is the length of the shortest closed cycle of which that vertex is a member. If v is in no cycle, then the girth at v is infinite.*

Look at the graphs used in Problem 659. These graphs have girth 4, and the girth at every vertex is 4. The Petersen graph $P_{5,2}$ has girth 5.

Definition 14.10 *A **(3, n)-cage** is a cubic graph with girth n and the smallest possible number of vertices. Examples of some of the known cages are given in Figure 14.10.*

The $(3, n)$-cages are also called the cubic cages or even just the cages, because the notion was first defined for cubic graphs. The cubic cages form an interesting example of a phenomenon that is widespread in mathematics: small examples are not representative. The cages shown in Figure 14.10 are unique and symmetric. Unique means that they are the only cubic graphs with their girth and size of vertex set. In this case, symmetric means that there is a way to permute the vertices that takes edges to edges such that any vertex can be taken to any other. The $(3, 7)$-cage is unique, but not symmetric. No other cage is symmetric in this sense. There are 18 different $(3, 9)$-cages, 3 different $(3, 10)$-cages, one known $(3, 11)$-cage, and a unique $(3, 12)$-cage. The $(3, 13)$-cage(s) is (are) not known.

Starting with beautiful, unique, symmetric graphs, the family of cages rapidly degenerates into fairly ugly graphs that are not unique. The ugliness means that cages will, in the future, probably mostly be worked on by stochastic search algorithms (though success here is not guaranteed at all). The current lower bound on the size of a $(3, 13)$-cage is 202 vertices, a number that Brendan McKay and Wendy Myrvold computed by a cleverly written exhaustion of all possibilities. The current best-known girth-13 cubic graph has 272 vertices and is given by an algebraic construction found by Norman Biggs.

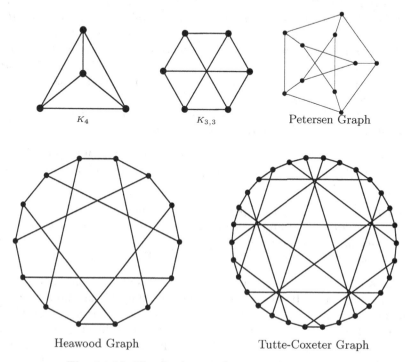

K_4 $K_{3,3}$ Petersen Graph

Heawood Graph   Tutte-Coxeter Graph

Fig. 14.10. The $(3, n)$-cages for $n = 3, 4, 5, 6$, and 8.

As before, the sticky wicket is writing a fitness function. The girth of a graph is the minimum of the girths at each of its vertices. Girth, however, would make a shockingly inefficient fitness function. At a given number of vertices, graphs of a smaller girth are far more common than those of a higher girth. At 30 vertices, it is possible to get girth 8, for example, only by the discovery of a unique and highly symmetric graph. Before we decide on a fitness function, let's perform a sampling experiment to see how much trouble we are in.

Experiment 14.22 *Write or obtain software for an algorithm that starts with the initial graph configuration from Figure 14.7 and executes editing commands, sampled uniformly at random, until the graph has 30 vertices. Generate 100,000 graphs in this fashion and make a histogram of their mean girth at each vertex and their girth. Report, also, the ratio of each girth to the most common girth. Were any graphs of girth 8 found?*

Experiment 14.22 should have verified the assertion about rarity of high-girth graphs and the problem with using girth directly as a fitness function. The mean girth at vertices is a much smoother statistic and will form the basis for herding graphs toward higher girth in the course of evolution.

Experiment 14.23 *Write or obtain software for a steady-state evolutionary algorithm that operates on a population of k graph-edit strings of length n generated uniformly at random. Use size-7 tournament selection, two-point crossover, and three-point mutation. Use the lexical product of girth and mean girth at each vertex, with girth being the dominant fitness function. Save and graph the mean and variance of both fitness functions and the maximum girth in each generation. Try all possible pairs of n and k, for k = 100, 500, 1000 and n = 30, 60, 90. For the length-30 strings, run through the strings once, twice, or three times. What girths do you obtain?*

In the past, we have tried various schemes for initializing populations to give evolution a boost. Combining Experiments 14.22 and 14.23 gives us a means of doing this.

Experiment 14.24 *Repeat Experiment 14.23, initializing each run with the following procedure. Generate 100,000 genes. Test the fitness of each graph. As each graph is tested, save its gene only if it is in the top k of the graphs tested so far. Compare the results with those of Experiment 14.23.*

This section gives only a small taste of what could be done with cellular encodings of graphs. It treats one possible encoding for an interesting but limited class of graphs. There are many other problems possible. It would be possible, for example, to have a "current vertex" editor like the ones used for finite state automata in Section 14.2. The editing commands might involve insertion of whole new subgraphs in place of the current vertex. They could also include commands to swap edges as in Chapter 13 (page 366).

Problems

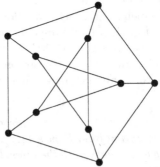

Problem 656. Find a derivation, using the editing commands given in this section, for the Petersen graph. Use the standard starting point, as in Example 30.

Problem 657. Find a sequence of editing commands that transforms K_4 into $K_{3,3}$ into the Petersen graph into the Heawood graph. (Extra Credit: find a sequence of commands that transforms the Petersen graph into the Tutte–Coxeter graph.)

Problem 658. Give a minimal derivation for a graph with girth 4 using the starting graph and editing commands given in this section.

Problem 659. The cube, derived in Example 30, is also called the 4-prism. Above are the 5-prism, the 6-prism, and the 7-prism. Find a sequence of editing commands, including a segment repeated some number of times, that can create the n-prism for any n. Say how many times the repeated fragment must be repeated to get the n-prism.

Problem 660. Define the graph IG(n) to be the result of applying the edit command **I1** to the initial configuration n times. Draw IG(0), IG(1), IG(2), and IG(3).

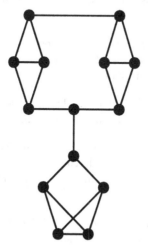

Problem 661. Make a copy of the graph above and label each vertex with its eccentricity.

Problem 662. Make a copy of the graph above and label each vertex with the girth at that vertex.

Problem 663. Prove that the set of editing commands for cubic graphs given in this section always produces a 3-connected graph. Do this by showing that the commands cannot produce a graph that can be disconnected by deleting one edge or by deleting any two edges.

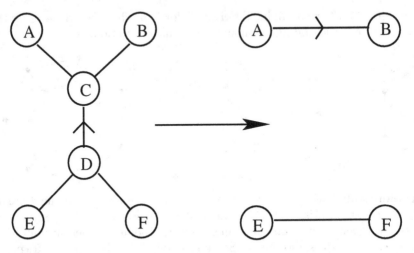

Problem 664. Suppose that we add the above deletion command D_1 to our editing commands. Assume that $A < B$ (so as to make the direction of the arrow unambiguous). Also assume, if you want, that there is a corresponding command D_2 involving the second graph bot. Prove that the language can now create disconnected graphs and graphs that are 1-connected or 2-connected but not 3-connected.

Problem 665. What restrictions do we need to place on the use of the deletion command(s) defined in Problem 664 if we want to avoid graphs with edges from a vertex to itself or multiple edges?

Problem 666. Reread Experiment 14.18. Compute the probability of any genes appearing in the initial population that have no insertion commands.

Problem 667. Design a short sequence of editing commands that if repeated will create a large-diameter graph. Estimate or (better yet) compute exactly the ratio of the diameter of the graph to the number of repetitions of your sequence of editing rules.

Problem 668. Prove that if we delete one of the insertion commands from the language, we can still make all of the same graphs.

Problem 669. Write or obtain software for expressing graph-edit genes. Take 100,000 samples obtained by running through random 120-character genes until a graph with 100 vertices is constructed. Make a histogram of the diameter and mean eccentricity of the graphs.

Problem 670. The edge swap operation used to generate random regular graphs in Chapter 13 is described in Appendix D (on page 549 in the definition of random regular graph). What restrictions would have to be placed on the positioning of the graph bots to permit adding a command that performed such an edge swap on the edges where the bots are? Assume that the new edges connect the former heads of the graph bots and the former tails of the graph bots, leaving the heads of the graph bots pointing toward the same vertex.

Problem 671. Give and defend a different representation for evolving graphs. It may be cellular or direct. If possible, make it more general than the representation given in this section.

Problem 672. Essay. In Chapter 10, we developed a GP automata representation for discrete robots performing the Tartarus task. Suppose that we were to put GP automata in charge of our graph bots. List and defend a set of input terminals for the deciders that would be good for the task used in Experiment 14.17.

Problem 673. Essay. In Chapter 10, we developed a GP automata representation for discrete robots performing the Tartarus task. Suppose that we were to put GP automata in charge of our graph bots. List and defend a set of input terminals for the deciders that would be good for the task used in Experiment 14.23.

Problem 674. Essay. In Chapter 12, we developed ISAc lists for discrete robots performing a variety of discrete robotics tasks. Suppose that we were to put ISAc lists in charge of our graph bots. List and defend a set of data vector entries that would be good for the task used in Experiment 14.17.

Problem 675. Essay. In Chapter 12, we developed ISAc lists for discrete robots performing a variety of discrete robotics tasks. Suppose that we were to put ISAc lists in charge of our graph bots. List and defend a set of data vector entries that would be good for the task used in Experiment 14.23.

Problem 676. Essay. Is the diameter or the average eccentricity a better measure of how dispersed or spread out a graph is, assuming that we are using the graph to control mating as in Chapter 13?

Problem 677. Essay. Generalize the editing method given in this section for cubic graphs to regular graphs of degree 4. Give details.

14.4 Context Free Grammar Genetic Programming

One of the problems with genetic programming is the disruptiveness of subtree crossover. Another problem is controlling the size of the parse trees created.

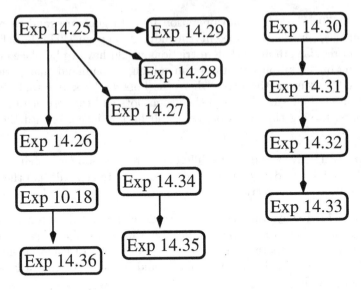

25 Context free grammar for PORS.
26 Variations on PORS experiment.
27 Excluding STO-STO.
28 Population seeding.
29 Adding a do-nothing rule.
30 The plus-times-half genetic programming problem.
31 PTH with context free grammar.
32 Exploring new grammars.
33 Exploring more new grammars.
34 Boolean parse trees.
35 The 3-parity problem.
36 Tartarus.

Fig. 14.11. The topics and dependencies of the experiments in this section.

In this section, we will use a cellular representation to take a shot at both problems by changing our representation for parse trees. We will do this by creating *context free grammars* that control the growth of parse trees. The grammar will form a cellular encoding for the parse trees. Grammar rules will be used the way editing commands were used for finite state automata and graphs in earlier sections of this chapter.

In addition to neatening crossover and controlling size, using grammatical representations for specifying parse trees solves the data typing problem. Using the old method, we could use multiple types of data in a parse tree only by encoding them in the data type of the tree. In Chapter 9, for example, the ITE operation took 3 real-number arguments, but the first was used as if it were a Boolean argument by equating negative with false. Since grammars can restrict what arguments are passed to operations, they can do automatic data type checking. There will be no need to write complex verification or repair operators that verify subtree crossover obeys data typing rules.

The use of grammars also permits us to restrict the class of parse trees examined by embedding expert knowledge into the grammar. For example, the grammar can exclude redundant combinations of operators, such as store a number in a memory and then store the exact same number in the same memory again. Both data typing and expert knowledge are embedded in the algorithm at *design* time rather than run time. So, beyond the overhead for managing the the context free grammar system, there is close to zero runtime cost for them. Cool, huh?

A *context free grammar* contains a collection of *nonterminal symbols*, a collection of *terminal symbols*, and a set of *production rules*. (This is a different use of the word "terminal" from the one we used in previous chapters on genetic programming. To avoid confusion, in this chapter, we will refer to the terminals of parse trees as "leaves.") In a context free grammar, a nonterminal symbol is one that can still be modified; a terminal symbol is one that the grammar cannot modify again. A sequence of production rules is called a *production*. A production starts with a distinguished nonterminal symbol, the *starting nonterminal*. It then applies a series of production rules. A production rule replaces a single nonterminal symbol with a finite string of terminal and nonterminal symbols. By the end of a production, all nonterminal symbols are resolved. Let's do an example.

Example 31. Recall the PORS language from Chapter 8. Here is a context free grammar for the PORS language:

Nonterminals: S
Starting nonterminal: S
Terminals: +, 1, Rcl, Sto

Production Rules:
Rule 1: **S** → (+ **S S**)
Rule 2: **S** → (Sto **S**)
Rule 3: **S** → **Rcl**
Rule 4: **S** → **1**

Starting with a single S let's see what parse tree we get if we apply the sequence of rules 121443.

Start:	S
Apply 1:	(+ S S)
Apply 2:	(+ (Sto S) S)
Apply 1:	(+ (Sto (+ S S)) S)
Apply 4:	(+ (Sto (+ 1 S)) S)
Apply 4:	(+ (Sto (+ 1 1)) S)
Apply 3:	(+ (Sto (+ 1 1)) Rcl)

The result is a correct solution to the Efficient Node Use problem for 6 nodes.

There are two things that you will have noticed in Example 31. First, there are sometimes multiple nonterminals to which a given production rule could have been applied. When applying productions in a cellular encoding, we must give a rule for which nonterminal to use when multiple nonterminals are available. In the example, we used the leftmost available nonterminal. Second, there is the problem of unresolved nonterminals. If we are generating random nonterminals, it is not hard to imagine getting to the end of a list of production rules before resolving them all. In order to use a string of context free grammar products as a cellular encoding, we must deal with this issue.

Definition 14.11 *The* **cauterization rules** *are a set of rules, one for each nonterminal in a context free grammar, that replace a nonterminal with a string of terminal symbols. These rules are used to finish a production when an evolved list of production rules is used and leaves nonterminals behind.*

A third issue that may be less obvious is that of rules that cannot be applied. If there is a production rule that cannot be applied, e.g., for want of an appropriate nonterminal symbol, then simply skip the rule.

We are now ready to formally state how to use context free grammars as a cellular representation for parse trees.

Definition 14.12 *A* **cellular representation for parse trees** *is a context free grammar together with a set of cauterization rules and rules for choosing which nonterminal will be chosen when multiple nonterminals are available. The representation is a string of production rules that the evolutionary algorithm operates on as a standard string gene.*

Let's perform some experiments.

Experiment 14.25 *Build or obtain software for a steady-state evolutionary algorithm for the PORS n-node Efficient Node Use problem using a context free grammar cellular encoding for the parse trees. Use strings of 2n production rules from the 4-rule grammar given in Example 31. The creatures being evolved are thus strings over the alphabet $\{1, 2, 3, 4\}$.*

Let the algorithm be steady-state and use tournament selection of size 11 on a population of 400 productions. Use two-point crossover and single-point mutation. When executing a production, do not execute production rules 1 and 2 if they drive the total number of nodes above n (count all symbols, +, 1, Rcl, Sto, and S, as nodes) If the tree still has nonterminal symbols after the entire string of production rules has been traversed, use rule 4 as the cauterization rule. When multiple nonterminals are available, use the leftmost one.

Perform 100 runs recording mean and standard deviation of fitness as well as time-to-solution for 30 runs for n = 12, 13, and 14. Cut off runs that take more than 1,000,000 mating events to finish. Also, save the final populations of productions in each run for later use.

Experiment 14.25 is our first try at context free grammar genetic programming. We used the simplest nontrivial genetic programming problem we have studied. Let's check the effect of tinkering with a few of the parameters of the system. Review the definition of nonaligned crossover, Definition 7.21.

Experiment 14.26 *Repeat Experiment 14.25 with 3 variations. In the first, expand the rightmost nonterminal, rather than the leftmost. In the second, replace one-point mutation with two-point mutation. In the third, make one-fourth of all crossover nonaligned. Document the impact of each of these variations.*

In the introduction to this section, we claimed that we can use the grammar to cheaply embed expert knowledge into our system. Let's give an example of this process. The tree **(Sto (Sto T))**, where **T** is a tree, wastes a node. Let's build a grammar that prevents this waste.

Definition 14.13 No Sto-Sto grammar

Nonterminals: S,T
Starting Nonterminal: S
Terminals: +, 1, Rcl, Sto

Production Rules:
Rule 1: **S** → **(+ S S)**
Rule 2: **S** → **(Sto T)**
Rule 3: **S** → **Rcl**
Rule 4: **S** → **1**
Rule 5: **T** → **(+ S S)**

Using the No Sto-Sto grammar will be a little trickier, because the use of a **T** forces the use of 3 nodes.

Experiment 14.27 *For the No Sto-Sto grammar, perform the variation of Experiment 14.25 that worked best. Do not let the number of symbols other than **T** plus 3 times the number of **T** symbols exceed the number of nodes permitted. For **T**, use the cauterization rule **T**→(+ 1 1). Compare your results with the other PORS Efficient Node Use experiments you have performed.*

Let's see how well population seeding works to generalize the result. It is time to use the productions we saved in Experiment 14.25.

Experiment 14.28 *Repeat Experiment 14.25 using the saved populations from Experiment 14.25 as the starting populations. Instead of n = 12, 13, and 14, do runs for n = 15, 16, and 17. Run 9 experiments, using each of the 3 types of populations saved (for n = 12, 13, and 14) to initialize the runs.*

Before you perform the runs, predict which initialization will help the most and least with each kind of run (n = 15, 16, and 17). Predict whether random initialization would be superior for each of the 9 sets of runs. Compare your predictions with your experimental results and explain the reasoning that led you to make those predictions.

This embedding of expert knowledge and evolved knowledge can be carried further, but additional development of killer PORS grammars and initializing populations is left for the Problems.

Let's take a look at the effect of a kind of rule analogous to biological *introns*. An intron is a sequence of genetic code that does not produce protein. In spite of "doing nothing," introns can affect (and in fact enable) crossover.

Experiment 14.29 *Repeat Experiment 14.26 using the best variation, but with a fifth production rule that does nothing. Perform 2 sets of runs that include the fifth rule. Make the strings of production rules 25% longer.*

We will now shift to a new problem: a maximum problem with two operations and a single numerical constant.

Definition 14.14 *A* **maximum problem** *is one in which the computer is asked to produce the largest possible result with a fixed set of operations and constants subject to some resource limitation.*

The PORS Efficient Node Use problem is a type of maximum problem in which the resource limitation was total nodes. Limiting parse trees by their total nodes is not traditional in genetic programming as it was originally defined by John Koza and John Rice. Instead, the depth of the tree from the root node is limited. The following maximum problem is a standard one for depth-limited genetic programming. We will start by building a standard depth-limited genetic programming system.

Experiment 14.30 *The* plus-times-half (PTH) *maximum problem uses the operations* + *and* * *and the numerical constant one-half (0.5). As with the PORS Efficient Node Use problem, the fitness of a parse tree is the result of evaluating it with the value to be maximized. Create or obtain software for a steady-state evolutionary algorithm that operates on PTH parse trees of maximum depth k, with the root node considered to be depth zero. Use size-7 single tournament selection. During reproduction, use subtree crossover 50% of the time. When subtree crossover creates a tree that exceeds the depth limit, prune it by deleting nodes that are too deep and transforming those nodes at depth k to leaves. For each tree, use a mutation operator that selects an internal node of the tree uniformly at random (if it has one) and changes its operation type. Run 50 populations until they achieve the maximum possible value for k = 4 (16) and for k = 5 (256). Cut a given run off if it has not achieved the maximum possible value in 1,000,000 mating events.*

With a standard baseline experiment in place, we can now try some experiments for the PTH problem with context free grammars. The following experiments demonstrate the way knowledge can be embedded in grammars.

Definition 14.15 Basic PTH Grammar

Nonterminals: S
Starting Nonterminal: S
Terminals: +, *, 0.5

Production Rules:
Rule 1: S → (* S S)
Rule 2: S → (+ S S)
Rule 3: S → 0.5

Experiment 14.31 *Repeat Experiment 14.30 with a context free grammar encoding using the basic PTH grammar. Expand the leftmost nonterminal first and use the obvious cauterization rule: rule 3 from the grammar. Do not execute any production that will make the tree violate its depth limit. Use two-point crossover and two-point mutation on your strings of production rules. Use strings of length 40 for k = 4 and strings of length 80 for k = 5. In addition to reporting the same statistics as those from Experiment 14.30 and comparing to those results, examine the genes in the final population. Explain why the system that exhibited superior performance did so.*

Let's see whether we can get better results using a more effective grammar.

Definition 14.16 Second PTH Grammar

Nonterminals: S,T
Starting Nonterminal: S
Terminals: +, *, 0.5

Production Rules:
Rule 1: S → (* S S)
Rule 2: S → (+ T T)
Rule 3: T → (+ T T)
Rule 4: T → 0.5

Experiment 14.32 *Repeat Experiment 14.31 with the second PTH grammar. Report the same statistics and compare. What was the impact of the new grammar? Cauterize all nonterminals remaining at the end of a production to 0.5.*

It is possible to build even more special knowledge into the grammar.

Definition 14.17 Third PTH Grammar

Nonterminals: S, T, U

Starting Nonterminal: S
Terminals: +, *, 0.5

Production Rules:
Rule 1: **S → (* S S)**
Rule 2: **S → (+ T T)**
Rule 2: **T → (+ U U)**
Rule 3: **U → 0.5**

Experiment 14.33 *Repeat Experiment 14.32 with the third PTH grammar. Report the same statistics and compare. What was the impact of the new grammar? Cauterize all nonterminals remaining at the end of a production to 0.5.*

The three grammars for the PTH problem contain knowledge about the solutions to those problems. This is a less-than-subtle demonstration of how to cook a problem to come out the way you want. We now turn to a grammar for Boolean parse trees and look to see whether we can extract generalizable knowledge from the system.

Definition 14.18 Boolean Parse Tree Grammar

Nonterminals: S
Starting Nonterminal: S
Terminals: AND, OR, NAND, NOR, NOT, T, F, $X_i (i = 1, \ldots, n)$

Production Rules:
Rule 1: **S → (AND S S)**
Rule 2: **S → (OR S S)**
Rule 3: **S → (NAND S S)**
Rule 4: **S → (NOR S S)**
Rule 5: **S → (NOT S)**
Rule 6: **S → T**
Rule 7: **S → F**
Rule 8: **S → X_1**
...
Rule 7+n: **S → X_n**

The Boolean parse tree grammar works on n-input variables and so has a number of rules that vary with n. First, let's repeat a familiar experiment and see how well the system performs.

Experiment 14.34 *Create or obtain software for a steady-state evolutionary algorithm that uses a context free grammar genetic programming representation for Boolean parse trees based on the Boolean parse tree grammar given* *above. Attempt to solve the odd parity problem: solutions should return true if there are an odd number of true inputs and false otherwise. Use a population of 400 productions of length 15 with two variables. Notice that your productions will be over a 9-letter alphabet. Cauterize by changing the first nonterminal to X_1, the second to X_2, and so on, cyclically. Fitness is the number of correct predictions of the parity of two binary variables for all possible combinations of Boolean values those variables can have. Use size-7 single tournament selection. Record time-to-solution for 100 runs and save the final population of productions from each run.*

The next step is to see whether productions that solve the 2-parity problem can help solve higher-order-parity problems.

Experiment 14.35 *Repeat Experiment 14.35 for the 3-parity problem with length 30 productions over the Boolean parse tree grammar. Perform one set of runs with random initialization and a second set in which you replace a random substring of length 15 in each initial string with one of the saved strings from Experiment 14.34. Load all the strings from all the final populations and select at random among the entire set of strings when initializing. Report times-to-solution and document the impact of the nonstandard initialization method.*

We now want to move on to demonstrate explicit data typing with context free grammar genetic programming. In the grammars used so far, there is something like data typing: e.g., look at the roles of **U** and **T** in the third PTH grammar. When we used GP automata for the Tartarus problem, the deciders were created with genetic programming. The decider's job was to reduce a large set of possible patterns to a single bit (the parity of an integer) that could drive transitions of the finite state portion of the GP automata. What we will do next is create a grammar for the deciders with the types **B** (Boolean) and **E** (sensor expression).

Definition 14.19 Tartarus Decider Grammar

Nonterminals: S, B, E
Starting Nonterminal: S
Terminals: UM, UR, MR, LR, LM, LL, LM, UL, ==, !=, 0, 1, 2

Production Rules:
Rule 1: **S → (AND B B)**
Rule 2: **S → (OR B B)**
Rule 3: **S → (NAND B B)**
Rule 4: **S → (NOR B B)**
Rule 5: **S → (NOT B)**
Rule 6: **B → T**
Rule 7: **B → F**

Rule 8: **B** → (**== E E**)
Rule 9: **B** → (**!= E E**)
Rule 10: **E** → **UM**
Rule 11: **E** → **UR**

...

Rule 17: **E** → **UL**
Rule 18: **E** → **0**
Rule 19: **E** → **1**
Rule 20: **E** → **2**

Experiment 14.36 *Rebuild Experiment 10.18 to use a cellular parse tree encoding with the above grammar for its deciders. Change "if even" and "if odd" to "if false" and "if true." For cauterization rules, use* **S**→(**== UM 1**), **B**→ **T**, *and* **E**→*UM. Use a string of 20 production rules for each decider. Compare the results with those obtained in Experiment 10.18.*

Problems

Problem 678. Give a string of context free grammar productions to yield an optimal tree for the Efficient Node Use problem for all n given in Figure 8.6.

Problem 679. The No Sto-Sto grammar encodes a fact we know about the PORS Efficient Node Use problem: a store following a store is a bad idea. Write a context free grammar that encodes trees for which all leaves of the parse tree executed before the first store are 1's and all leaves after the first store executed are recalls.

Problem 680. Using the four-rule grammar for PORS given in Example 31, do the following:

(i) Express the production **S=1212121443333**.
(ii) Express the production **T=1212144133133**.
(iii) Perform two-point crossover on the productions after the third and before the ninth character, coloring the production rules to track their origin in **S** and **T**. Express the resulting strings, coloring the nodes generated by productions from **S** and **T**.

Problem 681. Give the exact value of the correct solution to the PTH problem with depth-k trees. Prove that your answer is correct.

Problem 682. Suppose that instead of limiting by depth, we limited PTH by total nodes. First, show that the number of nodes in the tree is odd. Next, show that if $f(n)$ is the maximum value obtainable with n nodes, then

$$f(n+2) = \max(f(n) + 0.5, f(k) * f(m)),$$

where m and k are odd, and $n + 1 = m + k$. Using this fact, compute the value of $f(n)$ for all odd $1 \le n \le 25$.

Problem 683. Explicitly state in English the knowledge about solutions to the PTH problem embedded in the second and the third PTH grammars given in this section.

Problem 684. Is it possible to write a grammar such that if all the nonterminals are resolved, it must give an optimal solution to the PTH problem?

Problem 685. Essay. One problem we have in using genetic programming with real-valued functions is that of incorporating real constants. In the standard (parse tree) representation, we use ephemeral constants: constants generated on the spot. Describe a scheme for incorporating real constants in one or more production rules to use context free grammars with real-valued genetic programming.

Problem 686. Suppose we are using context free grammar genetic programming and that we have a mutation that sorts the production rules in the gene into increasing order. If this mutation were mixed in with the standard one at some relatively low rate, would it help?

Problem 687. Would you expect nonaligned crossover (see Definition 7.21) to help or hinder if used at a moderate-to-low rate in the PTH problem?

Problem 688. Could one profitably store the leaves in parse trees for the PTH problem as nil pointers?

Problem 689. Suppose, in Experiment 14.25, we permit only 1's and 2's in the production rules and simply allow the cauterization rule to fill in the leaves. What change would this make in the experiment?

Problem 690. Suppose we added a new terminal, **0.4**, and a fourth rule, S→0.4, to the basic grammar for the PTH problem. What would the maximum value for depths 4 and 5 now be? Would the problem become harder or easier to solve?

Problem 691. Give a cellular representation for parse trees yielding rational functions with operations **+**, **-**, *****, **/**, randomly generated ephemeral constants, and a variable x. (Recall that a rational function is a ratio of two polynomials.)

Problem 692. Essay. Considering the various grammars used for the PTH problem, address the following question: Do the grammars presented later in the section make the search space larger or smaller? Defend your view carefully.

Problem 693. Essay. Design and defend a better grammar for deciders for Tartarus than the one given in Definition 14.4.

15

Application to Bioinformatics

This chapter gives examples of applications of evolutionary computation to bioinformatics. The four sections all solve completely separate problems, and so again, the experiments are organized section by section. We will start with an application requiring only the very simple sort of evolutionary computation from Chapter 2. The fitness function will align binary strings with a type of genetic parasite called a *transposon*. The next application will evolve finite state automata to try to improve the design of polymerase chain reaction (PCR) primers. The third example will use evolutionary computation to locate error-correcting codes for DNA, useful in bar codes for genetic libraries. The technique involves using an evolutionary algorithm to control a greedy algorithm, a technique called a *greedy closure* evolutionary algorithm. The final application is a tool for visualizing DNA that uses a finite state automaton combined with a fractal technique related to iterated function systems.

15.1 Alignment of Transposon Insertion Sequences

A *transposon* is a form of genetic parasite. A genetic parasite is a sequence (or string) of DNA bases that copies itself at the expense of its host. It appears multiple times, possibly on different chromosomes, in an organism's genome.

In order to discuss transposons, we need a bit of molecular biology. Deoxyribonucleic acid (DNA) is the primary information storage molecule used by living organisms. DNA is very stable, forming the famous double helix in which complementary pairs of DNA sequences bind in a double spiral. This stability means that manipulating DNA requires a good deal of biochemical effort. Because there is a trade-off, in biochemical terms, between stability and usability, DNA is transcribed into a less stable but more usable form: ribonucleic acid (RNA). RNA is then sent to a subcellular unit called a *ribosome* to be converted into protein. Proteins are the workhorse molecules of life, performing much of the active biochemistry. The central dogma of molec-

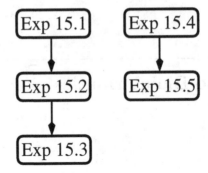

1   DNA alignment with an evolutionary algorithm.
2   Exploring the mutation rate.
3   Exploring population and tournament size.
4   Motif finding with an evolutionary algorithm.
5   Excluding complex motifs with latitude control.

Fig. 15.1. The topics and dependencies of the experiments in this section.

ular biology is that the information in DNA follows the path given in Figure 15.2.

Fig. 15.2. The central dogma of molecular biology.

The complementary binding of DNA bases does not only lend stability to the DNA molecule; it also enables the process of copying the information. There are four DNA bases: **C**, **G**, **A**, and **T**. The bases **C** and **G** bind to each other, as do the bases **A** and **T**. When DNA is copied to make RNA, the RNA that is made is the complement, with **C** copied as **G**, **G** copied as **C**, **A** copied as **T**, and **T** copied as **A**.

There are three kinds of transposons. Type I transposons are segments of DNA that cause the cell to transcribe RNA from them. This RNA is then transformed back into DNA by an enzyme called *reverse transcriptase* and integrated back into the genome. These transposons are thought to prefer specific positions to reintegrate their copies into the genome. Type II transposons simply copy themselves from DNA directly to DNA. Type III transposons are similar to type II, save that they are on average much shorter and use a different copy mechanism. Almost any text on molecular biology, e.g., *Genes VII* by Benjamin Lewin [41], contains a full description of the state of knowledge about transposons and their intriguing relationship with viruses.

In this section, we will be working with the problem of identifying the sequence that type I transposons use to integrate back into the genome. The data set available on the webpage associated with this text (click on data and then Chapter 15) was gathered by finding the point at which a particular transposon integrated into the genome. This is done by comparing a gene with no transposons to the same gene with transposons. Where the genes differ is a transposon site.

The problem is that while we know where the transposon integrated, we do not know into which strand of DNA it integrated. If there is a particular sequence of DNA bases required for integration, it appears on one strand, and its complement appears on the other. This means that if we want to compare the insertion sites, we must first decide into which strand the transposon integrated.

This is where the evolutionary computation system comes into play. It is used to decide whether to use a DNA sequence as found or to use its reversed complement. This is a binary problem, so the binary string evolvers we learned about in Chapter 2 will be useful.

We use the reverse complement instead of just the complement, because DNA strands have opposite orientations on opposite strands. This means that if the transposon integrated on the opposite strand, we should not only complement the DNA, but reverse it (turn it end-for-end).

Example 32. **Reverse complementation.** The DNA sequence
$$\textbf{CGATTACTGTG}$$
has reverse complementary sequence
$$\textbf{CACAGTAATCG.}$$
Not only do we apply the swaps **C⇔G** and **A⇔T**, but we also rewrite the sequence in reversed order.

Suppose that we have a sequence containing several hundred examples of transposon insertion sites. We delete the regions of the sequence without transposon sites and line up the regions with insertion sites so that the sites coincide. We then need to specify an orientation for each insertion site sequence, either forward or reverse-complement. This specification will be what we evolve. A binary string gene of length N can specify the orientation of a set of N sequences. For a data set with N sequences, we thus need a binary string evolver that operates on length-N strings. It remains to construct a fitness function.

In this case, we presume that there is a conserved motif at the point of transposon insertion. A *motif* is a set of characters, possibly with some wildcards or multibase possibilities. So,
$$\textbf{C, G, A or T, anything, C, C}$$
is an example of a motif that includes 8 sequences: **CGACCC, CGTCCC, CGAGCC, CGTGCC, CGAACC, CGTACC, CGATCC**, and **CGTTCC**. There may also be some other properties, like an above-average

fraction of **A**'s and **T**'s. Because of this, there is reason to believe that when the sequences are placed in their correct alignment, there will be a decrease in the total "randomness" of the base composition of the alignment.

Definition 15.1 *For a collection C of DNA sequences, define* P_X, $X \in \{\mathbf{C},$ $\mathbf{G}, \mathbf{A}, \mathbf{T}\}$, *to be the fraction of the bases that are X.*

```
01234567890123456789012345678
·----------·----------·--------
CACCGCACCGCACTGCATCGGTCGCCAGC
ACCCGCATCGGTCGCCAGCCGAGCCGAGC
CACCGCATCGGTCGCCAGCCGAGCCGAGC
CACTGCATCGGTCGCCAGCCGAGCCGAGC
GCTCGACACACGGGCAGGCAGGCACACCG
```

Fig. 15.3. A gapless alignment of 5 DNA sequences.

Definition 15.2 *A* **gapless alignment** *of a set of sequences of DNA bases consists in placing the sequences of DNA on a single coordinate system so that corresponding bases are clearly designated. An example of such an alignment appears in Figure 15.3. (Gapped alignments will be discussed in Section 15.3.)*

The transposon insertion data have to be trimmed to make the DNA sequences the same length. This means that the orientation, either forward, or reversed and complemented, is the only thing that can change about the way a sequence fits into an alignment. We now need a fitness function that will compute the "nonrandomness" of a given selection of orientations of sequences within an alignment.

Definition 15.3 *Assume that we have a gapless alignment of N DNA sequences, all of the same length M. View the alignment as a matrix of DNA bases with N rows and M columns. Let* X_i *be the fraction of bases in column i of the matrix of type X, for* $X \in \{\mathbf{C}, \mathbf{G}, \mathbf{A}, \mathbf{T}\}$. *Then, the* **nonrandomness** *of an alignment is*

$$\sum_{i=1}^{M} (X_i - P_X)^2.$$

The nonrandomness function is to be maximized. Lining up the motif at the point of insertion will yield less randomness. Notice that we are assuming that the DNA sequences are essentially random away from the transposon insertion motif. We are now ready to perform an experiment.

Experiment 15.1 *Write or obtain software for a steady-state evolutionary algorithm using single tournament selection with tournament size 7 that operates on binary genes of length N. Download transposon insertion sequences from the website associated with this book; N is the number of these sequences. Use two-point crossover and probabilistic mutation with probability $\frac{1}{N}$. Use a population of 400 binary strings for 400,000 mating events. Use the nonrandomness fitness function. Run the algorithm 100 times and save the resulting alignments. If an alignment specifies the reverse complement of the first sequence, reverse-complement every sequence in the alignment before saving it (this puts alignments that are the same except for reverse-complement status into the same orientation for comparison). How often do you get the same alignment? Are alignments that appear often the most fit, or are the most fit alignments rare?*

This experiment produces alignments and gives us a baseline notion of an acceptable fitness. With the baseline fitness in hand, let's perform a parameter sensitivity study for various algorithm parameters. We will start with mutation rate.

Experiment 15.2 *Modify the software from Experiment 15.1 as follows. Take the most common fitness you got in Experiment 15.1 and assume that any alignment with this fitness is "'correct." This lets us compute a time-to-solution. Now repeat the previous experiment, but for mutation rates $\frac{1}{2N}$, $\frac{1}{N}$, $\frac{3}{2N}$, and $\frac{2}{N}$. Report the impact on time-to-solution and the number of runs that fail to find a solution in 500,000 mating events.*

Now let's look at the effects of varying population size and tournament size.

Experiment 15.3 *Repeat Experiment 15.2 using the best mutation rate from Experiment 15.2. Use all possible pairs of population sizes 100, 200, 400, and 800 and tournament sizes 4, 7, and 15. Report the impact on time-to-solution and the number of runs that fail to find a solution in 500,000 mating events.*

Now that we have a way of aligning the transposon insertion sites, we need a way of finding the motif. A motif is a sequence of DNA bases with wildcard characters. A motif could thus be thought of as a string over a 15-character alphabet consisting of the nonempty subsets of $\{C, G, A, T\}$. We will encode this alphabet by letting $C=8$, $G=4$, $A=2$, and $T=1$ and by representing each subset with the sum of the numbers of its members. Thus, the number 9 is a partial wildcard that matches the letters C and T. We can use a string evolver to search for this encoding of a motif. As always, we need a fitness function.

Definition 15.4 *A kth-order Markov model of a collection of DNA sequences is an assignment to each DNA sequence of length k an empirical probability that the next base will be C, G, A, or T. Such a model is built from a collection of target DNA sequences in the following manner. For each* *length-k subsequence S appearing in the target DNA, the number of times the next base is a* **C**, **G**, **A**, *or* **T** *is tabulated. Then, the probabilities are computed by dividing these empirical counts by the number of occurrences of the subsequence S. For subsequences that do not appear, the first-order probabilities of each DNA base are used.*

Example 33. Let's take the target DNA sequence

```
AAGCTTGCAGTTTAGGGCCCCTGATACGAAAGAAGGGAGGTCCGACAGCCTGGGGCCGAC
TCTAGAGAACGGGACCCCGTTCCATAGGGTGGTCCGGAGCCCATGTAGCCGCTCAGCCAG
GTCCTGTACCGTGGGCCTACATGCTCCACCACCCCGTGACGGGAACTTAGTATCTAGAGT
TATAAGTCCTGCGGGTCCGACAACCTCGGGACCGGAGCTAGAGAACGGACATTAGTCTCC
TGGGGTGGTCCGGAGCCCGTACAGCCGCTCAGCCTAGTCCCGTACCATGGTCCTGCACGC
TCCACCGCCCTGTGACAAGTGTCCTAGTATCTAGAACCGCGACCCAAGGGGGTCCGGACA
AGCAACTTGGCCACCCGGACTAAAACCTGCAGGTCCCTAGCATGTATCAAAGGGCGACTA
ATGTCAGACGGAGAACCCTATGAGGTGTACTACTAACGCTTCCTAGCTAAAAGTTGTGTA
CAGATCCAGATCTCGGCGAGTTTGCCTCCCGAGGATTGTTGACAACCTTTTCAGAAACGC
TGGTATCCAACCTCAACACATCAAGCCTGCATCCGAGGCGGGGGGCCAGGTACTAAGGAG
AAGTCAACAACATCGCACATAGCAGGAACAGGCGTTACACAGATAAGTATTAAATACTGC
TTAGAAGGCATTATTTAATTCTTTACAAAAACAGGGGAAGGCTTGGGGCCGGTTCCAAAG
AACGGATGCCCGTCCCATAGGGTGGTCCGGAGCCTATGTGGCCGGTTAGCCTGGTTCCGT
ACCCAAAATCCTGCACACTCCACCGCTCTGTGGTGGGTGTCCTAGTATTTAAAACTAAAG
```

To build a second order ($k = 2$) Markov model of the DNA sequence, we need to tabulate how many times a **C**, **G**, **A**, or **T** appears after each of the possible 2-character sequences. This is the work computers were meant to do, and they have, yielding the following tabulation:

Sequence	N_C	N_G	N_A	N_T
CC	18	25	16	23
CG	9	17	9	8
CA	11	16	15	12
CT	12	14	19	8
GC	20	6	10	12
GG	13	26	17	20
GA	14	14	13	6
GT	17	13	14	10
AC	17	9	20	11
AG	16	19	16	13
AA	19	17	17	4
AT	10	8	7	7
TC	27	3	8	7
TG	10	14	5	13
TA	13	18	11	10
TT	6	7	12	7

Dividing through by the number of times each 2-character sequence occurs with another base following it yields the second-order Markov model for the target above.

Markov model $k = 2$				
Sequence	P_C	P_G	P_A	P_T
CC	0.220	0.305	0.195	0.280
CG	0.209	0.395	0.209	0.186
CA	0.204	0.296	0.278	0.222
CT	0.226	0.264	0.358	0.151
GC	0.417	0.125	0.208	0.250
GG	0.171	0.342	0.224	0.263
GA	0.298	0.298	0.277	0.128
GT	0.315	0.241	0.259	0.185
AC	0.298	0.158	0.351	0.193
AG	0.250	0.297	0.250	0.203
AA	0.333	0.298	0.298	0.070
AT	0.312	0.250	0.219	0.219
TC	0.600	0.067	0.178	0.156
TG	0.238	0.333	0.119	0.310
TA	0.250	0.346	0.212	0.192
TT	0.188	0.219	0.375	0.219

For each 2-character sequence, we have the probability that it will be followed by each of the 4 possible DNA bases.

What use is a kth-order Markov model? While there are a number of cool applications, we will use these Markov models to baseline the degree to which a motif is "surprising." In order to do this, we will use the Markov model to generate sequences "like" the sequence we are searching. A kth-order Markov model of a given set of target DNA sequences can be used to find more sequences with the same kth-order statistics. Let's look at the algorithm for doing this.

Algorithm 15.1 Moving Window Markov Generation Algorithm

Input: *A kth-order Markov model and a number m*
Output: *A string of length m*
Details: *Initialize the algorithm as follows. Select at random a sequence of k characters that appeared in the target DNA sequence used to generate the original Markov model. This is our initial window. Using the empirical distribution for that window, select a next base. Add this base to the end of the window and shift the window over, discarding the first character. Repeat this procedure m times, returning the characters generated.*

Algorithm 15.1 can be used to generate any amount of synthetic DNA sequences with the same kth-order base statistics as the original target DNA used to create the Markov model. This now puts us in a position to define a fitness function for motifs.

Definition 15.5 *Suppose we have a set of target DNA sequences, e.g., the set of aligned transposon insertion sequences generated in Experiments 15.1–15.3. The* **count** *for a motif is the number of times a sequence matching the motif appears in the target DNA sequences.*

Definition 15.6 *Suppose we have a set of target DNA sequences, e.g., the set of aligned transposon insertion sequences generated in Experiments 15.1–15.3. The* **synthetic count** *for a motif is the number of times a sequence matching the motif appears in a stretch of synthetic DNA, generated using Algorithm 15.1 with a Markov chain created from the target DNA or an appropriate set of reference DNA.*

Definition 15.7 *The p-***fitness** *of a motif is the probability that the count of a motif will exceed its synthetic count. The* p_N*-***fitness** *of a motif is the estimate of the p-fitness obtained using N samples of synthetic DNA. Compute the* p_N*-fitness of a motif as follows. Obtain target and reference DNA (they may or may not be the same). Pick k and generate a kth-order Markov model from the reference DNA. Compute the count of the motif in the target. Pick N and compute the synthetic count of the motif in N sequences of the same length as the target sequence generated with the Markov chain derived from the reference DNA. The fraction of instances in which the synthetic count was at least the count is the* p_N*-fitness.*

The p_N-fitness of a motif is to be minimized; the harder it is for the synthetic count to exceed the count of a motif, the more surprising a motif is. Notice that the p-fitness is a probability and so not only selects good motifs, but gives a form of certificate for their statistical significance. It is important to remember that this p-fitness is relative to the choice of reference DNA. The transposon insertion studies used in this chapter study insertion into a particular gene and are published in a paper by C. Dietrich et al [16]. A good set of reference DNA is thus the sequence of that gene, available on the website for this book. Let's go find some motifs.

Experiment 15.4 *Write or obtain software for a steady-state evolutionary algorithm using single tournament selection with tournament size 7 that operates on string genes over the motif alphabet described in this section. Download the glu18 gene sequence from the website for this text for use as reference DNA. Build a 5th-order (k = 5) Markov model from the glu18 code and use it to implement the* p_N*-fitness function (N = 2000) for motifs in aligned sets of transposon insertion sites from Experiments 15.1–15.3. Use two-point* *crossover and single-point mutation in the motif searcher. Use a population of motifs of length 8 for* 100,000 *mating events with a population size of* 400. *Perform* 100 *runs. Sort the best final motifs found in each population by their fitnesses. Report the number of times each motif was found. Are there cases in which the sequences specified by one motif were a subset of the sequences specified by another?*

If this experiment worked for you as it did for us, you have discovered a problem with this technique for finding motifs: what a human thinks of as a motif is a bit more restrictive than what the system finds. The system described in Experiment 15.4 managed to find several motifs with high fitness values, *but* appearing in the target sequence only once each. This means that our motif searcher can assemble a motif from rare strings that has a high *p*-fitness but is not of all that much interest. A possible solution to this problem is to insist numerically that the motifs be more like what people think of as motifs.

Definition 15.8 *A character in the motif alphabet stands for one or more possible matches. The* **latitude** *of a character in the motif alphabet is the number of characters it stands for minus one. The* **latitude** *of a motif is the sum of the latitudes of its characters.*

Experiment 15.5 *Repeat Experiment 15.4. Modify the fitness function so that any motif with a latitude in excess of d is awarded a fitness of 1.0 (the worst possible). Perform* 100 *runs for* $d = 5, 6, 7$. *Contrast the results with the results of Experiment 15.4.*

It would be possible to perform additional experiments with the motif searcher (you are urged to apply the searcher to other data sets), but instead we will move on to an application of evolutionary algorithms to a problem in computational molecular biology. If you are interested in further information on motif searchers, you should read [45] and look at the Gibbs sampler, a standard motif location tool [30].

Problems

Problem 694. Give a 20-base DNA sequence that is its own reverse complement.

Problem 695. The nonrandomness fitness function compensates for first-order deviation from uniform randomness in the DNA used by computing the fractions P_X of each type X of DNA base. Consider pairs of adjacent DNA bases. These will show up in certain proportions in a random DNA sample. Write a function that compensates for this second-order randomness.

Problem 696. Explain in a sentence or two why the function given in Definition 15.3 measures nonrandomness.

Problem 697. Give a motif, of the sort used in this section, that matches as few sequences as possible, but also matches each of the following:

AAGCTCGAC	CACGGGCAG	CGGGCAGGC	GGGGCAGGC
ACACAGGGG	CACTCCGCC	CTACCAAAG	GTCGCCAGC
ACCGGATAT	CACTGCATC	CTCCGTCTA	GTCGCCAGC
AGCCGAGCC	CCACCGGAT	CTGTCGATA	GTCGCCAGC
CACAGGGGC	CCCCAAATC	CTGTGTCGA	GTGCGGTGC
CACCCGCAT	CCCTCATCC	GAGTAGAGC	TCCTAGAAT
CACCGCACC	CCGCACCGC	GCTGCGCGC	TCCTGATGG
CACCGCATC	CGGCTCGGC	GGAGAGAGC	TTCACTGTA

Problem 698. Construct and defend a better fitness function for motifs than the p-fitness.

Problem 699. Give an efficient algorithm for checking the count of a motif in a sequence of DNA.

Problem 700. Essay. Explain why the order of the Markov model used in Experiments 15.4 and 15.5 must be shorter than the length of the motifs being evolved to get reasonable results.

Problem 701. Essay. Based on Experiments 15.1–15.3, make a case that the nonrandomness fitness function on the data set used is unimodal or polymodal.

Problem 702. Essay. Why is maximizing the nonrandomness fitness function the correct choice?

Problem 703. Essay. With transposon data, we have a simple method of locating where the transposon inserted itself: there is a transposon sequence where before there was none. This gives us an absolute reference point for our alignment and so leaves us to worry only about orientation. Describe a representation for gapless alignment where we suspect a conserved motif but do not know exactly where, laterally in the DNA sequence, that alignment is.

Problem 704. Essay. Taking the minimal description of transposable elements (transposons) given in this section, outline a way to incorporate structures like transposable elements into an evolutionary computation system. If you are going to base your idea on natural transposons, be sure to research the three transposon types and state clearly from which one(s) you are drawing inspiration.

15.2 PCR Primer Design

Polymerase chain reaction (PCR) is a method of amplifying (making lots of copies of) DNA sequences. DNA is normally double-stranded. When you heat

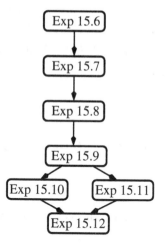

6   Evolving finite state primer predictors.
7   Balanced training data and lexical fitness.
8   Alternative fitness.
9   Hybridization.
10 Documenting the impact of hybridization.
11 Exploring the number of states in the predictor.
12 The impact of hybridization with optimized state number.

Fig. 15.4. The topics and dependencies of the experiments in this section.

the DNA, it comes apart, like a zipper, at a temperature determined by the fraction of **GC** bonds (**GC** pairs bind more tightly than **AT** pairs). Once they are apart, an enzyme called DNA-polymerase can grab individual bases out of solution and use them to build partial double strands. As the DNA cools, it re-anneals as well as being duplicated by the polymerase. A single PCR cycle heats and then cools the DNA with some of it being duplicated by the polymerase.

Primers are carefully chosen short segments that re-anneal earlier, on average, than the whole strands of DNA. If we start with a sample of DNA, add the correct pair of primers, a supply of DNA bases, and the right polymerase enzyme, then we will get exponential amplification (roughly doubling in each cycle of the reaction) of the DNA between the two primers. (The primers land on opposite strands of DNA.) A diagram of this process is given in Figure 15.5.

The length of each new strand is controlled by time and temperature. Typically, you let the strands grow, on average, *beyond* the annealing point for the complementary primers. Since the primers are on opposite strands, they amplify in opposite directions and only the DNA between them undergoes exponential growth. Primers are of length 17 to 23, typically, and the amplified DNA is a few hundred to several thousand bases. Evolutionary computation can help pick good primers.

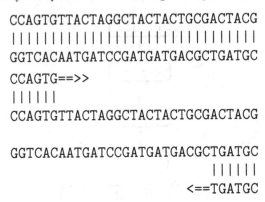

```
CCAGTGTTACTAGGCTACTACTGCGACTACG
|||||||||||||||||||||||||||||||
GGTCACAATGATCCGATGATGACGCTGATGC

CCAGTG==>>
||||||
CCAGTGTTACTAGGCTACTACTGCGACTACG


GGTCACAATGATCCGATGATGACGCTGATGC
                          ||||||
                       <==TGATGC
```

Fig. 15.5. Double-stranded DNA and single-stranded DNA undergoing replication.

Existing primer-picking tools make sure that the DNA biophysics of a pair of primers is correct. These tools match the melting temperature of the right and left primer, make sure that the primer does not anneal with itself or its partner, and check for other problems that can ruin a PCR reaction. (There are potential problems specific to the organism for which the primers are being designed.)

A problem many current primer-picking tools do not address is the problem of duplicate sequences. Given the size of genomes, 20-character DNA sequences (e.g., typical primers) should be unique. If genomes were generated at random, such sequences would be unique. However, many biological processes duplicate sequences within a genome. Transposons, discussed in Section 15.1, are an example of a source of duplicate sequences. What effect does this have on a PCR reaction?

If both members of a primer pair are inside a duplicated sequence in an organism, then they will amplify both copies. If the duplicates are identical, this isn't a problem for the PCR reaction (it may be one for the biologist). Often, though, duplicated sequences have diverged, and so many different sequences are amplified. These amplifications happen at slightly different exponential rates and diverge from each another exponentially. In practice, primer pairs in a duplicated sequence are unusable.

Another problem is created when one member of a primer pair is part of a duplicated sequence. If the number of copies of this sequence is small, then the exponential amplification of the paired sequence permits things to work properly. Some transposons have a ridiculous number of copies, and in this case, one-half of the primer pair anneals in so many places that the amplification of the region flanked by the primer pair goes nowhere. So what do we do?

The technique that follows was developed in the context of a large sequencing project in corn. Tens of thousands of primers were designed and tested. The results of these tests can be used to create a fitness function for evolved predictors that guess which primers will or will not work. (This data is available on the website associated with this book.)

The basic idea is this. Use primer-picking software to generate multiple primers for a given sequence target. Performance predictors, trained on past results, examine these primers and rate them. The more highly rated primers will have a better chance of working if the predictors are correctly generalizing about the sequence features that make primers work or fail. We also save a set of primers with known performance on which to test our predictors; this is called *cross validation*.

We need to chose a representation for our primer performance predictors. In this case, a finite state automaton is a natural choice, since it can process strings of characters and embed its opinion of them in its state space. Figure 15.6 shows a finite state automaton specialized for use on DNA. The automaton shown is a Moore automaton with states labeled with the automaton's output.

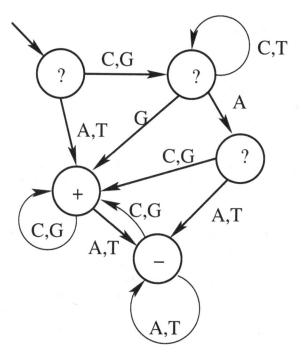

Fig. 15.6. A finite state automaton that can be driven by DNA and used as a performance predictor. (The input alphabet is **C, G, A ,T,** and the output alphabet is +, −,?, interpreted as "good," "bad," and "don't know.")

In order to train finite state automata to predict PCR performance, we need a fitness function. The data for training are in the format shown in Figure 15.7 and are available on the text website. We will divide the data into randomly selected training and cross-validation sets, with $\frac{1}{5}$ of the sequences (selected uniformly at random) in the cross-validation set. Primers in the training set are marked with a 0, 1, or 2; those marked with 0 are considered "good"; the others are considered "bad." The fitness function will select for automata that end in a + state on a "good" primer and a − state on a "bad" primer. (Note: we are treating primers that don't amplify their targets and those that amplify multiple targets as "bad.")

Definition 15.9 *The* **raw prediction fitness** *of a finite state automaton is computed as follows. Initialize the fitness to zero. Run each primer in the training set through the finite state automaton. If it ends in a + state for a good primer or a − state for a bad primer, add 1 to the fitness. If the automaton ends in a ? state, add nothing to the fitness. Otherwise, subtract 1 from the fitness.*

```
. . .
2 CTCCACTATAGCTGCCGTCG
2 TACAGGGACATCTGGATGGG
0 CTGCAGTACATCTACCACCACC
0 TGCAGAGCTTCGAGCACC
0 CGATCAGCATGTTCATTTGC
1 CAAGGAGGGAGTGATTCAGC
1 AAGAACAGCACTCAATCGGG
1 CAAGGAGGGAGTGATTCAGC
. . .
```

Fig. 15.7. Format for primer training data. (Numerical codes are 0 = primer works; 1 = primer amplifies multiple targets; 2 = primer does not amplify.)

Experiment 15.6 *Write or obtain software for evolving Moore automata with transitions driven by the DNA alphabet and state labels +, −, ?. Divide randomly the available primer data from the text website into training $\left(\frac{4}{5}\right)$ and cross-validation $\left(\frac{1}{5}\right)$ data. Treat the states of the automaton as indivisible objects with the string of states forming the basis of the crossover operator. Perform two-point crossover. Use three-point mutation. Each single-point mutation should change a transition destination, state label, or the initial state. Set the probabilities so that all arrows and labels in a given automaton have the same chance of being affected. Evolve a population of 400 primer predictors for 100,000 mating events using a steady-state algorithm with size-7 tournament selection. Let the predictors have 32 states. Run 30 populations.*

Report both the fitness tracks and a cross-validated final fitness for each run. This latter is the best fitness found when assessing the fitness of the entire final population on the cross-validation data. Also, report how often the most fit automaton according to cross validation is also the most fit according to the fitness on the training data. Report the density of each of the three state types (+, −, ?) in each population. What do these densities suggest? Be sure to save the best predictor from each run for later use.

The results of this experiment suggest a couple of modifications. Leaving the automata the option of saying "I don't know" gives the system flexibility, but is it happening too often? Also, since there is a finite number of examples of good and bad primers in the training set, there is a possibility of falling into a useless local optimum: the predictor could predict that all primers were of whatever type is most common. To avoid these pitfalls, let's improve the experiment.

Experiment 15.7 *Take the available primer data and divide them into good and bad primers. Randomly select, from whichever sort are more common, a number of examples equal to the number that are less common and then discard the excess of the more common type. This set of training data is now* balanced; *we won't get solutions which just guess whatever is most common. Modify the evolutionary algorithm from Experiment 15.6 to add a lexical fitness: the number of ? results, to be minimized. Rerun Experiment 15.6 both with and without the lexical fitness. Discuss the effect of balancing the training data and the impact of using the lexical fitness.*

At this point, we will redesign the fitness function. Insisting that the predictor get the fitness right at its final state is somewhat brittle. Perhaps there are automata that are on the right track, but get the final answer wrong.

Definition 15.10 *The* **incremental reward fitness function** *is computed in almost the same manner as the raw prediction fitness function. The difference is that the fitness is scored at each state transition. This yields more finely grained fitness information. As a good primer runs through the predictor, 1 is added for each + state and 1 is subtracted for each − state with ? still yielding a reward of 0. The opposite is done for bad primers.*

Experiment 15.8 *Using the nonlexical version of the software, repeat Experiment 15.7 with the incremental reward fitness function in place of the raw fitness function. Document the impact. Examine your best predictor: are there lots of ?'s near the initial state?*

Now we can try applying some other techniques from earlier chapters. Since there are many patterns in the training data (distinct possible sources of fitness), it follows that different runs will find different patterns. How do we combine patterns from distinct evolutionary runs?

Definition 15.11 *The practice of* **hybridization** *consists in initializing an evolutionary algorithm with superior genes from multiple populations that have already been evolved.*

Note that hybridization is a type of population seeding. Be careful; the term hybridization can also refer to using an algorithm that combines evolutionary and nonevolutionary techniques. A "hybrid optimizer" might, for example, use a hill climbing optimizer to evaluate the fitness of a gene in an evolutionary algorithm. This would permit the evolutionary algorithm to search the landscape for good hills and then use the hill climber to climb them. This sort of division of labor, while beyond the scope of this text, often pays big dividends.

Experiment 15.9 *Repeat Experiment 15.8, incorporating the 30 best-of-run automata saved during Experiment 15.8 into the initial population (in addition to random predictors). Does this affect the results?*

There is a problem with hybridization as performed in Experiment 15.9: it does not control for the effect of unmodified added evolution on the original populations. The 30 hybridized automata evolved through 200,000 mating events, while the others evolved through only 100,000. The next experiment will take far longer to run, but should yield a more meaningful test of the utility of hybridization.

Experiment 15.10 *Repeat Experiment 15.8 with the following modifications. Set the experiment to run for 100,000 mating events, but save the best-of-run automata (according to the cross-validation data) at mating event 50,000. Now initialize a new set of 30 runs with these best-of-run automata included in the initial population and run them for 50,000 mating events. We have two sets of runs, both run for 100,000 mating events, but with the second set benefiting from hybridization. In addition to reporting the other performance measures, discuss the impact of hybridization.*

The number of states used is a measure of the amount of information a finite state automaton can store. The experiments performed thus far yield a baseline for performance. Let's check the sensitivity of the system to the number of states.

Experiment 15.11 *Repeat Experiment 15.10 but with 48 and 64 states. What impact does this change have on the baseline and hybridized runs?*

For our last experiment, let's check the sensitivity to mutation rate.

Experiment 15.12 *Repeat Experiment 15.11 using only the number of states that performed best. Use 1-, 5-, and 7-point mutation and compare with the 3-point mutation used in Experiment 15.11. What impact does this change have on the baseline and hybridized runs?*

This section is a modest introduction to using machine learning to improve primer design. The technique of hybridization is a potentially valuable one. The incremental reward fitness function is an example of a redesign of a fitness function that makes the hill climbing functionality of an evolutionary algorithm more effective. There are a number of other possible technologies for this sort of machine learning: Markov modeling of good and bad primers, for example. There are also other EC techniques we could use, such as graph-based algorithms. We now leave primers for a much stranger application, DNA bar codes, with a new type of fitness function.

Problems

Problem 705. Is a predictor that has a ? on all its states in a local optimum or a big flat space with uphill paths at its edge? Defend your conclusions.

Problem 706. Would a real function optimizer benefit from hybridization? Explain.

Problem 707. Prove that there is a finite state automaton that can achieve maximal raw prediction fitness on the training data. Assume that no primer appears in the training set twice.

Problem 708. Explain why it is impossible to receive a reward on every state transition when computing incremental reward fitness, no matter what finite state automaton you use.

Problem 709. The system developed in this section runs primers through the finite state automaton one at a time. Come up with a fitness function that scores finite state automata on pairs of primers that are used together.

Problem 710. Is 32 states a reasonable number for the task in this section? Your answer should involve mathematics, probably counting arguments.

Problem 711. Essay. One of the advantages of GP automata is that deciders compress the bandwidth of the environment. Specify and defend a decider language that uses inputs of 3- or 5-base windows (instead of a single base at a time as the finite state automata do), permitting GP automata to evaluate primers.

Problem 712. Essay. Primers work or fail in pairs. That means that a primer with a bad score might have gotten a better one if it had been paired with a different partner. Given this, can we still hope to get useful results from the primer prediction system given in this chapter? Is it important that we are picking the best from among multiple primers when we use the system to select new primers?

Problem 713. Essay. Address the following statement. The finite state automaton whose existence was proved in Problem 707 would not perform well on the cross-validation set.

Problem 714. Essay. Would hybridization help more with the grid-robot tasks in Chapters 10 and 12 or with playing Iterated Prisoner's Dilemma?

Problem 715. Essay. In Chapter 10, several representations are used for Tartarus controllers: strings, parse trees, and GP automata. Rank them by the relative benefit you think they would get from hybridization.

Problem 716. Essay. One of the more controversial ideas in evolutionary computation is whether there are building blocks that can be brought together by crossover. The reason for the controversy is mostly failure to think on the part of various vociferous proponents and opponents of the idea. The "truth" is that some problems have neat easy-to-assemble building blocks, and others don't. Your topic: can the degree to which hybridization improves performance be used as an objective probe for the presence of building blocks?

15.3 DNA Bar Codes

13 Conway's lexicode algorithm.
14 Greedy fitness.
15 Greedy closure evolutionary algorithms.
16 Alternative crossover.
17 Exploring seed size.
18 Exploring population size.

Fig. 15.8. The topics and dependencies of the experiments in this section.

Our goal in this section is to find an algorithm for creating error-correcting codes for DNA libraries. These codes can be used to identify the source that contributed that DNA as part of a sequencing project. We will take some long detours and, along the way, invent a new type of evolutionary algorithm.

The DNA bar codes that we study in this text are short sequences of DNA used to mark genetic constructs so that they can be identified later. They should not be confused with another sort of naturally occurring DNA bar code that is used to identify animals. All animals tested so far have the gene *cytochrome oxidase C*. So far this gene is different, at the DNA sequence level, in every animal that has been tested. This means that this gene can be used to "bar code" animals. That is, it can be used as a unique identifier like the VIN number on an automobile or serial number on a DVD player. This quick identification of an animal with a small kit, usable by biologists in the field or customs agents, is both useful and convenient. It is also an example of convergent terminology. Both the DNA error correcting tags developed in this section and the unique genetic sequences found in animals are well described by the name "DNA bar code." Context should suffice to separate these two meanings of the term.

Greedy algorithms are familiar to people who study programming or discrete math. (We defined them in Chapter 7, on page 185.) A few, like the algorithms for finding a minimal-weight spanning tree, can be proven to yield optimal results. Other problems, like graph coloring and the Traveling Salesman problem, admit a plethora of greedy algorithms, all of which yield suboptimal results. While it would seem that the control of greedy algorithms is a natural target for evolutionary computation, relatively few methods have been devised. There are several possible approaches. The approach explored in this section seeks to deflect the behavior of a greedy algorithm by giving it a small hint. The hint is the target of our evolutionary computation, and we call the technique used to evolve good hints a greedy closure evolutionary algorithm.

Definition 15.12 *A* **greedy closure** *evolutionary algorithm is an evolutionary algorithm that uses a representation consisting of partial structures called* seeds. *The seeds are completed (closed) with a greedy algorithm. The quality of the complete structure, as finished by the greedy algorithm, is the fitness of the seed.*

The structures created from the seeds during evaluation of the fitness function will be said to have *grown* from those seeds. In order to understand the bioinformatic application in this section, DNA bar codes, we will need both a small amount of additional molecular biology and some basic theory of error-correcting codes. We begin with the error-correcting codes.

Error-Correcting Codes

An *error-correcting code* is a collection of strings to be sent over a possibly noisy communications channel. While any collection of strings is technically a code, the science of error-correcting codes seeks to create codes that permit us to correct some of the errors that occur during transmission. Thus, a complete error correction system contains not only the code but a decoding algorithm. Let's look at an example.

Example 34. Imagine a pair of neighbors one of whom is selling a car and the other of whom is contemplating purchasing the car. The neighbors live across a ravine from one another and must cross an arroyo to reach one another. The person selling the car has told the buyer that she must decide whether the price is acceptable by 5:00 p.m. Otherwise, the car goes to another buyer, who is offering a higher price but is not a friend and neighbor. At 3:50 p.m., a huge storm blows up and wipes out the bridge and the phone lines (and the cell tower for you high-tech types). The potential buyer must get a yes or a no to the seller. The neighbors walk out into the backyards of their houses and try to talk over the sound of the flood waters. The seller realizes that it is almost impossible to hear and yells something three times. What can happen?

Well, "yes" and "no" don't sound that similar, but with raging flood waters, there is a chance of mishearing what was said. Let's assume that there is a probability α of mishearing the result. Then, we get a simple binomial distribution (see Appendix B) which tabulates this way:

Answers misheard	Probability	$\alpha = 0.1$	$\alpha = 0.2$
0	$(1-\alpha)^3$	0.729	0.512
1	$3\alpha(1-\alpha)^2$	0.243	0.384
2	$3\alpha^2(1-\alpha)$	0.027	0.096
3	α^3	0.001	0.008

A code requires a decoding algorithm. In this case, we will take a majority vote on the answers heard. What does this do to the probability of error? Well, for $\alpha = 0.1$, the chance of error drops from 0.1 (with only one yell) to 0.028 for majority vote over three yells. This is about a 3.5-fold decrease in the chance of error. When $\alpha = 0.2$, the improvement is from 0.2 to 0.104, about a 1.9-fold improvement. Let's plot this "fold improvement:"

$$FoldImprovement(\alpha) = \frac{\alpha}{\alpha^3 + 3\alpha^2(1-\alpha)} = \frac{1}{3\alpha^2 - 2\alpha^3}.$$

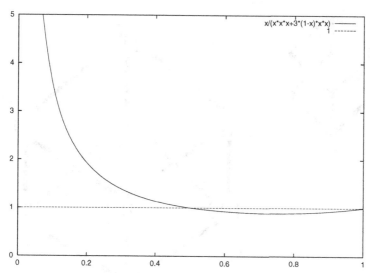

When α is small, fold improvement in the chance of understanding correctly with three yells is huge, with a vertical asymptote at zero. The technique ceases to help at $\alpha = 0.5$ (as one would expect). The behavior for $\alpha > 0.5$ is weird, but no one would use a communications channel with more that a 50% chance of miscommunication.

The code used in the example is called the *odd-length repetition code of length 3*. When working with error-correcting codes, the usual thing is to send bits; flipping a bit constitutes an error. If we repeat each bit an odd number of times, then the received bits can be decoded with a simple majority vote. This means that any communications channel that has the chance of flipping a bit $\alpha < 0.5$ can be used with any desired degree of accuracy. The more times you repeat the bit, the more likely you are to decode the bit correctly. What is the price? Repeating the bit uses up a lot of bandwidth.

A repetition code of length $2n + 1$ can decode n errors, but it is not very efficient. A code is a collection of strings, or *code words*. The code words of the length-3 repetition code are $\{000, 111\}$. Any code has a set of code words, and they are the words that are sent down the communications channel. The *received words* are the ones we try to correct. If we receive a code word, we assume that there were no errors. If we receive a word that is not a code word, then we try to find the code word closest to the received word. In this case, the notion of closest used is the *Hamming metric*, which defines the distance between two words to be the number of positions in which they disagree.

If we take the rate at which we can send bits on the channel times α, we get the *fundamental rate* of the channel. Claude Shannon proved that you can use a channel at any rate below its fundamental rate with any positive probability of error; i.e., you can get the error probability down to any level you like above zero. Shannon's theorem does not tell you how to construct the code; it only proves that the code exists. Most of the current research on error-correcting

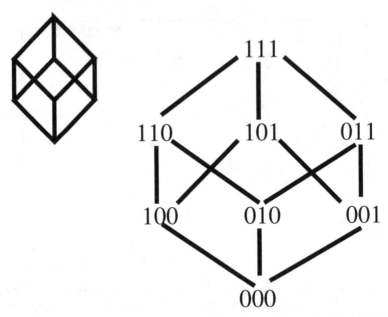

Fig. 15.9. A 3-cube formed by joining the words of length 3 over the binary alphabet with edges connecting words at Hamming distance 1.

codes amounts to finding constructions for codes that Shannon proved must exist decades ago.

At this point we change viewpoint a little to get a geometric understanding of error-correcting codes. The code words in the yelling-over-the-flood example, 000 and 111, are at opposite corners of the 3-hypercube shown in Figure 15.9. If we take the binary words of length n and join those that have Hamming distance 1, then we get an n-hypercube. This is the underlying space for standard error-correcting codes. Code words are, geometrically, vertices in the hypercube.

A *ball* is a collection of vertices at distance r or less from a distinguished vertex called the center. The number r is the radius of the sphere. *Hamming balls* are sets of vertices of a hypercube at Hamming distance r or less from a distinguished vertex called the center. If each word of a code is in a Hamming ball of radius r that is disjoint from the ball of radius r around any other code word, then any set of r errors during transmission leaves the received word closer to the transmitted word than to any other code word. This means that a code that is a set of centers of disjoint Hamming balls of radius r can decode up to r errors.

We call a Hamming ball of radius r an r-ball. A collection of centers of disjoint r-balls is called a *sphere packing* of radius r. The problem of finding good error-correcting codes is identical to that of packing spheres into a hypercube. A good introduction to error-correcting codes is [47]. A book that puts codes into an interesting context and continues on into interesting fundamental mathematics is [56].

This view of code words as sphere centers will be fundamental to understanding the algorithm that produces DNA bar codes. Another useful fact is left for you to prove in the Problems. We call the smallest distance between any two code words the *minimum distance* of the code. If the minimum distance between any two words in a code is $2r + 1$, then the code is a packing of radius-r spheres. We now know enough coding theory to continue on to the molecular biology portion of this section.

Edit Distance

DNA sequencers make errors. If those errors were always substitutions of one DNA base for another, we could correct them with a version of the binary error-correcting codes, upgraded to use the 4-letter DNA alphabet. Unfortunately, sequencing errors include finding bases that are not there (insertions) and losing bases that are there (deletions). These errors are called, collectively, *indels.* Our first task is to find a distance measure that can be used to count errors in the same way that the Hamming distance was used to count bit flips.

Definition 15.13 *The* **edit distance** *between two strings is the minimum number of single-character insertions, deletions, and substitutions needed to transform one string into the other.*

From this point on we will denote the Hamming distance between two strings x and y by $d_H(x, y)$ and the edit distance by $d_E(x, y)$. It is easy to compute Hamming distance, both algorithmically and by eyeball. In order to compute the edit distance, a more complex algorithm is required.

Algorithm 15.2 Edit Distance

Input: *Two L-character strings a, b*
Output: *The edit distance $d_E(a, b)$*
Details:

```
int dEdit(char a[L],char b[L]){//edit distance

int i,j,q,r,s,M[L+1][L+1];

  for(i=0;i<=L;i++){//initialize matrix
    M[i][0]=-i;
    M[0][i]=-i;
  }
  //fill in the dynamic programming matrix
  for(i=1;i<=L;i++)for(j=1;j<=L;j++){
    q=M[i-1][j-1];
```

```
    if(a[i-1]!=b[j-1])q--;
    r=M[i-1][j]-1;
    s=M[i][j-1]-1;
    if(s>q)q=s;
    if(r>q)q=r;
    M[i][j]=q;
}
```

```
return(-M[L][L]);  //the lower right corner is -(edit distance
```

```
}
```

The edit distance algorithm is a modification of a dynamic programming algorithm used to perform sequence alignment. If you are interested in the connections between sequence alignment and the computation of edit distance, read [33]. The edit and Hamming distances have a one-sided relationship. In the Problems, you will prove that Hamming distance is an upper bound on edit distance. We now do an example to show that the separation between Hamming and edit distance can be almost the length of the strings.

Example 35. Notice that

$$d_H(CACACACACA, ACACACACAC) = 10,$$

while

$$d_E(CACACACACA, ACACACACAC) = 2.$$

To see that the edit distance is two, delete the last character and insert it as the first.

Conway's Lexicode Algorithm

We will use *Conway's lexicode algorithm* as the greedy algorithm in our greedy closure evolutionary algorithm. It is a greedy algorithm that permits us to build error-correcting codes. A good discussion of its use for standard (binary Hamming) codes appears in [15].

Algorithm 15.3 Conway's Lexicode Algorithm

Input: *A minimum distance d, an alphabet A, and a word length n*
Output: *A code C with minimum distance d over A^n*
Details:

Place the list of all words of length n over A in lexicographical (alphabetical) order. Initialize an empty set C of words. Scanning the ordered collection of words, select a word and place it in C if it is at distance d or more from each word placed in C so far.

Conway's lexicode algorithm is a greedy algorithm that creates a code that is constructively of minimum distance d. As long as the space of words can be alphabetized, the algorithm produces a code, no matter what notion of distance is used. This turns out to be critical for finding error-correcting codes for the edit metric. The standard constructions for error-correcting codes relative to the Hamming metric don't seem to have versions over the edit metric. Briefly, the edit metric is far messier than the Hamming metric. Let's do an example.

Example 36. Suppose we run Conway's algorithm on the edit metric space for 5-letter DNA words. Then the resulting set of words at pairwise edit distance at least 3 is

AAAAA	AACCC	AAGGG
AATTT	ACACG	ACCAT
ACGTA	ACTGC	AGAGT
AGGAC	ATATC	ATTAG
CAACT	CAGTC	CATGA
CCCCA	CCGAG	CGCGC
CGTTG	CTAGG	CTCTT
CTTCC	GAAGC	GATCG
GCATT	GCTAA	GGCAG
GGGCT	GTGGA	TAATG
TAGCA	TCCTC	TCGGT
TGACC	TGTAT	TTCAA

DNA Bar Codes, At Last

We now have all the parts needed to create a greedy closure evolutionary algorithm to locate error-correcting codes for the edit metric over the DNA alphabet. We still lack, however, a motive for doing so. As we noted in Section 15.1, some organisms have a great deal of repeated sequences. The human genome project fragmented human DNA in several different ways, sequenced the fragments, and then fitted overlapping fragments together like a puzzle. In an organism like corn, with far more repeated sequences than humans, the step of fitting the puzzle together isn't possible. The repetitive nature of the sequences makes too many of the puzzle pieces look the same.

A related problem is that of locating the genes in an organism. A gene is a stretch of DNA that makes a protein. Most DNA is not part of a gene, rather it is "junk" DNA. Junk DNA may in fact be junk, or it may be a transposon sequence, or it may play a regulatory role. In any case, most applications of genomics need to know where the genes are. Genes can be located by sequencing their mRNA transcripts. While genes may be hard for humans to spot, an organism "knows" where its genes are; it can transcribe them. An *expressed sequence tag* (EST) is exactly an mRNA transcript. A complex biochemical process can be used to intercept transcribed genes, transform the mRNA into complementary DNA (cDNA). This cDNA is then placed in constructs in *E. coli* (a kind of bacteria). A collection of *E. coli*–carrying cDNA is called a *genetic library*. Which ESTs are present in a given bacteria is random, and, so, an EST sequencing project is a random sampling of the transcribed genes. The bacteria can be grown, increasing the amount of the cDNA. Primer annealing sites in the constructs placed in the *E. coli* permit selective amplification of the cDNA, providing enough DNA for sequencing. So what is the problem?

Most genes are not transcribed all the time. Heat shock genes in plants require the plants to be subjected to heat stress before they are transcribed. Genes that confer resistance to a parasite are typically transcribed only when the parasite is present. Genes used in development of a young organism cease being transcribed in the adult. There are thousands of genes in a given organism that are transcribed only in some weird circumstance.

In preparing a genetic library, samples are taken from as many organismal states as possible. An EST sequencing project in corn, for example, will use libraries prepared from different tissues, developmental stages, and different stress states (such as drought or disease). For economic reasons, these libraries are pooled before sequencing. A *DNA bar code* is a short sequence of DNA incorporated into the genetic construct placed in the *E. coli*. This bar code is used much the way bar codes are used in grocery stores: to identify the product. Each tissue, developmental stage, and stress type is assigned its own bar code. When a pooled library is sequenced, the bar codes allow the researchers to figure out which states stimulate which genes. If the bar codes happen to be drawn from an edit metric error-correcting code, then sequencing errors that hit the bar code might not prevent identification of the bar code. With this motivation, let's move on to the algorithm for finding sets of bar codes.

The primary attribute of a code, after its length and minimum distance, is its size. A large code is one that packs more spheres into the same string space. All codes found by the lexicode algorithm (Algorithm 15.3) have the property that they cannot accept any more words. However, they might not be as large as possible. Our evolutionary algorithm searches for larger codes within a fixed word length and minimum distance.

Experiment 15.13 *Implement Conway's lexicode algorithm for the edit metric over the DNA alphabet. Run the algorithm for the following parameter sets: length 5, distance 3; length 6, distance 3; length 8 distance 5. Verify both the sizes (from Table 15.1) and the membership in the (5, 3) case (from Example*

Code Sizes	Minimum Distance						
Length	3	4	5	6	7	8	9
3	4	-	-	-	-	-	-
4	12	4	-	-	-	-	-
5	36	8	4	-	-	-	-
6	96	20	4	4	-	-	-
7	311	57	14	4	4	-	-
8	1025	164	34	12	4	4	-
9	3451	481	90	25	10	4	4
10	*	1463	242	57	17	9	4
11	*	*	668	133	38	13	4

* denotes big.
- denotes empty.

Table 15.1. Size of DNA edit-metric lexicodes found with the unmodified lexicode algorithm.

36). Record the running time of the algorithm in all three cases. Now modify the algorithm to first check the Hamming distance. Since Hamming distance exceeds edit distance, if a word is too close to a word already in the code in the Hamming sense, then it is too close in the edit sense. This can be done in two ways: either (i) scan for Hamming rejection against all words in the code first, then scan for edit rejection, or (ii) check Hamming and then edit rejection of a potential new word against each word in the code. Try both possible modifications and report the impact on runtime.

Our evolutionary algorithm will search for a length-n minimum-distance d code. The structure we will evolve (our seed) is a set of 3 words at mutual distance d. Instead of starting Conway's algorithm with an empty code C, we will use a seed as the starting point. Let us now define our fitness function.

Definition 15.14 *The* **greedy closure fitness with Conway's algorithm**, *or* **greedy fitness** *for short, is computed as follows. Initialize the code in Conway's algorithm with a set S of words already at mutual distance d. Run the algorithm. The fitness of S is the size of the resulting code.*

A fact we have not yet established is that the size of codes produced by Conway's algorithm can vary when different seeds are used. A simple sampling experiment can settle this question.

Experiment 15.14 *Using the fastest version of the lexicode algorithm found in Experiment 15.13, implement the greedy fitness function. Evaluate this function on 20,000 sets of three words of length 6 with a minimum distance of 3 generated at random over the DNA alphabet. To get such sets of words: generate a first word; generate a second repeatedly until it is edit distance 3* *or more from the first word; generate a third word repeatedly until its edit distance from the first and second words is at least 3. Plot a histogram showing the number of codes of each size found. Compare your results with Figure 15.10.*

The result of the lexicode algorithm without a seed, 96, is slightly better than the mode code size of 95 for length-6 distance-3 codes in our version of the sampling experiment. The best, 103, contained just over 7% more words. Since longer bar codes are expensive in terms of biochemical success in creating libraries, squeezing in a few more bar codes at a given length is worth the trouble. From a mathematical perspective, getting some idea as to how large the codes can be is itself interesting. In any case, we see that using seeds does change the behavior of Conway's algorithm, and so using seeds can "control" the algorithm. But how?

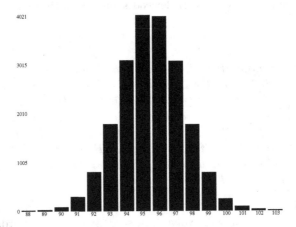

Fig. 15.10. A histogram showing the distribution of sizes of length 6, distance 3 edit-metric codes on the DNA alphabet located in 20,000 random samples of 3-word seeds. (The largest code located by sampling has 103 code words.)

A code with minimum distance d is made of words that are at least distance d apart in the string space from which the code is drawn. When we select a word to be in the code, we exclude all words within distance less than d of the selected word. By selecting a few words at the beginning, we exclude a large number of other words from consideration. This means that the control that using seeds has over the behavior of the lexicode algorithm is pretty substantial, but also quite unpredictable. A word chosen to be in the seed excludes a word the algorithm would otherwise have chosen. This in turn causes other words to be chosen, and a domino effect cascades through the code. Not only does the choice of seed change the size of the code, but it also changes the membership of the code far more than one might suppose given the size of the seed.

Since understanding the impact of seed choice on code size is difficult, choosing seeds is a sensible task for an evolutionary algorithm. An evolutionary algorithm does not require understanding to function. We have a representation, the 3-word seed, and we have a fitness function, the greedy fitness of seeds. We still need variation operators.

Definition 15.15 *For two seeds,* **uniform exclusive crossover** *is performed as follows. If two seeds have words in common, then we leave one copy in each seed. The words not in common are pooled and then randomly assigned during crossover. Uniform exclusive crossover is similar to uniform crossover for string genes but (i) it does not have positions the way a string does and (ii) it does not permit duplication of words by crossover.*

Definition 15.16 *We define* **seed point mutation** *to consist in changing one character in one uniformly selected word within a seed to a new character selected uniformly at random.*

Definition 15.17 *We define* **seed word mutation** *to consist in changing one word in a seed to a new word selected uniformly at random.*

A seed is a collection of words, so far three words, that obey the minimum-distance rule for the code the size of which we are trying to maximize. All three of the variation operators defined above have the potential to create seeds that violate this minimum-distance rule. To fix this, we extend the fitness function to award a fitness of zero to any seed that violates the minimum-distance criterion. We are ready to construct the first evolutionary algorithm.

Experiment 15.15 *Write or obtain code for the following steady-state evolutionary algorithm. Use size-7 tournament selection. Operate on a population of 200 seeds containing three words each of length $n = 6$ and minimum distance $d = 3$. Generate the initial population at random with code words at least distance 3 from each other. Use the greedy fitness function. Use uniform exclusive crossover 50% of the time and no crossover in the remainder of the mating events. Optionally, use seed point mutation or seed word mutation. Perform 100 runs using both mutation operators on each new seed and also 100 runs using one or the other mutation operator with equal probability.*

Save the maximum and population average fitness of those population members that do not have fitness zero. Also, save the number of zero-fitness seeds. Give histograms of the best final fitness for both sets of runs using different mixes of mutation operators. Does the appearance of a new best fitness have a subsequent impact on average fitness or the number of zero-fitness individuals? Which type of mutation turned in the best performance?

The above is our first implementation of a greedy closure evolutionary algorithm. In the Problems we explore other possible targets for this sort of algorithm. As a tool for locating bar codes, it avoids the problem of finding an encoding that stores an entire code. Selecting roughly 100 code words from 4^6 length-6 DNA words is a daunting problem, especially since the minimum-distance constraint creates a vast degree of interdependence among the words. The greedy closure algorithm we used fails badly to make a global search of the space of codes; instead, it searches some subset of those codes with great efficiency. It is also a completely new type of evolutionary algorithm, and so the "knobs," or operational parameters, will need to be explored.

Experiment 15.16 *Repeat Experiment 15.15, using the mutation operator(s) that turned in the best performance, but modify the crossover probability and perform runs with* $0\%, 25\%, 75\%,$ *and* 100% *chances of doing crossover. What is the impact?*

Another critical parameter is seed size.

Experiment 15.17 *Repeat Experiment 15.16, using the crossover rate that turned in the best performance. Change the algorithm so that it uses seeds of size* $1, 2,$ *and* 4*. What is the impact of varying seed size?*

Let us also check the impact of population size and sharpness of selection.

Experiment 15.18 *Repeat Experiment 15.17, using the seed size that turned in the best performance. Survey all possible combinations of population sizes* $100, 200,$ *and* 400*, and tournament sizes* $4, 7,$ *and* 15*. What is the impact?*

The structure of these experiments is not a sound one. Experiments 15.15–15.18 assume that once we have found an optimum for one parameter relative to the algorithm's current settings, it remains optimal. If we knew that there was no interaction between, say, the mutation operator(s) and the tournament size, then we would not have a problem. A complete factorial study, however, would take an inordinate amount of time. You may want to do a final project that is either a sparse factorial study or fills in parts of an ongoing one.

This section barely scratches the surface of both edit-metric error-correcting codes (note that decoding is left as an exercise) and of the application of greedy closure evolutionary algorithms. Other applications are suggested in the Problems. A natural thought is to attempt to apply the setup in this chapter to standard (binary Hamming) error-correcting codes. The author has done so and failed to improve on the known best codes for a given length and minimum distance. Given that the mathematical theory is far more beautiful for standard codes, it is not surprising that a messy technique like evolutionary algorithms cannot outperform it. Nevertheless, please contact the author if you manage a breakthrough.

Problems

Problem 717. Reread Example 34. Compute a general formula for the fold improvement caused by using a length-$(2n + 1)$ repetition code in the probability of misunderstanding a single-bit message when the probability of mis-understanding each individual bit is α. Plot the function for $2n + 1 = 5$ in a manner like that in the example for $2n + 1 = 3$.

Problem 718. Prove that if a collection C of code words has the property that for any $u, v \in C$, the Hamming distance from u to v is at least $2r + 1$, then the Hamming ball of radius r around any code word in C contains no other code word in C.

Problem 719. Suppose that we have a matrix M_k whose columns are every binary word of length k, *except* the all-zero word, in counting order. The matrix M_3 is shown below.

$$M_3 = \begin{bmatrix} 0\,0\,0\,1\,1\,1\,1 \\ 0\,1\,1\,0\,0\,1\,1 \\ 1\,0\,1\,0\,1\,0\,1 \end{bmatrix}$$

Let HC_k be the set of words that are the null space of the matrix, i.e., binary vectors $\mathbf{x}$ of length $2^k - 1$ such that $M_k * \mathbf{x} = \mathbf{0}$. For example, since

$$\begin{bmatrix} 0\,0\,0\,1\,1\,1\,1 \\ 0\,1\,1\,0\,0\,1\,1 \\ 1\,0\,1\,0\,1\,0\,1 \end{bmatrix} * \begin{bmatrix} 1 \\ 0 \\ 0 \\ 1 \\ 1 \\ 0 \\ 0 \end{bmatrix} = \begin{bmatrix} 0\,0\,0 \end{bmatrix},$$

we see that $\mathbf{x} = (1, 0, 0, 1, 1, 0, 0)$ is in HC_3. Prove that HC_k is a code with minimum Hamming distance 3 between any two words.

Problem 720. Enumerate (list the members of) HC_3, defined in Problem 719.

Problem 721. Let $d_H(x, y)$ be the Hamming distance between two strings x and y, and let $d_E(x, y)$ be the edit distance. Prove that

$$d_E(x, y) < d_H(x, y).$$

Problem 722. Compute the edit distance and show a minimal sequence of edits for all pairs of the following words: {**ACGTA, GCTAA, AAGGG**}.

Problem 723. Review Section 7.4. Outline a greedy closure algorithm for finding Costas arrays.

Problem 724. Outline a greedy closure algorithm for the Traveling Salesman problem. Is the Traveling Salesman problem a natural target or a poor one?

Problem 725. Prove that a code found with Conway's algorithm, using a seed or not, is maximal in the sense that no larger code with the same length and minimum distance contains it.

Problem 726. Using the edit distance lexicode algorithm, give a decoding algorithm for edit metric lexicodes. Assume that you are using DNA bar codes of length n and minimum distance $d = 2r + 1$. Given a received (sequenced) word, you should return either a member of the code C or an error message (if the received word is not closer to one code word than another).

Problem 727. Essay. A direct encoding of an error-correcting code would require a gene that picks out the members of the code from the space of words. Is such a direct encoding practical for the type of code located in Experiment 15.15?

Problem 728. Essay. Is Conway's algorithm specific to the Hamming or edit metric or can it be used with any notion of distance? With what kinds of notions of distance can it be used?

15.4 Visualizing DNA

In this section, we will make a substantial departure from applied bioinformatics and enter the realm of speculative bioinformatics. In the course of this, we will create a data-driven, evolvable fractal visualization tool. The starting point is a fairly well-known type of fractal algorithm called a chaos game.

Chaos Game Fractals

A *chaos game* is characterized as the process of generating a fractal by accumulating the positions of a moving point. This moving point is repeatedly displaced toward one of a fixed set of points, e.g., the vertices of an equilateral triangle. Figure 15.12 shows the Sierpiński triangle. It is generated by a chaos game in which a moving point is displaced, in each iteration, halfway from its present position toward a randomly selected vertex of a triangle. Figure 15.12 is plotted over 100,000 iterations of this process.

Algorithm 15.4 Simple chaos game

Input: *A set of fixed points in the real plane*
Output: *A set of points in the real plane*
Details:

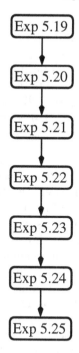

Fig. 15.11. The topics and dependencies of the experiments in this section.

A point, called the moving point, *is initialized to the position of one of the fixed points. An* updating *of the moving point's position is performed by*

Fig. 15.12. The Sierpiński triangle generated by a 3-cornered chaos game.

choosing one of the fixed points uniformly at random and then averaging the current position of the moving point with that of the fixed point. The moving point is moved halfway to the chosen fixed point.

A series of updatings is made to burn in *the moving point. This process permits the moving point to enter, up to the resolution of plotting, the fractal that is characteristic of the chaos game. Typically, a few hundred updatings are more than enough to burn in the moving point. During the burn-in, the chaos game just updates the moving point's position. In the post-burn-in period, the moving point is plotted.*

The character of the fractal resulting from a chaos game is controlled by the number of fixed points being used and the order in which those points are chosen to specify the direction of motion for the moving point. This latter point is key. The Sierpiński triangle is generated by using the vertices of a regular triangle, with the next point chosen uniformly at random. If instead, we choose points with some degree of nonuniformity, then the resulting fractal is a subset of the fractal obtained by driving with uniform random data.

If the 4 points at the vertices of a square are used as the fixed points, the chaos game produces a dense subset of the square. If the fractal is visualized, the square simply fills in. This represents an opportunity for the visualization of DNA or RNA data in a manner discussed in [35] and [18]. If we assign each corner of the square to one of the 4 DNA bases, then deviations from uniformity of the nucleic acid sequence will appear as gaps in the square filled in by the fractal process.

Figure 15.13, for example, demonstrates the results if we drive a chaos game on the square with sequence data from an HIV virus. As each base, **C, G, A, T**, is handed to the fractal process, the moving point is moved halfway from its current position to the corner of the square associated with the base. The averaging or halfway moves subdivide the square by sequence data, as shown in the left part of Figure 15.13. The resulting gaps indicate subsequences that do *not* appear in the HIV genome. In this case, many of the gaps can be attributed to the HIV virus's lack of methylization sites.

Interpretation of chaos game fractals such as those shown in Figure 15.13 requires a good deal of biological knowledge. The lack of methylization sites is obvious in Figure 15.13 only if you know the sequences for methylization sites and can picture where they are on the chaos game's square. This problem becomes more acute when an attempt is made to use these techniques to derive visual representations of protein sequences. Proteins are built out of 20 building blocks, called amino acids, rather than the 4 bases of DNA or RNA. In [52], both placing the 20 amino acids in a circle and extending the fractal into a third dimension are attempted. As one would expect, the interpretation difficulties grow.

A number of biological issues can be used to inform the choices made in designing a biological representation for a fractal. The map from nucleic acid to protein reads DNA in triplets, producing 64 *codons*. These codons are in

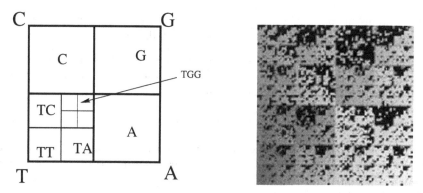

Fig. 15.13. The diagram in the left half of this figure shows how sequence data subdivide the square in a 4-cornered chaos game. Such a chaos game, driven by HIV sequence data, is displayed in the right half of the figure.

turn taken by a many-to-one map (the *genetic code*) onto the 20 amino acids as well as a *stop codon*. This stop codon indicates the end of transcription of a given sequence of DNA. The many-to-one map that forms the genetic code is the same in almost all organisms. The choice of which of several possible codons to use to specify a given amino acid, however, has a substantially organism-specific character. These biological considerations will factor into the design of evolvable fractals. Our next step is to generalize the chaos game.

Iterated Function Systems

Chaos games are a particular type of iterated function system [10]. In an *iterated function system* (IFS), a number of maps from the Cartesian plane to itself are chosen. These maps are then called in a random order, according to some distribution, to move a point in a manner similar to the chaos game. A *dynamical system* is a process that moves a point in space. The iterated function systems we use in this section are a stochastic (as opposed to deterministic) dynamical system; they move points in a random fashion constrained by the particular choice of contraction maps. A dynamical system breaks the space its points are in into two categories. The first category are points that the dynamical system will never reach again even if it hits them once. The second category are those that the dynamical system visits over and over with some positive probability. The second category of points is called the *attractor* (or attractors) of the dynamical system. Iterated function systems have a single attractor; some dynamical systems have several. Interested students should check out a text on dynamical systems, many of which are available in a typical college library. The attractor of an iterated function system is a fractal. [10] describes the properties of the fractal attractors of iterated function systems in some detail. In order to get a well-behaved fractal, the maps in the iterated function system must have the following property.

Definition 15.18 *Let $d(p, q)$ be the distance in the Cartesian plane between points p and q. A function $f : \mathbb{R}^2 \to \mathbb{R}^2$ from the plane to itself is called a* **contraction map** *if for any pair of points p, q,*

$$d(p, q) > d(f(p), f(q)).$$

An iterated function system made entirely of contraction maps has a bounded fractal attractor. A rich class of maps that are guaranteed to be contraction maps is that of similitudes.

Definition 15.19 *A* **similitude** *is a map that performs a rigid rotation of the plane, displaces the plane by a fixed amount, and then contracts the plane toward the origin by a fixed scaling factor. The derivation of a new point (x_{new}, y_{new}) from an old point (x, y) with a similitude that uses rotation t, displacement $(\Delta x, \Delta y)$, and scaling factor $0 < s < 1$ is given by*

$$x_{new} = s \cdot (x \cdot \cos(t) - y \cdot \sin(t) + \Delta x), \tag{15.1}$$
$$y_{new} = s \cdot (x \cdot \sin(t) + y \cdot \cos(t) + \Delta y). \tag{15.2}$$

To see that a similitude must always reduce the distance between two points, note that rotation and displacement are isometries (they do not change distances between points). This means that any change is due to the scaling factor, which necessarily causes a reduction in the distance between pairs of points. Let's look at a couple of iterated function system fractals.

Example 37. An iterated function system is a collection of contraction maps together with a distribution with which those maps will be applied to the moving point. In Figure 15.14 are a pair of fractal attractors for iterated function systems built with the 8 similitudes shown in Table 15.2. These similitudes are called uniformly at random.

The similitudes in this example were generated at random. The rotation factors are in the range $0 \le \theta \le 2\pi$ radians. The displacements are selected uniformly at random to move the origin to a point with $-1 < x, y < 1$. The scaling factor is chosen uniformly at random in the range $0 < s < 1$.

15.5 Evolvable Fractals

Our goal is to use a data-driven fractal, generalizing the 4-cornered chaos game, to provide a visual representation of sequence data. It would be nice if this fractal representation could work smoothly with DNA, protein, and codon data. These sequences, while derived from one another, have varying amounts of information and are important in different parts of cells operation. The

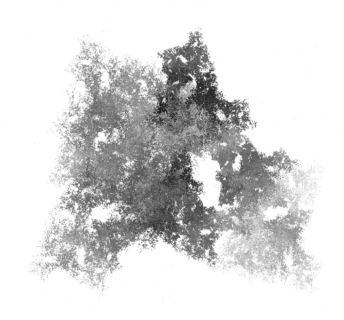

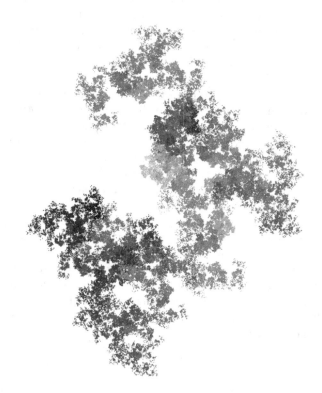

Fig. 15.14. The fractal attractors for the iterated function systems given in Example 37.

First IFS				Second IFS			
Map	Rotation	Displacement	Scaling	Map	Rotation	Displacement	Scaling
M1	4.747	(0.430, 0.814)	0.454	M1	2.898	(-0.960, 0.253)	0.135
M2	1.755	(−0.828, 0.134)	0.526	M2	3.621	(0.155, 0.425)	0.532
M3	3.623	(0.156, 0.656)	0.313	M3	5.072	(0.348,−0.129)	0.288
M4	0.207	(−0.362, 0.716)	0.428	M4	3.428	(−0.411,−0.613)	0.181
M5	2.417	(−0.783, 0.132)	0.263	M5	4.962	(−0.569, 0.203)	0.126
M6	1.742	(−0.620, 0.710)	0.668	M6	4.858	(−0.388,−0.651)	0.489
M7	0.757	(0.444, 0.984)	0.023	M7	5.953	(−0.362, 0.758)	0.517
M8	4.110	(−0.633,−0.484)	0.394	M8	1.700	(−0.696, 0.876)	0.429

Table 15.2. Similitudes, rotations, displacements, and scalings used to generate the iterated function systems in Example 37.

raw DNA data contains the most information and the least interpretation. The segregation of the DNA data into codon triplets has more interpretation (and requires us to work on DNA that is transcribed as opposed to other DNA). The choice of DNA triplet used to code for a given amino acid can be exploited, for example, to vary the thermal stability of the DNA (more **G** and **C** bases yield a higher melting temperature), and so the codon data contains information that disappears when the codons are translated into amino acids. The amino acid sequence contains information focused on the enzymatic mission of the protein. This sequence specifies the protein's fold and function without the codon usage information muddying the waters.

Given all this, we design an iterated function system fractal that evolves the contraction maps used in the system as well as the choice of which contraction map is triggered by what biological feature. For our first series of experiments, we will operate on DNA codon data, rich in information but with some interpretation. Our test problem is reading frame detection, a standard and much-studied property of DNA. Reading frame refers to the three possible choices of groupings of a sequence of DNA into triplets for translation into amino acids. Figure 15.15 shows the translation into the three possible reading frames of a snippet of DNA. Only the first reading frame contains the **ATG** codon for the amino acid Methionine (which also serves as the "start" codon for translation), and the amino acid **TAG** (one of the three possible "stop" codons).

The correct reading frame for a piece of DNA, if it codes for a protein, is typically the frame that is free of stop codons. Empirical verification shows that frame-shifted transcribed DNA is quite likely to contain stop codons, which is also likely on probabilistic grounds for random models of DNA. We remind you that random models of DNA must be used with caution; biological DNA is produced by a process containing a selection filter, and therefore contains substantial nonrandom structure. Figure 15.13 serves as an example of such nonrandom structure.

```
ATG GGC GGT GAC AAC TAG
Met Gly Gly Asp Asn Stp

A TGG GCG GTG ACA ACT AG
. Trp Ala Val Thr Ala ..

AU GGG CGG TGA CAA CTA G
.. Gly Arg Gly Gln Val .
```

Fig. 15.15. A piece of DNA translated in all 3 possible reading frames. (Amino acids are given by their 3-letter codes, which may be found in [50].)

A Fractal Representation

The data structure we use to hold the evolvable fractal has two parts: a list of similitudes and an index of DNA triplets into that list of similitudes. This permits smooth use of the fractal on DNA, DNA triplets, or amino acids by simply modifying the way the DNA or amino acids are interpreted by the indexing function. A diagram of the data structure is given in Figure 15.16. Each similitude is defined by 4 real parameters in the manner described in Definition 15.19. The index list is simply a sequence of 64 integers that specify, for each of the 64 possible DNA codon triplets, which similitude to apply when that triplet is encountered.

Interpretation	Contains
First similitude	t_1 $(\Delta x_1, \Delta y_1)$ s_1
Second similitude	t_2 $(\Delta x_2, \Delta y_2)$ s_2
	...
Last similitude	t_n $(\Delta x_n, \Delta y_n)$ s_n
Index	$i_1, i_2, \ldots, i_{64}$

Fig. 15.16. The data structure that serves as the gene for an evolvable DNA-driven fractal. (In this work, we use $n = 8$ similitudes, and so $0 \leq i_j \leq 7$.)

In order to derive a fractal from DNA, the DNA is segregated into triplets with a specific reading frame. These triplets are then used, via the index portion of the gene, to choose a similitude to apply to the moving point. The IFS is driven by incoming DNA triplets.

This representation permits evolution to choose both the shape of the maximal fractal (the one we would see if we drove the process with data chosen uniformly at random) and which DNA codon triplets are associated with the use of each similitude. Any contraction map has a unique fixed point. The fixed points of the 8 similitudes we use play the same role that the 4 corners of the square did in the chaos game shown in Figure 15.13.

We need variation operators. The crossover operator performs a one-point crossover on the list of 8 similitudes, treating the similitudes as indivisible objects, and also performs two-point crossover on the list of indices. We will used two mutation operators. The first, termed a *similitude mutation*, modifies a similitude selected uniformly at random. It picks one of the 4 parameters that define the similitude, uniformly at random, and adds a number selected uniformly in the interval $[-0.1, 0.1]$ to that parameter. The scaling parameter is kept in the range $[0, 1]$ by reflecting the value at the boundaries so that numbers $s > 1$ are replaced by $2 - s$ and values $s < 0$ are replaced by $-s$. The other parameters are permitted to move outside of their initial range. The second mutation operator, called an *index mutation*, acts on the index list by picking the index of a uniformly chosen DNA triple and replacing it with a new index selected uniformly at random.

Aside from a fitness function, we now have all the machinery required to evolve fractals. For our first experiment, we will attempt to tell whether DNA is in the correct reading frame. The website associated with this text has a file of in-frame and out-of-frame DNA available. We will drive the IFS alternately with these two sorts of data and attempt to get the IFS to plot points in different parts of the plane when the IFS is being driven by distinct types of data.

Definition 15.20 *The* **separation fitness** *of a moving-point process P, e.g., an IFS, being driven by two or more types of data is defined as follows. Compute the mean position (x_i, y_i) when the IFS is being driven by data type i. The fitness is*

$$SF(P) = \sum_{i \neq j} \sqrt{(x_i - x_j)^2 + (y_i - y_j)^2},$$

.

Experiment 15.19 *Write or obtain code for evolving iterated function systems with the representation given in Figure 15.16. Use the crossover operator. The evolutionary algorithm should be generational, operating on a population of 200 IFS structures with size-8 single tournament selection. In each tournament, perform a similitude mutation on one of the new structures and an index mutation on the other.*

To perform fitness evaluation, initialize the moving point to $(0, 0)$ and then drive the IFS with 500 bases of in-frame data and 500 bases of out-of-frame data before collecting any fitness information; this is a burn-in as was used in the chaos game. After burn-in, compute the mean position of the moving point for each type of data while alternating between the two types of data using 100 to 400 triplets of each data type. Select the length, 100 to 400, uniformly at random. The mean position data for each of the two data types may be used to compute the separation fitness.

*Perform 30 runs of 500 generations. Report the fitness tracks and esti-
mate the average number of generations needed to reach the approximate final
fitness. If you have skill with graphics, also plot the fractals for the most fit
IFSs using different colors for points plotted while the IFS is being driven by
different data types. Report the most fit IFS genes.*

Experiment 15.19 should contain some examples that show that there is
a very cheap way for the system to generate additional fitness. If we take an
IFS of the type used in Experiment 15.19 and simply enlarge the whole thing,
the separation fitness scales with the picture. This suggests that we may well
want to compensate for scaling.

Definition 15.21 *The **diameter** of a moving-point process is the maximum
distance between any two plotted points generated by the moving-point process.
For an IFS, the diameter should be computed only after the IFS has been
burned in.*

Definition 15.22 *The **normalized separation fitness** of a moving-point
process P, e.g., an IFS, being driven by two or more types of data is the
separation fitness divided by the diameter of the moving-point process.*

Experiment 15.20 *Repeat Experiment 15.19 using the normalized separation fitness instead of the separation fitness. Also, reduce the number of generations to 120% of the average solution time you estimated in Experiment
15.19. Comment on the qualitative differences of the resulting fractals.*

There is a second potential problem with our current experimental setup.
This problem is not a gratuitous source of fitness as was the scaling issue. This
issue is an aesthetic one. A very small scaling factor moves the moving point
quite rapidly. Using tiny scaling factors in well-separated regions results in
high fitness and good separation of the data, but not in a good visualization.

Experiment 15.21 *Repeat Experiment 15.20, but modify both initialization
and similitude mutation so that scaling factors are never smaller than a. Perform runs for a = 0.5 and a = 0.8. What impact does this modification have
on the fitness tracks and on the pictures generated by the most fit IFS?*

Chaos Automata

The IFS representation we have developed has a problem that it shares with
the chaos game: it is forgetful. The influence of a given DNA base on the
position of the moving point is diminished with each successive execution
of a similitude. To address this problem we introduce a new representation
called *chaos automata*. Chaos automata differ from standard iterated function
systems in that they retain internal state information. This gives them the
ability to associate events that are not nearby in the sequence data.

The internal memory also grants the fractals generated a partial exemption from self-similarity. The IFS fractals generated thus far have parts that look like other parts. When driven by multiple types of input data, a chaos automaton can "remember" what type of data it is processing and use distinct shapes for distinct data. Two more-or-less similar sequences separated by a unique marker could, for example, produce very different chaos-automata-based fractals by having the finite state transitions recognize the marker and then use different contraction maps on the remaining data.

```
Starting State:6

      Transitions:                  Similitudes:
   If  C  G  A  T   Rotation    Displacement    Contraction
   -----------------------------------------------------------
   0)  3  2  3  3  : R:0.678 D:( 1.318, 0.606) S:0.905
   1)  5  3  5  3  : R:1.999 D:( 0.972, 0.613) S:0.565
   2)  7  7  2  3  : R:0.521 D:( 1.164, 0.887) S:0.620
   3)  3  0  0  3  : R:5.996 D:( 0.869, 0.917) S:0.805
   4)  0  0  0  5  : R:1.233 D:( 0.780,-0.431) S:0.610
   5)  5  5  5  7  : R:1.007 D:(-0.213, 0.706) S:0.623
   6)  3  7  3  4  : R:3.509 D:( 0.787, 0.767) S:0.573
   7)  1  5  5  2  : R:0.317 D:( 0.591, 0.991) S:0.570
```

Fig. 15.17. A chaos automaton evolved to visually separate two classes of DNA. (The automaton starts in state 6 and makes state transitions depending on inputs from the alphabet {**C, G, A, T**}. As the automaton enters a given state, it applies the similitude defined by a rotation (R), displacement (D), and shrinkage (S).)

Chaos automata are modified finite state automata. An example of a chaos automaton, evolved to be driven with DNA data, is shown in Figure 15.17. Each state of the chaos automaton has an associated similitude, applied when the automaton enters that state. The finite state automaton supplies the memory, and the similitudes supply the contraction maps. A chaos automaton is an IFS with memory. Note that we have made the, somewhat arbitrary, choice of associating similitudes with states rather than transitions. We thus are using "Moore" automata rather than "Mealy" automata. Algorithm 15.5 specifies how to use a chaos automaton as a moving-point process.

Algorithm 15.5 Using a chaos automaton

Input: *A chaos automaton*
Output: *A sequence of points in the plane*
Details:

Set state to initial state.
Set moving point (x,y) to (0,0).

Repeat
  Apply the similitude on the current state to (x,y).
  Process point (x,y).
  Update the state according to input with the transition rule.
Until (out of input).

In order to use an evolutionary algorithm to evolve chaos automata, we need variation operators. We will use a two-point crossover operator operating on the string of states (each state treated as an indivisible object with the integer identifying the initial state attached to the first state). There are three kinds of things that could be changed with a mutation operator. Primitive mutation operators are defined for each of these things and then used in turn to define a master mutation operator that calls the primitive mutations with a fixed probability schedule. The first primitive mutation acts on the initial state, picking a new initial state uniformly at random. The second primitive mutation acts on transitions to a next state. It selects one such transition uniformly at random and then selects a new next state uniformly at random. The third primitive mutation applies a similitude mutation to a similitude selected uniformly at random. The master mutation mutates the initial state 10% of the time, a transition 50% of the time, and a similitude 40% of the time. For our first experiment, we will test our ability to evolve chaos automata to solve the reading frame problem.

Experiment 15.22 *Modify the software from Experiment 15.21, including the lower bound on the scaling factor for similitudes, to use chaos automata. What impact did this have on fitness?*

Let's now test chaos automata on a new problem. In a biological gene, there are regions called *exons* that contain the triplets that code for amino acids. There are also regions between the exons, called *introns*, that are spliced out of the mRNA before it is translated into protein by ribosomes. We will use chaos automata to attempt to visually distinguish intron and exon data.

Experiment 15.23 *Repeat Experiment 15.22 but replace the in-frame and out-of-frame DNA with intron and exon sequences downloaded from the website for this text. Report the fitness tracks. Do the chaos automata manage to separate the two classes of data visually? Report the diameter of the best fractal found in each run as well as the fitness data.*

Now, let's tinker with the fitness function. We begin by developing some terminology. To efficiently describe new fitness functions, we employ the following device: the moving point, used to generate fractals from chaos automata driven by data, is referred to as if its coordinates were a pair of random variables. Thus (X, Y) is an ordered pair of random variables that gives the position of the moving point of the chaos game. In working to separate several types of data $\{d_1, d_2, \ldots, d_n\}$, the points described by (X, Y) are partitioned into $\{(X_{d_1}, Y_{d_1}), (X_{d_2}, Y_{d_2}), \ldots, (X_{d_n}, Y_{d_n})\}$, which are the positions of the moving points of a chaos automaton driven by data of types $d_1, d_2, \ldots, d_n$, respectively. For any random variable R, we use $\mu(R)$ and $\sigma^2(R)$ for the sample mean and variance of R. Using this new notation, we can rebuild the separation fitness function of a moving-point process P, with d_1 and d_2 being the in-frame and out-of-frame data:

$$SF(P) = \sqrt{(\mu(X_{d_1}) - \mu(X_{d_2}))^2 + (\mu(Y_{d_1}) - \mu(Y_{d_2}))^2} \qquad (15.3)$$

The problem of having fractals made of sparse sets of points is only partially addressed by placing a lower bound on the scaling factor within the similitudes. Our next function will encourage dispersion of the points in the fractal while continuing to reward separation by multiplying the separation by the standard deviation of the position of the moving point.

Definition 15.23 *The* **dispersed separation fitness** *for a moving-point process P is given by*

$$F_3 = \sigma(X_{d_1})\sigma(Y_{d_1})\sigma(X_{d_2})\sigma(Y_{d_2})SF(P).$$

Experiment 15.24 *Repeat Experiment 15.23 with dispersed separation fitness in place of separation fitness. In addition to the information recorded previously, track the diameter of the resulting fractals over the course of evolution. Compare this with the diameters recorded in Experiment 15.23. Also, check to see whether the fractals visually separate the data.*

If your version of Experiment 15.24 worked the way ours did, then you got some huge fractals. The dispersed separation fitness function overrewards dispersion. This too can be fixed.

Definition 15.24 *The* **bounded dispersed separation fitness** *for a moving-point process P is given by*

$$F_4 = \tan^{-1}(\sigma(X_{d_1})\sigma(Y_{d_1})\sigma(X_{d_2})\sigma(Y_{d_2}))SF(P).$$

Experiment 15.25 *Repeat Experiment 15.24 using bounded dispersed separation fitness in place of dispersed separation fitness. Did the new fitness function help the dispersion problem? As before, report whether the fractals visually separate the data.*

We have not made a study of the sensitivity of the evolution of chaos automata to variation of the algorithm parameters. This is not the result of laziness (though the length of this chapter might justify some laziness), but rather because of a lack of a standard. The meaning of the fitness values for chaos automata is quite unclear. While the fitness functions used here did manage to visually separate data during testing, higher fitness values did not (in our opinion) yield better pictures. The very fact that the metric of picture quality is "our opinion" demonstrates that we do not have a good objective fitness measure of the quality of visualizations of DNA. If you are interested in chaos automata, read [4] and [5]. You are invited to think up possible applications for chaos automata. Some are suggested in the Problems.

Problems

Problem 729. The dyadic rationals are those of the form

$$q = \sum_{i=-n}^{\infty} x_i 2^{-i}.$$

Run a chaos game on the square with corners $(0,0), (0,1), (1,1)$, and $(1,0)$. Prove that the x and y coordinates of the moving point are always dyadic rationals.

Problem 730. Is the process "move halfway from your current position to the point (x, y)" a similitude? Prove your answer by showing that it is not or by identifying the rotation, displacement, and contraction.

Problem 731. When the chaos game on a square is driven by uniform random data, it fills in the square. Suppose that instead of moving halfway toward the corners of the square, we move 40% of the way. Will the square still fill in? If not, what does the resulting fractal look like?

Problem 732. Consider the following modification of the chaos game on a square. Number the corners 0, 1, 2, 3 in the clockwise direction. Instead of letting the moving point average toward any corner picked uniformly at random, permit it to move only toward a corner other than the next one (mod 4) in the ordering. What does the resulting fractal look like?

Problem 733. Prove that chaos games are iterated function systems.

Problem 734. For the 8 similitudes associated with the first IFS in Example 37, compute the fixed point of each similitude to 4 significant figures. Plot these fixed points and compare with the corresponding fractal.

Problem 735. For the 8 similitudes associated with the second IFS in Example 37, compute the fixed point of each similitude to 4 significant figures. Plot these fixed points and compare with the corresponding fractal.

Problem 736. What variation of the chaos game on the square produced the above fractal?

Problem 737. Prove that a contraction map has a unique fixed point.

Problem 738. True or false? The composition of two contraction maps is a contraction map. Prove your answer.

Problem 739. Suppose that the HIV-driven chaos game in Figure 15.13 is of size 512×512 pixels. How many DNA bases must pass though the IFS after a given base b to completely erase the influence of b on which pixel is plotted?

Problem 740. When evolutionary algorithms are used for real function optimization, the number of independent real variables is called the dimension of the problem. What is the dimension of the representation used in Experiment 15.19?

Problem 741. When evolutionary algorithms are used for real function optimization, the number of independent real variables is called the dimension of the problem. What is the dimension of the representation used in Experiment 15.22?

Problem 742. What problems would be caused by computing the diameter of an IFS without burning it in first?

Problem 743. Assume that we are working with k different types of data and have k disjoint circles in the plane. Create a fitness function that rewards a moving-point process for being inside circle i when plotting data type i.

Problem 744. Suppose that instead of contracting toward the origin by a scaling factor s in a similitude, we had distinct scaling factors s_x and s_y that were applied to the x and y coordinates of a point. Would the resulting modified similitude still be a contraction map? Prove your answer.

Problem 745. Essay. Create a parse tree language for genetic programming that always gives a contraction map from the real line to itself.

Problem 746. Essay. Would two chaos automata that achieved similar fitness values on the same data using the bounded dispersed separation fitness produce similar pictures?

Problem 747. Essay. Suppose we had a data set consisting of spam and normal e-mail. Outline a way to create a fractal from the character data in the e-mail. Assume that you are working from the body of the e-mail, not the headers, and that the number of recipients of an e-mail has somehow been concealed.

Problem 748. Essay. When trying to understand the behavior of evolutionary algorithms, we have used the metaphor of a fitness landscape. Describe, as best you can, the fitness landscape in Experiment 15.19.

Problem 749. Essay. When trying to understand the behavior of evolutionary algorithms, we have used the metaphor of a fitness landscape. Describe, as best you can, the fitness landscape in Experiment 15.22.

Problem 750. Essay. Suppose that we have a black and white picture. Construct a fitness function that will encourage the type of fractal used in Experiment 15.19 to match the picture.

Problem 751. Essay. Define *chaos GP automata* and describe a problem for which they might be useful.

Glossary

This glossary includes terms used in the text and terms that connect the text to the broader literature. Since evolutionary computation is a new and rapidly evolving field , it is necessarily incomplete. Because good ideas are often rediscovered, you may run across different terms for some of the concepts in this glossary; in some cases arbitrary choices were made; in many other cases we were simply unaware of the other terms. Please send us any suggestions or additions you may have for future editions of the glossary.

adjacency matrix A square matrix all of whose entries are zero or one that gives the connectivity of a combinatorial graph. For an undirected graph, the matrix is symmetric; for a directed graph, asymmetry gives the direction of the edges.

Always Cooperate A strategy for playing Prisoner's Dilemma. The player always cooperates.

Always Defect A strategy for playing Prisoner's Dilemma. The player always defects.

argument An argument of a function is one of its input values. The argument of a node in a parse tree is one of the subtrees that provides input to it or the root node of such a subtree.

artificial life In the context of this book, artificial life is any algorithmic simulation of life or any algorithm that owes its inspiration to a living system.

artificial neural net A simple simulation of a system of connected neurons inspired by biological networks of neurons. Artificial neural nets are programmed by adjusting connection strength between various artificial neurons. If these connections contain closed, directed loops then the artificial neural net is recurrent, otherwise it is feed-forward.

atomic A synonym for indivisible. In an EC context we call the smallest elements of a chromosome that cannot be cut by whatever crossover operators are being used the *atomic* elements of the representation of that chromosome. Varying which elements are atomic can substantially change the behavior of a system.

attractor See basin of attraction. A basin of attraction is called an attractor.

automatically defined function A genetic programming term denoting a distinct parse tree that can be called from the "main" parse tree. A given genetic programming system may use no ADFs, one, or several. The ADF is analogous to a subroutine or procedure in standard programming and, properly used, yields the benefits of encapsulation and reuse that subroutines do.

basin of attraction A feature of a dynamical system. A dynamical system moves points in a space. A basin of attraction is a set of points that have the property that once a point being moved has entered the basin it is difficult or impossible for it to escape. In evolutionary computation, the set of points that will reach a given optima if pure hill climbing is performed are said to be the basin of attraction for that optima.

Bernoulli trial A random experiment with two possible outcomes. The standard example is flipping a coin.

bijection A function that is one-to-one and onto. Bijections preserve cardinality and hence are useful in enumeration.

binary variation operator (also called a *crossover operator*) A system for taking two parents and generating children each of which incorporates elements of each parent. Examples include one- and two-point string crossover and subtree crossover from genetic programming.

binomial coefficient An entry in Pascal's triangle; the coefficient of x^k in the binomial expansion of $(1 + x)^n$. The binomial coefficient $\binom{n}{k}$ counts the number of different ways that k objects can be chosen from n objects. See Appendix B.

binomial probability model, theorem The binomial probability model is used to understand sums of multiple Bernoulli trials. The binomial theorem is a closely related algebraic result that also gives the form of powers of binomials. Both these results are discussed in Appendix B.

biodiversity The total diversity of an ecosystem. As in real-world biological ecosystems, biodiversity or population diversity is an important quantity in evolutionary computation. Since evolutionary computation systems typically use very small, highly inbred populations tracking the biodiversity of a population is one way of telling whether it has become stuck (or has converged).

bioinformatics The application of information science to biological data. Typically, bioinformatics uses algorithmic and statistical methods to pull useful information out of large sets of biological data.

biology Life and living things or the sciences that study life and living things. Evolutionary computation has a nontrivial intersection with theoretical biology.

bloat A type of growth of structure size in an evolving population of variable-sized structures. To be bloat, the increase in structure size has to be unneeded from a perspective of fitness. Rather than contributing to fitness, bloat is caused by a secondary selection for being able to pass on a quality. The bloated material gives protection against disruption by crossover, mutation, or other variation operators. Bloat is best known to occur in genetic programming.

Boolean Two-valued, usually the logical values true and false.

burn in This is the process of running a dynamical model for some number of steps before using it. Such models are often initialized at random to an impossible state. Running the model often moves its state variables into the realm of the possible. An iterated function system is an example of a system that may require burn in.

cardinality A high-tech word for size. For finite sets, the cardinality of a set is simply the number of elements in the set.

Catalan Nnumbers The Catalan numbers count the number of binary trees with n leaves. They have the closed form

$$C_n = \frac{1}{n}\binom{2n-2}{n-1}$$

cellular encoding The process of using directions about how to build a structure as the chromosome type in a an evolving population instead of a direct specification of the parameters of the structure itself. Cellular encodings are indirect, and come in many varieties. Cellular encoding is a type of representation.

chaos automata Finite state machines with a similitude (simple contraction map) associated with each state. Chaos automata are typically data-driven and act to map data onto chaotic dynamics. Chaos automata are typically used to visualize data. The name is derived from the so-called chaos game.

chaos game A very simple discrete stochastic dynamical system used to generate a fractal, the Sierpiński triangle. Three points in the plane, usually the vertices of an equilateral triangle, are fixed. A moving point is initially placed at one of the fixed points chosen at random. The moving point is then repeatedly moved halfway to one of the fixed points chosen uniformly at random. The position of the moving point is plotted. An example of a Sierpiński triangle is shown in Figure 15.12.

chopping A variation operator used on parse trees to control their size. When a parse tree exceeds an external size limit, the chop operator is used to reduce its size. An argument of the root node of the tree is chosen at random, and the subtree rooted at that argument becomes the entire tree. Chopping is repeated until a tree does not exceed the external size limit.

chromosome The data structure used to store each member of the evolving population. The string evolver uses a string chromosome; simple genetic programming uses a parse tree chromosome.

coevolution A special type of evolution in which the fitness of an individual is not measured purely by the fitness function, but also depends on the other members of the population. The most natural place for coevolution to arise is in evolutionary algorithms in which fitness is assessed by competition between population members. Niche specialization and other techniques that take population structure (e.g., crowding) into account as well as the normal figure-of-merit from the fitness function are also coevolutionary. (In biology, coevolution is the tendency of interacting populations to adapt to one another.)

coparent In an evolutionary algorithm using crossover, the participants in a given instance of crossover are called parents. When the parent selection is not symmetric, as in graph based evolutionary algorithms, the second parent is referred to as the coparent.

combinatorial graph A collection of points, called vertices, and specification of pairs of those vertices joined by edges. Appendix D defines combinatorial graphs and gives some of their properties.

complete set of operations A set of operations on a space is said to be complete if concatenation of its members can generate any operation on that space. For Boolean logic, for example, AND and NOT form a complete set of operations.

computational fluid dynamics (CFD) The body of techniques used to compute, model, and numerically simulate the behavior of fluids or gasses. While nice differential equations (the Naiver–Stokes equations) can be written to describe fluid dynamics, these equations are unsolvable for a number of real-world problems. This in turn forces the use of large numerical simulations in the solution of fluid dynamics problems.

connection topology (In this text, this term refers to artificial neural nets.) The choice of which neurons in the network will be connected. The strength of the connections is not part of the connection topology.

context free grammar A collection of rules and symbols used to specify a context free language. The rules expand only single symbols and hence are context free. The language of a grammar is the set of all strings that can be obtained by applying the rules a finite number of times in a manner that removes all so called nonterminal symbols. See Section 14.4 for a more complete definition.

contingent evolution The notion that evolution is unrepeatable. The biological version of contingent evolution was first advanced by Stephen Jay Gould. It hypothesizes that if we could somehow rerun the experiment of "life on earth," we would get very different results each time. Evolutionary computation tends to support this view in that many EC experiments have a large number of possible outcomes, and restarting a simulation with a different random number seed is likely to produce a different outcome. Unlike biology, however, EC experiments are repeatable: be sure to record your random number seeds.

contraction map A map from a metric space to itself that has the property that for any pair of points those points in the image under the map are closer together than the points themselves were. Contraction maps are used in iterated function systems to generate fractals. An interesting fact about contraction maps is that they have a single fixed point.

connection weights Numbers associated with the connections between neurons in a neural net. Adjusting these weights is the primary method of programming a neural net.

Conway's lexicode algorithm Named after its inventor, John Conway, a method of picking out a set of points at guaranteed minimum mutual distance. Such points are often used as an error-correcting code. The lexicode algorithm orders a space of points. A minimum distance is specified and an empty collection of points is initialized. Traversing the set of points in order, a point is added to the collection if it is at least the minimum distance from the points already chosen.

copy number The number of copies we make when we allow a solution to reproduce.

Costas array A square array with one dot in each row and column such that the vectors connecting pairs of dots each occur at most once.

crossing number The number of edges that cross one another in a drawing of a combinatorial graph. The smallest crossing number of any drawing of a given graph is the crossing number of the graph itself.

crossover The process of blending two structures to make one or more new structures. There are many types of crossover; see the following entries in the glossary.

crossover, adaptive Crossover that changes in a manner that permits selection to enhance the effectiveness of crossover. There are many potential ways to implement adaptive crossover, e.g., with a crossover template or by having variable strength for possible crossover points.

crossover, conservative (also called *pure* crossover) Crossover that has the property that crossing over two copies of a single parent yields copies of that parent as the resulting children. The standard one- or two-point crossover for strings is conservative, subtree crossover is not conservative.

crossover, multi-point A binary variation operator in which several points are chosen randomly in a linear genome, and then the material between those points is contributed alternately from each parent to the children. See crossover, one-point and crossover, two-point.

crossover, nonaligned Nonaligned crossover relaxes the constraint placed on most string-genome binary variation operators that the material exchanged between copies of parents to make children be from corresponding parts of the same genome. In non aligned crossover, substrings of the parental genome with distinct starting points are exchanged. Typically, the lengths of the strings exchanged agree.

crossover, nonconservative A crossover operator for which the result of applying it to a pair of parents that are identical to one another need not produce children that are copies of the parents.

crossover, null This is a formal name for not having a crossover operator. Null crossover exchanges no material between parents.

crossover, one-point A binary variation operator in which a point is chosen randomly in a linear genome and then suffixes of the parents starting at the chosen point are exchanged to obtain the children. See crossover, two-point, and crossover multipoint.

crossover operator A variation operator that blends parts of two parent structures to make one or more new child structures. A major part of the design of a representation for a problem is selecting the crossover operator. One possible choice is no operator, and if one is used, it is the source (or obstacle) of inheritance of parental characteristics.

crossover, permutation Permutations are ordered lists of a set of elements in which each element appears once. This means that standard string crossover operators are useless (they often produce nonpermutations). As a result, there are many techniques from repair operators to random key encodings for performing crossover in permutations. See Chapter 7 for examples.

crossover, pure See crossover, conservative.

crossover, single-point The same as one-point crossover.

crossover, subtree A binary variation operator for tree-structured chromosomes that picks a subtree at random in each parent and then exchanges those subtrees. It is the standard example of nonconservative crossover, enables bloat in many cases, and has been called a macromutation with some justification.

crossover, two-point A binary variation operator in which a pair of points are chosen randomly in a linear genome, and then the material between those points is exchanged to obtain the children. See crossover, one-point and crossover, multipoint.

crossover, uniform A binary variation operator in which the contribution of each parent to each child is determined at random at each locus. For a string chromosome, for example, uniform crossover would flip a coin at each character to decide which parental character went to each child. The probability distribution of this binary operation can control the degree of habitability versus blending of parental characteristics.

cross-validation The practice of reserving some data to later validate a model or predictor generated from data. If, for example, a least squares fitness function was used to fit a model to data, then the most fit models are sometimes very bad. This is because they have, in effect, memorized the training data without picking up any real underlying data model. Reserving some data to check the model's effectiveness on data not used in creating the model is a sensible method of detecting and rejecting such "over training."

cycle type, of a permutation An unordered list of the lengths of cycles in a permutation given in cycle notation. See Chapter 7 for a definition of cycle notation. The cycle type is a list of positive integers summing to n, e.g., the cycle type of $(0\ 1\ 2)(3\ 4)(5\ 6)$ is 3 2 2.

data mining The transformation of data into information. A common goal of data mining is to take voluminous data, e.g., every purchase that customers make at a chain of convenience stores, and produce useful information, e.g., putting beer on sale will increase diaper sales. Evolutionary computation can be used to do data mining, but it is a relatively minor player. Evolutionary computation also provides a good target for data mining. Genetic programming produces interesting bits of code. Finding them in the morass of output can be tricky and is a type of data mining. Data mining has a second meaning in the social sciences: overuse of a single data set. Don't be surprised if a social scientist has a negative reaction to the term data mining.

deceptive fitness function A fitness function is *deceptive* if outside of a small neighborhood of the global optimum, movement toward the optimum actually reduces fitness.

decider A parse tree used in GP automata to drive finite state transitions. The role of the decider is to process the input data down to a small amount of information.

developmental biology The field of biology that studies how the information in the genome that specifies the growth and development of an organism is expressed. In evolutionary computation, developmental biology is the process of transforming the chromosome into the structure evaluated for fitness. Many evolutionary computation systems have no developmental biology. The Sunburn system and cellular encoding are examples of EC systems that use a nontrivial artificial version of developmental biology.

direct encoding Using a direct encoding means to store, as the structures you are evolving, exactly the structures used by your fitness function.

Divide the Dollar A game in which both players bid. If the bids total a dollar or less then both players are paid their bid; otherwise, they receive nothing. This game was constructed by John Nash as an extreme example of a game with many Nash equilibrium. Any pair of bids X, $100 - X$ forms a Nash equilibria and so knowing the Nash equilibria of the game does not predict the game's behavior well.

discontinuity A place in a function where it makes an abrupt jump in value. A standard example of such a function is the cost of a package as a function of its weight. The cost jumps abruptly at certain weights.

DNA Abbreviation for deoxyribonucleic acid. DNA is the primary information-storage molecule in most known living creatures. (Arguably some living creatures, e.g., RNA viruses, use RNA to store information.) The chromosomes, or representation, used in an evolutionary computation system are sometimes colloquially called the system's "DNA". Please avoid this usage, it confuses potential biological collaborators.

DNA bar code In this text, a DNA bar code is an error-correcting code using the edit metric over the four-letter DNA alphabet. The term also means a stretch of DNA that is unique to an organism and can be used to identify it.

dynamic programming A method of tracing the shortest path between two points in a metric space. Using in this text to find the edit distance between DNA strings. See Section 15.3 for details. Dynamic programming can also be used for tasks such as robot path planning and enumerating the number of paths between two points in a graph.

dynamical system A process that moves points in a space. It can to so deterministically or stochastically, and the motion may be continuous or discrete. Most evolutionary algorithms are discrete stochastic dynamical systems. Evolution moves the points making up the population around the gene space.

Many other sorts of dynamical systems exist but they are beyond the scope of this text.

echo machine A single-state finite state machine that returns its input as output. Used as the starting point for editing in the cellular representation for finite state machines given in Section 14.2.

edit distance The distance computed by the edit metric. See edit metric.

edit metric A distance measure on the space of strings over a given alphabet. The distance between two strings in the edit metric is the minimal number of single-character insertions, deletions, or substitutions required to turn one string into the other. The edit metric is a formal metric and obeys the three metric space axioms.

Efficient Node Use problem A type of maximum problem for the PORS (plus-one-recall-store) environment. Given a fixed number of nodes the Efficient Node Use problem seeks the PORS tree that generates the largest possible number.

ephemeral constant A type of constant used in genetic programming. In a genetic programming system operating on real values or integer data it is usual to have terminals that are constants. Either the system will maintain a list of constants to be used as terminals, or constants will be generated randomly whenever a terminal constant is needed in a parse tree. These randomly generated constants that appear and disappear with the trees that contain them are called ephemeral constants. The genetic programming in this text is all of the sort that uses ephemeral constants.

error correcting code A collection of strings (code words) that are chosen so that many errors are required to transform one into another. If only code words are transmitted, then the large number of errors criterion means that small numbers of errors can be corrected by changing the word received into the code word it most closely resembles.

Euclidian distance The square root of the sum of the squares of the differences of the coordinates of two points. In the plane, with each point having two coordinates, the Euclidian distance from (x_1, y_1) to (x_2, y_2) is

$$d = \sqrt{(x_1 - x_2)^2 + (y_1 - y_2)^2}$$

evolution Biologically, evolution is the variation of allele frequencies in a population over time. In abstract evolution is the sort of change that happens over time when a population that reproduces inaccurately is subject to some form of selection.

evolution strategies A type of evolutionary computation usually applied to parameter estimation, e.g., model fitting. They were developed by I. Rechenberg, H. P. Schwefel, and others in Germany in the 1970s. Evolution strategies use a very sharp type of selection and employ self-adaptation. A small number of best solutions are chosen in each generation, and many variations of each of these are generated. The solutions include not only the parameter values that are being optimized but also parameters that control how to vary those parameters. They are similar to evolutionary programming, though evolutionary programming uses more gentle selection than evolution strategies use.

evolutionarily stable strategy (ESS) An evolutionarily stable strategy is defined in the context of a population evolving to play a game, e.g. Prisoner's Dilemma. An ESS is a strategy so that if the entire population is made of that strategy then no single individual can survive if they appear in the population. Colloquially, a population consisting of a single ESS cannot be invaded.

evolutionary programming A type of evolutionary computation originally conceived by Lawrence J. Fogel in 1960. Like other forms of evolutionary computation, it operates on populations of solutions. The original form does not use binary variation operators like crossover, but does employ sophisticated forms of self-adaptation in which individuals in the population have evolving parameters that control their own variation operators. In some ways, evolutionary programming is similar to evolution strategies but typically uses far less sharp forms of selection.

fake bell curve A function with a single maximum whose basin of attraction is all of $\mathbb{R}^n$. It is shaped like the famous bell curve but is a simpler function. The univariate fake bell curve is

$$f(x) = \frac{1}{x^2 + 1}$$

The general fake bell curve in n dimensions is

$$f(x_1, x_2, \ldots, x_n) = \frac{1}{\left(\sum_{i=1}^{n} x_i^2\right) + 1}$$

fan-out (used in logic circuits) The fan-out of a logic gate is the largest number of other gates its output can drive. In logical theory fan out is not an issue but when designing real circuits there is a limit to the number of other devices a logic gate can drive.

feature selection This term originates in pattern recognition. A feature is something that can be computed from each member of a data set. When using evolutionary computation as a machine learning technique, or when using any sort of machine learning or automatic classification, selecting good features from the input data can enhance performance. Evolutionary computation can be used to perform feature selection, as in the hyperspectral example in Chapter 1.

finite state machine A device that takes inputs and looks up appropriate outputs from an internal table in a manner dependent on the machine's internal state. The internal state permits the machine to look up different outputs depending on the history thus far. This internal state is thus a form of memory. The outputs may either be associated with the transitions to a next state or may be associated with the states themselves. These two different types of finite state machines are called Mealy and Moore machines, respectively. See Chapter 6.

finite state predictor A finite state machine used to predict an input sequence.

finite state transducer A finite state machine that is used to transform a string or sequence of symbols. Rather than recognizing something the machine outputs symbols as new symbols come in.

fitness biased reproduction This is the analogue of "survival of the fittest" in evolutionary computation. It is the practice of making better solutions more likely to reproduce.

fitness function A heuristic measure of the quality of solutions. The fitness function is used to tell which solutions are better than others. There may be several fitness functions that are appropriate to a given problem.

fitness landscape The graph of the fitness function. The dependent variable drawn from the set of fitness values. The independent variables are dawn from the same space that the population is drawn from. The evolving population can be thought of as moving on this landscape. This landscape metaphor is useful for thinking about evolutionary computation but in the case of complex representation or high-dimensional problems, can be a little hard to picture.

fitness trial Used when there are many possible cases of the problem of interest, e.g., the Tartarus problem with its many possible starting boards. A fitness trial is a single case in a fitness evaluation, which typically uses the sum or average of many fitness trials to estimate fitness.

fixed point For a function f, a point such that $f(x) = x$.

fractal An object whose dimension is not a whole number. Clouds, ferns, and snowflakes are examples of natural objects that have good fractal models.

fractal attractor See basin of attraction. An iterated function system has a set of points that its moving point approaches as the system is iterated. This is the sole basin of attraction or attractor of the iterated function system when it is viewed as a dynamical system. This attractor is typically a fractal, a fact that makes iterated function systems of interest to people who study fractals.

function optimizer An algorithm or piece of software that can find points in the domain of a function that yield the largest or smallest values possible for the dependent variable. Evolutionary computation can be used for function optimization, but many other techniques exist as well.

game theory The study of the theory and strategy of games. The material in this text on Prisoner's dilemma is an example of game theory. Evolutionary stable strategies and Nash equilibria are game-theoretic ideas.

Gaussian distribution A probability measure on the real numbers given by the famous bell curve,

$$f(z) = \frac{1}{\sqrt{2\pi}} e^{-z^2/2}$$

It is also called the normal distribution. See Appendix B for more information.

Gaussian mutation A mutation operator (unary variation operator) for real variables representations based on the Gaussian distribution; see same. A Gaussian mutation can make changes of any size but has a strong tendency to make small changes.

gene duplication In biology, gene duplication is the copying of the DNA containing a gene so that an organism has one or more copies of the gene. Gene duplication enables discovery of new gene function because a gene with multiple copies can have mutations to one copy while retaining critical function in another copy. Some analogy to gene duplication may take place when using the GP automaton or MIPs net representations.

generation An updating of the entire population via breeding. In its most extreme form, a generation is the replacement of the entire population of an evolutionary algorithm by new structures. An evolutionary algorithm that does this is said to be *generational* The population-updating techniques of an evolutionary algorithm lie on an axis from generational to steady state. Steady-state algorithms use mating events in which a minimal amount of breeding takes place before a new creature is placed into the population.

generational See generation.

genetic algorithm A form of evolutionary computation typified by using a string representation, typically a string of bits, crossover, and mutation. Genetic algorithms lack self adaptation in their original form. Many variations exist that attempt to dynamically locate correct representation for a given problem. Invented by John Holland and substantially generalized by his students.

genetic programming A form of evolutionary computation featuring variable-sized genomes that encode formulas or pieces of computer code. The most common representation for genetic programming is a tree-structured chromosome, but there are many other structures (linear structures such as ISAc lists, complex structures such as GP automata). Invented by John Koza and John Rice.

gladiatorial tournament selection A form of tournament selection in which pairs of individuals compete until two winners are found. These winners are permitted to breed, and their children replace the winners.

global optimum An optimum that takes on the maximum possible fitness value. It need not be unique, but multiple global optima must take on the same fitness value. See also optimum, local optimum.

GP automata A representation used to evolve small computer programs. A GP automaton is a modified finite state machine. Each state has a parse tree associated with it that serves to abstract information from the environment to drive transitions. The parse trees are called *deciders* Examples of application domains for GP automata are grid robot tasks such as Tartarus, optical character recognition, and control of simulated chemical reactions. See Chapter 10.

Graduate School Game A two-player simultaneous game with two moves: cooperate and defect. This game is a modification of iterated prisoner's dilemma in which the highest score comes from taking turns defecting. The game is supposed to evoke a couple that take turns putting one another through graduate school. See Section 6.3.

graph based evolutionary algorithm (GBEA) A type of evolutionary algorithm that uses a combinatorial graph as a population structure that limits the spread of information via breeding.

greedy algorithm An algorithm that tries to satisfy an objective function (fitness function) by making choices that pay the largest possible immediate benefit. Examples include Kruskal's algorithm for a minimal-cost spanning tree and Conway's lexicode algorithm.

greedy closure evolutionary algorithm An evolutionary algorithm in which a greedy algorithm is part of the fitness evaluation. The representation for this type of evolutionary algorithm is a partial structure that is completed by a greedy algorithm. Fitness of the partial structure is measured by the quality of the complete structure after it is built by the greedy algorithm. An example of this type of evolutionary algorithm appears in Section 15.3.

greedy packing (used in the bin-packing problem in Chapter 7) Given a set of bins and an order in which to consider goods, goods are placed in the first bin with sufficient room to hold them. In order to pack the bins, the order in which the goods are considered is evolved.

grid robot A form of virtual robot that exists in a world built on a Cartesian grid. See Chapters 10 and 12.

Hamming ball A ball in the Hamming metric. It is the set of all strings no more than some fixed Hamming distance from a given string (the center).

Hamming distance The distance computed by the Hamming metric. See Hamming metric.

Hamming metric A distance measure for distances between strings that are the same length. The Hamming distance between two strings is the number of positions in which their characters are not the same. The Hamming metric is a formal metric obeying the three metric space axioms.

Hamming sphere A sphere in the Hamming metric. It is the set of all strings at some fixed Hamming distance from a given string (the center).

headless chicken crossover A binary variation operator resembling standard crossover. Instead of performing crossover between two population members a single population member is selected and then crossed over with a new, randomly generated structure. Typically the random structure is generated in the same fashion as a member of the initial population. Headless chicken crossover is most often used in genetic programming. See macromutation.

Herbivore task A grid robot task that is used to model foraging by herbivores. (A herbivore is an animal that gets its energy by eating plants.) The usual grid robot actions of turning left or right and moving forward are augmented by a fourth action called eating. When an eat action is executed any box in front of the grid robot vanishes. See Chapter 12.

hexer A game with dice used to help people gain an intuitive understanding of Markov chains. A number of dice, usually 4, are used. On the first move all the dive are thrown by the player. If there are no sixes then the player loses. If all the dice are sixes the player wins. If some sixes are thrown then the sixes are put in a *six pool* and the game enters an iterative phase. All dice not in the six pool are thrown. If there are no sixes then a die is removed from the six pool and the player throws again. If no sixes are available in the six pool when no sixes are thrown the player loses. If all the dice are placed in the six pool the player wins. It is an odd feature of hexer that using four dice gives the worst chance of winning.

hill climbing An optimization technique that operates on a function or fitness landscape. It operates by moving in the direction of increasing fitness or functional value (decreasing value if minimization is the goal. Hill climbing serves as a baseline technique for evolutionary algorithms, as a potential type of helpful mutation, and as an optimization technique in its own right. Hill climbing may be highly sophisticated, e.g., following the gradient of a differentiable function, or quite simple, e.g., repeatedly generate new examples, saving them if they are better than the current best.

hybridization (of algorithms) When a problem is solved by combining different algorithmic techniques the results is a hybrid algorithm. The greedy closure evolutionary algorithm in Chapter 15 is an example of a hybrid of Conway's lexicode algorithm with an evolutionary algorithm.

hybridization (of populations) An evolutionary algorithm technique that seeks to improve results by combining creatures from distinct populations. In this text it is used in evolving grid robots. Many populations are evolved for some number of generations. The best creature from each populations are then copied and combined to create a new starting population. This technique does improve the results for some grid robot problems.

hyperspectral data Hyperspectral data is spectral data (light intensity or reflectances) that involve hundreds to thousands of frequencies of light. The adjective "hyper" means "far more than used previously."

hypercube A generalization of the usual cube in three dimensions. The usual cube is a 3-hypercube. A square is a 2-hypercube (the number gives dimension). The tesseract is a 4-hypercube. Hypercubes are interesting in that they are abstract descriptions of the partial order of subsets of a set, a structure of adjacency for the Hamming metric on binary strings of a given length, a combinatorial graph, a geometric polytope, and many other things.

injection A function that one-to-one.

intron In biology, an intron is a piece of a gene that does not code for protein and that is not on the ends of the gene. Introns are thought to sometimes have a regulatory function or play a role in managing gene crossover but their DNA is not subject to the same selection pressure as DNA that actually codes for protein. By analogy, unused parts of a data structure in a member of a population in evolutionary computation are called introns. The term is most commonly used in genetic programming. See bloat.

iterated function system A collection of functions used to create a fractal. Each of the functions must be a contraction map. A moving point is chosen, and then the functions are applied to it with the next function applied chosen uniformly at random with replacement from all available functions. For graphics applications, the moving point is plotted, yielding a fractal. This requires that the system undergo burn in. The set of all places the moving point could reach is the full fractal, called the attractor of the iterated function system.

ISAc (If-Skip-Action) list A representation for evolving programs. An ISAc list uses a data vector of inputs. The list is a list of ISAc nodes. Each node contains two pointers into the data vector, an action, and a pointer to some other part of the ISAc list. A node is executed by comparing the two data items, e.g., is the first object larger than the second? If the comparison is true, then the node's action is executed. The nodes are executed in order, except in the case of a jump action, which moves execution to the other node pointed to by the current node's pointer into the ISAc list. The available actions that could be placed in a node's action field include jump, a null operation called NOP, and actions relevant to the environment in which the ISAc list is operating. In a Tartarus environment, for example, the possible actions are NOP, jump, left, right, forward. See Chapter 12.

least squares fit The practice of fitting a model to data by minimizing the sum of squared error of the model on the data. This minimization may be performed by adjusting parameters of the model, as in the least squares fit of a line, or may be performed by a procedure that also searches a space of possible models such as genetic programming.

indirect encoding Using an indirect encoding means that the objects stored in the evolving population are interpreted or developed before they are passed to your fitness function.

k-**max function** A fitness function on strings over an alphabet with k letters. It returns the count of whatever character appears the most often. The function has k modes, the k strings composed of a single character. It is an example of a polymodal fitness function for string representation evolutionary algorithms. Except for its polymodality, it is structurally similar to the one-max function.

Lamarckian evolution Lamarck held an incorrect, but in his times plausible, view that acquired characteristics could be inherited. Standard evolution would hold that a giraffe's long neck was the result of selection for the longest available necks in the current population together with slow production of even longer necks by mutation. Lamarck's theory held that necks lengthened because the giraffes were stretching up to reach high leaves and that in addition, children would start with the neck their parents had achieved after years of stretching. The molecular mechanisms of inheritance in biology cannot support Lamarck's idea for the most part but it can be used to build software. The most common method is to create a mutation operator that performs some sort of local search of the best structure in the area of the search space near the population member being mutated.

Lamarckian mutation A mutation operator that searches the space of all genes within some fixed number of mutations of the gene being mutated and returns the best result. See Lamarckian evolution.

Lambda (λ) A string containing no characters; the empty string.

lambda-transition (λ-transition) A transition in a finite state machine that produces no output or response. Used in GP automata as the "think" action to permit multiple deciders to contribute to a single action.

Law Of Averages (LOA) Strategy A strategy for playing the game Rock Paper Scissors that assumes that your opponent believes in the law of averages. The law of averages, which is false, states that a random experiment will display a bias that tends to return it to its average behavior. A coin that has flipped many heads would, according to the law of averages, flip an excess of tails until it got back to somewhere near 50/50 heads and tails. The nice thing about the law of averages is that many people believe it. For Rock Paper Scissors, the strategy plays to beat the move used the least so far by an opponent.

lazy evaluation A technique from artificial intelligence that involves failing to evaluate expressions until they are required to yield some value that leads to a visible result. In a spreadsheet, for example, the cells of the sheet contain formulas that depend on one another. The user sees some portion of the sheet. When new values are entered the sheet could recompute every cell but needs only compute the ones the user is able to see and those that the cells the user can see depend on. This is more than not recomputed cells that do not change value at all. Cells that should change value are not recomputed until the are looked at or until the spreadsheet is saved.

least squares fit A test for how well a model fits a collection of data. The values of the model for the independent variables of the available data points are computed to predict the independent values. The squared error of the predicted and actual values is computed by summing the squared of the differences of the actual and predicted values. Fitting a model is done by minimizing the squared error of the model from its data. Squared error is used as a fitness function (to be minimized) in fitting a model with evolutionary computation. See Section C.4 for the derivation of a least squares linear model.

lexical fitness The practice of using a lexical partner fitness function. This is a second fitness function that serves as a tie-breaker.

lexical partner A second fitness function used to break ties in fitness values is a *lexical partner* for the fitness function it is supposed to aid. The name comes from lexical or dictionary ordering. Since the second fitness function, the lexical partner, can only break ties it is infinitely less important than the first fitness function, just as the first letter of a word is infinitely more important than the second in determining it position in an alphabetical list.

local mating rule In graph-based evolutionary algorithms, the method of choosing which neighbor of a creature already chosen to breed is to be the second parent together with the method of placing the offspring in the population.

local optimum An optimum that does not take on the best possible fitness value. It is the best only in its "local neighborhood." See also optimum, global optimum.

logic gate An electrical circuit that implements a Boolean function. In this text these circuits are simulated as artificial neural networks.

macromutation A unary variation operator that makes a large change in the structure it operates on. The term comes from the debate over the status of subtree crossover in genetic programming. In the view of its inventors, subtree crossover blends characteristics of the parents. Others think that subtree crossover often makes novel structures substantially unrelated to either parent. Thus, the question, "is subtree crossover a macromutation?" There is experimental evidence supporting both sides of the argument, and the true status of subtree crossover appears to depend on the details of the genetic programming system being used and the problem being solved. There is also evidence that some amount of macromutation is not a bad thing. See headless chicken crossover.

Markov chain A state-conditioned probability space in which the chance of an event occurring in the next sample of the space depends on the current state. See Section B.2.

Markov model A model with a space of events that has a state space and probabilities of moving between the states. Used to summarize knowledge about depending probabilities, e.g., as in the model of DNA given in Section 15.1. A *hidden* Markov model is one in which the probabilities in the model are estimated from data.

mating event A single act of reproduction. In an algorithm that uses crossover, a mating event is the selection of two parents, the process of copying them, possibly crossing the copies over, mutating the copies, and then possibly placing the copies back in the population.

maximum problem A type of problem in genetic programming. A maximum problem seeks to generate the maximum possible value with a given set of operations and terminals and some restriction on size. The size restriction might be on the depth of the tree or on its total number of nodes.

Mealy machine A finite state machine in which the outputs of the machine are associated with the transitions. See Moore machine.

mean A high-tech word for average.

metric space Any space or collection of points that has a distance function. A distance function takes pairs of points to the non negative reals in a manner that computes the distance between those points. The three defining properties of distance are as follows: the distance from a point to itself is zero; the distance between two distinct points is positive; and the distance around two sides of a triangle is at least as far as the distance along the remaining side. An example is standard Euclidian distance in the plane.

minimal description problem A task in the plus-one-recall-store programming test bed. Given a number n find a PORS tree that computes n and that has as few nodes as possible. This is a task for which it is challenging to write a fitness function.

MIPs net A representation used for genetic programming. MIPs stands for multiple interacting programs. A MIPs net is a collection of parse trees that have terminals that hold the output of the other trees in the net. A MIPs net can have a strict ordering so that MIPs nets can see only trees before them in the order. There are called *feed-forward* MIPs nets, in analogy to the corresponding neural net notion. MIPs nets can also permit all trees to potentially access all other trees, in which case they are said to be recurrent. See artificial neural net.

mode A local maximum, typically in a probability distribution or fitness function.

model of evolution The process whereby population members are selected to breed and chosen to die. See Chapter 2 for a discussion.

monotone function A univariate function that either only increases or only decreases as the independent variable increases. Monotone functions either preserve or completely reverse the order of their inputs while possibly changing their values.

Monte Carlo integration A numerical integration technique that randomly samples the function begin integrated. Used for high-dimensional functions as a way of avoiding the explosion in size of a regular grid of samples as dimension increases.

Monte Carlo method A slightly archaic name for simulations that involve sampling or random numbers. Evolutionary algorithms can be seen as a type of Monte Carlo method. Named after a famous casino.

Moore machine A finite state machine in which the outputs of the machine are associated with the states. See Mealy machine.

motif A short pattern with wild cards, usually over the protein, DNA, or RNA alphabets.

moving point When generating a fractal with an iterated function system or chaos automaton, the moving point is a point being operated on by the contraction maps in the iterated function system or chaos automata. The moving point is the point that sweeps out the fractal and that is plotted to generate an image of the fractal.

multicriteria optimization Optimizing with more than one fitness function. Since there are often multiple desirable properties, e.g., durability and cost, that fight with one another, multicriteria optimization often comes up in applications.

Multiple Interacting Programs See MIPS net.

Multiplexing problem A logic gate specification problem. A multiplexer has n data inputs and k address inputs, $n = 2^k$. The binary number given at the k inputs selects which of the n inputs will be the output of the multiplexer.

mutation A variation operator that makes random changes in a single structure in the population is called a mutation. There are a vast number of different mutation operators. Selection of the mutation operator(s) is a fundamental part of the design of an evolutionary algorithm. A mutation operator is supposed to make small changes in the structure it operates on, providing an ongoing source of variation in the population.

mutation rate The rate at which mutations happen. Usually measured in number of mutations per new structure.

Nash equilibrium A collection of strategies for playing a game. It requires that if a group of players are using the strategies then a unilateral change by any one player will lower his score. Everyone defecting is an example of a Nash equilibria of the Prisoner's dilemma.

neighborhood A small region in a space. Typically a neighborhood is "of a point," which means that the point is at the center of the neighborhood.

neural net A network of connected neurons. The connections have associated numbers called weights. These weights establish the strength of the connections and, together with the pattern of connections, control the behavior of the net. Neural nets are programmed by adjusting the strengths of the connections. See neuron.

neuron The units out of which neural nets are built. A neuron is connected to other neurons with weights, accepting the sum of the weights times their outputs as its input. The neuron also has a transfer function that transforms its input into its output.

neutral mutation A mutation that does not affect fitness. Neutral mutations may happen by accident, for instance mutating an unused state in a finite state machine. It is also possible to construct intentional neutral mutations that enhance population diversity. In the k-max problem one might change the identity of the characters, e.g., at $k = 4$ exchange all ones for zeros and twos for threes. This will not change fitness but will greatly change the results of crossing over with other members of the population.

neutral selection Running an evolutionary algorithm with a fitness function that returns a constant value, with no selection. Neutral selection experiments are used to characterize the rate at which information spreads in a selection scheme. As an example, examine the neutral selection experiments in Chapter 13 that characterize the selection behavior of graphs.

niche specialization (in evolutionary computation) The practice of reducing the fitness of an organism if there are many other similar organisms. There are many types of niche specialization depending on the details of how similarity of organisms is measured and how the number of similar organisms is transformed into a fitness penalty.

niche specialization, domain A type of niche specialization that measures similarity by comparing the chromosomes (data structures) of population members. An evolutionary algorithm using a string chromosome that compared the strings via Hamming distance would be an example of domain niche specialization. The problem of finding a reasonable similarity measure on chromosomes limits the utility of this technique for some representations.

niche specialization, range A type of niche specialization that measures similarity by comparing fitnesses. It is simple but can take a similar creatures that are very different in their underlying data structures.

node An operation or terminal in a parse tree. Both operations and terminals are stored in the same sort of structure and are distinguished by the fact that operations take arguments (inputs), while terminals do not.

nonaligned crossover See crossover, nonaligned.

nonconservative crossover See crossover, nonconservative.

North Wall Builder A grid robot task. The robot is placed next to a single box. The usual moves, left, right, forward, and rules for pushing boxes from the Tartarus task apply. If the square that initially held the box is ever empty then a new box is placed there. Fitness is computed, after a fixed number of moves by the robot, by proceeding from the north side of the board to the south until obstructed by a box. Grids visited during this process are subtracted from the total number of grids in the board to give the score. For details and variations see Chapter 12.

one-max function A fitness function on binary strings that counts the number of ones. Is is a standard example of a unimodal fitness function for string representation evolutionary algorithms. It is also a case of the string evolver with an all-ones reference string.

one-to-one A property a function can have. A function is one-to-one if no point in the range is the value of more than one point in the domain. The function $y = 3x + 2$ is one-to-one; each y value is associated with one x value. The function $y = x^2$ is not one-to-one, since, for example, both 2 and -2 are taken to 4.

onto A property a function can have. A function is onto if every value in the range actually occurs. The function $y = \tan(x)$ is onto because every real number is the tangent of some angle. The function $y = x^2$ is not onto the set of real numbers, because no negative number occurs as a y-value.

operation A node in a parse tree that both accepts and returns values.

optimum (of a fitness function or fitness landscape) A point with a better fitness value than all the other points near to it. See also local optimum, global optimum.

optional game A game in which refusing to play is one of the possible moves.

Packing problem The problem of placing a collection of goods, of integer size, into a collection of bins, also of integer size. A packing is represented as an assignment of goods to bins such that the total size of the goods in each bin is not more than the size of the bin. The problem is used in Chapter 7 in a hybrid algorithm. The order in goods are presented to a greedy packing algorithm is evolved.

parity problem (function) (gate) Discovering an implementation of the parity function. The parity function is a Boolean function reporting true if an odd number of its inputs are true. A parity gate is a logic gate that computes the parity function. The parity problem and its negation are, in a sense, the hardest functions to implement because they depend on all their inputs in any situation. Most logic functions have some set of inputs where, given the value of all but one of the inputs, the remaining input doesn't affect the output if you change it. For parity all inputs always matter. This makes parity a "hardest" problem.

parse tree A dynamically allocated data structure, composed of individual nodes, that can store mathematical or symbolic formulas. The basic data structure stores a single operation, constant, or input value and pointers to nodes storing its arguments. A parse tree is also a diagram of a formula that shows the order in which its parts are evaluated. The data structure is derived from the diagramming technique.

Pascal's triangle (see Appendix B) A method of displaying the binomial coefficients. Except for the ones bordering the triangle, each entry is the sum of the two above it. Each row, numbered with zero-base counting, contains the entries $\binom{n}{0}, \binom{n}{1}, \binom{n}{2}, \ldots, \binom{n}{n}$, whereby n is the second entry in the row. Pascal's triangle is named after Blaise Pascal, who did foundational work in enumeration and probability.

Pavlov In the context of this text, a strategy for playing Iterated Prisoner's Dilemma. The player cooperates initially, and thereafter cooperates with its opponent when they both made the same move in the previous play. Pavlov is an error-correcting strategy, in that it can recover from a noise event when playing against a copy of itself.

PCR primer An abbreviation for polymerase chain reaction primer; see Section 15.2. A primer, in this context, is a short stretch of DNA that anneals to a longer strand of DNA, initializing a polymerase reaction. Polymerase is an enzyme that can take single DNA bases and add them to a strand of DNA, making a complementary copy. Run over and over with the correct primers the polymerase reaction can exponentially amplify a particular type of DNA.

penalty function A function that gives a valuation to the violation of some condition or rule. Penalty functions are used to build up or modify fitness function by reducing the fitness of a member of a population by some function of the number of undesirable features it has.

Perato frontier The set of all Perato-optimal objects is the Perato frontier for a problem with two or more quality measures. The frontier exhibits the trade-offs between the quality measures. See Perato optimal.

Perato optimal If we are comparing objects with two or more quality measures then one dominates another if it is better in all quality measures. A strategy that cannot be dominated is said to be perato optimal.

permutation An ordering of a set of distinct objects. Also, a bijection of a finite set with itself. Permutations are discussed extensively in Chapter 7.

permutation matrix A square matrix with a single one in each row and column and zeros elsewhere. Permutation matrices are associated with corresponding permutations by allowing the ones to designate a "this row goes to this column" relationship that specifies a permutation. Permutation matrices are discussed in Chapter 7.

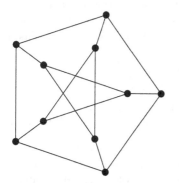 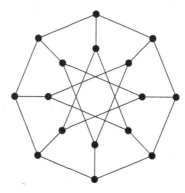

Petersen graph The Petersen graph, shown above left, is a famous combinatorial graph. Many theorems in graph theory are of the form, "if a graph is not (does not contain) the Petersen graph, then ...," and so the graph is well know, even having had a book named after it. The Petersen graph can be generalized as two cycles joined by spokes. A generalized Petersen graph is shown above right, the $(8, 3)$-Petersen graph with two eight-cycles, the inner of which proceeds by jumps of size three. A careful definition of the Petersen graph appears in Appendix D.

phase change A frontier where the character of some quantity changes. The standard example of distinct phases are the ice, liquid, and steam phases of water. In a function a phase change is an abrupt change in some characterization of the function's behavior, e.g., a function that oscillated for positive inputs and remained constant for negative inputs might be said to have a phase change at zero.

Plancha eco-stove A design of wood-burning stove. One type was designed via evolutionary computation; see Chapter 1.

plus-one-recall-store (PORS) A very simple genetic programming language with a single binary operation (plus, integer addition), a single unary operation that returns its arguments value and also stores that value in a memory (store), and two terminals, one (the integer) and recall, which returns the value in the memory. There are a number of problems that use this language. See Chapter 8.

plus-times-half (PTH) A maximum problem used in genetic programming. The operations plus and times are available together with the constant $\frac{1}{2}$. This problem is typically used with tree-depth-limited genetic programming rather than the tree-size-limited genetic programming common in this text. Producing a maximal value requires that the addition operator add constants to a number bigger than one and that that number then be multiplied by itself.

point mode A mode (high or low point) in a function or search space that has no other points at the same height (depth) in a small neighborhood. The optimum of $f(x) = 5 - x^2$ is an example of a function with a point mode. The optima of

$$g(x,y) = \frac{\sin(x^2 + y^2)}{x^2 + y^2 + 1}$$

are examples of nonpoint modes. They form a circle in which each highest point has an infinite number of highest points within any positive radius of it.

point mutation In biology, a point mutation is a change to a single base of DNA. In evolutionary algorithms a point mutation is a minimal mutation. In a bit string representation the analogy to biology is very plain: a point mutation flips the value of a single bit. When the representation is an array of real numbers the analogy becomes less exact because a distribution is required to describe what changes can be made at one point in the array. For more complex representations the meaning of point mutation is less clear. Point mutations are also called one-point mutations and may be used to build up other operations such as two-point, n-point, and so on.

pointer mutation A mutation used in ISAc lists that modifies one of the two pointers into the data vector in an ISAc node.

Poisson distribution A discrete distribution that appears as the limiting case of the binomial for a very large number of trials with a very small probability of success. See binomial distribution, Bernoulli trial. The Poisson distribution is useful in writing an efficient implementation of uniform mutation in a string representation. Rather than checking the probability of mutating each character of a string the Poisson distribution is used to generate a number of characters that will be mutated. These are then selected and mutated, substantially reducing the quantity of random numbers needed.

population The collection of solutions on which an EC system operates. The term is draw from biology. Population size and structure are both critical design parameters for an evolutionary algorithm.

polymerase chain reaction See PCR primer.

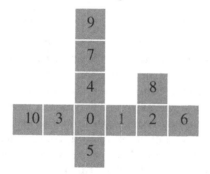

polyomino (generalization of domino) A collection of squares arranged so that all the squares that meet, meet full face to full face and so that the shape, as a whole, is connected. An example, with numbered squares, is shown above.

polysymbot A representation in which a single structure contains specifications for multiple symbots.

population seeding An ordinary evolutionary algorithm generates an initial population at random. Population seeding is the practice of adding superior genes to the initial population, or, possibly, creating an entire population from superior genes. These genes can be from previous evolution, designed according to heuristics, or created with expert knowledge.

PORS See plus-one-recall-store.

predator–prey model A mathematical model of a predator and a prey species. An example is given in Chapter 1.

primitive (mod p) A number n is said to be primitive (mod p) if p is prime and the powers $n^0, n^1, n^2, \ldots$ include all the nonzero numbers (mod p). In this next such numbers are used to construct Costas arrays.

Prisoners Dilemma A simple two-player simultaneous game that serves as a mathematical model of cooperation and conflict. The game may be played once or many times (iterated). This game is discussed in the later sections of Chapter 6 and Section 14.2. The game is described and applications to the real world discussed in the book *The Evolution of Cooperation*, by Robert Axelrod.

probabilistic mutation See uniform mutation.

probability The probability of an event is the limiting fraction of the number of times the event will occur if the experiment in which the event can occur is repeated an infinite number of times. The meaning of probability is a subject of deep and ongoing debate, and that debate is beyond the scope of this text.

PTH See plus-times-half.

Public Investment Game A multiplayer game. Each player is given some amount of money. They may keep any or all of the money. The money they choose not to keep is placed in a common pool. The common pool money is then multiplied by some factor larger than one and the resulting payout divided evenly among all the players. The game is intended to model public investment, e.g., in building roads with funding via taxation. The game is interesting because the maximum total payout comes when all money is placed in the common pool, but the highest payout goes to the player who invests the least in the common pool. To place this in the context of road building, a citizen that evades his taxes keeps his money and has the use of the roads. The roads, however, were built with less money and are hence presumably not as good as they could have been. In this text, the game is modified by adding a required minimum investment (law) and penalty for not making the minimum investment (fine).

random key encoding A representation for storing permutations. A sorting key is a field in a database that is used to sort the records of the database. If we place random numbers in an array then those numbers can be vied as encoding the permutation that would sort them into ascending or descending order. This is called a random key encoding for permutations. It permits evolutionary computation software that works with real valued genes to encode and evolve permutations. Random key encoding is discussed in Chapter 7.

Random (Strategy) In the context of this text, a strategy for playing prisoner's dilemma in which the player chooses uniformly at random between cooperation and defection in each step.

random variable The numerical outcome of an experiment with more than one possible outcome. Flipping a coin and assigning one to heads and zero to tails is an example of such an experiment. The value actually obtained in a particular flip of the coin is an example of a random variable.

range abstraction (operator) A statistical summary of some contiguous part of a data set. Range abstractions are summaries of parts of a data set that may have more predictive value than the raw data because they filter noise or are simpler then the raw data. Range abstractions are used in an example in Chapter 1.

range niche specialization See niche specialization, range.

range operator See range abstraction.

rank selection A method of choosing members of a population to reproduce or die. Population members are ranked and then chosen in proportion to their rank. When selecting for reproduction, the best creature has rank n and the worst has rank one. The reverse holds when selecting creatures to die. This method is similar to roulette selection, but buffers against the impact of very compact or very broad spreads in the population fitness.

recurrent For the neural net definition of this term, see neural net. A state in a Markov chain is recurrent if the chain never permanently leaves the state. A Markov chain is said to be recurrent if all its states are recurrent. The property of recurrence is more generally ascribed to a system that never permanently abandons any of its possible configurations. Uniform mutation, thus, can be said to be recurrent because it can recreate any possible structure if the correct random numbers come up.

repair operator A repair operator is used to compensate for destructive crossover. When a crossover operator creates an invalid structure a repair operator is used to make a small number of changes to restore validity. This text avoids repair operators for the most part, though some of the crossover techniques used with permutations could be viewed as incorporating repair operators.

representation The method of coding and manipulating potential solutions in the computer is the representation. Representation includes not only data structure but also variation operators. Changing the representation can completely transform system behavior and so choice of representation is critical in evolutionary computation.

Ripoff A strategy for playing the Iterated Prisoners Dilemma. The strategy defects initially. If the opponent ever defects, Ripoff cooperates on the next move and plays Tit-for-Tat thereafter. If the opponent does not defect then Ripoff alternates defection and cooperation thereafter. Ripoff was discovered by an evolutionary algorithm in which the population contained immortal Tit-for-Two-Tats players.

Rock Paper Scissors A two-player simultaneous game with three possible moves. The moves are rock, paper, and scissors. Rock beats scissors, scissors beats paper, and paper beats rock. If the players make the same move, they tie. Rock paper scissors is discussed in Chapter 14.

root The topmost node in a parse tree. Its output is the output of the tree.

roulette selection A method of choosing members of a population to reproduce or die. Population members are in proportion to their fitness or, in some cases, to a monotone strictly positive function of their fitness. Roulette selection is a general technique for selecting according to an empirical distribution. It is used in this way for efficient implementation of probabilistic (or uniform) mutation, operating on a tabulation of the Poisson distribution. See Section 2.7. Roulette selection is named in analogy to a Roulette wheel.

roulette wheel selection Same as roulette selection.

round robin tournament A tournament in which each possible pair of players compete.

Royal Road function A fitness function on string genes in which fitness is given for getting entire blocks correct. The original Royal Road function used a length-64 genes with eight disjoint blocks of eight bits. A fitness of $+8$ was awarded if all bits in a block were 1's. The function can be generalized by changing the length and block size. The function is valuable because of its explicit blocks which permit analysis of an algorithms ability to discover and preserve blocks.

SAW Abbreviation for self-avoiding walk. See self-avoiding walk.

selection This is the process that picks population members to breed or to survive or both. An evolutionary algorithm uses selection and requires it to be biased in favor of more fit organisms. The model of evolution incorporates the selection procedure.

self-avoiding walk An evolutionary computation problem using a string gene representation over the alphabet up, down, left, right. The members of the alphabet represent moves on a grid from a fixed starting point. The fitness function is the number of grids visited. The length of the string of moves is chosen so that there are exactly enough moves to visit each square once. A perfect solution is thus a *self-avoiding walk*, or walk that repeats no square.

sequence An infinite list or string drawn from some set.

Shannon entropy The Shannon entropy of a probability distribution P is the expected value of the negative log of the probabilities of the elements of the distribution. The Shannon entropy counts the number of bits needed, on average, to report the outcome of an experiment that has the distribution P.

Shannon's theorem A result proved by Claude Shannon that says that there is a way to transmit information on a noisy channel so that the information lost to the noise can be, with arbitrarily high but nonunit probability, be arbitrarily close to that which would be lost if the pieces of information trashed by noise were known in advance. Shannon's proof is nonconstructive and so does not say how to find a system that transmits the information in the maximally noise-resistant fashion. The mathematical discipline of coding theory grew out of attempts to find a system for defeating noise that Shannon's theorem proved must exist.

shape evolution A system for evolving polyominos given in Chapter 14. Shape evolution serves as a simple example of cellular encoding.

shortest path problem The problem of finding the shortest path connecting two points. In this text we explore a very simple version of this problem, finding a path from $0,0)$ to $(1,1)$ across a featureless landscape.

Sierpiński triangle A fractal, shown above, see Section 15.4 for a discussion. The Sierpiński triangle is interesting in that there are several different fractal algorithms that generate representations of it. In this texts it is generated with a simplified iterated function system called a chaos game.

similarity radius The distance inside of which to things are considered similar for the purpose of niche selection.

similitude A map from a metric space to itself that displaces, rotates, and then contracts the space. Similitudes are all contraction maps and are used in chaos automata as well as iterated function systems.

similitude mutation A mutation of a chaos automaton that modifies one of the parameters of a similitude within the automaton. See Section 15.5.

simple evolutionary fit Selecting a model for a data set and then finding the parameters of the model with an evolutionary algorithm. See Chapter 9 for comparisons with other methods.

simplexification An operation on a vertex of a combinatorial graph or the entire graph. When applied to a single vertex, the vertex is replaced by a number of vertices equal to its degree. All of the new vertices are adjacent to one another (thus forming a simplex). Each former neighbor of the replaced vertex is made adjacent to one of the new vertices so that each of the new vertices has one of the replaced vertices former neighbors as a neighbor. The choice of new vertex adjacent to old vertex is irrelevant, since all choices yield isomorphic graphs. Simplexification of an entire graph is simplexification of each vertex in the graph.

simultaneous game A game in which the players move at the same time, or at least without knowledge of the other players current move. Examples include Prisoner's Dilemma, Divide-the-Dollar, and Rock Paper Scissors.

single tournament selection See tournament selection.

Sombrero ———

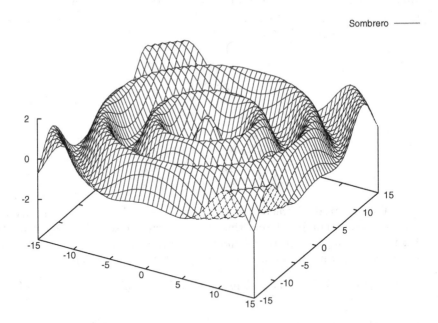

Sombrero function Shown above, the cosine of the distance from the origin. A function with a single-point optimum and an infinite number of nonpoint optima.

standard normal distribution A Gaussian distribution with a mean of zero and a standard deviation of one.

steady state An evolutionary algorithm that performs a minimal amount of breeding (a single mating event) before placing a new structure in the population. See generation.

stochastic hill climber An optimization algorithm that generates random variations of a current best configuration and retains the variation if it is better or no worse than the current best. Since an evolutionary algorithm has unary variation operators available it is easy to build a stochastic hill climber along the way to building an evolutionary algorithm. The stochastic hill climber can be used to select among potential mutation operators and to yield baseline performance for the evolutionary algorithm. See hill climbing.

stochasticity A five-dollar word for randomness.

string evolver A string representation evolutionary algorithm. Mostr string evolvers have a very simple fitness function: match to a fixed reference string. The One-max problem is a string evolver with the all ones string as the reference. String evolvers with this simple fitness function are not intrinsically interesting but can be used to understand the impact of parameter variation, e.g. mutation rate, tournament size, on simple evolutionary algorithms. The Royal Road function is a somewhat more challanging fitness function for a string evolver. The self-avoiding walk problem in Chapter 2 supplies an example of a string evolver with a fairly complex fitness function. The transposon insertion site alignment problem in Chapter 15, while apparently fairly simple, is an example of an applied problem solver with a string evolver.

subtree A tree inside a parse tree, rooted at one of its nodes. Strictly speaking a whole parse tree is a subtree rooted at the root node of the tree. Subtrees that are not the whole tree are called *proper* subtrees.

sum of squared error See least squares.

Sunburn An evolutionary algorithm for designing very simple star fighters that fight in a one-dimensional arena. See Chapter 4 for details. Sunburn uses a nonstandard model of evolution called gladitorial tournament selection.

subtree crossover A form of crossover used on parse trees. A node in each parse tree is chosen and the sub-trees rooted at those nodes are exchanged. Subtree crossover is highly disruptive and nonconservative. It is sometimes called a macromutation.

surjection A function that is onto.

symbolic regression Using genetic programming to perform least squares fit. The suggestion is that because it searches the space of models, that genetic programming is performing symbolic computations to fit the model.

symbot A type of very simple virtual robot. Described in Chapter 5.

Tartarus A grid robot task. Tartarus places a single grid robot into a world with walls and some number of boxes. The grid robots task is to move the boxes into the corners or against the walls of the world. The typical Tartarus task has a 6 × 6 grid with 6 boxes. See Chapter 10.

terminal A node in a parse tree that returns a value. It may be a constant or a means of passing values to the parse tree from outside.

Tit-for-Tat A strategy for playing Iterated Prisoner's Dilemma. The player cooperates initially and returns its opponent's last move thereafter. Tit-for-Tat is a bad strategy in a noisy environment but a very good one when noise is absent.

Tit-for-Two-Tats A strategy for playing Iterated Prisoner's Dilemma. The player cooperates unless the opponent has defected against it in the last two moves.

total order A partial order with the additional property that for any two objects a and b, either a is related to b or b is related to a.

tournament A small group selected from a population. See tournament selection.

tournament selection Any model of evolution is which a small group, the tournament, is selected and its members compared to decide who will breed and who will die. There are several variations. Single tournament selection permits the two best within a single tournament group to breed and replace the two worst. Double tournament selection uses two tournament groups, and the best in each group is chosen to breed. Replacement may be within the group or in the population using some other replacement method.

transposition Exchanging two elements of an ordered list or permutation. Any permutation may be turned into any other by applying the correct sequence of transpositions.

transposition mutation A mutation operator for ordered genes (permutations).

transposon A biological term. A piece of genetic code capable of moving from one part of a creature genome to another. There are many kinds of transposons. In Chapter 15 an evolutionary algorithm is used to align sites where a transposon inserted into the genome of corn.

Traveling Salesman problem A difficult optimization problem using ordered genes. A case of the problem consists of a list of cities together with the cost of travel between each pair of cities. The objective is to find a cyclic order to visit the cities that minimizes the total cost. This problem is explored in Chapter 7.

tree-depth limited In genetic programming, subtree crossover can cause the size of the parse trees to explode. Tree-depth-limited genetic programming controls size by pruning away any nodes beyond a certain depth from the root node of the tree after crossover. Compare with tree-size-limited genetic programming. Tree-depth-limited genetic programming is the most widespread and original type.

tree-size limited In genetic programming, subtree crossover can cause the size of the parse trees to explode. Tree-size-limited genetic programming controls size by promoting an argument of the root node of the tree to become the whole tree (chopping) after crossover whenever the number of nodes in the tree becomes too large. Compare with tree-depth-limited genetic programming.

triangle inequality The observation that going around two sides of a triangle is at least as far as going along the third side. The triangle inequality is one of the three properties a function must satisfy to be a metric.

truth table A tabulation of the output of a logic function given its inputs. Examples appear in Chapter 11.

unary variation operator Another term for mutation. A unary variation operator takes a single structure and creates a variation on it.

uniform exclusive crossover A crossover operator on sets intended to preserve the set property that elements are not repeated. When two sets undergo uniform exclusive crossover two children are produced. Any elements in common to the parent sets are placed in both children. The remaining elements are divided uniformly at random between the children. Uniform exclusive crossover is used for the the error-correcting code seeds in Section 15.3.

uniform mutation A unary variation operator typically used for string representations. Uniform mutation has a fixed, low probability of changing each character in the string. More generally, uniform mutation is a operator that can change any or all parts of a structure. One nice theoretical property of uniform mutation is that it is recurrent and hence ensures eventual convergence.

uniform real point mutation A unary variation operator that makes a point mutation with low fixed probability in each entry of a real array representation. The mutation itself is with a uniform distribution on a small interval symmetric about zero.

unimodal Having a single mode or optimum.

unimodal problem A problem having a single optimum, i.e., one global optimum and no local optima other than the global one.

unpenalized local population size A term related to niche specialization. In niche specialization, the fitness of individuals is penalized if too many accumulate "close" to one another. The unpenalized local population size of a niche specialization scheme is the maximum number of creatures that can be close before the penalties kick in.

useful mating Mating (crossover) between dissimilar creatures. A heuristic notion used to count wasted crossovers that turn to similar parents into two children similar to those parents and also to count potentially useful crossover between dissimilar parents that may explore new territory.

Vacuum Cleaner task A grid robot task. In a featureless world the Vacuum Cleaner task asks the robot to visit every square of the grid.

variation operators An umbrella term for operations such as crossover and mutation that produce variations of members of an evolving population. A variation operator is called unary, binary, etc. depending on how many different members of the population it operates on. Mutation operators are unary variation operators. Crossover operators are binary variation operators.

VCC fitness function Abbreviation for violations of the Costas condition. The VCC function accepts a permutation and returns the number of times the associated permutation matrix violates the Costas condition, that no vector joining two nonzero entries of the matrix be repeated between any other two entries.

vector A one-dimensional array of real numbers.

VIP A VI(rtual) P(olitician). A nonviolent variation of the type of evolutionary algorithm used in Sunburn. See Section 4.4.

virtual robotics Robotics simulated inside the computer as opposed to performed with physical hardware. Both symbots and grid robots are examples of virtual robots.

weighted fitness functions Combining two or more measures of quality into a single fitness function by taking a sum of constants times the quality measures yields a weighted fitness function. The constants are called the weights and establish the relative importance of the different fitness measures.

walk A sequence of vertices in a graph so that adjacent members of the sequence are also adjacent in the graph. A walk may repeat vertices where a path in a graph may not.

A

Example Experiment Report

This appendix gives an example report for Experiment 2.1. It shouldn't be followed slavishly. Individual instructors will have their own expectations for what makes a good lab report and will make them known. We suggest that lab reports contain the following:

- **Overview** Explain what the experiment is and why it is being done.
  - Which features of evolutionary computation are under examination?
  - How are they being examined?
  - What do you hope to find out? *State* your hypothesis.
- **Methods** Describe the methods used.
  - Give a description, possibly pseudocode, for the algorithm(s) you used.
  - State the source of your code (instructor, self, help from whom).
  - State the number of runs and the type of runs you made.
- **Results.** Clearly describe your results.
  - Give a summary, usually in the form of table(s) or graph(s), of your results. Present those results as clearly as possible.
  - Organize the results so that items that belong together are together.
- **Conclusions and Discussion.** Explain what happened and why you think it happened.
  - State what significance the results have in terms of the test environment and the features of evolutionary computation under investigation.
  - State problems you encountered in gathering data.
  - Explain your results, including alternative explanations if you see them and give logical support for the explanations.

The Impact of the Model of Evolution on String Evolver Performance

John Q. Student

October 17, 2004

Overview

This experiment looks at the effect of changing the model of evolution for a string evolver. The performance of a string evolver is defined as the number of generations required to find the first instance of a perfect match to the reference string. For each of seven models of evolution, 100 runs of a string evolver were performed. The models of evolution used were (**1**) single tournament selection with tournament size 4, (**2**) roulette selection and locally elite replacement, (**3**) roulette selection and random replacement, (**4**) roulette selection and absolute fitness replacement, (**5**) rank selection and locally elite replacement, (**6**) rank selection and random replacement, and (**7**) rank selection and absolute fitness replacement.

The issues discussed in class before we performed the experiment were as follows. Some needed characters for a string evolver may be present in low-fitness individuals. If this is the case, then lower selection pressure will help, avoiding the delay needed for mutation to place the character in a high-fitness individual. Of the methods examined, single tournament selection has the lowest selection pressure, roulette selection the highest. Saving the best results so far (elitism) may also play a critical role. All the techniques except random replacement save the best gene so far (or one superior to it, as in random elite replacement).

Methods

The experiments for all seven models of evolution operated on a population of 60 strings. The reference string was 20 characters long ("the reference string") over the ASCII printable characters, codes 32–127. In each generation, one-half the population was replaced to make the other six models of evolution agree with single tournament selection. One hundred evolutionary runs were performed for each model of evolution. The number of generations until a solution was located were recorded.

If a run required in excess of 10,000 generations to find a solution, that run was ended without a solution and recorded as a failure.

I wrote the code used for the experiment; it is given in Appendix 1 of this report. The code uses a single numerical switch to change the model of evolution being used. In the *generation* routine in the code there are a pair of switch statements turning on the model of evolution. The first performs selection (including all of tournament selection); the second performs replacement (except for tournament selection, which has replacement tightly integrated with selection).

I used the random number generator supplied by the instructor.

Results

Mean and standard deviation of time to solution for the various models of evolution, together with data on failures, are given in the table below:

Model of Evolution	Mean	Std. Dev.	Failures
Single tournament, size 4	258.5	82.8	0
Roulette selection, locally elite replacement	316.5	89.8	0
Roulette selection, random replacement	n/a	n/a	100
Roulette selection, absolute fitness replacement	394.6	157.8	0
Rank selection, locally elite replacement	361.1	84.6	0
Rank selection, random replacement	n/a	n/a	100
Rank selection, absolute fitness replacement	389.2	82.1	0

The results clearly show that random replacement is not a good idea. For the five models of evolution not involving random replacement, the following table gives 95% confidence intervals on the mean time to solution. This confidence interval uses the fact that the mean of a repeated experiment is normally distributed. The 95% confidence interval is

$$\mu \pm \frac{1.96s}{\sqrt{n}},$$

where μ is the mean, s is the sample standard deviation, n is the number of trials performed, and 1.96 is the value of $Z_{\alpha/2}$ for confidence $\alpha = 0.05$ in the normal distribution. The table is given in increasing order of mean time to solution:

Model of Evolution	95% C.I.
Single tournament, size 4	(242.4, 274.6)
Roulette selection, random elite replacement	(298.9, 334.1)
Rank selection, random elite replacement	(344.5, 377.7)
Rank selection, absolute fitness replacement	(373.1, 405.3)
Roulette selection, absolute fitness replacement	(363.7, 425.5)

Conclusions and Discussion

The initial hypothesis that saving the best result so far is important, i.e., using a fitness ratchet, seems to be correct. Random replacement apparently overwrote good results often enough to prevent the algorithms from functioning inside the permitted time horizon. Clearly, both models of evolution using random replacement were far worse than all the others.

Single tournament selection was a clear winner, with a significantly better mean time to solution, using the confidence intervals in the second table. Absolute fitness replacement seems worse than locally elite replacement, but the effects are somewhat confounded. In second place after tournament selection was roulette selection with random elite replacement; it is significantly better than the other methods. The table seems to give weak evidence that roulette is better than rank selection and that random elite is better than absolute fitness replacement.

The evidence suggests that preservation of diversity is a good thing, but a good deal less conclusively than the evidence in favor of a fitness ratchet. The winner in overall performance was the method that preserved diversity the best, as stated in class, but the second-best diversity preserver, rank selection, came in pretty much last. Increasing the number of trials may help resolve the question.

Rank selection does preserve diversity better than roulette selection in the face of a large fitness gradient. It destroys diversity better when there is a very small fitness gradient. Given the small size of the population, there may have been a relatively small fitness gradient for much of the time in each run. I thus conclude that the evidence in favor of diversity preservation from the tournament selection results should be given greater weight. A series of experiments varying tournament size (a larger tournament implies less diversity preservation) may provide valuable independent evidence.

Appendix 1: Code Listing

```cpp
/*  Experiment 2.1 for Evolutionary Computation
 *  for Modeling and Optimization
 *
 *  Models of evolution:
 *     0 - Size 4 single tournament selection
 *     1 - Roulette selection and random elite replacement
 *     2 - Roulette selection and random replacement
 *     3 - Roulette selection and absolute fitness replacement
 *     4 - Rank selection and random elite replacement
 *     5 - Rank selection and random replacement
 *     6 - Rank selection and absolute fitness replacement
 *  use these as the first argument to "generation" in the main program
 *  loop to select which model of evolution you are using.
 */
#include <iostream>
#include <cmath>

using namespace std;

#include "Random378.h"

//model of evolution to use, see above comment for codes
#define modevo 6
//characters in the string for the string evolver
#define string_length 20
//population size
#define population_size 60
//number of trials to run
#define trials 100
//Timeout number
#define timeout 10000
//reference string
#define ref "The reference string"

                         /* Function Prototypes */
//set initial algorithm parameters
void initialize_algorithm();

//compute the fitness of a string
int fitness(char population_member[string_length]);

//create a new random population
void initialize_population(
     char population[population_size][string_length],//population members
        int fit[population_size],                    //fitness
        int &bestfit                                 //best fitness tracker
     );

//proportional selection, used for roulette and rank selection
//selects a population index in proportion to the values
//all values must be nonnegative!
int proportional(int values[population_size]);

//run a generation of evolution with the given model of evolution
void generation(
        int model_evo,                               //model of evolution
        char population[population_size][string_length],//population members
        int fit[population_size],                    //fitness
        int &bestfit                                 //best fitness tracker
     );
```

```
main(){//main program

char population[population_size][string_length];  //population
int fit[population_size];                          //fitnesses
int trial_counter,gen_counter;                     //loop indices
int bestfit;                                       //best fitness tracker
int times[trials];                                 //record time to solution
double mean,variance,value;                        //statistical scratch variables
int failures;                                      //counts number of timeouts
int successes;                                     //trials-failures

  initialize_algorithm(); //seed the random number generator
  failures=0;
  for(trial_counter=0;trial_counter<trials;trial_counter++){//loop over trials
    bestfit=0;  //initialize fitness tracking variable
    initialize_population(population,fit,bestfit);  //new population
    gen_counter=0;  //initialize generation counter
    while((bestfit<string_length)&&(gen_counter<timeout)){//generation loop
      generation(modevo,population,fit,bestfit);
      gen_counter++;
    }
    cout << "Trial " << trial_counter+1 << " solution or timeout in "
         << gen_counter << " generations." << endl;
    times[trial_counter]=gen_counter;
  }
  failures=successes=0;
  mean=variance=0.0; //initialize statistical accumulators
  for(int i=0;i<trials;i++){//sum fitness, squared fitness
    if(times[i]<timeout){
      value=((double)times[i]); //prevent integer wrap around
      mean+=value;
      variance+=(value*value);
      successes++;
    } else failures++;
  }
  if(successes>2){//report mean and varience if they mean anything
    mean/=successes; //compute mean
    //compute variance
    variance/=successes;
    variance-=(mean*mean);
    cout << "With N=" << successes << " trials that terminated." << endl;
    cout << "Mean generations to solution " << mean << "." << endl;
    cout << "Standard deviation " << sqrt(variance) << "." << endl;
  }
  cout << "There were " << failures << " failures." << endl;

}
```

```
//set initial algorithm parameters
void initialize_algorithm(){

  seedMT(9120782);  //change the random number seed to get
                    //a different sample of the outcomes

}

//compute the fitness of a string
int fitness(char population_member[string_length]){

int i,cnt;  //loop index, fitness counter

  cnt=0; //intialize counter
  for(i=0;i<string_length;i++){//traverse the string
    if(population_member[i]==ref[i])cnt++;  //check for match in each position
  }
  return(cnt);  //return the number of agreements
}

//create a new random population
void initialize_population(
    char population[population_size][string_length], //population members
        int fit[population_size],                    //fitness
        int &bestfit                                 //best fitness tracking
    ){

int i,j;  //loop index variables

  for(j=0;j<population_size;j++){//loop over population
    //The following fills each position of the population member with
    //a printable ASCII string, codes 32-127
    for(i=0;i<string_length;i++)population[j][i]=lrandMT()%96+32;
    //compute the initial fitnesses
    fit[j]=fitness(population[j]);
    if(fit[j]>bestfit)bestfit=fit[j];
  }
}
```

```
//proportional selection, used for roulette and rank selection
//selects a population index in proportion to the values
//all values must be nonnegative!
int proportional(int values[population_size]){

int total,dart;  //total value and the proportional selection dart
int i;           //loop index

  total=0;//zero the value acucmulator
  for(i=0;i<population_size;i++)total+=values[i];  //sum the values

  //if there is nothing to select on, select randomly
  if(total==0)return(lrandMT()%population_size);

  //otherwise
  dart=lrandMT()%total; //throw the dart

  for(i=0;i<population_size;i++){//figure out where the dart landed
    dart-=values[i];  //subtract the ith value
    if(dart<0)return(i);  //if it goes negative, there we are
  }

  //warn the user
  cerr << "Selection failure" << endl;

  //return an in-range value
  return(lrandMT()%population_size);

}
```

```
//run a generation of evolution with the given model of evolution
void generation(
        int model_evo,                          //model of evolution
        char population[population_size][string_length], //population members
        int fit[population_size],                //fitness
        int &bestfit                            //best fitness tracker
        ){

int sortindex[population_size];  //Sorting index and value buffer
int i,j;                         //loop indices
int rv,sw;                       //sorting and unsorting variables
int cp;                          //crossover point
int rp;                          //random position for mutation
int p1,p2;                       //population index of first and second parents
int flag;                        //rank stop flag

//new population for nontournament models of evolution
char newpop[population_size/2][string_length];  //genes
int newfit[population_size/2];                   //fitness data

  switch(model_evo){//selection code and tournament selection
  case 0: //tournament selection, size four

    for(i=0;i<population_size;i++)sortindex[i]=i; //initialize sortindex

    for(i=0;i<population_size;i++){//randomize the population order
      rv=lrandMT()%population_size;  //get a random position
      //swap current and random sorting indices
      sw=sortindex[i];sortindex[i]=sortindex[rv];sortindex[rv]=sw;
    }

    for(i=0;i<population_size;i+=4){//loop by tournaments
      //Knuth-hardsort the tournament to decreasing fitness order
      if(fit[sortindex[i]]<fit[sortindex[i+3]]){//compare 0:3
        sw=sortindex[i];sortindex[i]=sortindex[i+3];sortindex[i+3]=sw;
      }
      if(fit[sortindex[i+1]]<fit[sortindex[i+2]]){//compare 1:2
        sw=sortindex[i+1];sortindex[i+1]=sortindex[i+2];sortindex[i+2]=sw;
      }
      if(fit[sortindex[i]]<fit[sortindex[i+1]]){//compare 1:1
        sw=sortindex[i];sortindex[i]=sortindex[i+1];sortindex[i+1]=sw;
      }
      if(fit[sortindex[i+2]]<fit[sortindex[i+3]]){//compare 2:3
        sw=sortindex[i+2];sortindex[i+2]=sortindex[i+3];sortindex[i+3]=sw;
      }
      if(fit[sortindex[i+1]]<fit[sortindex[i+2]]){//compare 1:2
        sw=sortindex[i+1];sortindex[i+1]=sortindex[i+2];sortindex[i+2]=sw;
      }
      //Sorted

      //reproductions: copy the parents over the dead to create children
      //performing the crossover as you do so
      cp=lrandMT()%string_length; //compute crossover point
      for(j=0;j<cp;j++){//copy parents to crossover point
        population[sortindex[i+2]][j]=population[sortindex[i]][j];
        population[sortindex[i+3]][j]=population[sortindex[i+1]][j];
      }
      for(j=cp;j<string_length;j++){//copy parents from crossover point to end
        population[sortindex[i+2]][j]=population[sortindex[i+1]][j];
        population[sortindex[i+3]][j]=population[sortindex[i]][j];
      }

      //Now perform mutation in both children
      rp=lrandMT()%string_length; //select position of mutation
      population[sortindex[i+2]][rp]=lrandMT()%96+32;  //put in new character
      rp=lrandMT()%string_length; //select position of mutation
```

```
      population[sortindex[i+3]][rp]=lrandMT()%96+32;   //put in new character

      //update fitness information for both children
      fit[sortindex[i+2]]=fitness(population[sortindex[i+2]]);
      if(fit[sortindex[i+2]]>bestfit)bestfit=fit[sortindex[i+2]];
      fit[sortindex[i+3]]=fitness(population[sortindex[i+3]]);
      if(fit[sortindex[i+3]]>bestfit)bestfit=fit[sortindex[i+3]];
    }
    break; //end of tournament selection

  case 1: //roulette selection
  case 2:
  case 3:
    for(i=0;i<population_size/2;i+=2){//select pairs of parents
      //roulette selection of parents
      p1=proportional(fit);
      p2=proportional(fit);

      //reproductions: copy the parents to new populations
      //performing the crossover as you do so
      cp=lrandMT()%string_length; //compute crossover point
      for(j=0;j<cp;j++){//copy parents to crossover point
        newpop[i][j]=population[p1][j];
        newpop[i+1][j]=population[p2][j];
      }
      for(j=cp;j<string_length;j++){//copy parents crossover to end
        newpop[i][j]=population[p2][j];
        newpop[i+1][j]=population[p1][j];
      }

      //Perform mutation of the children
      rp=lrandMT()%string_length;       //select position of mutation
      newpop[i][rp]=lrandMT()%96+32;    //new character
      rp=lrandMT()%string_length;       //select position of mutation
      newpop[i+1][rp]=lrandMT()%96+32; //new character

      //compute new fitnesses
      newfit[i]=fitness(newpop[i]);
      if(newfit[i]>bestfit)bestfit=newfit[i];
      newfit[i+1]=fitness(newpop[i+1]);
      if(newfit[i+1]>bestfit)bestfit=newfit[i+1];
    }
    break; //done with roulette selection
  case 4: //rank selection
  case 5:
  case 6:
    //create the rank array
    for(i=0;i<population_size;i++)sortindex[i]=i; //initialize sortindex
    do {//bubble sort the population into descending fitness order
      flag=0;
      for(j=0;j<population_size-1;j++){
        if(fit[sortindex[j]]>fit[sortindex[j+1]]){
          sw=sortindex[j];sortindex[j]=sortindex[j+1];sortindex[j+1]=sw;
          flag=1;
        }
      }
    }while(flag);
    //sort is 0..populationsize-1, need to add one to get ranks
    for(i=0;i<population_size;i++)sortindex[i]++;
    //rank array is created

    for(i=0;i<population_size/2;i+=2){//select pairs of parents
      //rank selection of parents
      p1=proportional(sortindex);
      p2=proportional(sortindex);
      //reproductions: copy the parents to new populations
      //performing the crossover as you do so
```

```
      cp=lrandMT()%string_length; //compute crossover point
      for(j=0;j<cp;j++){//copy parents to crossover point
        newpop[i][j]=population[p1][j];
        newpop[i+1][j]=population[p2][j];
      }
      for(j=cp;j<string_length;j++){//copy parents crossover to end
        newpop[i][j]=population[p2][j];
        newpop[i+1][j]=population[p1][j];
      }

      //Perform mutation of the children
      rp=lrandMT()%string_length;        //select position of mutation
      newpop[i][rp]=lrandMT()%96+32;   //new character
      rp=lrandMT()%string_length;        //select position of mutation
      newpop[i+1][rp]=lrandMT()%96+32; //new character

      //compute new fitnesses
      newfit[i]=fitness(newpop[i]);
      if(newfit[i]>bestfit)bestfit=newfit[i];
      newfit[i+1]=fitness(newpop[i+1]);
      if(newfit[i+1]>bestfit)bestfit=newfit[i+1];
    }
    break; //done with rank selection
}

switch(model_evo){//replacement code - except for tournament selection
case 0: //do nothing - tournament selection is finished
    break;
case 1: //random elite replacement
case 4:
    for(i=0;i<population_size;i++)sortindex[i]=i; //initialize sortindex
    for(i=0;i<population_size;i++){//randomize the population order
      rv=lrandMT()%population_size;  //get a random position
      //swap current and random sorting indices
      sw=sortindex[i];sortindex[i]=sortindex[rv];sortindex[rv]=sw;
    }

    //perform the random elite replacement
    for(i=0;i<population_size/2;i++){//for all new strutures
      //compare with a population member selected randomly without replacement
      if(newfit[i]>=fit[sortindex[i]]){//child is at least as good
        //replace creature and fitness information
        for(j=0;j<string_length;j++)population[sortindex[i]][j]=newpop[i][j];
        fit[sortindex[i]]=newfit[i];
      }
    }
    break; //done with random elite replacement
case 2:
case 5:
    for(i=0;i<population_size;i++)sortindex[i]=i; //initialize sortindex
    for(i=0;i<population_size;i++){//randomize the population order
      rv=lrandMT()%population_size;  //get a random position
      //swap current and random sorting indices
      sw=sortindex[i];sortindex[i]=sortindex[rv];sortindex[rv]=sw;
    }
    //perform the random replacement
    for(i=0;i<population_size/2;i++){//for all new strutures
      //replace creature and fitness information
      for(j=0;j<string_length;j++)population[sortindex[i]][j]=newpop[i][j];
      fit[sortindex[i]]=newfit[i];
    }
    //since we may have overwritten the best creature we now need
    //to recompute the bestfit variable
    bestfit=0;
    for(i=0;i<population_size;i++)if(fit[i]>bestfit)bestfit=fit[i];
    break; //done with random replacement
case 3: //random elite replacement
```

```
case 6:
  //Order the population by fitness
  for(i=0;i<population_size;i++)sortindex[i]=i; //initialize sortindex
  do {//bubble sort the population into increasing fitness order
    flag=0;
    for(j=0;j<population_size-1;j++){
      if(fit[sortindex[j]]>fit[sortindex[j+1]]){
        sw=sortindex[j];sortindex[j]=sortindex[j+1];sortindex[j+1]=sw;
        flag=1;
      }
    }
  }while(flag);

  //perform the absolute fitness replacement
  for(i=0;i<population_size/2;i++){//for all new strutures
    //replace creature and fitness information
    for(j=0;j<string_length;j++)population[sortindex[i]][j]=newpop[i][j];
    fit[sortindex[i]]=newfit[i];
  }
  break; //done with absolutereplacement
}

}
```

B

Probability Theory

This appendix reviews some terms and mathematical notions from probability theory used in this book that may not have appeared in your program of study or which you may have forgotten. Ubiquitous in the theory of artificial life is the notion of a *Markov chain*, a set of repeated trials that are not independent. On the way to the elementary parts of the theory of Markov chains, we will review a good deal of basic probability theory.

B.1 Basic Probability Theory

A *distribution* D is a triple (Q, E, P) consisting of a set of *points* Q, a collection of *events* E that are subsets of Q, and a function $P : E \to [0, 1]$ that assigns probabilities to events. How would we represent the familiar example of flipping a fair coin in this notation?

Example 38. **Flipping a fair coin** When D represents flipping a fair coin, we have point set $Q =$ {heads, tails}, events $E = \{\{\},\{\text{heads}\},\{\text{tails}\},\{\text{heads, tails}\}\}$, and probability assignment

$$P(\{\}) = 0,$$
$$P(\{\text{heads}\}) = 0.5,$$
$$P(\{\text{tails}\}) = 0.5,$$
$$P(\{\text{heads, tails}\}) = 1.$$

Probabilities are real numbers in the unit interval. There is one additional requirement to make a triple (Q, E, P) a distribution. As long as the set Q is finite or countably infinite, we demand that

$$\sum_{q \in Q} P(\{q\}) = 1. \tag{B.1}$$

In the event that Q is uncountable, we demand that

$$\int_{q \in Q} P(\{q\}) = 1. \tag{B.2}$$

Typically, we confuse singleton sets with their sole member so that we define $P(q) := P(\{q\})$ for each $q \in Q$. You may wonder why we have points *and* events. Since events are built out of points, their presence seems redundant. There are two reasons. First, events consisting of many points in the distribution are often the actual objects of interest. Second, in the case in which Q is an uncountably infinite set (like the real numbers), the probability of singleton point events is typically zero. This forces us to deal with multi-point events to get anything done.

Example 39. **The uniform distribution on [0, 1]** A *uniform distribution* is one in which all points are equally likely. Notice that the distribution in Example 38 was uniform on two points. On an uncountable set, we achieve a uniform distribution by insisting that events of the same size be assigned the same probability by P. Two events A and B are the same size if

$$\int_{a \in A} dx = \int_{b \in B} dx.$$

A little work will show that for the uniform distribution on $[0, 1]$, we may take

$$P(x) = 1.$$

We compute the probability of an event by computing the integral of $P(x)$ on that event. Notice that we have been vague about specifying what E is in this example. Events that are built from intervals by the operations of intersection, union, and complementation are safe. For a better treatment, a course in measure theory is required.

A *trial* is the result of sampling a point from a distribution, flipping a coin, for example. A way of looking at the probability of an event is that it is the chance that a point in the event will be chosen in a trial. A *set of repeated trials* is a collection of trials taken one after the other from the same distribution or a sequence of distributions. We can place a *product distribution* on repeated trials by letting the points in the product distribution be the possible sets of outcomes of a repeated trial and then inducing the events and their associated probabilities in the natural manner.

Example 40. **A product distribution** Suppose we flip 3 coins. We then have an example of 3 repeated trials sampled from the distribution given in Example 38. The set of 3 trials forms a single trial in a (3-fold) product distribution. The points of this distribution are

$$\{\{H, H, H\}, \{H, H, T\}, \{H, T, H\}, \{H, T, T\},$$

$\{T, H, H\}, \{T, H, T\}, \{T, T, H\}, \{T, T, T\}\}.$

The set of events consists of all 256 subsets of the set of points. Each single-point event has probability 1/8, and the probability of an event is the number of points in it divided by 8.

Two events A and B are said to be *independent* if

$$P(A \cap B) = P(A) \cdot P(B).$$

An example of two independent events is as follows. If we flip two coins and put a product distribution on the 4 possible outcomes, then the events "the first coin comes up heads" and "the second coin comes up tails" are independent. If you want to know the probability of several independent events all happening, then you multiply their probabilities. The probability of getting 3 heads on 3 flips of a fair coin, for example, is $\frac{1}{2} \cdot \frac{1}{2} \cdot \frac{1}{2} = \frac{1}{8}$. (Each of 3 independent flips has probability $\frac{1}{2}$ of producing a head. Multiply them to get the probability of 3 heads in a row.)

If two events are not independent, then they are said to be *dependent*. Suppose, for example, we have a pot containing 5 black and 5 white balls, and we have two trials in which we draw balls out of the pot at random. If we do not replace the first ball before drawing the second, then the probability of drawing a black or white ball is dependent on what we drew the first time. In both trials, the events are $\{black\}$ or $\{white\}$, but the distribution of the second draw is changed by the first draw. The events "first ball is white" and "second ball is white" are *dependent* in the product distribution of the two trials.

If two events are such that either one happening completely precludes the other happening, then the events are said to be *disjoint*. Mathematically, A and B are disjoint if

$$P(A \cup B) = P(A) + P(B).$$

If you want to know the probability of one of several disjoint events happening, then you simply sum their probabilities. Each of the faces of a fair 6-sided die has probability 1/6 of being rolled, and all 6 events are disjoint. The probability of rolling a prime number on a 6-sided die is $P(2) + P(3) + P(5) = 1/6 + 1/6 + 1/6 = 1/2$. (Try asking a friend to call "prime" or "nonprime" on a die instead of "heads" or "tails" on a coin. A humorous argument often ensues, especially in the presence of those who believe 1 to be prime.)

If a distribution is on a set of numbers, then a distribution has an *expected value*. One computes the expected value of a distribution on a set of numbers by summing the product of the numbers with their respective probabilities. Take, for example, the numbers 1 through 6, as generated by a fair die. The probability of each is 1/6, and so the expected value of the roll of a single die is $\frac{1}{6} \cdot 1 + \frac{1}{6} \cdot 2 + \frac{1}{6} \cdot 3 + \frac{1}{6} \cdot 4 + \frac{1}{6} \cdot 5 + \frac{1}{6} \cdot 6 = 3.5$. The notion of expected value is a mathematical generalization of the more familiar notion of *average*. Formally, if $D = (Q, E, P)$ is a distribution for which $Q \subseteq \mathbb{R}$, then the expected value $E(D)$ is given by

$$E(D) = \sum_{q \in Q} q \cdot P(q). \tag{B.3}$$

Many introductory probability classes deal largely with sets of independent repeated trials or sets of disjoint events, because they are far easier to work with mathematically. The modus operandi of evolution is to have strongly *dependent* trials. Rather than maintaining the same distribution by replacing balls in the pot between trials, we throw away most of the balls we draw and produce new balls by combining old ones in odd fashions. This means that dependent probability models are the norm in artificial life. The independent models are also useful; they can, for example, be used to understand the composition of the initial population in an evolutionary algorithm.

B.1.1 Choosing Things and Binomial Probability

The symbol $\binom{n}{k}$, pronounced "n choose k" is defined to be the number of different sets of k objects that can be chosen from a set of n objects. There is a simple formula for the choice numbers:

$$\binom{n}{k} = \frac{n!}{k!(n-k)!}. \tag{B.4}$$

In choosing k objects out of n there are n choices for the first object, $n - 1$ choices for the second, and so on until there are

$$n \cdot (n-1) \cdots (n-k+1) = \frac{n!}{(n-k)!} \tag{B.5}$$

ways to choose the set. These choices, however, have an implicit order, and so in choosing k objects, there are $k!$ distinct orders in which we could choose the same set. Dividing by $k!$ yields the desired formula. Since choosing and failing to choose objects are dual to one another, we obtain the useful identity

$$\binom{n}{k} = \binom{n}{n-k}, \tag{B.6}$$

which also clearly follows from algebraic manipulation of the formula B.4. The choice numbers are also called the *binomial coefficients*, because of their starring role in the binomial theorem.

Theorem 7. *(Binomial theorem)*

$$(x+y)^n = \sum_{k=0}^{n} \binom{n}{k} x^k y^{n-k}.$$

A *Bernoulli trial* is a trial from a distribution $D = (Q, E, P)$ for which $|Q| = 2$. These two events happen with probability p and $1 - p$. One of the events is typically called a *success*, and the other is called a *failure*. The probability of success is p. The *binomial probability model* is used to compute the probability of seeing some number of successes in an independent set of repeated Bernoulli trials.

Theorem 8. *(Binomial probability model) If we are doing a set of n independent Bernoulli trials with probability p of success, then the probability of obtaining exactly k successes is*

$$\binom{n}{k} p^k (1 - p)^{n-k}.$$

The binomial probability model looks like a piece sliced out of the binomial theorem with p and $(1 - p)$ taking the place of x and y. This is the result of identical counting arguments producing the binomial probability model and the terms of the binomial theorem. If we are to have k successes, then we also have $n - k$ failures. Since the events are independent, we multiply the probabilities. Thus, any given sequence of successes and failures with k successes has probability $p^k(1 - p)^{(n-k)}$. Since the successes form a k-subset of the trials, there are $\binom{n}{k}$ such sequences. We multiply the probability of a single sequence with k successes by the number of such sequences to obtain the probability of getting k successes: the binomial probability model.

Example 41. Suppose that we have a population of 60 strings of length 20 that were produced by choosing characters "0" or "1" with a uniform distribution. What is the largest number of 1's we would expect to see in a member of the population? Answer this question by finding the number of 1's such that the expected number of creatures with that many 1's is (i) at least 1 and (ii) as small as possible.
Answer:
The expected number of creatures with k 1's is just the population size times the result of plugging $p = 1/2$, $n = 20$ into the binomial probability model. For 60 creatures, the expected number of creatures with 14 1's is 2.217. The expected number of creatures with 15 1's is 0.8871. So, 14 is a reasonable value for the largest number of 1's you would expect to see in such a population.

A quick way of generating binomial coefficients is to use *Pascal's triangle*, the first 11 rows of which are shown in Figure B.1. It is left to you to deduce how the triangle was generated and how to find a given binomial coefficient in the triangle.

B.1.2 Choosing Things to Count

In this section, we will use cards as our probability paradigm. We will use the machinery developed to learn something about single tournament selection.

```
                                    1
                                 1     1
                              1     2     1
                           1     3     3     1
                        1     4     6     4     1
                     1     5    10    10     5     1
                  1     6    15    20    15     6     1
               1     7    21    35    35    21     7     1
            1     8    28    56    70    56    28     8     1
         1     9    36    84   126   126    84    36     9     1
      1    10    45   120   210   252   210   120    45    10     1
   1    11    55   165   330   462   462   330   165    55    11     1
1    12    66   220   495   792   924   792   495   220    66    12     1
1   13    78   286   715  1287  1716  1716  1287   715   286    78    13     1
1  14    91   364  1001  2002  3003  3432  3003  2002  1001   364    91    14     1
1  15   105   455  1365  3003  5005  6435  6435  5005  3003  1365   455   105    15     1
```

Fig. B.1. Pascal's Triangle from $n=0$ to $n=15$.

Some familiarity with poker is assumed; consult Hoyle or a friend if you are unfamiliar with this game.

Example 42. What is the number of 5-card poker hands that can be dealt?
Answer:
    Compute the number of ways to *choose* 5 out of 52 cards, that is,

$$\binom{52}{5} = \frac{52!}{5! \cdot 47!} = 2{,}598{,}960.$$

    To get the probability of a given type of poker hand, you simply divide the number of ways to get the hand by the number of total hands. The next three examples illustrate this.

Example 43. What is the probability of getting three of a kind?
Answer:
    First let's solve the problem, "how many different poker hands are there that count as three of a kind?" Three of a kind is a hand that contains 3 cards with the same face value and 2 other cards with 2 other distinct face values. To get 3 cards the same, we choose the face value, choose 3 of the 4 cards with that face value, and then choose 2 of the other 49 cards, i.e., there are

$$\binom{13}{1} \cdot \binom{4}{3} \cdot \binom{49}{2} = 61{,}152$$

poker hands that contain 3 cards with the same face value.
    We are not done yet! This counting includes hands with 4 cards the same ("four of a kind") and with 3 cards with one face value and the other 2 with

another face value (a "full house"). Both of these are *better* than three of a kind and so do not count as three of a kind.

To get the correct count, we must therefore count the number of ways to get four of a kind and a full house and subtract these from the total. Four of a kind is quite easy: simply choose a face value, choose all 4 cards of that face value, and then choose one of the 48 other cards. There are

$$\binom{13}{1} \cdot \binom{4}{4} \cdot \binom{48}{1} = 624$$

ways to get four of a kind.

A full house is a little harder: choose 1 of the 13 face values to be the "three the same," choose 3 of those 4 cards, then choose 1 of the 12 remaining face values to be the "two the same," and then choose 2 of the 4 cards with that face value. In short, there are

$$\binom{13}{1} \cdot \binom{4}{3} \cdot \binom{12}{1} \cdot \binom{4}{2} = 3744$$

different ways to get a full house.

Putting this all together, there are

$$61{,}152 - 624 - 3744 = 56{,}784$$

ways to get three of a kind.

To get the probability of getting three of a kind, we divide by the total number of poker hands.

$$P(\text{three-of-a-kind}) = \frac{56{,}784}{2{,}598{,}960} \approx 0.02185.$$

Example 44. How many ways are there to get two of a kind?
Answer:

Again, we start by counting the number of hands that are two of a kind: 2 cards with the same face value and the other 3 with distinct face values. Since a large number of different types of good poker hands contain 2 cards with the same face value, it would be laborious to follow the count-and-subtract technique used in Example 43. We will, therefore, compute directly.

First, we select 1 of the 13 face values for our "two the same" and then choose 2 of those 4 values. This leaves 12 face values from which we must select 3 distinct face values to fill out the hand. Once we know these 3 face values, it follows that we must choose 1 of the 4 cards within each of these face values. This gives us

$$\binom{13}{1} \cdot \binom{4}{2} \cdot \binom{12}{3} \cdot \binom{4}{1}^3 = 1{,}098{,}240$$

ways to get two of a kind.

Dividing by the total number of poker hands, we get

$$P(\text{two-of-a-kind}) = \frac{1{,}098{,}240}{2{,}598{,}960} \approx 0.42256903.$$

One odd fact about poker hands is that the more valuable ones are easier to count. This is because they are not themselves related to still more valuable hands above them. The *flush*, a hand in which all 5 cards have the same suit, is quite easy to count, especially since a royal flush or a straight flush are, via linguistic technicality, still flushes.

Example 45. What is the probability of getting a flush?
Answer:
    First count the number of flush hands. We must choose 1 of 4 suits and then pick which 5 of the 13 cards in that suit we want. Thus, there are

$$\binom{4}{1} \cdot \binom{13}{5} = 5148$$

different ways to get a flush, yielding

$$P(\text{flush}) = \frac{5148}{2{,}598{,}960} \approx 0.001981.$$

Now, with the mental machinery all charged up to count things using *choose*, we can explore an issue concerning single tournament selection with tournament size 4. What is the expected number of children a creature participating in single tournament selection will have in each generation? First, let us agree that when two parents have two children, each incorporating some fraction of each parent's gene, this counts as one child per parent. This means that, in single tournament selection, the expected number of children of a parent is one times the probability that that parent will be placed by the random selection in a tournament in which it is one of the two most fit. Clearly, this probability can be computed from a creature's rank in the population in a given generation. (We will assume that when there are ties in fitness, they do not lead to ties in rank, but rather, rank is selected among equally fit creatures uniformly at random.)

Theorem 9. *The expected number of children of a creature with rank k out of a population of n creatures using single tournament selection as the model of evolution is*

$$\frac{\binom{n-k}{3} + \binom{n-k}{2}\binom{k-1}{1}}{\binom{n-1}{3}}.$$

Proof:
    There are two disjoint events that together make up the event in which we are interested, a creature being one of the 2 most fit creatures in its group of

4. Either it can be the top creature, or it can be the second in its group of 4. The number of choices of other creatures that leave the creature in question at the top is simply the number of creatures less fit than it choose 3, $\binom{n-k}{3}$. If it is the second creature, then we choose 2 creatures from those less fit, $\binom{n-k}{2}$, and 1 from those more fit, $\binom{k-1}{1}$. Since these events are disjoint, they add. Finally, divide by the number of possible ways to choose 3 creatures to obtain a probability. Finally, notice that in tournament selection, this probability is equal to the expected number of children. □

To give a feel for how the expected number of children is distributed, we show the probabilities for a population of size 24 in Example 46. It is interesting to note that the probability of death is exactly one minus the probability of having children in this model of evolution when the tournament size is 4. As an exercise, you could compute the probability based on rank of becoming a parent or of dying for tournament sizes other than 4.

Example 46. Probability of tournament selection

Rank	Expected Children	Rank	Expected Children
1	1	13	0.4658
2	1	14	0.3981
3	0.9881	15	0.3320
4	0.9656	16	0.2688
5	0.9334	17	0.2095
6	0.8927	18	0.1553
7	0.8447	19	0.1073
8	0.7905	20	0.0666
9	0.7312	21	0.0344
10	0.6680	22	0.0119
11	0.6019	23	0
12	0.5342	24	0

B.1.3 Two Useful Confidence Intervals

In many of the experiments in this book, we record the time until success, in generations or mating events, for a large number of populations. When there are variations in the evolutionary algorithms used to produce those times, we can ask which variation of the algorithm worked better. Let us imagine that we are studying the difference between single-point and probabilistic mutation in a string evolver of the sort used in Chapter 2. Figure B.2 gives a graph of the fraction of populations that contain a copy of the reference string as a function of the number of generations. The graphs show that single-point mutation outperforms probabilistic mutation at whatever rate it was performed. The question remains, "is the difference significant?"

Answering the question of significance can be done precisely. What you do is compute the probability that two experiments could be as different as they are by chance. To do this we construct confidence intervals. A *confidence interval with a given p-value for a quantity q* is a range of values $q_l \le q \le q_h$ such that the probability that the true value of q is between q_l and q_h is p. A general treatment of confidence intervals is given in any mathematical statistics book. We will treat two different sorts of confidence intervals: for the value of the probability of success for a Bernoulli trial and for the mean of a sampled random variable.

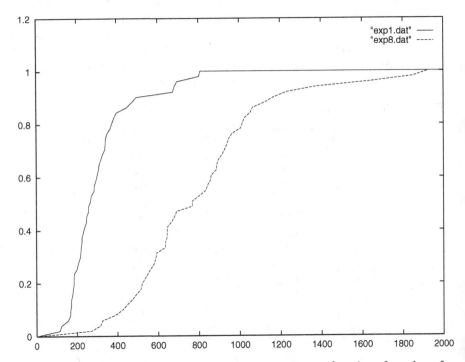

Fig. B.2. Fraction of populations with a correct answer as a function of number of generations. (The solid line graphs the data from a string evolver using single point mutation. The dotted line graphs the data from a string evolver using probabilistic mutation.)

Definition B.1 *A **random variable** X with distribution $D = (Q, E, P)$ is a surrogate for choosing a point from Q with probability as specified by P.*

A random variable X associated with flipping a coin has the distribution given in Example 38. It has two possible outcomes: "heads" and "tails." A random variable can be thought of as an instance of its distribution.

There are two important quantities associated with a random variable over a set of numbers: its mean and variance. The mean of a random variable is just its expected value (see Equation B.3). We denote the mean of a random variable X by the symbol μ_X. Restating Equation B.3 for a random variable X with distribution $D = (Q, E, P)$, we have

$$\mu_X = E(X) = \sum_{q \in Q} q \cdot P(\{q\}),\tag{B.7}$$

or

$$\mu_X = E(X) = \int_Q q \cdot P(q) \cdot dq.\tag{B.8}$$

The variance of a random variable is the degree to which it tends to differ from its mean. It is denoted by σ_X^2. Formally, the variance of a random variable X is given by

$$\sigma_X^2 = E((X - \mu_X)^2) = E(X^2) - \mu_X^2.\tag{B.9}$$

The variance is denoted by σ_X^2 in part because the square root of the variance is also a commonly used quantity, the *standard deviation*.

Definition B.2 *The* **standard normal distribution,** *denoted by* $N(0,1)$, *is a distribution with* $Q = \mathbb{R}$ *and*

$$P(E) = \frac{1}{\sqrt{2\pi}} \int_E e^{-x^2/2} \cdot dx.$$

The mean of this distribution is 0 and the variance is 1. The normal distribution with mean μ *and standard deviation* σ, *denoted by* $N(\mu, \sigma)$, *is a distribution with* $Q = \mathbb{R}$ *and*

$$P(E) = \frac{1}{\sqrt{2\pi}} \int_E e^{-\frac{(x-\mu)^2}{2\sigma^2}} \cdot dx.$$

α	z_α
0.05	1.64
0.025	1.96
0.01	2.33
0.005	2.57

Table B.1. Some useful values of the z-statistic.

The area under the standard normal for various values of the independent variable is called the *z-statistic*. If α is a probability then z_{alpha} is the value such that, for a standard normal random variable X, $P(X > z_\alpha) = \alpha$. Notice that, because the standard normal is symmetric, we might as well have said that z_α is the number for which $P(X < -z_\alpha) = \alpha$. Some values of the z-statistic are given in Table B.1.

We now have the pieces we need to construct confidence intervals.

Theorem 10. *Suppose we have a random variable with mean μ and standard deviation σ. Then if we compute the mean m and standard deviation s of n samples from the distribution there is a probability α that the true mean is in the interval:*

$$\left(m - z_{\alpha/2} \frac{s}{\sqrt{n}}, m + z_{\alpha/2} \frac{s}{\sqrt{n}}, \right)$$

Theorem 11. *If we perform n Bernoulli trials and obtain k successes. Then there is a probability α that the probability of success is in the interval*

$$\left(\frac{k}{n} - z_{\alpha/2} \sqrt{\frac{\frac{k}{n}\left(1 - \frac{k}{n}\right)}{n}}, \frac{k}{n} + z_{\alpha/2} \sqrt{\frac{\frac{k}{n}\left(1 - \frac{k}{n}\right)}{n}} \right)$$

Theorem 11 not only permits you to estimate the true probability of success for a binomial variable but permits you to compare different types of Bernoulli trials by comparing their confidence estimates.

B.2 Markov Chains

To analyze a series of trials that are not independent, the first mathematical technology to try is Markov chains. A *Markov chain* is a set S of states together with transition probabilities $p_s(t)$ of moving from state t to state s for any two $s, t \in S$. When you use a Markov chain, you start with an initial distribution on the states of the chain. If you know in which state you are starting, then the initial distribution will have probability one of being in that starting state. If your starting state is the distribution of an initial random population yet to be created, then you may have some initial probability of being in each state. The examples in this section should help clarify this notion.

We will be dealing only with Markov chains that have *fixed transition probabilities*. In this sort of Markov chain, the numbers $p_s(t)$ are fixed constants that have no dependence on history. We restrict our focus for clarity's sake and warn you that stochastic models of evolution, a topic beyond the scope of this text, will involve Markov chains with history-dependent transition probabilities.

Example 47. Suppose we generate a sequence of integers by the following rule. The first integer is 0 and subsequent members of the sequence are generated by flipping a coin and adding 1 to the previous number if the coin came up heads. The states of this Markov chain are $S = \{0, 1, 2, \ldots\}$. The transition probabilities are

$$p_s(t) = \begin{cases} 0.5 & \text{if } s = t \text{ or } s = t + 1, \\ 0 & \text{otherwise,} \end{cases}$$

and the initial distribution of states is to be in state 0 with probability 1.

It is easy to see that the integers generated are in some sense random, but the value of a member of the sequence is strongly influenced by the value of the previous member. If the current number is 5, then the next number is 5 or 6, with no chance of getting a 7, even though it is very likely that we will eventually get a 7. Here is a more complex example.

Example 48. Suppose we play a game, called Hexer, with 4 dice as follows. Start by rolling all 4 dice. If you get no 6's, you lose. Otherwise, put the 6's aside in a "six pool" and reroll the remaining dice. Each time you make a roll that produces no 6's, you pick up a die from the six pool to be used in the next roll. If you roll no 6's with an empty six pool, you lose. When all the dice are in the six pool, you win. In all other cases, play continues.

Hexer is a Markov chain with states $\{s_0, s_1, s_2, s_3, s_4, L\}$ corresponding to losing or the number of dice in the six pool. Attaining state s_4 indicates a win. The initial distribution is to be in state s_0 with probability 1. The transition probabilities are summarized in the *transition matrix*:

	$p_s(t)$	s_0	s_1	s_2	s_3	s_4	L
	s_0	0	0.3858	0.1157	0.0154	0.0008	0.4823
	s_1	0.5787	0	0.3472	0.0694	0.0046	0
t	s_2	0	0.6944	0	0.2778	0.0278	0
	s_3	0	0	0.8333	0	0.1666	0
	s_4	0	0	0	0	1	0
	L	0	0	0	0	0	1

Hexer transition matrix

A *transition matrix* for a Markov chain is a matrix $[a_{i,j}]$ indexed by the states of the Markov chain with $a_{i,j} = p_j(i)$.

Example 48 gives conditions for the game Hexer to end. The *terminal states* in which the games ends are s_4 and L. The definition of Markov chain we are using doesn't have a notion of terminal states, so we simply assign such states a probability of 1 of following themselves in the chain and then explain separately whether a state ends the chain or is repeated indefinitely whenever we reach it. The name for such states in Markov chain theory is *absorbing states*.

If we have a Markov chain M with states S, then a subset A of S is said to be *closed* if every state that can follow a state in A is a state in A. Examples of closed subsets of the state space of Hexer are $\{L\}$, $\{s_4\}$, and the entire set of states.

If S does not contain two disjoint closed subsets, we say that M is *indecomposable*. If for two states $x, y \in S$ it is possible for x to follow y and for y to follow x in the chain, then we say that x and y *communicate*. A subset A of S is a *communicating class of states* if any two states in A communicate. The set $\{s_0, s_1, s_2, s_3\}$ is a communicating class in the Markov chain for Hexer.

If there is a distribution d on the states such that for any initial distribution the limiting probabilities of being in each of the states converges to d, then we say that M is *stable* and we call d the *limiting distribution*. (The limiting probability of a state is just the limit as the number of steps goes to infinity of the number of times you have been in the state divided by the number of steps you have run the Markov chain.)

Notice that for Hexer there are two different "final" distributions as the number of steps goes to infinity: probability 1 of being in state L and probability 1 of being in state s_4. So, the Hexer Markov chain is not stable.

A *stable initial state* is a distribution d such that if you start with the distribution d on the states, you keep that distribution. If M is the transition matrix of a Markov chain and $\mathbf{d}$ is the row vector of probabilities in d, then d is a stable initial distribution if

$$\mathbf{d} \cdot M = \mathbf{d}.$$

It is not hard to show that the limiting distribution of a Markov chain, if it exists, is also a stable initial distribution. The following theorems, offered without proof, go a long way toward characterizing a very nice class of Markov chains.

Theorem 12. *An indecomposable Markov chain has at most one stable initial distribution.*

Theorem 13. *Stable Markov chains are indecomposable.*

If there is a partition of the set of states of a Markov chain

$$\{A_0, A_2, \ldots, A_{k-1}\}, \quad k \geq 2,$$

such that the only states that can follow the states in A_i are the states in A_{i+1} (addition mod k), then we say that a Markov chain is *periodic with period k*. The largest k for which a Markov chain is periodic is called the *period* of the Markov chain, and if there is no $k \geq 2$ for which a Markov chain is periodic, then we call the Markov chain *aperiodic*.

Theorem 14. *If a Markov chain is indecomposable, aperiodic, and has states that constitute a single communicating class, then either*
    (i) The Markov chain has no limiting distribution and the limiting probabilities of each state are zero, or,
    (ii) The Markov chain has a limiting distribution and is stable.

The next two examples are Markov chains that fit (i) and (ii) of Theorem 14 respectively.

Example 49. Suppose we modify Example 47 as follows. Roll a 6-sided die instead of flipping coins. Add 1 for a 5 or a 6, subtract 1 for a 1 or a 2, and otherwise leave the number unchanged. The states are now $S = \{\ldots, -2, -1, 0, 1, 2, \ldots\}$ and the transition probabilities become

$$p_s(t) = \begin{cases} 1/3 & \text{if } s = t - 1, \ t, \text{ or } t + 1, \\ 0 & \text{otherwise.} \end{cases}$$

It is not hard to see there is a single closed set of states, the whole state space, and that every state communicates with every other state. A bit of thought also shows that this Markov chain is aperiodic. This implies that Theorem 14 applies. Since we could choose our initial distribution to have probability one on any state, it follows that each state must have the same limiting probability as each other state. Since you cannot divide 1 into infinitely many equal pieces, there cannot be a limiting distribution, and so we are in case (i) of the theorem.

Example 50. Suppose we have a 4-state Markov chain with states $\{a, b, c, d\}$ and transition matrix

$$\begin{bmatrix} 0.25 & 0.25 & 0.25 & 0.25 \\ 0.25 & 0.25 & 0.25 & 0.25 \\ 0.25 & 0.25 & 0.25 & 0.25 \\ 0.25 & 0.25 & 0.25 & 0.25 \end{bmatrix}.$$

It is obvious from inspection that this Markov chain satisfies the hypothesis of Theorem 14. Since there are finitely many states, the limiting probability cannot be zero, and so this chain is of the type described by (ii). It is in fact easy to see that the limiting distribution is $(0.25, 0.25, 0.25, 0.25)$.

If a Markov chain has a stable limiting distribution, it is called the *stationary distribution* of the Markov chain. It isn't hard to approximate the stationary distribution of a Markov chain with a few states *if you know it has one.* Suppose M is a Markov chain with n states and transition matrix T. Pick an initial distribution d and then compute the sequence $\{\mathbf{d}, \mathbf{d} \cdot T, \mathbf{d} \cdot T^2, \ldots\}$. If M is stable, this sequence will converge to the stationary distribution. Essentially, repeated multiplication by the transition matrix will turn any initial distribution into the stationary distribution in the limit. For many choices of d (but not all), the sequence obtained by repeated multiplication by T will exhibit approximate periodicity if M is periodic.

Let us conclude with a simple Markov chain example that solves an estimation problem in artificial life. While reading this example, keep in mind that there are assumptions and estimates involved; do not accept these blindly. Any assumption, no matter how much you need it to cut down the problem to manageable size, should be repeatedly examined. With that caveat, let us proceed.

Example 51. Suppose we are running a string evolver on the alphabet $\{0, 1\}$ that uses an evolutionary algorithm with tournament selection and tournament size 2. If we have 60 creatures of length 20 and use single-point mutation, what is the expected time-to-solution?

Answer:

Assume that the point mutation must change the value of the locus it mutates. Also, assume that the reference string is "11111111111111111111." (Problem 11 showed that the choice of reference string is irrelevant to the solution time. This choice lets us use the results of Example 41.) With this reference string, the creature's fitness is the number of 1's in its gene.

The first step in solving this problem is to figure out how good the best creature in the population is. (If, for example, we had 2^{20} creatures, there would be an excellent chance that the solution would exist in the initial random population.) We solved this problem in Example 41; the answer is that the best creature has an expected fitness of 14.

The model of evolution (tournament selection) breaks the population into randomly selected sets of two creatures, copies the better over the worse in each group, and then performs a (bit flip) point mutation on the copy. This means that all creatures are following the same path to the reference string at the same rate. (Imagine how hard this example would be if we allowed crossover.) We therefore assume that the time-to-solution can be computed by following the best creature.

Let M be the Markov chain whose states are $\{0, 1, \ldots, 20\}$ representing the fitness of the best creature. The model of evolution ensures that the best creature will survive and that improvement always comes in the form of a single 0 being transformed into a 1. From this we can compute the transition probabilities to be

$$
p_s(t) = \begin{cases} (20 - t)/20 & \text{if } s = t + 1, \\ t/20 & \text{if } s = t, \\ 0 & \text{otherwise.} \end{cases}
$$

Our current guess at an initial distribution is

$$
(0, 0, 0, 0, 0, 0, 0, 0, 0, 0, 0, 0, 0, 0, 1, 0, 0, 0, 0, 0, 0)
$$

(that is, the best creature has fitness 14). Our expected time, in generations, to improve the best creature would be the reciprocal of the probability that it will improve (why?). Summing over the needed improvements, this gives us an estimate of

$$
\sum_{t=14}^{19} \frac{20}{20 - t} = 49
$$

generations.

We actually have the information needed to build the transition matrix, and the true initial distribution of the population is available; it is $\mathbf{d} = (p_0, p_1, \ldots, p_{20})$, where

$$p_i = \binom{20}{i}\left(\frac{1}{2}\right)^{20}.$$

We could get a much better estimate of time-to-solution by taking the true initial distribution and multiplying it by the transition matrix (with a computer) until the generation in which the probability of beginning in state 20 is at least $1/60$. Keep in mind that instead of following the best creature, we are now tracking the whole population of 60 creatures, so $1/60$ of a chance of being in state 20 gives us an expectation of one creature in state 20.

If we do this, our estimate becomes 21 generations, less than half of the far cruder and easier estimate above. In any case, an estimate of this sort should never be seen as a precise answer, but rather as a ballpark figure that tells you whether your simulation needs to be run for a few minutes, an hour, overnight, or on a generation of hardware not available until slightly after the FAA certification of a craft capable of interstellar travel.

C

A Review of Calculus and Vectors

In the real-function optimization treated in Chapter 3, an obvious choice for a Lamarckian mutation operator for a continuous function is to head uphill (when maximizing) or downhill (when minimizing). However, when you're optimizing a function of 25 variables, it becomes very hard to know which way *is* uphill using limited human intuition. When testing a real function optimizer, it can be quite embarrassing to make a test function, constructively place a large number of optima in the test function, and then find that the roots of the derivative of the function corresponding to your test optima have migrated into the parts of the complex plane off of the real axis (if you don't get the joke, review your calculus). A taste of this sort of test problem design appears in Problem 48. It's also nice, when making trivial initial tests of a real-function optimizer, to know the true optima of your test function.

Calculus is a key skill needed to do any of these things. It also is the key to steepest-descent techniques, important benchmarks that an alife system *must* beat to be worth the attention of optimizers. This appendix reviews a few selected features of calculus useful for problems in this book. We urge you to become skilled with calculus if you are not already.

C.1 Derivatives in One Variable

If you have a simple function, $f(x) = x^2$ for example, you can compute how fast $f(x)$ is changing on an interval $[a, b]$ by computing $\frac{f(b)-f(a)}{b-a}$. On the interval $[1, 3]$, our example function changes by a total of $9 - 1 = 8$. Since the interval is $3 - 1 = 2$ units wide, the rate of change on the interval is $8/2 = 4$. This is all very well and good for applications like issuing traffic tickets to red sports cars or figuring the profit you made per apple during the class fruit sale, but it's not enough to point the way uphill in 10 dimensions.

To do this, we need to be able to answer the question "What is the rate of change of $f(x)$ on the interval $[1, 1]$." Applying the average change technique causes you to divide by zero, not an easy feat and one prohibited by law in some mathematical jurisdictions. The way you avoid dividing by zero is to compute the average change on a whole sequence of intervals that get smaller and smaller and always include the point at which you want to know the change of $f(x)$. These average rates of change will, if $f(x)$ is a nice (differentiable) function, start looking alike and will converge to a reasonable value.

For the function $f(x) = x^2$, this reasonable value is always $2x$. We call this reasonable value by many names including *the instantaneous rate of change* of $f(x)$ and *the derivative of f(x)*. The formal notation is

$$D_x f(x) = 2x, \text{ or } \frac{d}{dx} f(x) = 2x, \text{ or } f'(x) = 2x.$$

If you want to be able to compute derivatives in general, take an honors calculus class (one with proofs). If you want to be able to compute derivatives for the usual functions that appear on a pocket calculator, the rules are given in this appendix in two tables: *Derivative Rules for Functions* and *Derivative Rules for Combinations of Functions*. These tables are not an exhaustive list, but they include, in combination, every continuously differentiable function used in this book.

The most important and confusing rule is the *chain rule*, which lets you nest functions: $D_x(f(g(x))) = D_x f(g(x)) \cdot D_x g(x)$. Here are a few examples to illustrate the rules.

Example 52. Compute : $D_x \cos(x^2 + 1)$.
Answer:
The form for $\cos(u)$ says that $D_u \cos(u) = -\sin(u)$. The derivative of $x^2 + 1$ is $2x$ (use the scalar multiple rule, the sum of functions rule, and the powers of a variable rule). Combining these results via the chain rule (set $u = x^2 + 1$) tells us that

$$D_x \cos(x^2 + 1) = -\sin(x^2 + 1) \cdot 2x.$$

Derivative Rules for Functions	
Powers of a variable	$D_x x^n = n \cdot x^{n-1}$
Trig. functions	$D_x \sin(x) = \cos(x)$ $D_x \tan(x) = \sec^2(x)$ $D_x \sec(x) = \sec(x)\tan(x)$ $D_x \cos(x) = -\sin(x)$ $D_x \cot(x) = -\csc^2(x)$ $D_x \csc(x) = -\csc(x)\cot(x)$
Log and exponential	$D_x ln(x) = \frac{1}{x}$ $D_x e^x = e^x$
Hyperbolic Trig.	$D_x \sinh(x) = \cosh(x)$ $D_x \tanh(x) = \operatorname{sech}^2(x)$ $D_x \operatorname{sech}(x) = -\operatorname{sech}(x)\tanh(x)$ $D_x \cosh(x) = \sinh(x)$ $D_x \coth(x) = -\operatorname{csch}^2(x)$ $D_x \operatorname{csch}(x) = -\operatorname{csch}(x)\coth(x)$
Inverse Trig.	$D_x \arcsin(x) = \dfrac{1}{\sqrt{1-x^2}}$ $D_x \arctan(x) = \dfrac{1}{1+x^2}$ $D_x \operatorname{arcsec}(x) = \dfrac{1}{|x|\cdot\sqrt{x^2-1}}$
Inverse Hypertrig.	$D_x \operatorname{arcsinh}(x) = \dfrac{1}{\sqrt{x^2+1}}$ $D_x \operatorname{arctanh}(x) = \dfrac{1}{1-x^2}$ $D_x \operatorname{arcsech}(x) = \dfrac{-1}{x\cdot\sqrt{1-x^2}}$

Example 53. Compute: $D_x \sqrt{x^2 + 2x + 3}$

Answer:

The first step is to rephrase the square root as a power so that the rule for powers of a variable may be used on it (the rule is stated in terms of the nth power, but n may in fact be any real number, e.g., $1/2$). Doing this transforms the problem to $D_x(x^2 + 2x + 3)^{1/2}$. Now, the powers of a variable rule tells us that $D_x u^{1/2} = \frac{1}{2}u^{-1/2}$, and combining the scalar multiple, sum of functions, and powers of a variable rule tells us that $D_x(x^2 + 2x + 3) = 2x + 2$. So, the chain rule says that

$$D_x \sqrt{x^2 + 2x + 3} = \frac{1}{2}(x^2 + 2x + 3)^{-\frac{1}{2}} \cdot (2x + 2) = \frac{x+1}{\sqrt{x^2 + 2x + 3}}.$$

Derivative Rules for Combinations of Functions	
Scalar multiples	$D_x(C \cdot f(x)) = C \cdot D_x f(x)$, C a constant.
Sum of functions	$D_x(f(x) + g(x)) = D_x f(x) + D_x g(x)$
Product Rule	$D_x(f(x) \cdot g(x)) = D_x f(x) \cdot g(x) + f(x) \cdot D_x g(x)$
Quotient Rule	$D_x \frac{f(x)}{g(x)} = \frac{D_x f(x) \cdot g(x) - f(x) \cdot D_x g(x)}{g^2(x)}$
Reciprocal Rule	$D_x \frac{1}{f(x)} = \frac{-D_x f(x)}{f^2(x)}$
Chain Rule	$D_x(f(g(x))) = D_x f(g(x)) \cdot D_x g(x)$

Example 54. Compute $D_x \left(\frac{\cos(1-x)}{x^2+1} \right)$.

For this problem we need the quotient rule as well as the chain rule to resolve $\cos(1 - x)$. The chain rule says that $D_x \cos(1 - x) = -\sin(1 - x) \cdot D_x(1 - x)$, and since $D_x(1 - x) = -1$, we get $D_x \cos(1 - x) = \sin(1 - x)$ once we cancel all the minus signs. Putting this result into the quotient rule yields

$$D_x \left(\frac{\cos(1 - x)}{x^2 + 1} \right) = \frac{(x^2 + 1) \cdot \sin(1 - x) - 2x \cdot \cos(1 - x)}{(x^2 + 1)^2}.$$

C.2 Multivariate Derivatives

One of the goals of this appendix is for you to learn to point uphill in any number of dimensions. The last section contained useful building blocks, but only in one dimension. In order to work in more dimensions, we need multivariate functions and vectors.

The vectors, n-tuples of numbers drawn from $\mathbb{R}^n$, are simply formal ways of writing down directions and magnitudes. In $\mathbb{R}^2$, if we take the positive y-axis to be north and the positive x-axis to be east, then the vectors $(1, 1)$ and $(7, 7)$ both point northeast and encode distances of $\sqrt{2}$ and $7 \cdot \sqrt{2}$ respectively. There are a number of standard operations on vectors that will be handy. The *vector sum* or just *sum* of two vectors $\mathbf{v} = (r_1, r_2, \ldots, r_n)$ and $\mathbf{u} = (s_1, s_2, \ldots, s_n)$ in $\mathbb{R}^n$ is defined to be

$$\mathbf{v} + \mathbf{u} = (r_1 + s_1, r_2 + s_2, \ldots, r_n + s_n).$$

The *scalar multiple* of a vector $\mathbf{v} = (r_1, r_2, \ldots, r_n)$ by a real number c is given by

$$c \cdot \mathbf{v} = (c \cdot r_1, c \cdot r_2, \ldots, c \cdot r_n).$$

The *norm* or *length* of a vector $\mathbf{v} = (r_1, r_2, \ldots, r_n)$ is given by

$$\|\mathbf{v}\| = \sqrt{r_1^2 + r_2^2 + \cdots + r_n^2}.$$

A *unit vector* is a vector of length one. The vector

$$\frac{1}{\|\mathbf{v}\|} \cdot \mathbf{v}$$

is called the unit vector in the direction of $\mathbf{v}$. Such unit vectors are useful for specifying the *direction* of Lamarckian mutations when the size is found by other means.

The entries of the vectors that point up- or downhill are going to be *partial derivatives*, and the vector that points up- or downhill is the *gradient*. A partial derivative is a derivative taken with respect to some one variable in a multivariate function. When you take the partial derivative with respect to a variable u, you use the same rules as for single-variable derivatives, treating u as the sole variable and all the other variables as if they were constants. Since normal derivatives (with respect to x) are denoted by D_x or $\frac{d}{dx}$, partial derivatives are denoted by the symbol ∂, as shown in Examples 55 and 56.

Example 55. If $f(x, y) = \left(x^3 + y^2 + 3xy + 4\right)^5$, then

$$\frac{\partial f}{\partial x} = 5 \cdot \left(x^3 + y^2 + 3xy + 4\right)^4 \cdot (3x^2 + 3y),$$

and

$$\frac{\partial f}{\partial y} = 5 \cdot \left(x^3 + y^2 + 3xy + 4\right)^4 \cdot (2y + 3x).$$

Example 56. If $f(x, y) = \cos\left(\sqrt{x^2 + y^2}\right)$, then

$$\frac{\partial f}{\partial x} = -\sin\left(\sqrt{x^2 + y^2}\right) \cdot \frac{2x}{\sqrt{x^2 + y^2}},$$

and

$$\frac{\partial f}{\partial y} = -\sin\left(\sqrt{x^2 + y^2}\right) \cdot \frac{2y}{\sqrt{x^2 + y^2}}.$$

Notice that in complicated expressions, the variables that are held constant still appear extensively in the final partial derivative.

If $f : \mathbb{R}^n \to \mathbb{R}$ is a continuously differentiable function of n variables, $f(x_1, x_2, \ldots, x_n)$, then the *gradient of f* is defined to be

$$\nabla f = \left(\frac{\partial f}{\partial x_1}, \frac{\partial f}{\partial x_2}, \ldots, \frac{\partial f}{\partial x_n}\right).$$

The gradient points uphill. To be more precise, if we are at a point in n-dimensional space and are examining an n variable function f that has a gradient, then the direction in which the function is increasing in value the fastest is ∇f. The two vectors

$$\frac{\nabla f}{\|\nabla f\|} \quad \text{and} \quad (-1) \cdot \frac{\nabla f}{\|\nabla f\|}$$

are unit vectors in the direction of maximum increase and maximum decrease of the function.

Example 57.

$$f(x_1, x_2, \ldots, x_n) = \frac{1}{x_1^2 + x_2^2 + \cdots + x_n^2 + 1}$$

is an n-dimensional version of the fake bell curve discussed in Section 2.2. Examine the gradient

$$\nabla f = \left(\frac{-2x_1}{x_1^2 + \cdots + x_n^2 + 1}, \frac{-2x_2}{x_1^2 + \cdots + x_n^2 + 1}, \cdots, \frac{-2x_n}{x_1^2 + \cdots + x_n^2 + 1} \right).$$

For any point in $\mathbb{R}^n$, each coordinate of the gradient is minus twice the value of the point in that coordinate divided by a positive number that is the same for all coordinates. This means that the gradient always points back toward the origin in each coordinate. Closer examination will show that the gradient points toward the origin globally as well as coordinatewise and that the length of the gradient vector corresponds to the steepness of the slope of the fake bell curve. Sketching the gradient at various points when $n = 2$ is instructive.

C.3 Lamarckian Mutation with Gradients

If $f : \mathbb{R}^n \to \mathbb{R}$ is a continuously differentiable function, then the gradient gives us a mutation operator that can be used in place of the more cumbersome Lamarckian mutation described in terms of multiple point mutations in Section 2.3. For the real function optimizers described in Chapter 3, we used a notion of point mutation in which we chose a maximum mutation size per coordinate of ϵ and mutated our genes (lists of points in $\mathbb{R}^n$) by adding ϵ times a number uniformly distributed in the range $-1 \le x \le 1$ to the value of a randomly chosen coordinate. The new Lamarckian mutation operator consists in adding the vectors

$$\epsilon \cdot \frac{\nabla f}{\|\nabla f\|} \quad \text{or} \quad -\epsilon \cdot \frac{\nabla f}{\|\nabla f\|}$$

when maximizing or minimizing f, respectively.

C.4 The Method of Least Squares

One standard minimization problem done in multivariate calculus is model-fitting with the method of least squares. First, pick a function with unknown parameters (treat them as variables) that you think fits your data well. Then, compare that function to your data by subtracting the values at selected points, squaring that difference, and then summing those squares. Then, use calculus to minimize the result. Recall that a function of several variables has its maxima and minima at points where all of its partial derivatives are zero. So, simply solve the system of equations obtained by setting the derivative with respect to each parameter of the sum of squared error to zero. Unless you are already familiar with the method of least squares, the preceding discussion is quite likely an impenetrable fog, and so an example is in order.

We will do the standard example, fitting a line to data. A line has two parameters, its slope and intercept. So, the function being fitted is

$$y = ax + b,$$

where a and b are the slope and intercept, respectively.

Imagine we have n data points $\{(x_1, y_1), (x_2, y_2), \ldots, (x_n, y_n)\}$. Then the sum of squared error is

$$E^2(a, b) = \sum_{i=1}^{n} (ax_i + b - y_i)^2.$$

Extracting the partial derivatives with respect to a and b, we obtain the system of equations

$$\frac{\partial E^2}{\partial a} = \sum_{i=1}^{n} 2 \cdot (ax_i + b - y_i)x_i = 0,$$

and

$$\frac{\partial E^2}{\partial b} = \sum_{i=1}^{n} 2 \cdot (ax_i + b - y_i) = 0.$$

Applying the linearity property of the summation and a little algebra yields linear equations in standard form:

$$a \cdot \sum_{i=1}^{n} x_i^2 + b \cdot \sum_{i=1}^{n} x_i = \sum_{i=0}^{n} x_i y_i,$$

and

$$a \cdot \sum_{i=0}^{n} x_i + b \cdot n = \sum_{i=1}^{n} y_i.$$

With these formulas, we can find the a and b that minimize squared error with respect to the data set. It is often worth going a small extra distance and solving the linear systems in general to obtain formulas for a and b in terms of the sums of products of data elements. To this end we present

$$a = \frac{n \sum_{i=1}^{n} x_i y_i - \sum_{i=1}^{n} x_i \sum_{i=1}^{n} y_i}{n \sum_{i=1}^{n} x_i^2 - \left(\sum_{i=1}^{n} x_i\right)^2},\tag{C.1}$$

and

$$b = \frac{\sum_{i=1}^{n} x_i^2 \sum_{i=0}^{n} y_i - \sum_{i=1}^{n} x_i \sum_{i=1}^{n} x_i y_i}{n \sum_{i=1}^{n} x_i^2 - \left(\sum_{i=1}^{n} x_i\right)^2}.\tag{C.2}$$

If you are putting these into software, you will want to note that the denominators of Equations C.1 and C.2 are identical. If you are prone to using formulas in place of thought, you are warned not to use these formulas on a single data point. Here is a mathematical issue that you may wish to ponder: when doing multivariate optimization by finding points where the partial derivatives equal zero, it is often necessary to go through a number of contortions to show that the critical points found are in fact minima or maxima and not just saddle points. There was a unique critical point in the above minimization. Why is it a minimum? Hint: to what analytic class of functions does $E^2(a, b)$ belong?

While the computation of the least squares best fit line is both the most widely used and the simplest nontrivial example of the method of least squares, many other models can be fit with least squares. Perhaps of more interest to you is the fact that the least squares line fit can be transformed to fit other models.

Suppose that we wished to fit an exponential model of the form

$$y = b \cdot e^{ax}.$$

Simply note that if $y = b \cdot e^{ax}$, where b is the initial value of the function at $x = 0$ and a is the growth rate, then it is also the case that $\ln(y) = ax + \ln(b)$. This means that if we take the natural log of the y-coordinates of our data set and then do a least squares fit of a line to the resulting transformed data set, then the slope of the line will be the growth rate of the model and the intercept will be the log of the initial value.

A similar technique can be used to fit a model of the form

$$y = \frac{1}{ax^2 + b}.$$

The derivation is left as an exercise for you.

The method of least squares can also be used to compare the quality of models. When we evolve data interpolators, we use it in exactly this fashion, judging a data interpolator to have higher fitness if its function has lower squared error with respect to the data. In some applications, mean squared error can serve as a fitness function (it is minimized).

D

Combinatorial Graphs

D.1 Terminology and Examples

In this appendix, we will go over the terminology and some elementary theory of combinatorial graphs. An excellent and complete introduction to the topics appears in [58].

Definition D.1 *A **combinatorial graph** or **graph** G is a set $V(G)$ of **vertices** together with a set $E(G)$ of **edges**. Edges are unordered pairs of vertices and as such may be thought of as arcs connecting pairs of vertices. The two vertices that make up an edge are its **ends** and are said to be **adjacent**.*

An example of a graph appears in Figure D.1. The graph has 16 vertices and 32 edges. In spite of their simplicity, graphs have a boatload of terminology. Prepare to remember.

Definition D.2 *If a vertex is part of the unordered pair that makes up an edge, we say that the edge and vertex are **incident**.*

Definition D.3 *The number of edges incident with a vertex is the **degree** of the vertex.*

Definition D.4 *If all vertices in a graph have the same degree, we say that the graph is **regular**. If that degree is k, we call the graph k-**regular**.*

The example of a graph given in Figure D.1 is a 4-regular graph. It is, in fact, the graph of vertices and edges of a 4-dimensional hypercube.

Definition D.5 *A graph is said to be **bipartite** if its vertices can be divided into two sets, called a **bipartition**, such that every edge has an end in each set.*

Definition D.6 *A* **subgraph** *of a graph G is a graph H whose vertex and edge sets are both subsets of V(G) and E(G).*

Definition D.7 *A graph is said to be* **connected** *if it is possible to start at any one vertex and then follow a sequence of pairwise adjacent vertices to any other.*

Definition D.8 *A graph is k-**connected** if the deletion of fewer than k edges cannot disconnect the graph.*

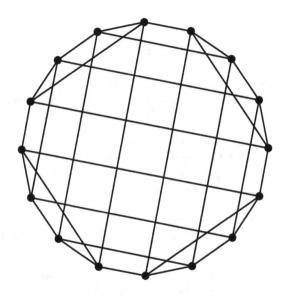

Fig. D.1. An example of a graph.

The example of a graph given in Figure D.1 is bipartite. Following is the so-called first theorem of graph theory.

Theorem 15. *The number of vertices of odd degree in a graph is even.*

Proof:

Count the number of pairs of incident vertices and edges. Since each edge is incident on two vertices, the sum is a multiple of two. Since each vertex contributes its degree to the sum, the total is the sum of all the degrees. A sum of integers with an even total has an even number of odd summands, and so the number of odd degrees is even. □

This theorem and its proof are included for two reasons. The first is to demonstrate the beautiful technique involved: count something two different ways and then deduce something from the equality of the two answers. The second is to show that even in a very general structure like graphs there are some constraints. Suppose, for example, that you have an evolutionary algorithm that is evolving 3-regular graphs. If you have a mutation that adds vertices, then it must add them in pairs, since a 3-regular graph has an even number of vertices. In some of the other examples, we will see other constraints on graphs. There are quite a lot of named families of graphs. Here are some that are used in this text.

Definition D.9 *The* **complete graph** *on n vertices, denoted by K_n has n vertices and all possible edges. An example of a complete graph with 12 vertices is shown in Figure D.2.*

Definition D.10 *The* **complete bipartite graph** *with $n + m$ vertices, denoted by $K_{n,m}$ has vertices divided into disjoint sets of n and m vertices and all possible edges that have one end in each of the two disjoint sets. An example of a complete bipartite graph with 8 (4+4) vertices is shown in Figure D.2.*

Definition D.11 *The* **n-cycle**, *denoted by C_n has vertex set $\mathbb{Z}_n$. Edges are pairs of vertices that differ by 1 (mod n) such that the vertices form a ring with each vertex having two neighbors. A* **cycle in a graph** *is a subgraph that happens to be a cycle.*

Definition D.12 *A* **path** *on n vertices is a graph with n vertices that results from deleting one edge from an n-cycle. A* **path in a graph** *is a subgraph that happens to be a path.*

Definition D.13 *The* **n-hypercube**, *denoted by H_n has the set of all n-character binary strings as its set of vertices. Edges consist of pairs of strings that differ in exactly one position. A 4-hypercube is shown in Figure D.2.*

Definition D.14 *The* $n \times m$ **torus**, *denoted by $T_{n,m}$ has vertex set $\mathbb{Z}_n \times \mathbb{Z}_m$. Edges are pairs of vertices that differ either by 1 (mod n) in their first coordinate or by 1 (mod m) in their second coordinate, but not both. These graphs are $n \times m$ grids that wrap (as tori) at the edges. A 12×6 torus is shown in Figure D.2.*

Definition D.15 *The* **generalized Petersen graph** *with parameters n and k is denoted by $P_{n,k}$. It has two sets of n vertices. The two sets of vertices are both considered to be copies of $\mathbb{Z}_n$. The first n vertices are connected in a standard n-cycle. The second n vertices are connected in a cycle-like fashion,* but the connections jump in steps of size k (mod n). The graph also has edges joining corresponding members of the two copies of $\mathbb{Z}_n$. The graph $P_{32,5}$ is shown in Figure D.2.

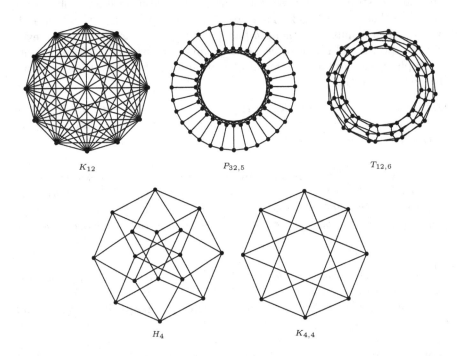

K_{12} $P_{32,5}$ $T_{12,6}$

H_4 $K_{4,4}$

Fig. D.2. Examples of complete, Petersen, torus, hypercube, and complete bipartite graphs. (These examples are all smaller than the graphs actually used, but are members of the same family of graphs.)

Definition D.16 *A sequence of pairwise adjacent vertices that is allowed to repeat vertices is called a* **walk**.

Definition D.17 *A graph that has no cycles as subgraphs is said to be* **acyclic**. *An acyclic connected graph is called a* **tree**.

Paths are examples of trees. There are a large number of constructions possible on graphs, a few of which are given here.

Definition D.18 *The* **complement** *of a graph G, denoted by $\overline{G}$, is a graph with the same vertex set but a complementary set of edges.*

The complement of a 5-cycle is, for example, another, different, 5-cycle; the complement of a 4-cycle is two disconnected edges.

Definition D.19 *If we take a vertex of degree* k *and replace it with a copy of* K_k *so that each member of* $V(K_k)$ *is adjacent to one of the neighbors of the replaced vertex, we say that we have* **simplexified** *the vertex. Simplexification of a graph is defined as simplexification of all its vertices.*

Simplexification is not a construction used much in elementary graph theory, but it is useful for the graph-based evolutionary algorithms discussed in Chapter 13. A picture of a graph and its simplexification are given in Figure D.3.

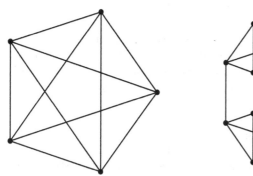 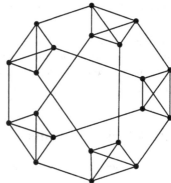

Fig. D.3. K_5 and K_5-simplexified.

Definition D.20 *A* **random graph** *is the result of sampling a particular graph from a random process that produces graphs.*

There are more types of random graphs than you can shake a stick at. We again give a few examples.

Definition D.21 *A* **random graph with edge probability** α *is generated by examining each possible pair of vertices and, with probability* α*, placing an edge between them. The number of vertices is determined in advance.*

Definition D.22 *A* **random regular graph** *can be generated by a form of random walk, as follows, with thanks to Mike Steel for the suggestion. Begin with a regular graph. A large number of times (think at least twice as many times as the graph has edges) perform the following edge swap operation. Pick two edges that have the property that (i) their ends form a set of 4 vertices and (ii) those 4 vertices have exactly two edges,* $\{a, b\}$ *and* $\{c, d\}$*, between them in the graph. Delete those two edges and replace them with the edges* $\{a, c\}$ *and* $\{b, d\}$*. Again, the number of edges is chosen in advance.*

Definition D.23 *To get a* **random toroidal graph with connection radius** β, *place vertices at random in the unit square. Connect with edges all pairs of vertices at distances at most* β *in the torus created by wrapping the edges of the unit square.*

Definition D.24 *A* **random simplicial graph** *is created by first choosing a number* n *of vertices, a number* k *of simplices, and a collection of allowed sizes, e.g.,* {3} *or* {7, 8, 9, 10}. *The graph is generated by performing the following move* k *times. A size* m *is selected at random from the list of allowed sizes. A set of* m *vertices is selected at random. All pairs of vertices in the selected set not already joined by edges are joined by edges.*

Definition D.25 *A* **simplexification-driven random graph** *is created by picking an initial graph and repeatedly choosing a vertex at random and simplexifying it. Since simplexification adds a number of vertices equal to the degree of the vertex it acts on less one, some planning is needed.*

D.2 Coloring Graphs

There is a plethora of problems that involve coloring the vertices or edges of a graph.

Definition D.26 *A* **vertex coloring** *of a graph is an assignment of colors to the vertices of a graph. A vertex coloring of a graph is said to be* **proper** *if no two adjacent vertices are the same color.*

Definition D.27 *The minimum number of colors required for a proper vertex coloring of a graph* G *is the* **chromatic number** *of the graph, denoted by* $\chi(G)$.

Bipartite graphs, for example, have chromatic number 2 (see if you can prove this in one or two lines).

Knowing the chromatic number of a graph is valuable, as can be seen in the following application. Suppose that we have a group of people from which are drawn several committees. Construct a graph with each committee as a vertex and with edges between two vertices if the committees in question share at least one member. Let colors represent time slots for meetings. A proper coloring of the vertices of this graph corresponds to a meeting schedule that allows every member of every committee to be present at each meeting of that committee. The chromatic number is the least number of slots needed for such a schedule.

Definition D.28 *An* **edge coloring** *of a graph is an assignment of colors to the edges of a graph. An edge coloring of a graph is* **proper** *if no two edges incident on the same vertex are the same color.*

Definition D.29 *The minimal number of colors required for a proper edge coloring of a graph G is the* **edge chromatic number** *of a graph, denoted by $\chi_E(G)$.*

Proper edge colorings are useful in the development of communications networks. Suppose we have a large number of sites that must send status or other information to all other sites. These sites are the vertices of the graph, and the edges represent direct communications links. If we assume that each site can communicate with only one other site at a time, then a proper edge coloring of the graph is an efficient algorithm for coordinating communications. If we have a proper edge coloring in n colors $0, 1, \ldots, n-1$, then processors talk over the edge colored i on each timestep congruent to i (mod n). Minimizing the number of colors maximizes usage of the communications links.

There are interesting coloring problems that do not involve proper colorings as well. In *Ramsey theory*, the goal is to color the edges of a complete graph with some fixed number k of colors, and then find some minimal number of vertices such that any edge coloring in k colors forces a monochromatic subgraph to appear that looks like K_m, $m < k$. For example, if we color the edges of a complete graph on 6 or more vertices red and blue, then there must be a red or a blue triangle (K_3). However, it is possible to bi-edge-color K_5 without obtaining any monochrome triangles. Formally, we say that the Ramsey number $R(3,3)$ is equal to 6. If you are interested, try to find a red-blue coloring of the edges of K_5 that avoids monochromatic triangles.

Very few Ramsey numbers are known, and improving lower bounds on Ramsey numbers is a very hard problem that one can attempt with evolutionary algorithms. Recently, Brendan McKay spent 4.3 processor years on Unix workstations showing that in order for a complete graph to have either a red K_4 subgraph or a blue K_5 subgraph forced no matter how it was red-and-blue edge colored, the graph must have at least 25 vertices. Formally, $R(4,5) = 25$. This is the hardest of the two-colored Ramsey numbers known so far. There is only one 3-colored Ramsey number known at the time of this writing, $R(3,3,3) = 17$ (neglecting the case in which monochromatic $K_2 s$ (edges) are forced). In other words, if we 3-color the edges of a complete graph in 3 colors, then no matter what coloring we use, we must have a monochromatic triangle if the complete graph has 17 or more vertices.

The proof that the Ramsey numbers are finite will appear in any good undergraduate combinatorics course, as will several more general definitions of Ramsey numbers and a plethora of Ramsey-style problems. The Ramsey numbers are pervasive in existence proofs in combinatorics and discrete math; so, additional information about a Ramsey number usually turns out to be additional information about many, many other problems as well.

D.3 Distances in Graphs

If we define the distance between two vertices to be the length of the shortest path between them (and define the distance to be infinite if no such path exists), then graphs become *metric spaces.*

Definition D.30 *A **metric space** is a collection of points, in this case the vertices of a graph, together with a function d (distance) from pairs of points to the real numbers, that has three properties:*

(i) For all points p, $d(p,p) = 0$;
(ii) For all pairs of points $p \neq q$, $d(p,q) > 0$; and
(iii) For all triples of points p, q, r, $d(p,q) + d(q,r) \geq d(p,r)$.

The third property is called the *triangle inequality.*

Definition D.31 *The **diameter** of a graph is the maximum distance between any two vertices of a graph.*

As we will see in Chapter 13, the diameter is sometimes diagnostic of the behavior of a graph-based evolutionary algorithm.

Definition D.32 *The **eccentricity** of a vertex is the largest distance from it to any other vertex in the graph.*

Notice that the diameter is then the maximum eccentricity of a vertex.

Definition D.33 *The **radius** of a graph is the minimum eccentricity (and it is not usually half the diameter, graphs aren't circles).*

Definition D.34 *The **center** of a graph is the set of vertices that have minimum eccentricity.*

Definition D.35 *The **periphery** of a graph is the set of vertices that have maximum eccentricity.*

Definition D.36 *The **annulus** of a graph comprises those vertices that are not in the periphery or the center.*

The several terms given above for different eccentricity-based properties are useful for classifying the vertices of network graphs in terms of their probable importance. Peripheral vertices tend to have lower traffic, while central vertices often have high traffic.

Definition D.37 *A **dominating set** in a graph is a set D of vertices with the property that every vertex is either in D or adjacent to a member of D.*

For graphs representing guards and lines of sight, or vital services and minimal feasible travel times to reach them, small dominating sets can be quite valuable. There may be reasons that we want dominating sets that are only in the periphery of a graph (imagine a town in which affordable land is only at the "edge" of town). Vertices in the center of the graph are more likely to be adjacent to lots of other vertices, and so it may be wise to choose them when searching for small dominating sets. The problem of locating minimal dominating sets is thought to be intractable, but evolutionary algorithms may be used to locate tolerably small dominating sets.

D.4 Traveling Salesman

It is possible to generalize the notion of distance in graphs by placing weights on their edges so that instead of adjacent vertices being at distance 1, they are at a distance given by the edge weight. In this case the edge weights may represent travel costs or distances.

Definition D.38 *The* **Traveling Salesman problem** *starts with a complete graph that has cities as its vertices and the cost of traveling between cities as edge weights. What we desire is an ordered list of all the cities that corresponds to a minimal-cost (total of edge weights) cycle in the graph that visits all the cities.*

Finding exact solutions to this problem is almost certain to be intractable (NP-complete for the computer science majors among you), but evolutionary algorithms can be used to find approximate answers (see Section 7.2). The Traveling Salesman problem is a standard test problem for evolutionary algorithms that operate on genes that are ordered lists without repetition (in this case the list is the salesman's itinerary).

D.5 Drawings of Graphs

Definition D.39 *A* **drawing** *of a graph is a placement of the vertices and edges of a graph into some space, e.g., the Cartesian plane.*

There are a number of properties of drawings that can be explored, estimated, or optimized with evolutionary algorithms. In Chapter 3, we discussed evolutionary algorithms that tried to minimize the crossing number of a graph when the edges were drawn as line segments.

Definition D.40 *The* **crossing number** *of a graph is the minimum number of times one edge crosses another in any drawing.*

Definition D.41 *A graph is said to be* **planar** *if it can be drawn with zero edge crossings in the Cartesian plane.*

Another property of a graph related to drawings is the thickness of a graph.

Definition D.42 *The* **thickness** *of a graph is the minimum number of colors in an edge-coloring of the graph that has the property that all the induced monochromatic graphs are planar.*

A planar graph thus has thickness 1. Thickness gives a useful measure of the complexity of a graph. An electrical circuit with a thickness of 3 might need to be put on 3 stacked circuit boards, for example. Many other problems concerning drawings of graphs exist but require a knowledge of topology beyond the scope of this text. If you are interested, look for books on topological graph theory that discuss the genus of a graph or the M-pire (empire) problem. The problem of embedding topological knowledge in a data structure that is to be manipulated by an evolutionary algorithm is a subtle one.

References

1. Dan Ashlock. GP-automata for dividing the dollar. In *Proceedings of the Second Annual Conference on Genetic Programming*, pages 18–26. IEEE Publications, Piscataway, New Jersey, 1997.
2. Dan Ashlock. Data crawlers for optical character recognition. In *Proceedings of the 2000 Congress on Evolutionary Computation*, pages 706–713. IEEE Publications, Piscataway, New Jersey, 2000.
3. Dan Ashlock and Mark Joenks. ISAc lists, a different representation for program induction. In *Genetic Programming 98, proceedings of the third annual genetic programming conference.*, pages 3–10, San Francisco, 1998. Morgan Kaufmann.
4. Daniel Ashlock and James B. Golden III. Computation and fractal visualization of sequence data. In *Evolutionary Computation in Bioinformatics*, chapter 11. Morgan Kaufmann, 2002.
5. Daniel Ashlock and James B. Golden III. Chaos automata: Iterated function systems with memory. *Physica D*, 181:274–285, 2003.
6. Daniel Ashlock and James Lathrop. A full characterized test suite for genetic programming. In *Proceedings of the 1998 Conference on Evolutionary Programming*, pages 537–546, New York, 1998. Springer-Verlag.
7. Robert Axelrod. *The Evolution of Cooperation.* Basic Books, New York, 1984.
8. Thomas Back, Ulrich Hammel, and Hans-Paul Schwefel. Evolutionary computation: Comments on the history and current state. *IEEE Transactions on Evolutionary Computation*, 1(1):3–17, 1997.
9. Wolfgang Banzhaf, Peter Nordin, Robert E. Keller, and Frank D. Francone. *Genetic Programming: An Introduction.* Morgan Kaufmann, San Francisco, 1998.
10. Michael F. Barnsley. *Fractals Everywhere.* Academic Press, Cambridge, MA, 1993.
11. James C. Bean. Genetic algorithms and random keys for sequencing and optimization. *ORSA Journal on Computing*, 2(2):154–160, 1994.
12. Randall D. Beer and John C. Gallagher. Evolving dynamical neural networks for adaptive behavior. *Adaptive Behavior*, 1:92–121, 1992.
13. Hans-Georg Beyer. *The Theory of Evolution Strategies.* Springer, New York, 2001.
14. Valentino Braitenberg. *Vehicles, experiments in synthetic psychology.* MIT Press, Cambridge, Mass., 1984.

15. Richard A. Brualdi and Vera Pless. Greedy codes. *Journal of Combinatorial Theory(A)*, 64:10–30, 1993.
16. C. Dietrich, F. Cui, M. Packila, D. Ashlock, B. Nikolau, and P.S. Schnable. Maize mu transposons are targeted to the 5' utr of the gl8a gene and sequences flanking mu target site duplications throughout the genome exhibit non-random nucleotide composition. *Genetics*, 160:697–716, 2002.
17. Theodosius Dobzhansky. Nothing in biology makes sense except in the light of evolution. *The American Biology Teacher*, 35:125–129, 1973.
18. Chitra Dutta and Jyotirmoy Das. Mathematical characterization of chaos game representations. *Journal of Molecular Biology*, 228:715–719, 1992.
19. David B. Fogel. The evolution of intelligent decsion making in gaming. *Cybernetics and Systems: An International Journal*, 22:223–236, 1991.
20. David B. Fogel. Evolving behaviors in the iterated prisoner's dilemma. *Evolutionary Computation*, 1(1):77–97, 1993.
21. David B. Fogel. On the relationship between the duration of an encounter and the evolution of cooperation in the iterated prisoner's dilemma. *Evolutionary computation*, 3(3):348–363, 1996.
22. David B. Fogel. *Evolutionary Computation, the Fossil Record*. IEEE Press, Piscataway, New Jersy, 1998.
23. L. J. Fogel. Autonomous automata. *Industrial Research*, 4:14–19, 1962.
24. L. J. Fogel. *On the organization of intellect*. PhD thesis, UCLA, UCLA, 1964.
25. L. J. Fogel. Artificial intelligence through a simulation of evolution. In *Biophysics and Cybernetics:Proceedings of the 2nd Cybernetic Sciences Symposium*, pages 131–155. Spartan Books, Washington, D.C., 1965.
26. R. M. Friedberg. A learning machine, part I. *IBM Journal of Research and Development*, 2(1):2–13, 1958.
27. R. M. Friedberg. A learning machine, part I. *IBM Journal of Research and Development*, 3(3):282–287, 1959.
28. Jonathan D. Gandrud, Dan Ashlock, and Elizabeth Blankenship. A note on general adaptation in populations of painting robots. In *Proceedings of the 2003 Congress on Evolutionary Computation*, pages 46–53. IEEE Publications, Piscataway, New Jersey, 2003.
29. David E. Goldberg. *Genetic Algorithms in Search, Optimization, and Machine Learning*. Addison-Wesley Publishing Company, Inc., Reading, MA, 1989.
30. Thore Graepel and Ralf Herbrich. The kernel Gibbs sampler. In *NIPS*, pages 514–520, 2000.
31. F. Gruau. *Neural Network Synthesis Using Cellular Encoding and the Genetic Algorithm*. PhD thesis, France, 1994.
32. Frederic Gruau. Automatic definition of modular neural networks. *Adaptive Behaviour*, 3(2):151–183, 1995.
33. Dan Gusfield. *Algorithms on Strings, Trees, and Sequences*. Cambrige University Press, New York, 1997.
34. W. Daniel Hillis. Co-evolving parasites improve simulated evolution as an optimization procedure. In Christopher Langton, editor, *Artificial Life II*, volume 10 of *Santa Fe Institute Studies in the Sciences of Complexity*, pages 313–324, Reading, 1991. Addison-Wesley.
35. H. Joel Jeffrey. Chaos game representation of gene structure. *Nucleic Acid Research*, 18(8):2163–2170, 1990.
36. Kenneth Kinnear. *Advances in Genetic Programming*. The MIT Press, Cambridge, MA, 1994.

37. Kenneth Kinnear and Peter Angeline. *Advances in Genetic Programming, Volume 2*. The MIT Press, Cambridge, MA, 1996.
38. John R. Koza. *Genetic Programming*. The MIT Press, Cambridge, MA, 1992.
39. John R. Koza. *Genetic Programming II*. The MIT Press, Cambridge, MA, 1994.
40. John R. Koza. *Genetic Programming III*. Morgan Kaufmann, San Francisco, 1999.
41. Benjamin Lewin. *Genes VII*. Oxford University Press, New York, 2000.
42. Kristian Lindgren. Evolutionary phenomena in simple dynamics. In D. Farmer, C. Langton, S. Rasmussen, and C. Taylor, editors, *Artificial Life II*, pages 1–18. Addison-Wesley, 1991.
43. Kristian Lindgren and Mats G. Nordhal. Evolutionary dynamics of spatial games. *Physica D.*, 75:292–309, 1994.
44. Alfred J. Lotka. *Elements of Mathematical Biology*. Dover Publications, New York, 1956.
45. Andrew Meade, David Corne, and Richard Sibly. Discovering patterns in microsatellite flanks with evolutionary computation by evolving discriminatory DNA motifs. In *Proceedings of the 2002 Congress on Evolutionary Computation*, pages 1–6, Piscataway, NJ, 2002. IEEE Publications.
46. James D. Murray. *Mathematical Biology*. Springer, New York, 2002.
47. Vera Pless. *Introduction to the Theory of Error-Correcting Codes*. John Wiley and Sons, New York, 1998.
48. T. S. Ray. An approach to the synthesis of life. In C. G. Langton, C. Taylor, J. D. Farmer, and S. Rasmussen, editors, *Artificial Life II*, pages 371–408. Addison Wesley, Reading MA., 1992.
49. Craig Reynolds. An evolved, vision-based behavioral model of coordinated group motion. In Jean-Arcady Meyer, Herbert L. Roiblat, and Stewart Wilson, editors, *From Animals to Animats 2*, pages 384–392. MIT Press, 1992.
50. Joao Setubal and Joao Meidanis. *Introduction to Computational Molecular Biology*. PWS Publishing, Boston, MA, 1997.
51. Neil J. A. Sloane. On-line encyclopedia of integer sequences.
52. Victor V. Solovyev. Fractal graphical representation and analysis of DNA and protein sequences. *Biosystems*, 30:137–160, 1993.
53. Brian L. Steward, Robert P. Ewing, Daniel A. Ashlock, Amy Kaleita, and Steve M. Shaner. Range operator enabled genetic algorithms for hyperspectral analysis. In *Intelligent Engineering Systems through Artificial Neural Networks*, volume 14, pages 295–300, 2004.
54. Gilbert Syswerda. A study of reproduction in generational and steady state genetic algorithms. In *Foundations of Genetic Algorithms*, pages 94–101. Morgan Kaufmann, 1991.
55. Astro Teller. The evolution of mental models. In Kenneth Kinnear, editor, *Advances in Genetic Programming*, chapter 9. The MIT Press, 1994.
56. Thomas M. Thompson. *From Error-Correcting Codes Through Sphere Packings to Simple Groups*. The Mathematical Association of America, Washington, 1984.
57. Gregory Urban, Kenneth Mark Bryden, and Daniel Ashlock. Engineering optimization of an improved plancha stove. *Energy for Sustainable Development*, VI(2):9–19, 2002.
58. Douglas B. West. *Introduction to Graph Theory*. Prentice Hall, Upper Saddle River, New Jersy, 2001.

59. Darrel Whitley. The genitor algorithm and selection pressure: why rank based allocation of reproductive trials is best. In *Proceedings of the 3rd ICGA*, pages 116–121. Morgan Kaufmann, 1989.

60. David H. Wolpert and William G. Macready. No free lunch theorems for optimization. *IEEE Transactions on Evolutionary Computation*, 1(1):67–82, April 1997.

Index